D1220077

World Survey of Climatology Volume 12

CLIMATES OF CENTRAL AND SOUTH AMERICA

World Survey of Climatology

Editor in Chief:

H. E. LANDSBERG, College Park, Md. (U.S.A.)

Editors:

H. ARAKAWA, Tokyo (Japan)
R. A. BRYSON, Madison, Wisc. (U.S.A.)
H. FLOHN, Bonn (Germany)
J. GENTILLI, Nedlands, W.A. (Australia)
J. F. GRIFFITHS, College Station, Texas (U.S.A.)
F. K. HARE, Ottawa, Ont. (Canada)
P. E. LYDOLPH, Milwaukee, Wisc. (U.S.A.)
S. ORVIG, Montreal, Que. (Canada)
D. F. REX, Boulder, Colo. (U.S.A.)
W. SCHWERDTFEGER, Madison, Wisc. (U.S.A.)
K. TAKAHASHI, Tokyo (Japan)
H. VAN LOON, Boulder, Colo. (U.S.A.)
C. C. WALLÉN, Geneva (Switzerland)

World Survey of Climatology Volume 12

Climates of Central and South America

edited by

Werner SCHWERDTFEGER

Department of Meteorology
University of Wisconsin
Madison, Wisc. (U.S.A.)

ELSEVIER SCIENTIFIC PUBLISHING COMPANY
Amsterdam-Oxford-New York 1976

ELSEVIER SCIENTIFIC PUBLISHING COMPANY
335 Jan van Galenstraat
P.O. Box 211, Amsterdam, The Netherlands

AMERICAN ELSEVIER PUBLISHING COMPANY, INC.
52 Vanderbilt Avenue
New York, New York 10017

Library of Congress Card Number 78-103353
ISBN 0-444-41271-9
With 106 illustrations and 264 tables

Printed in The Netherlands

World Survey of Climatology

Editor in Chief: H. E. LANDSBERG

Volume 1 General Climatology, 1
Editor: H. FLOHN

Volume 2 General Climatology, 2
Editor: H. FLOHN

Volume 3 General Climatology, 3
Editor: H. FLOHN

Volume 4 Climate of the Free Atmosphere
Editor: D. F. REX

Volume 5 Climates of Northern and Western Europe
Editor: C. C. WALLÉN

Volume 6 Climates of Central and Southern Europe
Editor: C. C. WALLÉN

Volume 7 Climates of the Soviet Union
by P. E. LYDOLPH

Volume 8 Climates of Northern and Eastern Asia
Editor: H. ARAKAWA

Volume 9 Climates of Southern and Western Asia
Editors: H. ARAKAWA and K. TAKAHASHI

Volume 10 Climates of Africa
Editor: J. F. GRIFFITHS

Volume 11 Climates of North America
Editors: R. A. BRYSON and F. K. HARE

Volume 12 Climates of Central and South America
Editor: W. SCHWERDTFEGER

Volume 13 Climates of Australia and New Zealand
Editor: J. GENTILLI

Volume 14 Climates of the Polar Regions
Editor: S. ORVIG

Volume 15 Climates of the Oceans
Editor: H. VAN LOON

List of Contributors to this Volume

M. A. ALAKA
Techniques Development Laboratory
National Weather Service, NOAA
Silver Spring, Maryland (U.S.A.)

A. M. JOHNSON
Cerro de Pasco Corporation
La Oroya (Peru)

H. H. LETTAU
Department of Meteorology
University of Wisconsin
Madison, Wisconsin (U.S.A.)

A. MILLER
Department of Meteorology
California State University-San José
San José, California (U.S.A.)

W. H. PORTIG
U.S. Army Tropic Test Center
Fort Clayton, Canal Zone

F. PROHASKA
Department of Geography
University of Wisconsin
Milwaukee, Wisconsin (U.S.A.)

L. R. RATISBONA
Servicio Meteorologico Nacional
Rio de Janeiro (Brazil)

J. W. SNOW
Department of Meteorology
University of Wisconsin
Madison, Wisconsin (U.S.A.)

Contents

Chapter 1. INTRODUCTION
by W. Schwerdtfeger

The atmospheric circulation over Central and South America 1
References 11

Chapter 2. THE CLIMATE OF ARGENTINA, PARAGUAY AND URUGUAY
by F. Prohaska

Introduction 13
Wind 13
    The westerlies of the mid-latitudes (Patagonia and Tierra del Fuego) 14
    Wind direction, 14—Wind velocity, 15
    The wind regime of the subtropical high-pressure systems 17
    Area of influence of the Pacific anticyclone with prevailing southerly winds, 17—
    Area of influence of the Atlantic anticyclone with prevailing northeasterly winds,
    and easterlies and southeasterlies, 18
    The wind regime of the continental subtropical low 21
    The wind regime of the cordillera and the Altiplano 22
Cloudiness 23
    Western Paraguay, northwestern Argentina: maximum in summer and minimum
    in winter 25
    Uruguay, Argentine mesopotamia, pampa and the eastern Chaco: small annual
    range with maximum in winter 27
    Central and eastern Paraguay, western Chaco, southern pampean sierras: double
    annual wave 28
    Cuyo: small annual range and low degree of cloudiness 28
    Patagonia 30
    Relation between cloudiness and number of clear and cloudy days 31
Duration of sunshine 32
    East puna 33
    The northwestern provinces 34
    Chaco and central Argentina 35
    Argentine–Uruguayan mesopotamia, La Pampa, Cuyo and Patagonia 35
    Tierra del Fuego 35
Global radiation 36

Temperature 36
    Annual averages 37
    Monthly averages 37
    Vertical temperature gradient 38
    Annual variation 43
    Daily variation 44
    Maximum temperatures 47
    Minimum temperatures 48
    Soil temperatures in Argentina 49
Humidity 50
    Vapor pressure 50
    Relative humidity 54
    Evaporation 56
Precipitation 57
    Tropical regime with a single maximum 60
    Continental summer rains of the subtropics 62
    Transitional zone to the maritime winter rains of the subtropics 63
    The winter rains of the temperate latitudes 64
    The uniform rainfall distribution throughout the year of the subpolar region . . 65
    Rainfall variability 66
    Days with precipitation 67
Thunderstorms 69
Appendix—Climatic tables 70
References 71
    General climatic descriptions, data, bibliographies 71
    Special regions and elements 72

Chapter 3. THE CLIMATE OF CHILE
by A. MILLER

Introduction 113
The arid north 117
Central Chile 122
Southern Chile 125
Appendix—Climatic tables 129
References 129

Chapter 4. THE CLIMATE OF PERU, BOLIVIA AND ECUADOR
by A. M. JOHNSON

Introduction 147
Precipitation 150
    General 150
    Eastern lowlands 155

Contents

Andes—eastern slopes 156
Andes—central 157
Andes—western slopes 164
Coastal plain and islands 165
Temperature 172
General 172
Eastern lowlands 175
Andes—eastern slopes 176
Andes—central 176
Andes—western slopes 178
Coastal plain and islands 179
Radiation, sunshine and cloudiness 182
Surface winds 185
Acknowledgements 188
Appendix I—Dynamic and energetic factors which cause and limit aridity along South
America's Pacific coast (by H. H. Lettau) 188
Appendix II—High thunderstorm frequency over the subtropical Andes during the summer;
cause and effects (by W. Schwerdtfeger) 192
Appendix III—The water-budget of Lake Titicaca (by W. Schwerdtfeger) 195
Appendix IV—Station pressure values vs. elevation of places in the high mountains (by
W. Schwerdtfeger) 199
Appendix V—Climatic tables 199
References 200

Chapter 5. THE CLIMATE OF BRAZIL
by L. R. Ratisbona

Introduction 219
Climatic factors 219
Latitude 220
Relief 220
Nature of the surface 221
Pressure centres 222
Air masses 225
Climatic elements 226
Winds 226
Cloudiness 229
Duration of sunshine 230
Distribution, 230—Annual variation, 232
Temperature 232
Average temperature, 233—Maximum temperatures, 236—Minimum temperatures,
237—Anomaly in temperature, 237—Annual variation, 238—Annual range, 240—
Continentality, 240—Daily variation, 240—Day to day variation, 241—Warm
days and nights, 242
Moisture 242

Vapour pressure, 242—Relative humidity, 244

Precipitation 244

Distribution, 245—Number of days, 245—Average intensity, 245—Maximum intensity, 248—Regimes, 248—Variability, 251

Thunderstorms 251

Fog and haze 253

Climatic types 255

Regional characteristics 260

Some problems 264

The droughts of the northeast 264

Rains in the interior of tropical Brazil 265

The rains of the east coast 266

Acknowledgements 267

Appendix—Climatic tables 267

References 267

Additional references 269

Chapter 6. THE CLIMATE OF NORTHERN SOUTH AMERICA
by J. W. SNOW

Preliminary remarks 295

The area and climate in general 295

Chapter outline and terminology 298

French Guiana 299

Geography 299

Climate 300

General features and controls, 301—Rainfall, 302—Other elements, 305

Surinam 305

Introduction and geography 305

Climate 307

Precipitation, 307—Diurnal distribution of rainfall, 309—Pressure and wind, 312—Storms, 313—Temperature, 314

Guyana 321

Geography 321

Climate 322

Rainfall, 322—Other climatic elements, 325

Venezuela 326

Introduction and geography 326

General climatic factors and climate types 327

Solar radiation, 327—Large scale circulation, 328—Climate types, 329

Precipitation 329

Basic annual distributions, 329—Annual totals, areal apportionment, 330—North and west section, 332—Llanos, 333—Lllanos pattern, 334—Guyana Highlands, 337

Temperature 338

Dependences, 338—Areal means, 340—Basic patterns, 341—Hourly distribution, 342

Rainfall producers 344

Percent of rainfall per rain-day, 344—Northern Hemisphere disturbances, 345—Southern Hemisphere disturbances, 346—Equatorial trough disturbances, 346

South Caribbean dry zone 348

Extent and characteristics, 348—Cause(s), 351—Dry zone résumé, 357

Colombia 358

Introduction 358

Caribbean lowland section 360

Andes section 360

Geography, 361—Rainfall, 362—Other climatic elements, 366

Llanos section 368

Amazon section 369

Pacific coast section 370

Excessive rainfall, 370—Geography, 371—Climate, 371—Factors effecting the raininess, 375

Recapitulation 376

Appendix—Climatic tables 376

References 376

Other references and additional data sources 379

Chapter 7. THE CLIMATE OF CENTRAL AMERICA
by W. H. PORTIG

The climatological elements 405

Wind 405

Barometric pressure 410

Cloudiness 412

Sunshine 414

Temperature 414

Annual variation of temperature, 415—Temperature variation with altitude, 416—Diurnal temperature variation, 417

Moisture 417

Rainfall 418

Introduction, 418—Sea, 419—Central America, 424—Islands near Central America, 432—Annual rainfall totals of islands other than those near Central America, 433—Puerto Rico and the Lesser Antilles, 438

Thunderstorms 445

Visibility 446

Weather 447

Introduction 447

"Grosswetterlagen" 447

Local developments 448

Appendix—Climatic tables 451

References 451

Chapter 8. CLIMATOLOGY OF ATLANTIC TROPICAL STORMS AND HURRICANES
by M. A. ALAKA

Classification of disturbances in the tropics 479
Sources of information 479
Hurricane formation 480
    The birthplace of Atlantic hurricanes 480
    Availability of latent energy, 483—Vorticity due to earth's rotation, 485
    Frequency of Atlantic hurricanes 489
    Seasonal variation, 489—Secular variations and trends, 495
Hurricane motion 496
    Types of storm movements 496
    Persistence and recurvature 497
    Tropical cyclones affecting the United States 498
Structural aspects and related phenomena 500
    The wind field 500
    The pressure field 500
    Rainfall and floods 501
    The storm surge 505
    Hurricane tornadoes 506
Acknowledgement 507
References 507

REFERENCE INDEX 511

GEOGRAPHICAL INDEX 517

SUBJECT INDEX 529

Chapter 1

Introduction

W. SCHWERDTFEGER

From steaming tropical rain forests to arid, desert-like coasts, from stormy peaks of the earth's longest mountain chain down to glaciers at sea level; truly, a great variety of climates can be found on the 13% of our planet's total land mass to which this volume refers.

Extreme climatic conditions, of course, are not at all propitious for that little outpost of civilized human activities about which we are directly concerned, that is, a properly installed meteorological station with white paint on its thermometer screen, continuously and diligently operated and maintained over several decades. Furthermore, the priorities of developing or changing economies, so typical of modern Latin America, do not favor the prompt and complete publication of existing observational records. Under such circumstances, more than usual emphasis has been given to indirect climatological evidence, from vegetation patterns to old travelogs. It also appeared of primordial importance to find as author for each region a true expert whose intimate personal knowledge of, and experience in, his area could help to correctly interpret the often inadequate information and to fill the many gaps. This is the reason why seven different authors were asked to contribute to the volume. If a full homogeneity of the entire text has thus not been achieved, it should be considered as the lesser of two possible deficiencies.

When K. Knoch wrote the introduction to his text "Klimakunde von Südamerika" for Köppen and Geiger's *Handbook of Climatology* (1930), he said: "Because reliable and comparable observational data always must be the principal basis for a description of the climate (of a large region), a climatology of South America can hardly satisfy high expectations." All the fast changes of the world we live in, and all progress of scientific activities notwithstanding, this statement is as true today as it was at the time it was written, more than forty years ago.

The atmospheric circulation over Central and South America

For an understanding of the large scale climatic features of Central and South America, a short comment on the predominant pattern of the atmospheric motions is in order. On a larger scale still, an up-to-date description of the meteorological characteristics of the entire Southern Hemisphere, its circulation, radiation budget, heat and water balance, and peculiar aspects of its synoptic meteorology, has been given by VAN LOON et al., 1972.

The seasonal sea level pressure maps for the southern summer (XII–II; Fig.1) and the

1

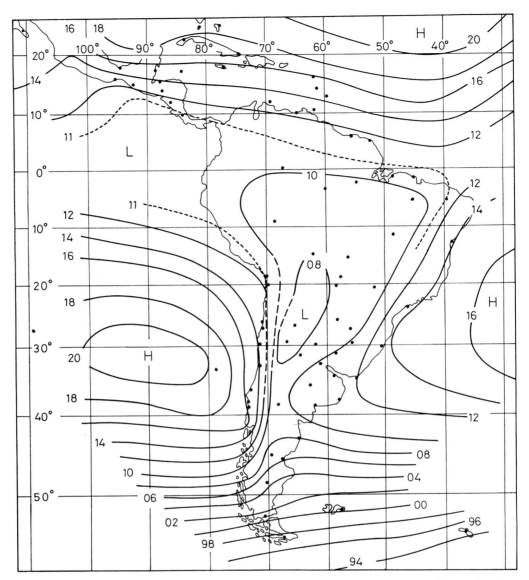

Fig.1. Atmospheric pressure at sea level, average of the three months December–February (southern summer). 20 = 1,020 mbar.

southern winter (VI–VIII; Fig.2) are essentially self-explaining. Only a few features deserve a brief comment because some earlier climatological descriptions have expressed differing opinions.

(*a*) In the equatorial belt, 10°S to 10°N, the pressure over the East Pacific ("downwind" when considering the main, although weak, tropospheric flow) is noticeably lower than over the West Atlantic ("upwind"). The same is true, incidentally, for the two other seasons; that is, it appears to be a perennial feature.

(*b*) In the southern summer, the center of the continental heat low is clearly located between 20° and 30°S, over the relatively high and dry terrain east of the Andes. For this region, of course, as for the major part of the continent, the reduction of station pressure to sea level is a problematic procedure. Therefore, it is important to note that an analysis

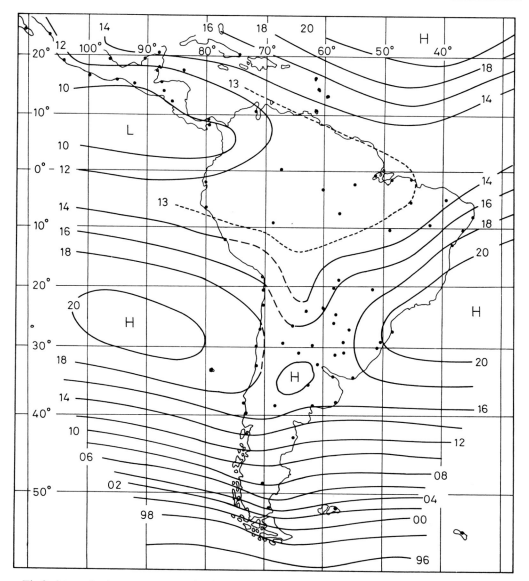

Fig.2. Atmospheric pressure at sea level, average of the three months June–August (southern winter). 96 = 996 mbar.

for the 1,000-m level (SCHWERDTFEGER, 1954) as well as surface wind frequency statistics (as given in the climatic tables of Chapter 2) confirm this location.

(*c*) Also in the southern summer, the high pressure cell over the eastern South Pacific is centered at about 32°S (and 90°W), that is, at a latitude a few degrees higher than the comparable high pressure area of the western South Atlantic. The Pacific cell also appears to be the stronger one.

(*d*) On the other hand, in the southern winter, the two cells show about equal strength and equal latitudinal extension. Of the two very weak troughs on both sides of the small winterly continental high pressure cell (centered at 34°S, 64°W), only the eastern one can be confirmed by surface wind frequencies.

(*e*) There has been some controversy, recently, regarding the sea level pressure pattern along the northern coast of South America and over Venezuela. In this regard it may be stated, therefore, that surface wind observations as well as the available aerological evidence militate against a high pressure cell which would have to be positioned over the interior of Venezuela nearly year-round if the published pressure and station-elevation data were accepted.

The map of the change of sea level pressure from the southern summer to the southern winter (Fig.3) has at first been elaborated as a consistency-check for the not-always-reliable original data. It must also be kept in mind that for the tropical and for the oceanic regions the maxima and/or the minima in the average annual march of the sea level pressure do not necessarily occur in these two seasons. For example, the subtropical high of the North Atlantic undergoes a semi-annual oscillation in the value of its central

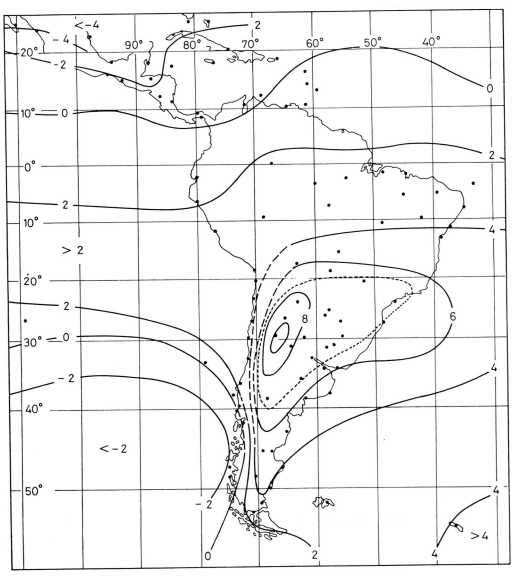

Fig.3. Average change (mbar) of atmospheric pressure at sea level, from southern summer to southern winter.

pressure, attaining maxima in February and July. Nevertheless, a map of the summer to winter change reveals some interesting aspects.

(*a*) At low latitudes, the zero-line clearly is in the Northern Hemisphere, close to the "meteorological equator", whereas along the geographic Equator itself values of about +2 mbar are found.

(*b*) The area of maximum positive change, over northwest Argentina, tends to confirm the concept of the thermal character of the summer low, just east of the Andes.

(*c*) There is a noticeably different pattern of change-values over the western South Atlantic and the eastern South Pacific. Some of the evidence is based solely on ship-observations (U.S. Navy, 1958, 1959), especially in the Southeast Pacific where no island station exists south of Juan Fernández (34°S), and west of Evangelistas (75°W); this may have led to some bias toward higher pressure during the winter season. Nevertheless, the few existing island- and coastal stations with long and relatively reliable series leave no

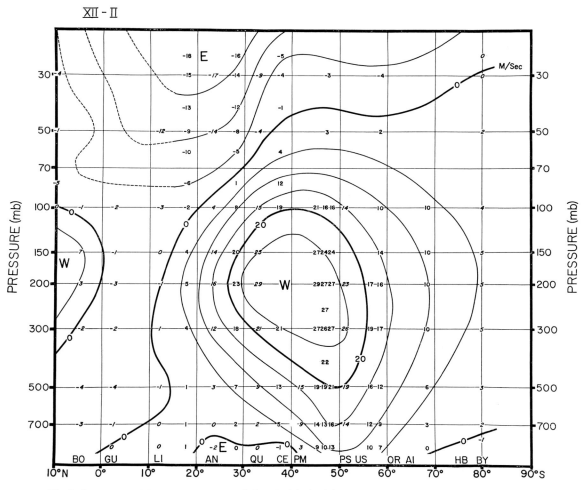

Fig.4. Average west to east component of the wind (m/sec) between 10°N and 80°S latitude, 1,000 to 30 mbar, over South America. The two-letter abbreviations above the 1,000-mbar line denote the radiosonde stations whose data were used. Three months period December–February. (For details, see SCHWERDT-FEGER and MARTIN, 1964).

doubt that the patterns are indeed different: the changes over the two ocean areas are of opposite sign.

The average zonal flow of the atmosphere between surface and the 25-mbar level (approx. 25 km in the Tropics, 25.5 and 22 km in the Antarctic summer and winter, respectively), roughly along the meridian 70°W, is shown in Fig.4. In the tropical regions, the main zonal flow (from the east) is weak, and the same is true for the meridional motion. Correspondingly, tropospheric air masses will have a long residence time in this area and can be efficiently modified according to the characteristics of the underlying surface. At all latitudes south of about 30°S, during a major part of the year even south of 25°S, the tropospheric flow above the level of the average Andes crest-line is moderate to strong from the west. At these latitudes and above the average crestline, then, new air masses from the Pacific are almost continuously carried through, and fast changes of the conditions in the upper half of the troposphere can affect the weather and climate in the lower layers.

It might be interesting to note that in the middle latitudes, between 40° and 50°S approxi-

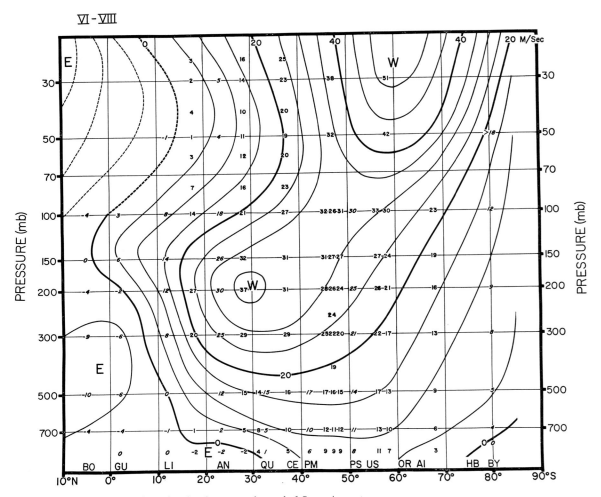

Fig.5. Same as Fig.4, for the three months period June–August.

mately, there is a belt in which the mean zonal flow of the troposphere is faster in the summer than it is in the winter. As VAN LOON (1964) has shown, this is a phenomenon of the *Southern* Hemisphere atmospheric circulation, related to the annual variation of the mean meridional temperature gradient in the troposphere. It has no counterpart in the Northern Hemisphere where the relative abundance of land area tends to minimize the summertime meridional temperature gradient.

In higher latitudes, south of 50°S, there is another important feature of the atmospheric circulation which cannot be seen from the two cross-sections given in Figs.4 and 5. A month-to-month analysis (SCHWERDTFEGER and PROHASKA, 1956; SCHWERDTFEGER, 1960, 1962a, 1963) shows that the zonal flow in the subpolar belt is characterized by a predominantly *semi-annual* variation, with the maxima occurring in March and September. This can be proven from the surface and upper air observations of Port Stanley (52°S), Ushuaia (55°S) and Argentine Islands (65°S), and from over 77,000 ship-observations taken primarily during the decades prior to the opening of the Panama Canal (SCHWERDTFEGER, 1962b, 1963). It is likewise reflected in the annual march of various weather elements near the southern tip of the continent, most clearly in the equinoctial maxima of the westerly winds over Tierra del Fuego and the Drake Passage. In the area around Cape Horn, the old sailors' adage of the equinoctial storms is very real indeed.

Between 10°N and 40°S, the wind data represented in Figs.4 and 5 refer to the west-coast of the continent. As far as can be deduced from vertical cross-sections of the mean zonal geostrophic wind south of 10°S, along the meridians 70° and 40°W during the four mid-season months (VAN LOON et al., 1971), the longitudinal variations in the tropospheric zonal flow across the continent do not appear to be large. It might be significant, however, that the strength of the average jet-stream, at about 200 mbar (11 to 12 km) between 30° and 45°S, is greater over the east-coast than over the west-coast in January and April,

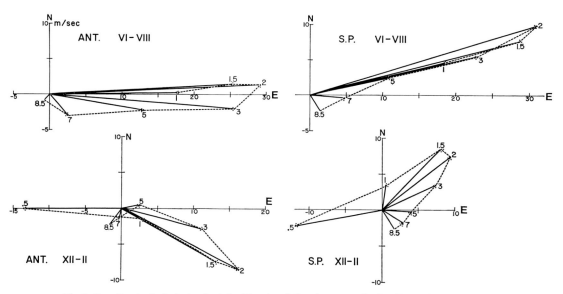

Fig.6. Resultant winds (m/sec) at the Tropic of Capricorn, at the Pacific and near the Atlantic coast of South America, in the southern winter (upper two graphs) and southern summer (lower two graphs). *ANT* = Antofagasta; *S.P.* = São Paulo. The solid lines represent the vector average of the winds at the levels 850, 700, 500, 300, 200, 150, and 100 mbar. The dashed lines show the thermal wind for the respective atmospheric layers.

while the opposite is the case in July and October. This could be related to the intense heating, in particular the release of latent heat, over the continent north of the Tropic of Capricorn during the southern summer and early fall.

Regarding the mean vector wind, a well-founded west versus east comparison can be made by means of the wind-soundings of Antofagasta (70.5°W) and São Paulo (46.5°W), both stations being almost exactly at the latitude of the Tropic. As Fig.6 shows, there are remarkable, characteristic differences. They are less pronounced in winter than in summer. In the former season (upper two graphs of Fig.6), the resultant wind vectors (full lines) over Antofagasta in all layers of the troposphere above the level of the mean crest of the Andes (about 580 mbar) are practically due west to east, and so is the thermal wind vector (dashed lines), at least up to 300 mbar. Over São Paulo the average wind structure is quite similar, only with a turn of 15° to 20° toward the north, that is, with a significant meridional component from the south.

In the summer (lower two graphs of Fig.6) a striking difference becomes evident. The thermal wind vectors (parallel to the mean isotherms, the warmer air to the left of the arrow) between 500 and 200 mbar at Antofagasta, and between 850 and 200 mbar at São Paulo, clearly indicate the presence of much warmer air over the continent than over the surrounding oceans, and this warm air's effect on the wind pattern itself. It thus appears that at the subtropical latitudes the thermal regime of the continents affects the general atmospheric stream system much more than the "dynamic" processes induced by a mountain chain, even when the mountains are as formidable and nearly perpendicular to the main flow as the Andes (GUTMAN and SCHWERDTFEGER, 1965).

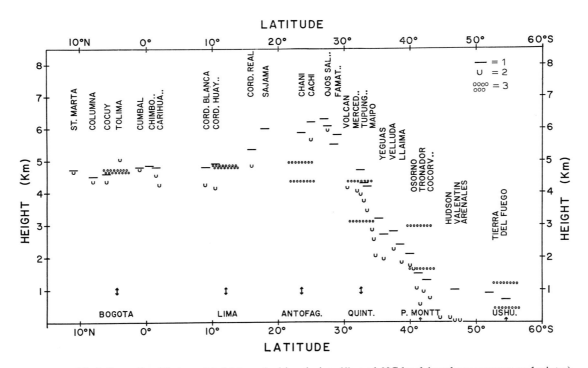

Fig.7. Snow line (*1*), lowest height reached by glaciers (*2*), and 0°C level (southern summer and winter) (*3*) in the Andes of South America.

In the region of the Andes themselves, the effects of the predominant tropospheric stream-pattern and its variation from north to south become immediately visible in the varying height of the snowline and the state of glaciation of the higher mountain massifs (Fig.7). North of about 12°S (latitude of Lima and Huancayo, Peru) the easterly flow persists up into the mid-troposphere, wherein the mountains penetrate, throughout the year. South of 35°S, at the height of the peaks and above, the westerly flow strongly increases poleward. In the intermediate subtropical region, a well defined snow-line cannot always be found even where the mountain tops exceed the 6 km level (PROHASKA, 1962); where such a limit has been observed, it clearly occurs at a higher elevation than the snow line in the tropical region (HASTENRATH, 1967). The change of the glaciation of the mountains from the subtropical regime, with its pronounced seasonal variation of wind, precipitation, cloudiness and other climatic parameters, to the belt of the perennial westerlies is quite spectacular. The lowest level to which glacier tongues descend drops from 4 km at 32°S to 500 m at 42°S, and by 46.6°S the glacier San Rafael extends all the way down to the waters of the Pacific Ocean (LLIBOUTRY, 1956).

The subtropical mountain climate of the Andes combines all conditions for a bizarre and beautiful phenomenon of ablation by evaporation, the *penitents*, which are narrow, stalagmite-like walls of ice and firn (Fig.8). The necessary conditions are: (*a*) considerable snowfall in winter, no precipitation in summer; (*b*) little cloudiness and the presence of dry air in late spring and summer; (*c*) as a consequence of (*b*) and the high elevation (above 3,000 m), high intensity of direct incoming radiation. Indeed, the most spectacular fields of penitents are formed between about 20° and 35°S. In this region firn-spikes or ice-prongs can easily reach a height of 3 m or more wherever the original depth of the

Fig.8. Penitents in the high Andes (photo A. E. Corte, 1954).

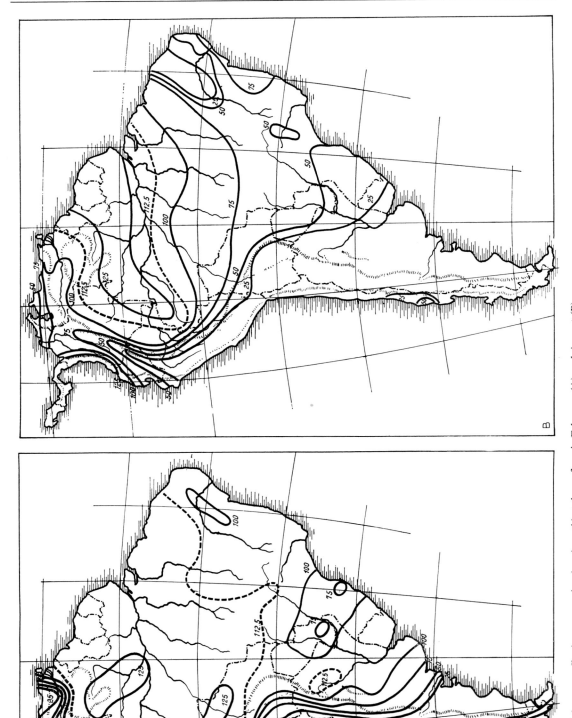

Fig.9. ALBRECHT's (1965) maps of the mean effective evaporation (mm/month) at the surface, in February (A) and August (B).

winter snow accumulation on wind-protected slopes or troughs of the terrain was adequate. The most detailed description of the phenomenon was given by TROLL, 1942 (also see LLIBOUTRY, 1954).

The penitents in the high mountains are a phenomenon of great aesthetic appeal, but of no importance for the application of climatological data and analysis to the more practical needs of mankind in agriculture and hydro-economy. To give information usable in these two fields is, of course, a main concern of the following regional descriptions. There is, however, one important climatic factor on which very few observing stations in Latin America provide adequate information, that is, effective or natural evaporation. In order to fill this gap somewhat summarily, at least for *South* America, Fig.9 contains the reproduction of two maps, for February and August, of the average natural evaporation in mm/month, as computed by ALBRECHT (1965). For the arid parts of the continent, the natural evaporation has been assumed as equal to the precipitation. Outside the arid zones, Albrecht's evaporation values are based upon heat budget computations together with precipitation and run-off records.

References

ALBRECHT, F., 1965. Untersuchungen des Wärme- und Wasserhaushalts der südlichen Kontinente. *Ber. Dtsch. Wetterdienstes*, 14(99): 54 pp.

GUTMAN, G. J. and SCHWERDTFEGER, W., 1965. The role of latent and sensible heat for the development of a high pressure system over the subtropical Andes, in the summer. *Meteorol. Rundsch.*, 18(3): 69–75.

HASTENRATH, ST. L., 1967. Observations on the snow line in the Peruvian Andes. *J. Glaciol.*, 6(46): 541–550.

KNOCH, K., 1930. Klimakunde von Südamerika. In: W. KÖPPEN and R. GEIGER (Editors), *Handbuch der Klimatologie*, Borntraeger, Braunschweig, Vol.2, Pt.G, 349 pp.

LLIBOUTRY, L., 1954. The origin of penitents. *J. Glaciol.*, 2(15): 331–338.

LLIBOUTRY, L., 1956. *Nieves y Glaciares de Chile*. Ediciones de la Universidad de Chile, Santiago, 471 pp.

PROHASKA, F., 1962. Algunos aspectos del clima de la alta cordillera y de la Puna Argentina. *Inst. Nacl. Tecnol. Agropecuaria, Buenos Aires, Publ.*, 79:21–30.

SCHWERDTFEGER, W., 1954. Análisis sinóptico y aspecto climatológico de dos distintos tipos de depresiones báricas en el Norte de la Argentina. *Meteoros*, IV(4): 301–323.

SCHWERDTFEGER, W., 1960. The seasonal variation of the strength of the Southern Circumpolar Vortex. *Mon. Weather Rev.*, 88(6): 203–208.

SCHWERDTFEGER, W., 1962a. Die halbjährige Periode des meridionalen Temperaturgradienten in der Troposphäre und des Luftdrucks am Boden im Südpolargebiet, ihre Erscheinungsform und kausalen Zusammenhänge. *Beitr. Phys. Atmosph.*, 35(3/4): 234–244.

SCHWERDTFEGER, W., 1962b. *Meteorología del Area del Pasaje Drake*. Publicación H 410 del Servicio de Hidrografía Naval, Secretaría de Marina, República Argentina, Buenos Aires, 78 pp.

SCHWERDTFEGER, W., 1963. The Southern Circumpolar Vortex and the Spring Warming of the Polar Stratosphere. *Proc. Int. Symp. on Stratospheric and Mesospheric Circulation, August 1962, Berlin—Meteorol. Abhandl. Inst. Meteorol. Geophys. Freie Univ. Berlin*, XXXVI: 207–224.

SCHWERDTFEGER, W. and PROHASKA, F., 1956. Der Jahresgang des Luftdrucks auf der Erde, und seine halbjährige Komponente. *Meteorol. Rundsch.*, 9, pp.33–43; pp.186–187.

SCHWERDTFEGER, W. and MARTIN, D. W., 1964. The zonal flow of the free atmosphere between 10°N and 80°S, in the South American sector. *J. Appl. Meteorol.*, 3(6): 726–733.

TROLL, C., 1942. Büsserschnee in den Hochgebirgen der Erde. *Petermanns Geogr. Mitt., Ergaenzungsh.*, 240, 68 pp.

U.S. Navy, 1958, 1959. *Marine Climatic Atlas of the World, IV. South Atlantic Ocean*, 50 - Cl - 531, 17 pp., 267 charts; *V. South Pacific Ocean*, 50 - Cl - 532, 15 pp., 267 charts. U.S. Govt. Printing Office, Washington, D.C.

VAN LOON, H., 1964. Mid-season average zonal winds at sea level and at 500 mb south of 25°S, and a brief comparison with the Northern Hemisphere. *J. Appl. Meteorol.*, 3: 554–563.

VAN LOON, H., TALJAARD, J. J., JENNE, R. L. and CRUTCHER, H. L., 1971. *Climate of the Upper Air, Southern Hemisphere, II. Zonal Geostrophic Winds.* National Center for Atmospheric Research, Boulder, Colo., NAVAIR 50 - IC - 56 and NCAR TN/STR - 57, 40 pp.

VAN LOON, H., TALJAARD, J. J., SASAMORI, T., LONDON, J., HOYT, D. V., LABITZKE, K. and NEWTON, C. W., 1972. Meteorology of the Southern Hemisphere. *Meteorol. Monogr.*, 13(35): 263 pp.

WELCH, M. M., 1921. Bibliography on the climate of South America. *Mon. Weather Rev. Suppl.*, 18: 42 pp.

Chapter 2

The Climate of Argentina, Paraguay and Uruguay

FRITZ PROHASKA[1]

Introduction

The territory comprising Argentina, Paraguay and Uruguay has an area of almost 3.4 million km² and extends from 19° to 55°S, i.e., from the tropics to the subpolar region of the archipelago of Tierra del Fuego. The subtropical section is essentially a large plain (Gran Chaco and the pampa) which rises gradually from east to west, interspersed by the pampean sierras and sheltered from the Pacific Ocean by the Cordilleras de los Andes, the most important climatic divide in South America. Uruguay and eastern Paraguay consist of rolling plains, while Patagonia shows a typical meseta landscape. The dominant climatic characteristic of the subtropical eastern (maritime) part is its high degree of atmospheric humidity, whereas in the remaining regions the dryness of the air is the decisive factor (except in the southern cordillera). In the subtropical continental parts this dryness is coupled with intense solar radiation, while in Patagonia it is accompanied by the well-known high velocity of the prevailing westerly winds, emphasizing, thus, the autochthonous or allochthonous (advective) features in the climate of the respective areas. Despite the absence of orographic divides these three main climatic regions can be clearly delimited: the transition zone between the moist and the dry subtropical regions is located along the Rio Paraguay and Rio Parana and the transition from the subtropic climate towards the west wind zone of the mid-latitudes takes place over the middle and lower courses of the Rio Negro and Rio Colorado, at about 39°S. The climate of the higher parts of the Andes is controlled mainly by the atmospheric circulation of the East Pacific Ocean and is discussed in Chapter 3.

Wind

The average wind conditions of the territories under discussion are controlled by three circulatory systems: the circulation within and between the subtropical high-pressure cells (the Pacific anticyclone controls the extreme west, the Atlantic anticyclone the eastern part of the subtropical region), the wind system which develops according to the intensity of the quasi-stationary low in the Gran Chaco, and the prevailing westerlies of the middle latitudes. Local winds (pampero, sudestada, zonda, land and sea breeze, etc.) are temporarily decisive and may assume climatic importance for certain regions. In ad-

---

[1] Fritz Prohaska died at Milwaukee, Wisconsin, U.S.A., December 22, 1970.

dition, the Andes and the Altiplano develop their own wind regimes in areas that do not extend into the prevailing westerlies of the lower and middle troposphere.

The westerlies of the mid-latitudes (Patagonia and Tierra del Fuego)

In few parts of the world is the climate of the region and its life so determined by a single meteorological element, as is the climate of Patagonia by the constancy and strength of the wind.

The region is situated between the southern flank of the semipermanent, subtropical high-pressure belt or cells, whose direct influence extends to about 40°S throughout the year, and the intensive subpolar low-pressure trough, which is centered approximately on the Antarctic Circle. Since these pressure systems undergo only small seasonal and spatial variations and show little changes in their intensity, west winds prevail in Patagonia during the entire year. This fact, thus, provides the best criterion for the delimitation of Patagonia as a uniform climatic region.

Also geographically the meseta landscape of Patagonia with its typical grass and shrub steppe is a unit and extends from the Rio Colorado and Rio Negro, i.e., from 39°–40°S, to the southern tip of the continent (52°S) and includes the northern part of Tierra del Fuego.

Wind direction

The annual frequency distribution of the wind directions of individual stations shows that west winds count for 50–70% of all observations (including calms). Within the western sector an annual variation is well expressed due to small seasonal displacements of the predominant pressure systems. In winter, the isobars parallel the latitudes, since a ridge over the continent connects the Atlantic and the Pacific high-pressure centers both at approximately the same latitude and with similar intensities. In the summer, on the other hand, the two high-pressure cells are separated over the continent by a low-pressure trough. Moreover, the center of the Pacific high-pressure area is then farther south and close to the South American coast, while the center of the Atlantic cell is situated far off the coast to the east and extends only a ridge to South America. Therefore pressure gradient and wind systems are better pronounced over western than over eastern Patagonia.

As a consequence, winter shows a uniform west wind circulation whereas summer has a relatively weak zonal component superimposed on the meridional gradient, a fact that gives the west winds a southerly component (west-southwest to southwest winds). At 45°S between 75° and 65°W the zonal gradient amounts to 6 mbar in January and disappears in July. Winds from the north and east occur only in winter and spring with certain regularity. In these seasons the "blocking action" is a characteristic but not very frequent pressure pattern (GRANDOSO and NUÑEZ, 1955); it can have far reaching consequences, since air masses from the Atlantic may penetrate up to the eastern slope of the Andes, producing sometimes intense precipitations.

Wind velocity

The west winds are characterized not only by their prevalence during the entire year but also by their intensity. For this reason Patagonia is one of the stormiest of any permanently inhabited regions. A geostrophic wind of about 7 m/sec corresponds to the mean annual pressure gradient between 40° and 60°S at 75°W, which rises to 8 m/sec over southern Patagonia. This figure is approached to within 10–20% by many stations as a yearly average, and is reached and exceeded at some of them.

As a matter of fact, these meridional pressure gradients and the resulting wind velocities exhibit small but distinct temporary changes which are caused by the annual variation in intensity and extension of the above mentioned pressure centers (SCHWERDTFEGER and PROHASKA, 1955). The annual variation of the pressure gradients and wind speed are shown in Table I. The mean values of wind speed have been calculated, taking into account all stations with records during the decade 1951–1960 (14 in total) in the three southernmost Argentine provinces irrespective whether they are freely exposed (Rio Gallegos), protected from the wind (Lago Argentino), subject to locally caused intensification (Perito Moreno), or orographically caused deviations of the westerlies (Ushuaia). All of Patagonia has wind speeds above the annual average from October through February (summer). At the southern coast the increase starts in September and lasts until March, up to which month higher wind speeds are also observed in the interior of Patagonia. In winter, the monthly average of wind speed is substantially lower in the interior than at the coast, except at orographically protected harbors (Puerto Santa Cruz, Ushuaia). This is the result of a high frequency of calms produced by a strong thermic

TABLE I

AVERAGE WIND SPEED (IN M/SEC; 1951–1960) IN PATAGONIA AND MEAN MERIDIONAL PRESSURE GRADIENT IN 70°W BETWEEN 45° AND 55°S

Month	Wind speed		Pressure grad. per degree lat. (mbar)
	coast[1]	inland[2]	
Jan.	7.0	7.8	1.5
Feb.	6.7	6.8	1.4
Mar.	5.9	6.4	1.3
Apr.	5.5	4.8	1.2
May	5.4	4.0	1.1
June	4.8	3.2	1.1
July	5.7	4.3	1.2
Aug.	6.0	5.1	1.2
Sept.	6.3	5.4	1.3
Oct.	6.8	6.8	1.5
Nov.	7.8	7.9	1.4
Dec.	7.1	7.7	1.4
Ann.	6.2	5.8	1.3

[1] Nine stations: Camarones, Comodoro Rivadavia, Trelew, Faro Cabo Virgenes, Puerto Deseado, Rio Gallegos, San Julian, Puerto Santa Cruz, Ushuaia.
[2] Five stations: Esquel, Sarmiento, Gobernador Gregores, Lago Argentino, Perito Moreno.

inversion during night time in central Patagonia. During day time, when the west winds are blowing, they are stronger inland than at the coast. On the other hand the coastal winds are more permanent, since local winds become operative when the west drift dies down (see Table II).

TABLE II

RELATIVE FREQUENCY (%) OF THE ANNUAL DISTRIBUTION OF THE DIRECTION OF THE WIND AND THE CORRESPONDING AVERAGE WIND SPEED (M/SEC) AT GOBERNADOR GREGORES AND AT PUERTO DESEADO (1951–1960)

N		NE		E		SE		S		SW		W		NW		C (%)	Mean speed
%	sp.	%	sp.	%	sp.	%	sp.	%	sp.	%	sp.	%	sp.	%	sp.		

Gobernador Gregores (48°47′S 70°15′W; 358 m):

| 5 | 7.8 | 1 | 4.4 | 1 | 5.0 | 1 | 3.3 | 5 | 5.6 | 8 | 8.1 | 33 | 8.6 | 14 | 10.3 | 32 | 6.1 |

Puerto Deseado (47°44′S 65°55′W, 79 m):

| 17 | 7.8 | 5 | 6.1 | 6 | 5.3 | 4 | 6.1 | 8 | 6.9 | 11 | 7.2 | 35 | 7.8 | 10 | 8.6 | 4 | 7.2 |

Gobernador Gregores, located practically midway between the cordillera and the coast, observes west winds with the same frequency but with greater intensity than the coast, and has at the same time a much larger number of calms. At Puerto Deseado, on the other hand, the frequency of calms is insignificant and all wind directions show a high wind speed, that result in a higher average.

The daily variation of the wind speed is generally strongly pronounced, above all in summer (i.e., in Rio Gallegos it increases from 5 m/sec during the night to 10 m/sec in the afternoon). Due to the dryness of the air the heat loss by radiation produces a stable stratification very rapidly after nightfall and thus, a decrease of wind velocity.

The published values of wind speed in the interior of Patagonia are certainly not representative. The older meteorological stations have been established in settlements that developed in the most sheltered places. This can be seen from a comparison of wind speed statistics before and after 1950, a year in which several new meteorological stations started to operate at airfields in less protected locations and with anemometers at a height of 10 m above the ground. The increase is in some places in excess of 100%.

Along the entire coast of Patagonia, land and sea breezes are superimposed on the prevailing westerlies. Therefore, the west winds have their greatest frequency in winter (when the sea breeze is hardly developed) but the maximum wind speed is observed in summer. In this season the sea breeze blows immediately when the westerlies lull. To speak of monsoon at the Patagonia coast (KNOCHE and BORZACOV, 1946) is unjustified since this circulation is limited to the coastal regions. Besides there is no seasonal change in the prevailing wind direction and neither cloudiness nor rainfall show a "monsoon-like" behavior.

The northern limit of the west wind regime runs from the Andean valleys in northern Neuquén at about 37°S across the Rio Negro to the Golfo San Matias at 41°S. The northwest–southeast position of this boundary is determined by both geographical and meteorological factors. Northwards of 37°S, the average height of the cordillera increases markedly (Aconcagua, 6,959 m, is at 33°S) and the continent widens considerably. It is therefore impossible for the west winds to blow down to the surface, and also impossible to reach as far east as the Atlantic coast throughout the year. However, the main reason why the prevalence of the westerlies ceases at these latitudes is that during the summer half year the meridional pressure gradient changes to a zonal gradient. A low-pressure trough located on the eastern side of the Andes splits the zonal high-pressure ridge. Hence, a change in wind direction from west to south takes place on the western side of the trough, and to the north or northeast on its eastern side. Consequently, the prevalence of the westerlies throughout the year is interrupted and the typical Patagonian wind conditions end.

The wind regime of the subtropical high-pressure systems (Cuyo, the pampas, mesopotamia, eastern Paraguay and Uruguay)

The transition from the prevalence of a zonal component of the wind, to a perennial dominance of a meridional component takes place in a small zone without any orographic divide, situated along the upper Rio Colorado and lower Rio Negro. Here, a zonal flow predominates in winter and a meridional flow in summer. The latter becomes dominant north of 38°S throughout the year.

The above mentioned low-pressure trough separating the south from the north winds, extends from the Chaco to northern Patagonia at about 67°W. Although it disappears in the mean-pressure field of the winter months, the daily weather maps show that also in this season the meridional component of the wind is stronger than the zonal.

As the center of the Pacific high is nearer to the South American continent as that of the Atlantic high, the pressure gradient to the west coast is stronger than to the east coast (Table III).

The difference of the zonal-pressure gradients from the two coasts to the low-pressure trough are, however, not reflected in the wind velocities since they are more than offset by the frictional effect of the Andes. Due to their protection the lowest wind velocities of Argentina are observed in their piedmont area (Cuyo) with annual averages of 1–2 m/sec.

TABLE III

PRESSURE DIFFERENCE (MBAR) AT 35°S BETWEEN THE CONTINENTAL TROUGH-LINE AT 67°W AND THE PACIFIC AND ATLANTIC COASTS RESPECTIVELY

	Jan.	Apr.	July	Oct.
77°–67°W	9	3	0	6
57°–67°W	3	2	1	2

Area of influence of the Pacific anticyclone with prevailing southerly winds (Cuyo)

The entire year south winds are predominant to the east of the Andes as far as 67°W between 27° and 37°S (or the rivers Fiambala, Desaguadero, Salado).

This is an arid region of long and wide longitudinal valleys and short and narrow transverse valleys, of bolsons, and of gently rolling plains, known as Cuyo (a geographical
unit). In the southern part of this zone south winds prevail and in the northern part southeast winds. These directions correspond exactly to the course of the isobars, since the
mean annual position of the center of the Pacific anticyclone is between 30° and 35°S. In
summer a greater frequency and a higher speed of the south winds is registered, corresponding to the stronger pressure gradient towards the continental low. In addition, an
easterly component in summer and a westerly component in winter are superposed on
this southerly wind (Table IV). However, as already mentioned, the wind speed in this
zone is generally very low throughout the year.

TABLE IV

RELATIVE FREQUENCIES OF THE MONTHLY DISTRIBUTION OF THE DIRECTION OF THE WIND AT MENDOZA,
1951–1960

	J.	F.	M.	A.	M.	J.	J.	A.	S.	O.	N.	D.	Y.
SE + S	32	26	21	17	12	13	13	15	21	29	28	28	21
SW + W	8	7	12	25	39	33	40	39	25	13	7	8	21
NW + N + NE + E	21	23	23	24	23	22	21	24	23	21	24	25	23
Calms	39	44	44	34	26	32	26	22	31	37	41	39	35

Area of influence of the Atlantic anticyclone with prevailing northeasterly winds (pampas, Argentine
mesopotamia and southwestern Uruguay), *and easterlies and southeasterlies* (eastern Paraguay,
northeastern Argentina, and northeastern Uruguay)

To the east of the low-pressure trough, the winds depend primarily on the position and
intensity of the Atlantic anticyclone. Therefore, in the Argentinian pampas and southwestern Uruguay north and northeast winds prevail throughout the year. In winter the
northward displacement of the subtropical high-pressure belt gives way in the southern
transition zone (southern part of the provinces Buenos Aires, La Pampa) to westerlies
while in summer the northern transition zone (southwestern Uruguay, Entre Rios, and
southern Santa Fe) shows the distinct east to southeast flow.
The average annual wind velocity presents an extremely uniform picture in this region.
Because of the weak pressure gradient the velocities are between 2 and 3 m/sec, rising to
about 4 m/sec only at the coast. Therefore, also in regard to the wind velocity exists a
clear difference between the wind regime in Patagonia and the pampas. It decreases
sharply in the transition zone and whereas in Patagonia the maximum speed occurs in
summer, starting from October, in the transition zone the increase of velocity begins in
September, just as in the previously described region controlled by the Pacific anticyclone,
and in the pampas and the connected regions in August. The end of the period with wind
speeds above the annual average shows a similar trend. In southern and middle Patagonia, the period of strongest winds ends in February or March, in northern Patagonia,
the piedmont area and the transition zone to the pampas in January or February, while

in the pampas itself it finishes as early as December. This shift of time of maximum wind velocity in north–south direction from early spring to summer is due to the differences of the annual variation of pressure in the continental part of Argentina (rapid warming and pressure decrease in spring) in contrast to the conditions over the Atlantic.

Higher wind speeds can be found in the mountain regions of the provinces of Córdoba, San Luis and Buenos Aires as well as in the exposed parts of the Atlantic littoral (Punta del Este). In these places monthly averages may approach those found in Patagonia. In this respect the wind observations of a lightship (Pontón Faro Recalada) deserve special interest by its exceptional location at the outer Río de la Plata, ca. 50 km to the south-west of Montevideo at the entrance of the shipping canal to Buenos Aires (W. Schwerdt-feger, personal communication, 1965). The data (Table V) show wind speeds throughout the year nearly as high as at the Patagonian coast (see Table I) with a maximum in October and a minimum in June. Furthermore, the predominance of the westerlies is well pronounced in winter (April through August); in fact, the zonal component is always stronger than the meridional component, except in the transition months April, August and September. In spite of the relative high wind velocities the resultant wind speed and the steadiness of the wind are rather low, that is only another expression for the very active weather development in this transitional region between subtropical (Brazil current) and subpolar (Falkland current) influences on the one hand and between continental and oceanic influences on the other.

TABLE V

SURFACE WIND CONDITIONS (KNOTS) AT PONTON FARO RECALADA (RIO DE LA PLATA)[1]

	J.	F.	M.	A.	M.	J.	J.	A.	S.	O.	N.	D.	Y.
V_S	12.5	12.4	12.5	13.1	12.2	11.3	11.7	12.7	12.5	13.6	12.8	13.2	12.5
V_R	3.2	3.1	3.5	1.3	0.8	1.8	2.1	1.2	3.7	4.3	3.3	3.5	1.7
q	0.26	0.25	0.28	0.10	0.06	0.16	0.18	0.10	0.30	0.32	0.26	0.27	0.13
D_R	102	087	117	200	261	253	273	210	142	121	118	113	130
w	−3.2	−3.1	−3.1	+0.4	+0.8	+1.7	+2.1	+0.6	−2.3	−3.6	−2.9	−3.3	−1.3
n	−0.7	+0.2	−1.6	−1.2	−0.1	−0.5	+0.1	−1.0	−3.0	−2.2	−1.6	−1.4	−1.1

[1] 35°10′S 56°19′W; 8 m above water; 1951–1960.
Explanation of symbols:
V_S = average wind speed; V_R = resultant wind velocity; q = steadiness (V_R/V_S); D_R = resultant wind direction; w = west component (+); n = north component (+).

Whereas the annual variation of wind speed is insignificant (within 1 m/sec), the daily variation of wind velocity is well expressed over the pampas, as over all large plains, and a pronounced land and sea breeze can be observed along the Atlantic coast. As an example for the latter, the data of Mar del Plata are shown in Fig.1. According to the direction of the coastline, winds from the northeast to southwest represent sea winds and west through north land winds. Since this circulation modifies only slightly the general wind pattern at the western flank of the Atlantic anticyclone the sea breeze blows mainly from the northeast and the land breeze from the northwest. The daily movement takes place

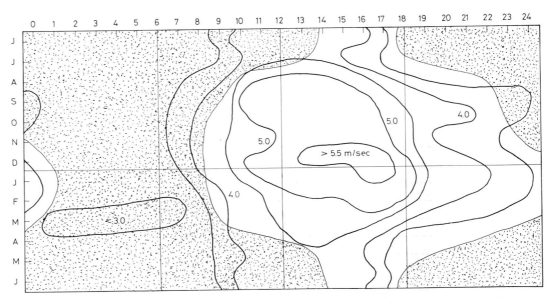

Fig.1. Land and sea breeze at Mar del Plata (m/sec). Shaded area: land breeze (W + NW + N).

counterclockwise, i.e., through south in the morning and through north in the afternoon, which agrees with the deflection of the winds to the left in the Southern Hemisphere. From October to March, the sea breeze sets in between 08h00 and 09h00 with a simultaneous increase in speed and maintains its direction until about midnight or even 01h00 in January. However, by this time, the velocity has almost reached its minimum. In midwinter the predominance of the sea breezes is reduced to about 3 h in the afternoon.

Over eastern Paraguay and Misiones, as well as over Corrientes and northeastern Uruguay, a weak but prevalent eastern to southeastern current throughout the year is observed. According to the annual variation of the pressure over the Atlantic and over the center of the continent the easterlies are more pronounced in spring and summer. Only the southern parts of this zone exhibit an increase in the frequency of northeasterly winds in winter, when the high-pressure belt has its northernmost position. The wind velocity of the whole region is somewhat higher than in the region previously described and shows practically no annual variation except an increase between August and October. These months are characterized by an increase in the meridional-pressure gradient with a cor-

TABLE VI

ANNUAL VARIATION OF THE MERIDIONAL PRESSURE GRADIENT AT 55°W BETWEEN 25°S AND 35°S AND FREQUENCY AND SPEED OF THE WIND OF THE EAST-SECTOR AT CORRIENTES, 1951–1960

	J.	F.	M.	A.	M.	J.	J.	A.	S.	O.	N.	D.	Y.
Press.													
gradient (mbar)	3.9	4.5	4.2	3.1	2.3	1.9	1.6	3.3	5.0	5.4	4.6	4.3	3.5
Wind (NE + E + SE)													
Frequency (%):	48	48	45	41	45	42	45	50	50	53	54	46	50
Speed (m/sec):	2.8	2.5	2.2	2.5	2.5	2.5	2.8	3.1	3.3	3.1	2.8	2.5	2.8

responding increase in the easterly component of the wind. This is due to the fact that in eastern Paraguay (about 25°S) a strong fall in pressure sets in as early as July, while in the area of the Río de la Plata (35°S) the pressure falls markedly only between October and November and shows a much smaller annual range (pressure differences between July and October: 5 mbar for Puerto Presidente Franco and 2 mbar for Punta del Este). See Table VI.

The wind regime of the continental subtropical low (Gran Chaco and pampean sierras)

Over the Paraguayan–Argentinian Chaco and the pampean sierras and bolsons a quasi-stationary low-pressure system is situated. The causes for the development of this cyclone are a combination of thermal and dynamic effects with one or the other predominating in the different seasons (SCHWERDTFEGER, 1951b). During the summer, when the low is particularly well developed, the thermal effect is decisive; during winter the dynamic or lee effect prevails, produced by the west winds blowing perpendicular to the Andes.

The wind field depends on the annual variation of position and intensity of the low and the corresponding pressure gradient in direction to the Atlantic anticyclone. Hence, *in winter*, when the low is located over the Chaco itself (western Paraguay until the northern limit of Santiago del Estero), frequent southerly winds occur in Santiago del Estero, Catamarca and La Rioja (extension to the north and east of the southerlies described on p.20) which farther north change into easterlies over the eastern part of the provinces of Chaco and Formosa, and subsequently turn northeasterly over central Paraguay. On the west side of this cyclone the air currents from the cordillera to the Chaco are not affected in the same degree by the low-pressure system, but rather dependent on other factors, such as high-level westerlies, thermal equilibrium currents, etc.

In summer, the more southerly position of the low, centered over the piedmont area of the pampean sierras (La Rioja–Catamarca), leads to north and northwest winds over the Paraguayan Chaco and to northeast and east winds over the Argentinian Chaco. The corresponding south winds of its west side can be detected only in La Rioja, while further north they are masked by the mountain and valley wind circulation.

The annual average of the wind velocity is not very different from that in mesopotamia, but the seasonal variation is pronounced and shows a maximum between August and November or December. It is of purely thermal origin due to the different rates of heating and cooling over the continent and over the Atlantic. Rapid heating noticeable as early as August in the Paraguayan Chaco (Mariscal Estigarribia) and Formosa (Las Lomitas) goes hand in hand with the likewise pronounced fall in pressure, which sets in just one month later in the Argentinian Chaco (Presidencia Roque Saenz Peña). On the other hand, in the Río de la Plata area the annual variation of pressure is comparatively weak and similar to that over the adjacent Atlantic, i.e., the pressure fall sets in only in November. Between August and November there is, thus, a well expressed southeast–northwest pressure gradient with a consequent intensification of the northeast winds. The opposite is observed during the first half of the year, the autumn and early winter, when the generally gentle winds show a low degree of uniformity due to the insignificant pressure gradient from the Atlantic into the continent.

The wind regime of the cordillera and the Altiplano

In the transition zone between temperate and subtropical latitudes the main ridges of the Andes rise more rapidly from south to north than does the lower limit of the westerlies. Consequently, the central range with the highest elevations (border between Argentina and Chile) is situated within the planetary westerlies (PROHASKA, 1957). Their high speed the year round is becoming even higher in winter. It exceeds sometimes the possibility to be registered with the currently used anemographs. For example, at Corrida de Cori (5,100 m) the few winter months for which observations exist, have a monthly average of 15 m/sec. The transition zone to the wind regime in the southern Andes with a summer maximum lies approximately between 30° and 35°S.

The mountains to the east (unless their peak regions extend into the planetary westerlies), the Altiplano and the longitudinal and transverse valleys have their own wind systems. The mountain and valley wind circulation is exceptionally well developed and in places rises regularly to storm strength for several hours, since the mountains rise sharply from the plains. To this diurnal circulation a seasonal one is superimposed in a corresponding sense, i.e., a predominance of winds from the mountains to the plains in the cool season and the reverse direction in the summer. This seasonal circulation is often so pronounced that it sometimes substantially enhances the prevailing diurnal circulation or counterbalances it to such a degree that very slight breezes or calms result.

Such predominant "mountain winds" in winter and "valley winds" in summer can be found in almost all large longitudinal and lateral valleys and slopes leading to the Altiplano, whatever their direction. This is clearly demonstrated by the wind observations of two stations only 220 km apart in longitudinal valleys running from their confluence in opposite directions, La Poma in the Calchaquí valley to the north and Santa Maria in the valley of the same name to the south. In both valleys the prevailing wind directions are also opposite, in summer as well as in winter (PROHASKA, 1961b). Similar conditions are observed in Punta de Agua, La Rioja or La Quiaca, Jujuy (PROHASKA, 1962) as is shown in Fig. 2. From November through March an advection of tropical air masses from the northeast is evident and from May through September the mountain winds predominate (from the south for topographic reasons) together with the westerlies of the general circulation.

The reasons for this pronounced seasonal change in the wind direction may be found in the extraordinarily great difference in height of 3,000–4,000 m between the Gran Chaco and the Altiplano over a short horizontal distance, and also in the fact that these high elevations are not peaks, as in other mountains, but rolling plains and smooth ridges; the summits of the main mountain chains are still 1,000–2,000 m higher. Furthermore, the Altiplano is practically bare of vegetation and situated at the Tropic of Capricorn so that it represents a marked heat source in summer at an altitude of 3,000–5,000 m (SCHWERDT-FEGER, 1961).

Foehn winds are observed in the Andes as well as in other mountains. Here they are called zonda, after a locality in the valley of the Rio San Juan (province of San Juan). Although not very frequent, they are particularly strong in the subtropical part of the Andes because of the great differences in height. The zonda is more frequent in winter when the pressure on the west side of the Andes reaches the maximum in the subtropical latitudes. When the zonda is blowing temperature rises of more than 30 °C can occur with

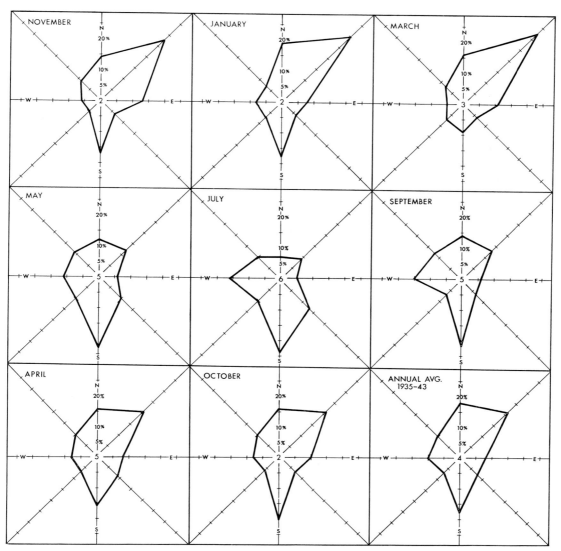

Fig.2. Frequencies of hourly values of wind directions at La Quiaca (%); period: 1935–1943.

a simultaneous fall in the relative humidity to below 10% and even below 5%. It differs from the Alpine foehn only in one respect. While the foehn in the Alps produces very transparent air and excellent visibility, the zonda brings about an extraordinary degree of turbidity. Since the cordillera in these latitudes is practically bare of vegetation or substantial snow fields or glaciers, any strong winds give rise to dust storms. After a powerful zonda several days are often required until all the dust has sedimented, and the atmosphere regains its normal clarity.

Cloudiness[1]

Advection is the main control of cloudiness in this part of the continent, except for very

[1] Only the total amount of cloud cover will be discussed since no statistics are available for the different cloud genera, in spite of their observation.

limited areas. Consequently a similar regional division exists as for the wind regimes. The boundaries are somewhat displaced, since the wind field and thus the cloud systems at higher levels are different from those near the ground. However, in subtropical latitudes the influence of subsidence is well expressed as can be seen in the annual averages (Fig.3).

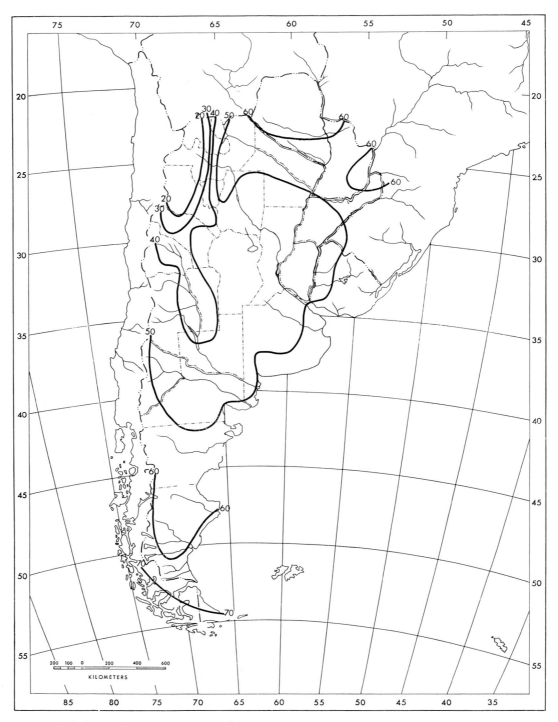

Fig.3. Annual isonephs in per cent of sky cover.

By far the largest region has values of only 40–60%. The isoneph of 50% circumscribes the subtropical continental part. Higher values are found in the maritime-influenced coastal regions of Buenos Aires and Uruguay, in Misiones and Paraguay and in the tropically influenced eastern part of Jujuy, Salta and Tucumán. From Chubut on southward, viz. south of 40°S, the mean cloudiness increases above all at the coast and in the Andes, and reaches 60% south of 50°S, 77% at Ushuaia. Less than 40% are found only in the Andean provinces from Mendoza on northward and particularly in the region to the west of the pre-cordillera which climatically belongs to the desert area of northern Chile (20% at Punta del Agua, 28°38′S 68°39′W, 2,600 m).

Winter shows the simplest pattern of the different seasons. Cloudiness increases uniformly from northwestern Argentina with values below 20% to the east and south. Values between 60% and 70% and locally higher are observed only on the coasts of Buenos Aires and Uruguay, in the cordillera south of 40°S, and in the extreme south of Patagonia. In summer the zone with the lowest degree of cloudiness is displaced towards the central and eastern parts of Argentina in a belt between 30° and 40°S. Higher amounts are found south of 45°S, in western Paraguay and the eastern half of Salta and Jujuy.

The annual variation of cloudiness is mainly a function of the north–south displacement of the subtropical anticyclone. Therefore, some regions are under the influence of different cloud regimes throughout the year. The following four characteristic annual variations exist.

(*1*) Tropical cloud regime in summer (maximum) and subtropical cloud regime in winter (minimum): northwestern Argentina and western half of Paraguay.

(*2*) Subtropical cloud regime throughout the year with higher values in winter and increase of cloudiness from the west to the east due to the increase of maritime influence: central Argentina, Uruguay and eastern Paraguay.

(*3*) Subtropical cloud regime in summer (minimum) and cloud systems of the westerlies in winter (maximum): northern Patagonia.

(*4*) Cloud systems of the westerlies throughout the year with higher values in summer and increase of cloudiness from north to south: southern Patagonia, south of 45°S.

Consequently in nearly every month the opposite extremes can be observed since the transitional zones have the maxima and minima in the intermediate seasons. Great differences are apparent also in the annual range. High values in narrowly delimited regions in northwestern Argentina, in Neuquén, and in western Rio Negro are contrasted by uniform cloud conditions throughout the year over the major part of the region with annual ranges less than 20% and even below 10% in certain zones (Cuyo and central Patagonia). The following description is based on geographical units as it is not possible to define the limits of cloud systems the same way as those of wind systems for obvious reasons and the fact that transition zones influence large regions.

Western Paraguay, northwestern Argentina (including La Rioja): maximum in summer and minimum in winter

The extreme differences in elevation of this region (200–6,000 m) are decisive for the absolute amount of cloud cover but not for the type of annual variation. The annual average of cloudiness is between 50 and 60% over the piedmont area and on the eastern (windward) slopes of the pre-Andean sierras as far as the southern limit of Tucumán.

The averages decrease to 20–30% on the leeward side of these mountains, in the pre-puna and puna and over the pampean sierras of Catamarca and La Rioja, which according to the prevailing northeast flow are at the lee side of the pre-Andean sierras. The lowest annual means are registered in the all-sides protected longitudinal valleys and bolsons of moderate elevation. The uniform annual variation shows the minimum in winter due to the subtropical high-pressure belt, and the maximum in summer due to advection of moist air masses of tropical origin. This is therefore the zone with the greatest annual range of cloudiness; for example, La Quiaca has a range of 51% which is high by any standards. In areas with such extreme ranges an annual average is an almost meaningless value.

This type of annual variation can be seen in Fig.4,*1.1*. The values are averages of 5 representative stations in the pre-Andean and pampean sierras. It is the typical variation to be found in such tropical areas as southeastern Bolivia and the Brazilian Mato Grosso; it has to be valid also for western and northern Paraguay where no observations exist. The transition from the small amount of cloudiness in winter to the large amount of cloud

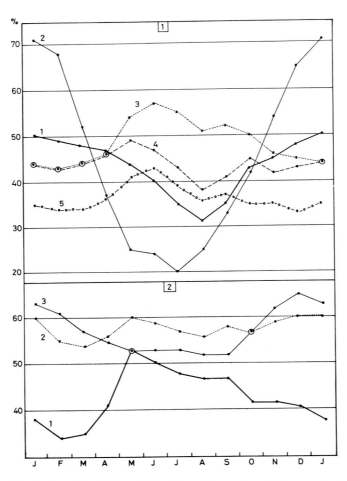

Fig.4. Annual variation of cloudiness (percentage of sky cover). *1.1* = average of Salta, Tucumán, Andalgalá, La Rioja, Catamarca; *1.2* = La Quiaca; *1.3* = average of Azul, Casilda, Victorica, Paso de los Libres, Ceres, Punta del Este; *1.4* = average of Córdoba, San Luis, Santiago del Estero; *1.5* = Mendoza, San Juan; *2.1* = Maquinchao; *2.2* = Sarmiento; *2.3* = Gobernador Gregores.

cover in summer takes place between September and October over the whole region; likewise uniformly observed characteristics are the maximum cloud cover in January, the slow decrease in autumn, the sharp decrease up to June and the minimum in August. Hence, the eastern Andean sierras have four months of predominantly clear sky and eight months with a high amount of cloud cover. In the puna itself and particularly in its western part the relationship is reversed; here the period of strong cloudiness is reduced to the three midsummer months (Corrida de Cori, La Casualidad). An extreme range of cloud amount appears at the northeastern verge of the puna (e.g., La Quiaca, Fig.4,*1.2*). In summer (November through March) the heavy cloudiness is produced by advection of warm and moist air masses of tropical origin, whereas few clouds are observed during the rest of the year, particularly as they are mainly high clouds drifting with the west winds (monthly means of 20–30% increasing towards the north). Therefore, the winter shows only small local differences in the whole zone and the summer big differences in the absolute amount of cloudiness, above all in the cordillera where it is due to orographic effects. Only the eastern slopes of the pre-Andean sierras and the western part of the Chaco have uniformly high mean values which equal those of southern Patagonia.

This pronounced annual variation is also clearly expressed in the number of clear and cloudy days[1]. In northwestern Argentina the three winter months together have an average of more than 30 clear days (La Quiaca 41, Punta del Agua (Catamarca) and La Casualidad (Salta) more than 60 days) and less than 20 cloudy days (La Quiaca only 2). In summer very few clear days are observed (total of the three summer months: La Quiaca 3, Tucumán 9, La Rioja 17), but not very many cloudy days either (an average of 20–30), since the prevailing convective clouds have a very pronounced daily variation and stable stratus clouds rarely occur.

Uruguay, Argentine mesopotamia, pampa and the eastern Chaco: small annual range with maximum in winter

The annual variation of cloudiness in this region is just the inverse of the preceding (Fig.4,*1.3*). In addition it is characterized by a small annual range (10–15%) and uniform annual averages (between 40 and 50%). On the coasts these values are somewhat higher but remain below 60%. The stabilizing influence of the subtropical circulation is conspicuous throughout the year, despite the very active secondary circulation at times.

As in all subtropical and maritime influenced regions the season of maximum cloudiness is winter due to the fact that the cloud systems of the west wind circulation extend farthest Equator-wards at this time of the year. This relatively strong winter cloudiness (about 60%) starts in May, reaches a well defined maximum in June, and ends in October. A weak but characteristic secondary minimum is found in August. It shows that the influence of the main minimum of cloudiness of northwestern Argentina and Paraguay extends southeastward to the Atlantic coast. August is the month where the high-pressure ridge over the continent begins to degrade, but the advection of tropical air masses has not yet started. The time of the lowest cloudiness (40–50%) is between

---

[1] The figures for clear and cloudy days are taken from the *Atlas Climático de la República de Argentina* (1901–1950). For their definition, see the Appendix on p.70. In view of the differences in the definitions Uruguay has not been included in the comparisons.

November and April and the main minimum is indistinctively found in one of the first three months of the year. The summer convective clouds are not nearly as frequent and intense as in the region previously discussed; this is due to the direct influence of the Atlantic high-pressure cell (see map of average sea-level pressure for the three summer months December–February, Fig.1, Chapter 1).

From winter to summer the number of cloudy days decreases from 25 to 10 and on the coast from 30 to 15 days for the three-month totals. In contrast, the number of clear days varies little throughout the year increasing only from winter to summer in the corresponding three-month intervals from 15–20 to 20–25 days inland and from 10 to 15 days on the coast. The influence of the subtropical high-pressure system on cloudiness is so conclusive that apart from a narrow coastal strip the annual sums of clear days (more than 80) actually outnumber the overcast ones.

Central and eastern Paraguay, western Chaco, southern pampean sierras: double annual wave

The broad transition zone between the regions with maximum cloudiness in summer and those having the maximum in winter extends from the Paraguayan–Brazilian boundary south- and westward through the Chaco, the pampean sierras (except those of Catamarca and La Rioja) to the border between Mendoza and San Juan. The mean annual cloudiness decreases in the same direction from 60% in northern Paraguay to 35% in San Luis. The southern part of this region, and the Cuyo region as well, have the smallest amounts of cloudiness of the whole area east of the Andes, for there are neither the clouds of tropical origin in summer nor those of Atlantic or Pacific origin in winter. Consequently, it is also the region with the smallest annual range which amounts to only 10% in the central and western part. The existing double wave shows a main maximum in May and a secondary one in October, a main minimum in August and a secondary one in February (Fig.4,*1.4*). In Paraguay, the maxima shift to June and January, while the main minimum remains in August with the secondary minimum weakly defined in the autumn.

Comparing Fig.4, *1.4* with *1.1* and *1.3*, it can be seen that in this region the annual variation between January and May is parallel to that of the maritime subtropical regime (summer minimum) while in winter it is parallel to that of the tropical regime (winter minimum). This transitional zone shares, thus, always the characteristics of the adjacent regimes at the time when these have their minimum cloudiness. The results are the low values of cloud cover throughout the year, and a total of 100 clear against only 60 cloudy days. In Paraguay this relationship is slightly reversed because of the closeness to the tropical zone; there the cloudy days slightly outnumber the clear days (70 against 60 days).

Cuyo: small annual range and low degree of cloudiness

Cuyo and particularly its western part (San Juan and Mendoza) has a subtropical continental cloud regime, since neither the maritime influence of the Atlantic (because of its great distance) nor that of the Pacific (because of the great height of the cordillera in these latitudes) affects the amount of the average cloudiness. The annual variation is typical for a subtropical region with the maximum in winter (Fig.4,*1.5*). However, the range and the absolute values are much lower than on the Pacific or Atlantic coast, and the extremes occur somewhat earlier (maximum in May–June and minimum in November–December). The mean values amount to only 30–40% with an annual range of about 10%. Mendoza, San Juan and the western parts of the provinces farther north are, as already mentioned, the region with the lowest values of cloud cover in the three countries. Substantially more than 100 clear days per year are matched by only 40 or fewer cloudy days.

This is therefore also a region which is not affected by either the intense summer cloudiness to the north or the intense winter cloudiness to the south. As neither north winds nor south winds are generally associated with an increase in cloudiness, the few overcast days are observed only during strong eastern to southeastern flow, i.e., during advection of Atlantic air masses in combination with orographic lifting; similar effects are also evident on the eastern slopes of the pampean sierras.

The cloudiness decreases with elevation as can be seen on a cross-section at the latitude of the city of Mendoza (Table VII). This is not in contradiction to the above-mentioned windward effect since east and southeast winds are rather infrequent and their influence on cloud cover is partly compensated by the foehn effect during west wind situations in the overall average. The mountain pass Cristo Redentor (3,832 m) which is freely exposed to the Pacific cloud regime shows once more an increase in cloudiness and has approximately the same annual amount as Mendoza. A similar variation of cloud cover with increasing height can also be seen farther north at the latitude of Salta (Table VII) only here with a reversed seasonal variation (summer maximum and winter minimum). The highest parts of the cordillera are therefore characterized by a stronger cloud cover than either the corresponding eastern, or the western flanks. In other words, the low degree of cloudiness observed on the Chilean and Argentine side of the Andes are not connected one with the other, but interrupted by the higher degree of cloudiness observed in the central part of the cordillera. Here, the limit between the regimes with winter and summer maximum, respectively, coincides probably with the boundary between prevalent summer and winter precipitation, found at about 27°S (to the north of Cerro Ojos del Salado). This line lies farther north than in the piedmont area where it coincides fairly clearly with the provincial border between San Juan and La Rioja (30°S) where the records of the station Jachal evidence a double wave with slightly more cloudiness in winter.

TABLE VII

CLOUDINESS AS A FUNCTION OF ELEVATION AT THE EAST SIDE OF THE ANDES IN PERCENT OF SKY COVER

At latitude of Mendoza, 33°S 1941–1960:	Jan.	Jul.	Year	At latitude of Salta, 25°S 1951–1960:	Jan.	Jul.	Year
Mendoza 827 m	38	51	43	Salta 1,226 m	70	27	45
Villavicencio 1,789 m	27	38	38	San Antonio de los Cobres 3,776 m	61	16	32
Puente del Inca 2,720 m	22	48	33	La Casualidad 4,000 m	18	14	15
Cristo Redentor 3,829 m	28	60	41	Corrida de Cori 5,100 m	50	31	34

Patagonia

Two types of annual variation in cloudiness running almost opposite one to another can be detected in Patagonia; northern Patagonia shows a winter maximum and southern Patagonia a summer maximum. The annual variation over the latter region corresponds to the intensification of the westerlies in summer, while during the same season the decrease of cloud cover over northern Patagonia must be ascribed to the southward movement of the subtropical anticyclone. An extremely narrow transition zone exists between 45° and 46°S (see climatic tables of Comodoro Rivadavia and Sarmiento, pp. 100, 101). It shows a double wave with a small annual range and maxima at the solstices and minima at the equinoxes. Moreover, this narrow zone is characterized by a more intense cloud cover than that of both adjacent zones; this may be due to an overlap of the two cloud regimes. Farther to the south, however, the cloud cover increases rapidly again. Fig.4,2 shows these three entirely different annual variations as well as the differences in their absolute values. They are represented by Maquinchao (41°15′S; Fig.4,2.1), Sarmiento (45°35′S; Fig.4,2.2) and Gobernador Gregores (48°47′S; Fig.4,2.3). All three stations are located on the central mesetas and have a total north–south distance of 850 km. The double wave in the transitional region (Sarmiento) is even better expressed on the coast (Comodoro Rivadavia; see the climatic tables).

Fig.4,2 shows particularly well the increase of summer cloudiness in southern Patagonia, and the monthly averages reach values of over 80% cloud cover in Tierra del Fuego. The period during which these large amounts of summer clouds are observed lengthens from north to south in a similar manner as its intensity. South of the transitional zone, at 46°S, it starts in November and continues until February, while south of 50°–52°S it begins as early as October and lasts until March. In winter the monthly averages are much more uniformly distributed and the increase towards the south is less pronounced. This leads to an increase in annual range from the border between Chubut and Santa Cruz (46°S) to Tierra del Fuego. These characteristics of cloudiness in Patagonia are represented in Table VIII.

In winter, the strong cloudiness of the westerlies is observed as far north as the limit between Neuquén and Mendoza. At this time of the year no great differences can be found apart from a slight decrease from the cordillera towards the coast. In summer, on the other hand, and as already mentioned, cloudiness decreases considerably northward, so that the greatest annual range of this region (almost 35%) is observed in Neuquén, the transition zone bordering the previously described cloud regime in Cuyo, where the degree of cloudiness is low the year round. In the Andean and pre-Andean part the increase of cloudiness sets in one month earlier (April) than in the rest of northern Patagonia (May), which at the same time is the month with the maximum cloudiness. The cloudy period ends in September. The minima occur almost uniformly in February. For obvious reasons, the cloudiness is less over the mesetas than in the cordillera or on the coast.

At the northern border of Patagonia the annual averages of clear and cloudy days are almost the same (between 60 and 80). In accord with the marked increase in cloudiness towards the south, the number of cloudy days at Ushuaia increases to 165, while that of clear days decreases to 10. Due to the different annual variation in cloudiness, northern Patagonia has its maximum of cloudy days in winter and southern Patagonia in summer.

TABLE VIII

CHARACTERISTIC FEATURES OF THE CLOUD REGIME IN PATAGONIA

Latitude °S:	37	39	41	43	45	47	49	51	53	55
Pre-Andean zone										
Maximum (%)	57	54	69	69	–	56	69	69	71	–
In months	V, VII	V	VI	V	–	I	I	I	XII	–
Minimum (%)	23	30	36	47	–	49	51	54	52	–
In months	II	I, II	II	II	–	III, IX	IX	IX	VII	–
Range	34	24	33	15	–	7	18	15	19	–
Ann. average (%)	42	41	55	57	–	53	57	61	62	–
Months above average	IV–IX	IV–IX	V–IX	IV–IX		XI–II, VI	XI–II	XI–II, V	XI–IV	
Meseta										
Maximum (%)	–	49	55	–	57	–	–	–	–	–
In months	–	V	V, VI	–	XI–I	–	–	–	–	–
Minimum (%)	–	20	35	–	49	–	–	–	–	–
In months	–	II	II	–	II	–	–	–	–	–
Range	–	29	20	–	8	–	–	–	–	–
Ann. average (%)	–	34	46	–	55	–	–	–	–	–
Months above average	–	V–IX	V, IX	–	XI–I, V–VI	–	–	–	–	–
Coast										
Maximum (%)	–	50	63	61	59	62	72	77	80	83
In months	–	VI	VI	V	I	I	I	I	XII	XII
Minimum (%)	–	32	40	48	49	49	56	55	54	63
In months	–	II	II	III	III	VII	VI	VI	VI	VI
Range	–	18	23	13	10	13	16	22	26	20
Ann. average (%)	–	41	52	55	55	55	62	64	66	73
Months above average	–	V–X	V–IX	V–IX	XI–I, V–VI	XI–II	XI–II	X–III	X–III	X–III

This different annual variation in north and south Patagonia, together with the increase in cloudiness southward leads to a rather uniform distribution of cloudy and clear days in winter, and to marked differences in summer. During the three winter months the entire Patagonia has about 25 cloudy days increasing in the extreme south to more than 30, and has about 10 clear days decreasing to half this number in the extreme south. In summer, on the other hand, the number of cloudy days increases from less than 10 in northern Patagonia to more than 40 in Tierra del Fuego, and the clear days decrease from 25 to less than 1.

Relation between cloudiness and number of clear and cloudy days

The relationship between average monthly cloudiness and the number of clear or cloudy days in Argentina is given in Table IX (comparative figures for Paraguay and Uruguay are as yet inadequate). Irrespective of the mean cloudiness, both clear and cloudy days occur more frequently in winter than in summer, another expression for the fact that the

TABLE IX

MEAN CLOUDINESS (N) AS RELATED TO THE NUMBER OF CLEAR AND CLOUDY DAYS IN
ARGENTINA AND COMPARATIVE VALUES FOR THE LAST ONES IN THE U.S.A.[1]

N (%)	Clear days		Cloudy days		Cloudy days U.S.A.	
	Jan.	July	Jan.	July	Jan.	July
20	17	17	0.5	1	2	3
30	12	13	2	3	3	4
40	8	10	4	5	5	6
50	4	7	7	8	8	10
60	1.5	4	9	11	10	13
70	0.5	2	13	15	12	17
80	0.1	–	18	–	–	21

[1] After LANDSBERG, 1958. Period 1941–1950; 158 stations.

vertical structure of the air masses is more stable in winter. Except for the extreme month-ly means of cloudiness (20 and 80%), neither the number of clear days nor the number of cloudy days is as much as half the number of days in the month. A brief comparison with the corresponding figures in the United States (LANDSBERG, 1958) shows that for the same degree of mean cloudiness the number of cloudy days is greater in the U.S.A., viz., at higher latitudes. Argentine data indicate the same trend when the scant Patagonian observations are considered separately. A corresponding comparison for clear days is not possible since the definition of a "clear day" is different in the U.S.A. and Argentina.

Duration of sunshine

In accordance to the latitudinal extension of the territory under discussion, from 20° to 55°S, the possible duration of sunshine increases from 13 h and 20 min to 17 h and 40 min in December and decreases in June from 11 h to 7 h. However, the different cloud regimes affect the annual variation of the duration of sunshine and its annual range in such a manner that the annual variation due to astronomic factors or its latitude-de-pendent amplitude becomes regionally reversed. Data of sunshine recorders are available for about 40 Argentine stations and for Montevideo, but the observation periods vary greatly. Only stations with at least five-year observations have been used for the follow-ing discussion. Since the data are not homogeneous only the main features can be out-lined.

The area under subtropical, maritime influence has an annual average of about 2500 hours, or 7 h/day, which corresponds to a relative duration of sunshine of about 55%. Towards the north, near the Paraguayan border, the values decrease to 2,200 h or 6 h/day and towards the extreme south (Tierra del Fuego) to 1,500 h or 4 h/day; the correspond-ing relative duration of sunshine is 50 and 33%, respectively. Only in the puna and the northwestern Andean provinces (as far as and including Mendoza) the annual average is above 3,000 h (8 h/day) or 65–75% of the maximum possible sunshine. La Quiaca has the highest average with 3,300 or 9 h/day, equal to a relative duration of sunshine of 75%. Towards the Chilean frontier, the strong decrease in summer cloudiness may lead to substantially higher values which are probably among the highest in the world.

Although the annual distribution is rather uniform, great differences can be seen with respect to the individual seasons. During *summer* an increase from 6–7 h/day in the north to 11 h/day in the province of San Juan, Mendoza and Neuquén is observed, an increase which is substantially greater than that which would be due to the increase in the latitude; thus the relative duration of sunshine increases from 50 to 75%. Duration of sunshine in these three provinces is the highest in South America east of the cordillera. South of the Rio Negro the duration of sunshine decreases to one-half of the above-mentioned value ($5\frac{1}{2}$ h/day in Ushuaia), i.e., only 35% of the astronomically possible amount because of the marked increase in cloud cover. In the latitudes of maximum sunshine, between 30° and 40°S, a decrease from 11 h/day to 10 h/day exists towards the coast. Farther north, on the other hand, where the tropical summer cloudiness is orographically enhanced at the east slopes of the cordillera, an increase in duration of sunshine can be observed towards the east. The other parts of the country show only small differences from this general pattern, not systematic enough to be discussed on the basis of the existing observations.

In *winter*, when the decrease in the astronomical duration of sunshine from north to south is intensified by the increase in cloudiness in the same direction, the actual duration of sunshine decreases almost regularly from the subtropical relatively cloud-free north with 9–10 h/day to the subpolar very cloudy south with only 1 h/day, which corresponds to a decrease in the relative duration of sunshine from 85% (La Quiaca) to 15% (Ushuaia). From the cordillera towards the coast, the duration of sunshine decreases in subtropical latitudes and increases in Patagonia because of the high cloudiness of the southern cordillera.

The interaction of latitude, different annual variation in cloudiness and local geographical factors (coast, mountains, etc.) leads to great differences in the annual range of the monthly averages. It increases with increasing latitude from a minimum of 2 h in the north to a maximum of 7 h in the zone of maximum duration of sunshine in summer (between 40° and 42°S). Farther south, the range diminishes rapidly to 4–5 h/day. If the observed annual ranges are compared with the astronomical amplitudes it can be seen that in subtropical latitudes up to and including northern Patagonia (the zone of maximum range) the former always have larger values than the latter, but farther south, in the zone of the westerlies, they decrease rapidly to one half the possible values (southern Patagonia and Tierra del Fuego). In the first case, this is caused by the fact that the cloudiness has its minimum in summer and the maximum in winter and in the latter case by the inverse variation of cloudiness and, at the same time, the increase of the astronomically possible range with latitude.

The main types of the annual variation of the duration of sunshine are the following (Fig. 5,*1* and *2*).

East puna (La Quiaca; Fig.5,*1.1*)

In accordance with the division of the year into a rainy and a dry season in the tropics and the small annual range in the length of the day ($2\frac{1}{2}$ h) the dry season, although in the astronomical winter, has more hours of sunshine than the rainy season. This reversal of the normal annual variation of the duration of sunshine explains the fact that the sunny winter is called "summer" and the rainy and cloudy astronomical summer is known as "winter" by the indigenous population. In La Quiaca the monthly averages vary around 9.7 h/day during the dry season (April through November) to reach a maximum of 10.1 h/day in November. At the time of maximum advection (rainy season) these values fall to 8.5 h/day in January which is 64% of the possible duration of sunshine. The annual range, therefore, amounts to only 1.6 h, which is the lowest value recorded at any of the available stations.

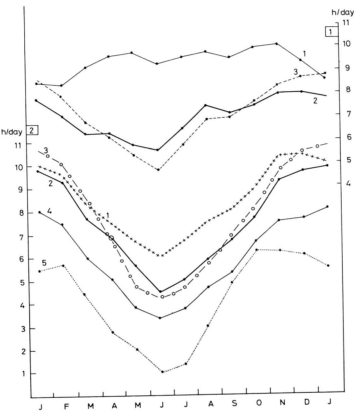

Fig.5. Annual variation of the duration of sunshine (hours per day). *1.1* = La Quiaca; *1.2* = average of Tucumán, La Rioja, Salta, Andalgalá; *1.3* = average of Córdoba, Santiago del Estero, Presidencia R. S. Peña; *2.1* = Cuyo; *2.2* = Río de la Plata; *2.3* = Neuquén and Rio Negro; *2.4* = southern Patagonia; *2.5* = Tierra del Fuego.

The northwestern provinces (Jujuy, Salta, Tucumán, Catamarca, La Rioja; Fig.5,*1.2*)

Although the absolute amounts increase from north to south, a uniform annual variation can be observed over the whole region. The annual curve has already the normal form of a summer maximum and a winter minimum, but with a very small annual range of 2–3 h, what approximately equals the astronomical value. Here again, a clear division of the year appears, but in this case into a relatively cloudy autumn and early winter (March–July) and a bright spring and summer (August–February). The maximum occurs as early as November–December, i.e., before the advective summer cloudiness sets in, while the minimum coincides with the astronomical minimum in June. A characteristic feature is the marked decrease from February to March and the sudden increase between July and August. The high amount of sunshine from August onwards, has a decisive influence on the temperature regime of this region and, thus, the early development of the thermic low.

Chaco and central Argentina (Fig.5,*1.3*)

The transition regime between tropical and subtropical cloud conditions (maximum April–May) is likewise visible in the duration of sunshine. It shows the same regular decrease in autumn, as the other subtropical regions of Argentina, but coincides with the northwestern provinces in respect to the increase in sunshine from July to August and in the marked predominance of hours of sunshine in spring and early summer over those of late summer and autumn and the clearly expressed maximum in December.

Argentine–Uruguayan mesopotamia, La Pampa, Cuyo and Patagonia (Fig.5,*2.1–4*)

The form of the annual variation is basically the same in the entire region only with different absolute values and annual ranges which vary between 4 and 7 h. Cuyo (Fig.5, *2.1*) evidences still high values in winter and represents, thus, a southward extension of type *1.2* (Fig.5) with the likewise early maximum in November or December. Neuquén and Rio Negro (Fig.5,*2.3*) exhibit the largest ranges. This is brought about by the particularly low amount of sunshine in winter (from May to July) caused by the intense cloudiness of the westerlies, whereas during the rest of the year the subtropical influence is predominant. The curves for southern Patagonia and the region around the Río de la Plata (*2.4* and *2.2* in Fig.5) are almost parallel, though at the former the absolute values are smaller and the summer depression starts as early as December. Representative curves of the pampa and the northern mesopotamia are not shown since they do not differ essentially from those already discussed (see the climatic tables in the Appendix). The annual variation in the pampa is similar to that in Neuquén and the northern mesopotamia resembles the Río de la Plata in this respect, except for somewhat higher values in spring.

Tierra del Fuego (Fig.5,*2.5*)

The annual variation in Ushuaia is not unlike that of northern Argentina since in both regions the duration of sunshine is reduced by a strong cloudiness in summer; only the absolute values are much lower. The main characteristics are the sharp increase in spring with the early maximum in October and November and the wide annual range. The last fact is due to the great variation in the length of the day at 55°S which amounts to 17 h and 23 min at the summer solstice and to 6 h and 43 min at the winter solstice or 10 h and 40 min of annual range, i.e., practically twice the variation in the northern part of Argentina; hence the annual range of the duration of sunshine is also nearly doubled.

Global radiation

Homogeneous series of global radiation measurements are available only for La Quiaca, Pilar (Córdoba) and Buenos Aires (Table X). Among these 3 stations only La Quiaca is remarkable for both the very high annual values and the very low annual range. The average annual sum of sun and sky radiation on a horizontal surface is 246 kly, a value that is probably more than 50 % higher than the amounts observed at lowland stations to the east according to scattered simultaneous observations. On the other hand, the annual range is 35 % less than that of the extra-terrestrial radiation. This equilibrated radiation regime is due to the extremely high winter values (500–550 ly/day) and relatively low summer values. This fact could be of great significance for a continuous utilization of solar energy as a power source. In the western part of the puna the conditions have to be even better, since La Quiaca lies at the cloudiest northeastern border. To the west greater elevations and therefore higher insolation, are combined with a minimum amount of water vapour and cloudiness all the year round (see climatic tables for La Casualidad and Corrida de Cori, pp.77, 135). It is possible that this combination of factors is unique at the earth's surface and that here the highest annual sums of solar radiation can be observed.

TABLE X

MEAN DAILY GLOBAL RADIATION (ly)

	J.	F.	M.	A.	M.	J.	J.	A.	S.	O.	N.	D.	Y.
La Quiaca (1945–1953)	765	695	690	630	560	515	530	600	700	810	805	785	665
Pilar[1] (1941–1943; 1945–1951)	650	570	435	355	255	205	240	315	410	535	625	665	440
Buenos Aires (1951–1960)	614	540	430	305	217	143	182	253	352	448	599	636	393

[1] 31°40′S 63°53′W; 338 m.

Temperature

Paraguay, Uruguay and the eastern half of Argentina as far as 40°S, rise only very locally to over 500 m above sea level. For this reason the general temperature distribution is only slightly affected by differences in height. Consequently, the isotherm-pattern, even without reduction to sea level, clearly exhibits the true thermic controls, that is, latitude and ocean. In Patagonia, isotherms are relevant only in certain zones along the coast where elevations are below 500 m; on the west–east downwards sloping mesetas they resemble nearly contour lines. In western Argentina, the almost abruptly rising Andes make the drawing of unreduced isotherms even more meaningless, since they follow directly the north–south contour lines.

Annual averages

The highest mean annual temperatures (between 26° and 27°C) are observed in the northernmost part of the Gran Chaco, i.e., near the Paraguayan–Bolivian boundary at about 20°S. This is the lowest part of the continent in these latitudes, partly even below 200 m, and not far from its geographical center (15°S 62°W). In the other parts of Paraguay and northwestern Argentina, the annual temperatures are still very high, so that the 20°C isotherm is not found until 30°S. South of this latitude, the temperature gradient steepens and a maritime influence becomes apparent (south of 35°S). Therefore, the isotherms have a northwest–southeast direction which changes to north–south in Patagonia, according to the above mentioned influence of elevation. The 10°C isotherm runs from Rio Limay, at 40°S, south-southeastwards to the mouth of the Rio Deseado (47°S). The lowest annual averages (5°–6°C) are observed at the southern tip of Patagonia and in Tierra del Fuego.

Over the 35° latitude (20°–55°S) the annual averages show a temperature difference of about 21°C or a decrease of 6°C for 10° latitude. The magnitude of this *total* decrease remains nearly the same the whole year, since the temperature regimes of the southernmost and the northernmost regions exhibit a similar and small annual variation with a range of about 8°C. The temperature differences between individual latitudinal zones, however, evidence large seasonal variations (Table XI). In summer, the meridional gradient increases from the subtropics to the west-wind zone from 3°C per 10° latitude to 8°C per 10° latitude. In winter it decreases from 11°C per 10° latitude to 4°C per 10° latitude. Towards the southern ocean—the Drake Passage—the temperature gradients are larger than farther north in both of these seasons (SCHWERDTFEGER et al., 1959). The annual variation of the meridional gradients is minimum between 35° and 40°S and maximum between 20° and 30°S. Absolute amount and annual variation of these gradients are the consequence of both the uniformly intense heating of the continent in summer as far south as 40°S, and the strong cooling in winter which extends northwards to about 30°S, caused by the advection of polar air masses, channeled by the Andes. The steepest gradient, i.e., the most rapid change of temperature with latitude is found between 50° and 60°S in summer and between 20° and 30°S in winter.

TABLE XI

MERIDIONAL TEMPERATURE GRADIENT ALONG THE CONTINENTAL AXIS[1] (°C PER 10° LATITUDE)

Latitude °S:	20–30	30–40	40–50	50–55	50–60	20–55
January	3	4	8	5	10	20
July	11	6	4	1	7	22
Year	7	6	5.5	2.5	7	21

[1] Elevations below 500 m NN.

Monthly averages

The highest monthly average temperatures (between 29° and 30°C) are found in the Paraguayan Chaco in January (Fig.6). During summer, this is the warmest region of the whole continent. However, the temperatures are not as high as they are at similar latitudes in North America or the other continents. The hot zone extends as far as the northern Argentine Chaco, where the highest monthly means are still above 28°C. Due to the continentality of the climate, mean January-temperatures above 24°C are normally found in the rolling plains between the pampean sierras and the Andes as well as in the western pampa and the piedmont areas of Cuyo. A continental influence on the temperature can be detected also on the mesetas of Patagonia (see 20°C isotherm). Because of the prevailing west winds, the cooling influence of the ocean (cold Falkland current) is not felt inland. On the other hand, the oceanic influence affects the entire province of Buenos Aires, southern Entre Rios, and southern Uruguay. The strongest effect of the ocean is observed in Tierra del Fuego, where the temperatures (9°–10°C) are only slightly above those of the surrounding sea. Summarizing, the isotherms for January show a tongue of high temperatures along the longitudinal axis of the continent to southern Patagonia and a characteristic bending towards the north near the coast and along the Andes.

The July isotherms show the temperature distribution in winter (Fig.7). Their pattern is much more regular then in summer and a general northwest–southeast direction can be seen, caused by the contrast continent–ocean. A bending of the isotherm northwards as in summer is evident only at the foot of the Andes due to the orographic effect and at the Atlantic slope of the Patagonia mesetas. Along the coastal strip, on the other hand, they bend directly southwards, indicating the sharp contrast between the cold continent and the relatively warm sea. As already stated, the total decrease in the July temperatures between northern Paraguay and Tierra del Fuego is almost the same as in January. The average of July in northern Paraguay is 22°C and in Tierra del Fuego about 1°C (Ushuaia, 1.6°C; Rio Grande, 0.4°C). The 18°C isotherm (Köppen's southern limit of the A-climate) coincides with the southern margin of the Paraguayan Chaco and bends northwards east of Asunción, where it runs along the eastern border of the broad basin of the Paraguay river. The effective and almost periodic advection of polar air masses as far as central Argentina is characterized by the 10°C isotherm. It runs along the northeast coast of the province of Buenos Aires (including Río de la Plata) in a northwesterly direction over northern Córdoba to La Rioja. The lowest monthly average temperature of the continent (apart from the mountains) is at Rio Gallegos (1.1°C).

The foehn effect, produced by the persistent strong west winds, is another reason for the relatively high winter temperatures on the coast of Patagonia. It is particularly pronounced where mesetas or mountains extend nearly to the coast. This can be seen comparing the temperatures of Sarmiento (260 m) and Comodoro Rivadavia (60 m), 150 km apart and separated by the Pampa del Castillo with an average elevation of 600–700 m. The resulting temperature differences (Table XII) suggest superadiabatic gradients in winter. But this strong gradient is also brought about by the special situation of Sarmiento which, protected against the wind, has a high frequency of nocturnal inversions and, thus, lower temperatures than the open mesetas. In summer, the cool ocean temperature counteracts an adiabatic temperature gradient as well as a strong heating of the Patagonian mesetas; for this reason and in spite of the foehn effect an almost normal temperature gradient prevails between meseta and coast.

Vertical temperature gradient

The temperatures in the cordillera, the pampean sierras, and the Altiplano are largely

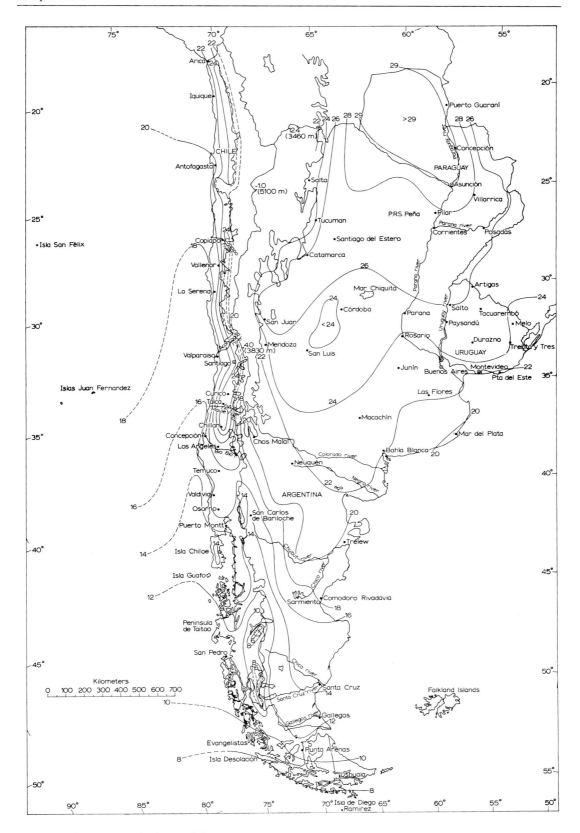

Fig.6. January isotherms (°C).

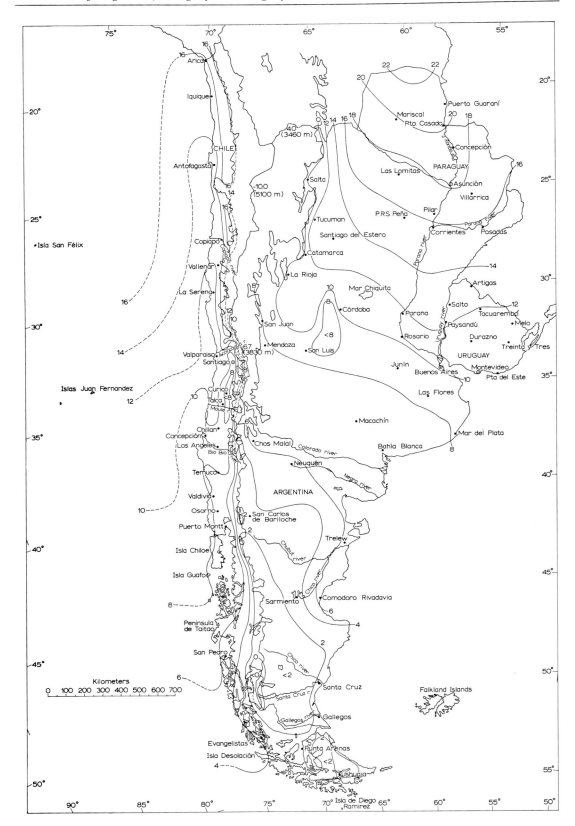

Fig.7. July isotherms (°C).

TABLE XII

TEMPERATURE GRADIENT BETWEEN SARMIENTO (260 m) AND COMODORO RIVADAVIA (60 m) (in °C/100 m)

Jan.	Feb.	Mar.	Apr.	May	June	July	Aug.	Sep.	Oct.	Nov.	Dec.	Yr.
0.63	0.72	0.72	0.89	1.01	1.42	1.38	0.99	0.72	0.63	0.63	0.53	0.86

affected by topography, exposure and nature of the ground surface, besides their dependence on elevation. Generally the vertical temperature gradients are similar to those found all over the world, both with respect to their absolute values and to their annual variation (BöGEL, 1956); in summer, they vary about 0.6°C/100 m and in winter about 0.4°C/100 m. Higher values are normally found in the arid zones, where even monthly means may reach dry adiabatic values, and smaller gradients are observed on humid slopes, exposed to orographic rains. But considerably lower temperature gradients result comparing low-level stations with those situated at the high plains or broad valleys of high elevation (PROHASKA and GARCíA, 1951). Here, the sparse vegetation cover of the soil together with the intense insolation in subtropical latitudes produces such a strong heating in summer that even the gradients in direction to the Gran Chaco (the "heat pole" of South America) assume values of only 0.35°C/100 m to 0.45°C/100 m. In the winter, local conditions play a still greater part in the absolute amount of temperature gradients because of the different intensity and duration of nocturnal cooling. Consequently, lapse rates vary widely between 0.5°C/100 m and negative values (inversions). The wide range of the values may also in part be due to the great distances between any two stations and their sometimes different observation periods.

To calculate the average height of the 0°C limit, a mean gradient of 0.55°C/100 m for the yearly average can, however, be used, as everywhere else. Using this mean value and the scattered high-level stations the 0°C levels have been calculated for various latitudes on the eastern side of the Andes (Table XIII). The increase in height of the 0°C level with decreasing latitude is surprisingly regular and even the expected sudden change in altitude where the transition takes place from the westerlies to the subtropics (35°C) and from the latter to the tropics (20°S), is hardly visible. The unusual height of the 0°C level at 20°S (5,400 m) may perhaps be the result of extrapolation from the overheated puna, as no mountain pass stations exist to the north of Corrida de Cori (25°S). In the free atmosphere over Antofagasta the 0°C level is rarely above 5,000 m as a monthly mean. Over the 35° latitude a mean value of 125 m per degree of latitude results for the increase in the height of the 0°C level. Comparison of calculated and observed values shows the mentioned coincidence.

TABLE XIII

ELEVATION OF THE 0° ISOTHERM OVER SEA LEVEL ON THE EASTSIDE OF THE ANDES OBSERVED (2nd line) AND CALCULATED (3rd line) ACCORDING TO A CONSTANT INCREASE OF HEIGHT AT A RATE OF 125 M PER DEGREE LATITUDE

Latitude °S	55	50	45	40	35	30	25	20
Elevation (km)	1.0	1.5	2.0	2.8	3.5	4.0	4.5	5.4
Calculated (km)	0.9	1.5	2.1	2.8	3.4	4.0	4.6	5.2

Annual variation

The region as a whole has a simple variation, with the maximum in January and the minimum in July. However, in northwestern Argentina the minimum occurs in June, while along the coast the extremes are shifted to February and August, respectively. The temperature differences between June and July inland and between July and August on the coast are, however, very small. The similarity of the annual variation even in very different climates is astonishing (compare in Fig.8 Bahía Negra at 20°S with Ushuaia at 55°S). Apart from different annual ranges, characteristic differences can be seen only between spring and fall values. Hence, the influence of latitude on the form of the annual variation is very small. On the other hand, the maritime and continental influence is clearly discernible.

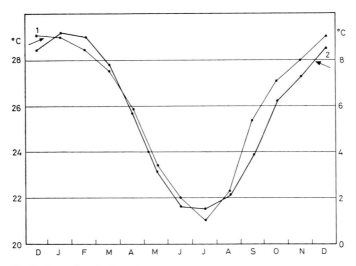

Fig.8. Annual variation of temperature at Bahía Negra (*1*) at 20°15′N 58°12′W, 100 m, and Ushuaia (*2*), Table XLV.

In the extreme oceanic region of Punta del Este and to a lesser extent of Mar del Plata the temperature in March is even higher than in December, and fall becomes 2°–3°C warmer than spring (Fig.9,*2.1* and *2.2*). Towards the interior of the continent, these higher autumn temperatures disappear very rapidly and the limit where spring and autumn values become equal follows approximately the course of the Rio Parana from Misiones to about the latitude of the city of Santa Fé, whence it runs southwest, and then directly south after reaching the western border of the province of Buenos Aires. In Patagonia, fall is warmer then spring only on the coast as far as 45°S and farther south just on the most exposed peninsulas such as Cabo Blanco, Punto Deseado, and Cabo Virgenes. If this limit is regarded as a division between oceanic and continental influence in the temperature regime, it evidences a continental influence almost to the coast of Patagonia and a not very deep penetration into the continent of the maritime influence in the subtropics.

The relatively high spring temperatures to the west of this limit and in Paraguay are caused by the small amount of cloudiness in late winter and spring coinciding with the end of the dry season, so that the temperature responds rapidly to the increasing intensity

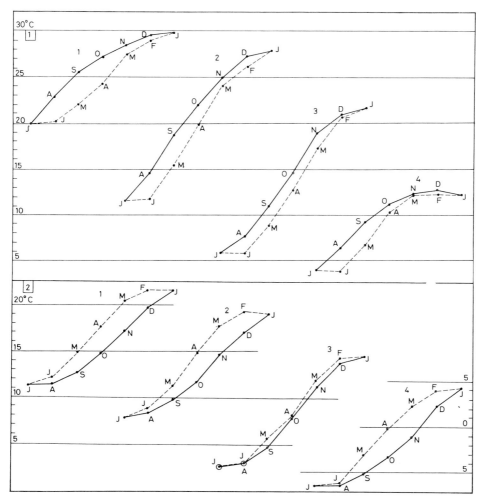

Fig.9. Annual variation of temperature in continental climate (*1*) and in maritime climate (*2*), in °C. *1.1* = Mariscal Estigarribia; *1.2* = Catamarca; *1.3* = Cipolletti; *1.4* = La Quiaca; *2.1* = Punta del Este; *2.2* = Mar del Plata; *2.3* = Bariloche; *2.4* = Cristo Redentor.

of solar radiation. The effect is most marked in the Chaco and on the rolling plains and in the valleys of the pampean sierras (Fig.9,*1.1* and *1.2*). In Mariscal Estigarribia and in Catamarca, the temperature in September is 3°C higher than in May and a difference remains throughout the spring although reduced to 1°C. This rapid rise in spring, which partly sets in as early as August, is the reason for the very early formation of the thermal low over the Chaco and for the response of the wind field to its position and intensity as already mentioned. In central Argentina (Fig.9,*1.3*) temperatures in spring are only a little higher than in fall. Spring has more sunshine than autumn, and moreover, the first outbreaks of cold polar air occur generally in May, marking thus an early beginning of winter.

A different annual variation is found in the eastern puna (Fig.9,*1.4*). At La Quiaca the temperature increases from July to November, remains practically constant until March (this five months period contributes only 0.6°C or 6% to the annual range of 9.6°C) and falls to its minimum in just three months. In other words the northeastern part of the

Argentine puna shares the tropical regime during summer (small changes from month to month) due to the strong advection of tropical air masses with the intrinsic strong cloudiness. The rest of the year it takes part in the temperature changes of the dry subtropics with a variation that corresponds perfectly to the annual change in net radiation. Cristo Redentor, the pass station of the central cordillera, exhibits normal high-altitude characteristics (Fig.9,*2.4*), which by the way are similar to those at the east coast at the same latitude. Likewise, the stations in the piedmont zone of the Patagonian Andes show an oceanic combined with lake influence with autumn warmer than spring (Fig.9,*2.3*). Consequently, the duration of the climatic seasons varies in the different regions. The intermediate seasons are becoming shorter with decreasing latitude and increasing maritime influence.

The highest range of the annual variation (16°–17°C) is observed in western Argentina between 30° and 40°S. Maximum continentality, according to this criterion, is therefore found in latitudes in which the width of the continent is rapidly decreasing from 2,000 to 1,000 km. On the one hand the Andes protect this region against maritime influences, on the other they channel tropical air masses in summer and polar air masses in winter regularly into this region. From this area the annual range decreases in all directions—most rapidly eastwards, towards the coast, where it amounts to 10°–11°C (Punta del Este, 10.2°C; Mar del Plata, 11.6°C). With increasing elevation to the west it falls to almost the same value (Cristo Redentor, 11.4°C). The decrease towards tropical and subtropical latitudes is much slower. The lowest annual ranges of 8°C are found in northern Paraguay and in Tierra del Fuego. However, the greater parts of Argentina and Uruguay (including the major part of Patagonia) have annual ranges greater than 12°C. Only the northern provinces of Argentina (Salta, Jujuy, Formosa, Misiones) and Paraguay show the decrease that is a characteristic of the approach to the tropical zone.

Daily variation

The magnitude and annual change of the periodic and aperiodic daily variation are functions of the autochthonous and allochthonous features of the climate. According to whether they reinforce or counterbalance one another, the mean monthly aperiodic daily variation may assume values of 24°C or of 5°C only. Likewise it depends on weather and season, when the extreme values are observed.

Aperiodic variation is the difference between mean monthly maximum and minimum (mean daily range, cf. the climatic tables of the Appendix). For obvious reasons, in the extra-tropical region the minimum of the aperiodic daily range occurs in winter, predominantly in June, and the maximum in summer (December or January). In the dry western and northwestern Argentina, both extremes appear earlier, the minimum in autumn (March to May) and the maximum in spring (September to November). In La Quiaca, located in the tropical transition zone, the annual variation of the aperiodic range is reversed and shows a maximum in winter (July) and a minimum in summer, i.e., a seasonal change that is essentially a function of the rainy and dry seasons and not of the insolation and the changing pattern of advection during the year. The combination of both factors explains the values observed in northwestern Argentina, since the end of the dry season coincides with the period of already intense insolation and at the end of the cloudy summer the intensity of the solar radiation is strongly declining.

The spatial distribution of the aperiodic temperature range (cf. the climatic tables) shows the highest values in the puna and in the dry west. In the former region, they are coupled with a marked annual variation (La Quiaca: 24.1°C in the dry season and 15.3°C in the rainy season), while in the latter region the aperiodic daily range is large throughout the year. It amounts to about 16°C in the Andean provinces (except Patagonia) and rises locally to 19°C, e.g., in late summer in Neuquén and in winter in Salta. The smallest daily ranges are observed on the coasts, e.g. in Punta del Este (between 5.2° and 7.4°C), Mar del Plata (between 7.0° and 11.4°C) and Ushuaia (between 6.0° and 9.2°C). In Paraguay, Uruguay, the Argentine mesopotamia and in Patagonia, the maximum range is found in summer with 12°–14°C; the minimum in winter, 8°–10°C. In central Argentina, including the province of Buenos Aires, the aperiodic daily range varies between 12° and 14°C all the year round. A comparison of the mean annual range of the aperiodic daily variation with that of the annual variation of the monthly averages reveals that the former is only slightly smaller than the latter. Only towards the north of Argentina and Paraguay the annual range of the monthly means shows the characteristic decrease in direction to the tropics, but the range of the aperiodic daily variation retains its value of about 12°C.

Periodic daily variation is the difference between the extremes of the average hourly values. This variation of the temperature shows no peculiarities and differs in the various climatic zones only in its amplitude which, however, varies over a wide range. The July values at La Quiaca (20.6°C) and at Ushuaia (2.8°C) can be regarded as the two limits in this entire region (Fig.10); but they represent just their equally extreme geographic situations at the puna and on the subpolar sea, respectively. In general, the periodic daily variation lies between 10° and 15°C in summer and between 7° and 12°C in winter. In summer these values rise to 17°C in the southern part of Cuyo, Neuquén and western Rio Negro, since long days coincide with intense radiation at 35°–40°S, but decrease to 8°C on the coast. In winter, they decrease further, to 6°C north of 40°S (in July, La Quiaca has the same noon temperature as Montevideo) and to 4°C on the Patagonian coast.

The influence of longitude and latitude on the occurrence of the extremes of the periodic daily variation is evident. The standard time of the three countries refers to 60°W, but as they extend from 53° to 73°S, they have a maximum local time difference of 1 h and 20 min, which is clearly visible by the time the extreme values are registered. In summer, the minima occur in the east at about 05h00, in the central region between 05h00 and 06h00, and in the west at about 06h00. South of 45°S, however, the influence of latitude is more evident and the minima are at 05h00 near the cordillera and in Ushuaia as early as 04h00. In winter, with a less pronounced daily variation, the minima occur generally at about 07h00 but advance to 06h00 in the extreme northeast and are at 08h00 in Patagonia. All these times correspond to local sunrise, or just after.

The time of the occurrence of the maxima depends more on local influences. In summer, they are generally registered at about 15h00, due to the strong insolation in the subtropics. In the west, as in Cuyo, the maximum is as late as 16h00 or even 17h00. At the coast it is normally observed at 14h00 except at places subject to strong sea breezes where it occurs not later than noon (Mar del Plata or Camarones). In winter the time of the occurrence of the maxima is more uniformly at about 15h00 even in the west. In Patagonia it is at 14h00 because of the higher latitude and strong winds and cloudiness.

The relationship of the aperiodic to the periodic daily variations gives a useful indication

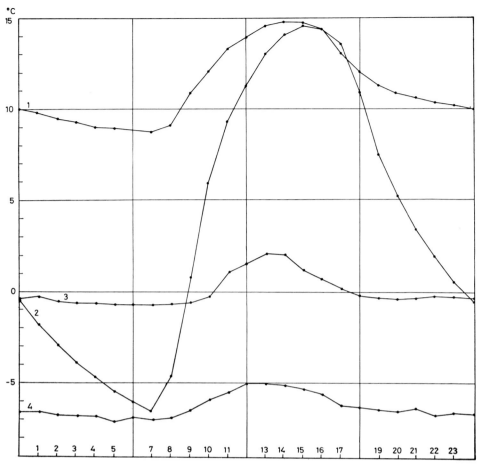

Fig.10. Diurnal variation of temperature in July (in °C) at Montevideo (*1*), La Quiaca (*2*), Ushuaia (*3*), Cristo Redentor (*4*).

of the perturbation of the autochthonous climate. This criterion leads to the same climate divisions as described in connection with the wind regime, viz., the zone of the prevailing westerlies, the zone under the influence of the subtropical high pressure area, and the transitional zone to the tropics with the characteristic division of the year into a rainy and dry season.

In the first of these regions, in Patagonia, this ratio is between 1.6 and 1.8 in winter and rises to 2.0 only over the Andean lakes openly exposed to the west, such as Nahuel Huapi or Lago Argentino, and reaches a value of 2.4 in Tierra del Fuego. But here the periodic daily variation amounts to scarcely 3°C in July. In summer, these values decrease to 1.3°C, particularly on the mesetas, but are not so low on the coast and at the piedmont of the cordillera. The reasons are the greater periodic daily variation and the stabilizing influence of the subtropical high which extends into Patagonia at this time of the year. For the same reasons, similar seasonal variations are found over the southern pampa, but with smaller absolute values (1.3°–1.4° in winter and 1.2° in summer). The other parts of Argentina show ratios between 1.2° and 1.3°, rising to 1.5° only on the coast and falling below 1.2° in Cuyo which is normally little affected by advective influences. A small annual variation can be seen in the northern pampa and mesopotamia, that is similar to the extra-tropical type, and in northern and northwestern Argentina, similar to the tropical weather type. This latter influence (rainy and dry seasons) is reflected, as an extreme case, in the data of La Quiaca with a ratio of 1.42° in January and 1.16° in July. A still bigger range is exhibited in Cristo Redentor, but with the maximum and minimum in the opposite seasons: in July 3.6° and in January 1.5°; this minimum value is higher than the maximum at the same time in La Quiaca which has approximately the same elevation. This reflects the strong influence of the subtropical high in summer and of the westerlies in winter in these latitudes and altitudes.[1]

Interdiurnal variations have been calculated only for Buenos Aires from 1906–1937, but the values are probably representative for the whole pampa. The annual average amounts to 2.0°C with a very small annual variation (2.2°C in December, 1.9°C in July). The greater variability in summer is the result of the more active weather pattern. This causes greater temperature contrasts at this time of the year. The pamperos have a stronger influence on the temperature field in summer than in winter and the same holds for advection of tropical air masses. In the annual average, large negative changes (inter-diurnal variation $\geq 5°C$) are four times as frequent as positive changes. This is also expressed in the annual averages of all negative and all positive changes, respectively; the former amounting to 2.3°C with a frequency of 13 days per month and the latter to 1.8°C with a frequency of 17 days per month. The averages of the greatest interdiurnal changes per year are $+6.6°C$ and $-9.2°C$.

Maximum temperatures

The combination of high monthly averages with a large daily variation makes the Argentine Chaco the region with the highest temperatures in South America. Almost every year temperatures of 45°C are observed and the highest values ever measured are between 49° and 50°C. The 50°C mark itself (which has been reached in all other continents, even in Europe) has so far never been registered at a meteorological station in South America. However, temperatures of 49°C have been occasionally recorded over a very wide region stretching from the tropic to approximately 30°S. Therefore it can be assumed that the 50°C limit is reached or exceeded somewhere in the southern Gran Chaco. It must be borne in mind that this very thinly populated area has only a small number of meteorological stations, located generally in, or in the neighborhood of, settlements which have irrigation systems; by the resulting oasis effect the observed temperatures are not entirely representative of this extremely arid climate (PROHASKA, 1959). Temperatures of 40°C are observed almost every year in Paraguay, northern and central Argentina and northern Uruguay. Only the most exposed coastal strips of Uruguay and Patagonia south of the Valdés peninsula never reached this temperature, and neither was it observed on the Patagonian mesetas south of 40°S. The level of 30°C has been registered over the entire mainland except at its southernmost tip at the Strait of Magellan. In Tierra del Fuego, the highest reported values are between 25° and 28°C. The absolute maxima measured at the stations in the cordillera and the puna are as follows: Corrida

---

[1] A more detailed discussion of this relationship is not justified since the periods of observation of the periodic and aperiodic daily variations are not identical and since the first ones have been available for only relatively short periods (at most 10 years).

de Cori (5,100 m) 14°C; La Casualidad (4,000 m), 25°C; La Quiaca (3,450 m), 28°C; Cristo Redentor (3,830 m), 20°C.

The mean monthly maxima of the warmest months generally exceed 30°C everywhere north of the Rio Negro except on the Atlantic coast and on the Río de la Plata. In the Gran Chaco 35°–39°C is reached, while in southern Patagonia the maxima are only 20°C and in Tierra del Fuego 15°C.

Minimum temperatures

The lowest temperature in South America is generally quoted as −33°C, observed at Sarmiento. Even if this value must be regarded as questionable, it can be assumed that the lowest temperatures of the continent, excluding the cordillera, are found in the wind protected places of the southern Patagonia mesetas. At these stations certainly −25°C or less can be reached. The lowest recorded values at the two stations on the main ridge of the cordillera are: Corrida de Cori (5,100 m) −24°C, and Cristo Redentor (3,830 m) −30.3°C. The annual minima of Patagonia and Tierra del Fuego (except the coasts), in Mendoza, and the western pampa are between −10°C and −15°C, and a bit lower values at the piedmont of the Andes. Minima of −5°C are observed almost every year as far north as southern Santiago del Estero (29°S). This temperature has been recorded as an absolute minimum in the center of the Paraguayan Chaco (Mariscal Estigarribia), in the sierras of Misiones, and in western and southern Uruguay. Thus, frost (0°C) occurs nearly everywhere in the three countries. According to available records Punta del Este constitutes an exception. Therefore it can be assumed that also the Atlantic coast of Uruguay never has temperatures below 0°C. The capital of Paraguay, Asunción, reports +2.3°C as the absolute minimum. It is quite possible that temperatures below 0°C never occur in the entire river basin of the Rio Paraguay. This is supported by observations at Corumbá at the Brazilian border (absolute minimum +0.8°C). With regard to the mean monthly minimum of the coldest month, the values are slightly below 0°C (up to −2°C) only on the Patagonian mesetas but not on the coast; in Tierra del Fuego they are between −2°C and −3°C, in Paraguay, Formosa, Misiones and northern Corrientes they remain over 10°C.

The difference between the mean monthly maxima of the warmest and the mean monthly minima of the coldest months exceeds in general 20°C over the continent, but is only 17°C in Tierra del Fuego. More than 25°C are observed in southern Chaco, the eastern part of the pampa and northern Patagonia and the differences exceed 30°C in Cuyo, and the western pampas as far south as the Rio Negro. Hence, the last mentioned zone is the most continental region of South America, according to its thermal characteristics.

The differences of the annual extremes show a broad band of 45°C centered at the longitudinal axis of the continent from Chaco to southern Patagonia, with a zone of maximum temperature differences of over 50°C stretching from southern Mendoza to northern Chubut. Only at higher elevations in the Andes, in the Argentine mesopotamia and in Paraguay and Uruguay is the absolute annual temperature variation below 40°C.

Ground frost is observed nearly everywhere since the above-discussed 0°C values refer to measurements in the meteorological screen. In general, frost may occur in the period between mid-May and mid-September in the agriculturally important regions of the three countries; this means that there is a frost-free period of about 250 days. In western

Patagonia and Tierra del Fuego, this value is reduced to less than 100 days, while in the Chaco, Corrientes, and Misiones more than 300 frost-free days are common. In the extreme north the risk of ground frost is restricted to the month of July. Since here frost is normally produced by advection of cold polar air with subsequent cooling by radiation, its occurrence is particularly variable and very late and early frosts may happen. If it happens it is especially injurious to plants, producing great damages to agriculture (BURGOS, 1963a).

Soil temperatures in Argentina

Although in Argentina soil temperatures have been measured at approximately 120 stations for about 20 years, averages are so far only available for few stations for the period 1941–1950. Since no recordings but only thrice daily readings are available, only those depths will be considered where the daily variation becomes so small that the time of observation is of no major influence on the computation of means; this is generally the case at depths below 40 cm. From this depth downwards, the daily changes and the changes from year to year are so small, that even different observation periods of the mean values are normally within the limits of error of the measurements (which are considerable in the case of mercury in glass thermometers).

As expected, the main climatic zones are likewise perceivable in the soil temperatures. The drier (and therefore the more bare of vegetation) and the warmer the climate, the more the annual average temperature of the soil exceeds that of the air (in the meteorological shelter); the moister and/or the cooler the climate and the denser the vegetation the smaller is the contrast (Table XIV). In northwestern Argentina this difference amounts to 3°C (3.5°C in the extremely dry and hot southern Chaco[1], and 2.5°C in Cuyo and the dry western pampa) and decreases to 1°C towards northeastern Argentina (Misiones) and towards Patagonia.

The time lag of the annual variation at 50 cm is still so small that the extreme values coincide with those of the air temperature (January and July); at a depth of 1 m exists already a displacement of phase, so that the extremes are delayed until February and to July or August, respectively. At 2 and 3 m an appreciable difference can be seen between the soil temperatures in the arid west and in the humid east. In the former the maxima occur in February at 2 m and March at 3 m, and the minima in August (2 m) and September (3 m), while in the latter region the maxima at 2 m are equally distributed between February and March and at 3 m are found in April, while the minima are in September (2 m) and October (3 m). In Buenos Aires, where measurements at greater depths are available, a displacement of phase by 6 months is observed at 5 m. The decrease of the annual range with depth depends on the properties of the soil (thickness and type of the soil horizons) and of the subsoil. As Table XIV shows, the decrease in range is almost normal, with the exception of two stations in the northwest of Argentina (Guemes and Andalgalá). Here the decrease is substantially less because of the particular soil components. The annual averages in Table XIV are valid for all depths considered, since the differences at different depths are within the accuracy of the measurements and show no definite trend.

---

[1] Here at 10 cm the temperature of January is above 30°C as an average of a 10-year period.

TABLE XIV

MEAN ANNUAL AIR AND SOIL TEMPERATURE AND MEAN ANNUAL RANGE (°C)[1]

Station	Air temperature		Soil temperature		Range at depth			
	mean	range	mean	soil-air difference	0.5 m	1 m	2 m	3 m
Guemes 24°36'S 64°59'W 655 m	19.7	12.3	22.6	2.9	10.4	8.5	6.0	–
Andalagala 27°35'S 60°19'W 1,072 m	18.1	15.4	20.7	2.6	–	11.4	7.8	–
Angel Gallardo 31°37'S 60°41'W 18 m	18.1	13.2	19.9	1.7	–	10.0	5.8	3.8
Pergamino 33°56'S 60°33'W 66 m	16.2	14.1	17.7	1.5	–	10.2	6.1	3.5
Col. Alvear 35°00'S 67°39'W 465 m	15.9	17.0	18.4	2.6	17.4	14.3	–	4.0
Jose C. Paz 34°31'S 58°45'W 30 m	15.9	13.6	17.7	1.8	–	9.6	6.0	3.6
Santa Rosa 36°34'S 64°16'W 189 m	15.8	16.7	18.3	2.5	–	13.9	7.2	4.1
Fortin Mercedes 39°31'S 62°38'W 25 m	14.9	15.4	16.8	1.9	15.5	12.9	7.7	4.6

[1] Period 1941–1950.

Humidity

Vapor pressure

The source regions of atmospheric moisture in extra-tropical South America are the humid tropical regions of the Brazilian–Bolivian lowlands, the Atlantic, and the catchment areas of the rivers Paraguay and Parana in the subtropics. The region is almost completely sheltered from the influence of the Pacific by the cordillera (the mean annual vapor pressure is 2 mbar at Corrida de Cori and 3 mbar at Cristo Redentor), and the air masses coming from polar regions have a very low humidity.

The most important factor is the advection of tropical air masses, for the direct influence of the Atlantic and of the two large rivers is mainly limited to the coastal region and the region lying between the rivers. In addition to the normal decrease of vapor pressure from north to south, there is therefore also a decrease from east to west. This is further accentuated by the gradual increase in elevation towards the Andes. The annual means of the vapor pressure actually observed therefore decrease from northeast to southwest. In Patagonia, where the moist tropical air masses do not normally penetrate, this becomes a pure east–west decrease, i.e., a more humid coastal strip contrasting with the dry interior. The influence of the enormous quantities of warm water flowing southwards in the rivers Paraguay, Parana, and Uruguay, with their numerous branches, swamps, and deltaic regions, on the water vapor content is well marked in the annual means, particularly in the wide basin of the Rio Paraguay, in Argentinian mesopotamia (Entre Rios, Corrientes and Misiones), and in western Uruguay. The highest values are therefore found in regions influenced by these river systems, as far south as the Río de la Plata. The contrast to lower values is particularly pronounced towards the west, towards the Chaco, and the pre-Andean plains. Table XV, where the values observed in the river regions are compared with those for the pre-Andean plains but reduced to the river levels, shows clearly that this contrast is not due to the increase in elevation.

TABLE XV

VAPOR PRESSURE (mbar) ON THE PRE-ANDEAN PLAINS AND THE PLAINS NEAR THE RIVERS PARAGUAY–PARANÁ–RIO DE LA PLATA

	Latitude °S: 20	25	30	35
Pre-Andean plains	20	17	13	10
The same values reduced to the level of the plains near the rivers	21	19	15	13
Observed on the plains near the rivers	22	21	17	15

The only exception to this is at the eastern slopes of the sierras in Tucumán, Salta and Jujuy, where as a result of orographic rainfall the tropical–subtropical jungle vegetation extends in a narrow strip southward and the vapor pressures even after reduction to the same height equal that at the Rio Paraguay. On the other hand, the annual mean vapor pressure in the southeastern half of Uruguay is somewhat lower. This is due to the decrease of temperature toward the coast. On the Patagonian plateau the annual mean vapor pressure decreases from 6.5 mbar in the north to 5.5 mbar in the south, while on the coast the annual mean vapor pressure decreases from 8.5 to 7 mbar in the same direction. This latter value is also found in Tierra del Fuego. A comparison between the Patagonian mesetas and the coast shows that here the influence of the Atlantic on humidity is slight, as more than 2/3 of the increase on passing from the plateau to the coast can be attributed to the difference in height, i.e., even the coast of Patagonia has a relatively dry atmosphere.

In the geographical and seasonal distribution of vapor pressure, the value of 19 mbar is of special significance. It is regarded as the threshold value for "sultriness". Annual means above this value are found in the whole of Paraguay (with the exception of the higher regions in the east of the country), in the eastern half of the province of Formosa, in the north of Corrientes, and in lower regions in Misiones.

Although the *annual variation* of the vapor pressure in the extra-tropical latitudes follows approximately the temperature variations, there are some differences, particularly regarding the amplitude. The greatest similarity between the two curves is found where there is a constant supply of water vapor throughout the year. This is the case in the above-mentioned river basins and on the Atlantic coast, except in Patagonia. The times of occurrence of the extremes indicate a normal time lag, so that in maritime-influenced regions the maxima and minima appear in February and in August, respectively. In general, however, the differences between January and February, and between July and August, are very slight. As a result, there is a humid late summer and autumn and a dry late winter and spring, and consequently the annual temperature curve shows a slow fall and a rapid rise.

In *winter*, the main source regions of the humidity are the Argentine–Paraguayan meso-potamia and the Atlantic. There is a simple spatial distribution of the vapor pressure, with a maximum (17 mbar) in northern Paraguay and a minimum (5 mbar) in western and southwestern Patagonia. The general direction of the isohumes is very similar to that of the isotherms. North of 35°S the difference between east (coast) and west (piedmont of the Andes) becomes in winter even more pronounced than in the annual means owing to the absence of advection of humid air masses of tropical origin. There is, thus, a clear division of the region into a dry west (from 7 mbar in Cuyo to 12–14 mbar in northern Chaco) and a humid east (from 11–12 mbar in Uruguay and Entre Rios to 16–17 mbar in eastern Paraguay). The narrow transition zone coincides nearly with the Paraguay and the Parana rivers, flowing almost in a north–south direction. The vapor pressure decrease from northeast to southwest is steady down to the 8-mbar isohume. This isohume runs from the southern coast of the province of Buenos Aires in an almost straight line over northern Córdoba to the sierras of Catamarca. To the west and south of this line there is only a slight further decrease, and the humidity distribution is consequently very uniform. Along the coast of Patagonia, on the other hand, a sharp decrease of vapor pressure is evident toward the west due to the influence of the dry west winds. The values are above 6 mbar at the coast and are already appreciably lower in bays penetrating deep into the continent, e.g., Comodoro Rivadavia (46°S) 5.3 mbar, Faro Cabo Blanco (47°S) 6.9 mbar, or Rio Gallegos (51.5°S) 5.5 mbar and Faro Cabo Virgenes (52.5°S) 6.3 mbar.

In *summer* the isohumes run more zonally, owing to the intensity and regularity of the advection of tropical air masses which can be traced from the vapor pressure as far south as northern Patagonia. At Bahía Negra (20°S) on the Paraguayan–Brazilian frontier, the January mean value is 28 mbar. This is a sign of pure tropical air masses, for such values are exceeded only occasionally even in the central parts of Amazonas. The vapor pressure indicates similar tropical air masses in a narrow strip running north–south on the eastern flank of the mountains in Jujuy and Salta. In the east, on the other hand, the values decrease to 20 mbar on the Río de la Plata, while in the west (Chaco, Cuyo, pampa) the values fall to 15 mbar. This shows the considerable drying-out of air moving towards the south over the continent in summer. An example is the very rapid decrease of vapor pressure from the Río de la Plata in the direction of the pampa. This decrease is so pronounced that it can be clearly traced within the network of stations in greater Buenos Aires. The corresponding January means run as follows: on the river bank (Aeroparque) 19.9 mbar, Observatorio Central (7 km inland) 18.8 mbar, Castelar (28 km inland) 18.4 mbar, and Ezeiza (50 km inland-international airport) 18.0 mbar. The latter value is representative for the eastern pampa at this latitude, showing that the direct influence of the warm and vast water surface of the Río de la Plata extends less than 50 km inland. This decrease is of great bioclimatic significance, since it includes the critical sultriness value (19 mbar). The parts of Buenos Aires (including the city center) lying within a coastal strip 10 km wide have a mean vapor pressure which is above this critical value for at least a month, whereas the western and southwestern districts of Greater Buenos Aires have mean vapor pressures below this critical value, even in the summer. This state of affairs is corroborated by the experience of the inhabitants, which also confirms the validity of the critical value of 19 mbar for sultriness. If we consider the diurnal variation of vapor pressure in Buenos Aires (Observatorio Central), we find, moreover, that in January and February the values are below the critical sultriness value only between 02h00 and 06h00, and that the means exceed the critical value between 10h00 and 15h00 still in March.

Inland, the 19-mbar isohume runs from the Río de la Plata, along the lower course of the Parana, toward the northwest over northern Córdoba, to the south of Tucumán. There the mean values of all four summer months are above the critical 19 mbar (December through March). The duration and intensity of sultriness increases fairly rapidly to the north and east of this line, so that on the Paraguayan–Brazilian frontier (Bahía Negra) there are 9 months in which the threshold value of sultriness is exceeded, while in eastern and southern Paraguay, and also in Misiones, it is exceeded during 6–7 months.

South of the 19-mbar isohume the direction of the gradient is maintained as far as the 10-mbar isohume, which runs northwest from the Valdés peninsula to Neuquén. In summer as well as in winter, the vapor pressure gradient over Patagonia is in a markedly east–west direction.

On the basis of the seasonal behavior, three regions with different annual ranges can be delineated: (*1*) the region with an ample supply of water throughout the year, so that the vapor pressure changes follow temperature changes with the usual time lag and show a range of 8–10 mbar; this region includes the subtropical Atlantic coastal area and mesopotamia; (*2*) the region where only a small amount of water vapor is available throughout the year (Patagonia); this is the amount of water vapor still remaining in the air masses of Pacific origin after crossing the southern cordillera; the range decreases from 5 mbar in northern Patagonia to 3 mbar in southern Patagonia and Tierra del Fuego; (*3*) the region to the west of the great rivers, where the main supply of water vapor comes from the tropics only during summer and where winter is the dry season characterized by low water vapor values; the annual range in this region is as high as 12 mbar on the eastern flanks of the mountains in Salta and Jujuy.

The *diurnal variation* is very small and shows the usual characteristics. In winter the diurnal curve has a simple wave form, with extreme values at approximately the same times as the extremes of temperatures. In summer the curve has a double-wave form due to the interaction of the diurnal variation of wind and temperature. The daily ranges in winter are around 1 mbar, somewhat lower (0.5 mbar) in the dry regions, and somewhat higher (1.5 mbar) in the humid northeast. In summer the ranges are much more affected by local conditions and vary between 1 and 3 mbar.

The decrease of vapor pressure with height is quite regular and normal during winter; only Corrida de Cori has rather low values for its elevation. However, the altitude of this station is known only approximately, and there are only a few observations for winter, at irregular intervals. On the other hand, during summer the vapor pressure is generally very low at the western Altiplano and the main range of the cordillera, due to the influence of the subtropical anticyclone. This results in a very small annual range of 2 mbar. Only the eastern and northeastern parts of the Altiplano are affected by pure tropical air masses in the summer, as shown by the vapor pressure during this season at La Quiaca (9 mbar) which, reduced to the level of the Bolivian lowland, gives 28–29 mbar agreeing with values actually observed there. For its elevation, this station has thus an unusually large annual amplitude of 6.4 mbar. A similar seasonal pattern also appears in the diurnal ranges which are greatest in the transition months March and November (2 mbar), when large changes of temperature are associated with advection of humidity. In the summer, during the rainy season, the range decreases to 1 mbar (owing to the small diurnal temperature variation), and in the dry season it amounts only to 1.4 mbar owing to the lack of water vapor, even though the diurnal range of temperature is then twice as large as in the rainy season.

Relative humidity

The combination of the annual average temperature gradient in direction from north to south with the northeast to southwest gradient of water vapor pressure results in a gradient of the relative humidity running from east to west. Misiones, the subtropical Atlantic coasts and the area of influence of the Río de la Plata have a high relative humidity of 75–80%, while the elongated zone extending from northern Chaco over central and western Argentina to central Patagonia has a low relative humidity of 55–60%. In the deeply indented gulfs of Patagonia the relative humidity decreases further to below 50%. In other words, the relative humidity decreases towards the coast due to the previously mentioned foehn effect. Even the highly oceanic influenced coast of the Valdés peninsula (Faro Punta Delgada 42°46′S, 63°38′W) has an annual average of only 67%, and it is only south of the Gulf of San Jorge, on the coasts jutting out towards the east, that the values become normal for such a situation (Faro Cabo Blanco, 47°12′S, 80%; Faro Cabo Vírgenes, 52°20′S, 81%); a likewise increase is observed at the foot of the southern cordillera. In the interior of southern Patagonia the relative humidity amounts to 70% and in Tierra del Fuego to 75%. In northern Argentina, values above 70% are found over a narrow zone parallel to the eastern slopes of the sub-Andean sierras (Tucumán and Salta). In the puna proper there is a sudden decrease to 50% and to 40% at the frontier with Chile.

The combination of the different types of annual variation of temperature and vapor pressure results in a great variety of the annual variation of the relative humidity in reference to amplitude, seasonal distribution and time of extremes. The largest annual ranges occur in northern Patagonia (Maquachín, 41°15′S 68°44′W, 888 m, 41%). Here a very pronounced annual variation of temperature is combined with a small range of vapor pressure due to the absence of readily available evaporation sources; but even in middle and southern Patagonia the annual range is above 25%. Similar values are found where the vapor pressure has its greatest range, namely in northwestern Argentina and in Paraguay, in spite of the already small annual range of temperatures. Advection of moist air masses in summer alternates with the autochthonous dryness in winter (La Quiaca 29%; Salta 27%). The lowest ranges (less than 15%) have the Argentine–Paraguayan mesopotamia up to the Río de la Plata and the coasts of Uruguay and Buenos Aires (Punta del Este 11%), since sufficient water vapor is present throughout the year. Similar small ranges are found at the exposed parts of the coasts of Patagonia and Tierra del

Fuego. A characteristic increase of range to over 20% evidences the center and eastern half of Uruguay and the adjacent parts of Entre Rios, due to the more pronounced annual variation in temperature.

The variety in the types of the *annual variation* of the relative humidity is illustrated by the fact that in every month of the year extreme values can be observed in one of the regions; but there is a pronounced seasonal division, since the maxima occur in the first half of the year (February to July) and the minima in the second half (August to January). Three main types can be detected: (*1*) the outer tropical zone with the summer maximum at the peak of the rainy season and the winter minimum towards the end of the dry season; (*2*) the extratropical regions with the normal variation of the relative humidity, i.e., inversely to that of the temperature; (*3*) the transitional type with extremes in the intermediate seasons, i.e., spring minimum due to a marked increase in temperature at a time when the advection of water vapor did not start, and an autumn maximum when an already pronounced fall in temperature is combined with a still high moisture content (Fig.11). The times of occurrence of the maxima and minima respectively are: February and August at the Argentine–Bolivian frontier; March and September in the central part of Jujuy and Salta; April and September in the western Chaco; May and September in the rest of the Chaco, northern part of Córdoba and western half of Santa Fé; May and October or November in Cuyo and the western pampa, and June or July and December or January in the entire mesopotamia (including eastern Paraguay), Uruguay and Patagonia. The spatial distribution of relative humidity in the individual months and its intermensual variation is therefore quite complex, since at the same time these quantities are increasing in one region and decreasing in another. August shows the simplest picture. During the

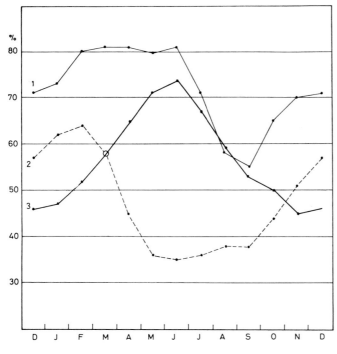

Fig.11. Annual variation of relative humidity (in %) at Tucumán (*1*), La Quiaca (*2*), Cipolletti (*3*).

month 70–80% are observed on the coast, in mesopotamia and in southern and western Patagonia, 60% in Chaco and Cuyo and 30–40% at the Altiplano; a weak south–north gradient is therefore superimposed to the pronounced east–west gradient. In spring the the minimum of the relative humidity is observed in the interior of the continent due to the marked increase in temperature. Only in the extreme northwest does the advection of moisture produce already a slight increase in the relative humidity. With the southward extension and intensification of this advection the relative humidity increases throughout central and western Argentina in late spring while it decreases in mesopotamia owing to an only slight increase in vapor pressure, but a pronounced increase in temperature. Thus, in spring, the contrast between east and west is evened out, to give place to a well expressed north–south gradient in summer. At this time of the year the highest values are observed in northwestern Argentina and the lowest in central and northern Patagonia (in the region of greatest annual temperature variation). Here, monthly averages are actually less than 40%. In autumn, as in spring, the distribution again becomes fairly regular over the whole region, the only differences being that the actual values are higher. Statistics of the *daily variation* of the relative humidity are available only for a small number of stations. The corresponding extreme values occur at the same time as those of the temperature variation. The range, however, depends to a large extent on the daily variation of vapor pressure, which is different in summer and winter. The mean daily ranges can be as little as 6% (June at Ushuaia) and as much as 54% (March at La Quiaca). Normally, however, the mean daily ranges are smaller than the corresponding annual ranges of the monthly averages and show little variation during the year. The greatest daily ranges occur approximately in the month of lowest relative humidity and vice versa. Only in the arid zones is this relationship reversed (Mendoza, La Quiaca). In the Argentine mesopotamia the daily ranges amount to about 20% in winter and about 35% in summer (25% on the coast); in the dry interior the figures are about 30% in winter and 40% in summer. In the puna, the summer rainy season exhibits the exceptionally large value of more than 50%, falling to 30% in the dry season. At the other extreme lies Ushuaia with a mean daily range of 24% in February and 6% in June.

Evaporation

It is difficult to estimate or determine the potential evaporation of large areas with sparsely scattered observations of this element. Besides, the few existing measurements have been obtained by different methods with their inherent sources of incontrollable error, which makes a comparison even more difficult. Attempts to calculate the evaporation from the available meteorological observations are likewise unsatisfactory. This applies particularly to the climatic conditions of these three countries, where different meteorological factors or different combinations of them are of decisive influence on evaporation in the different regions. Empirical or semi-empirical formulae which do not consider wind speed have to give erroneous results in Patagonia, those which do not include the influence of radiation produce erroneous results in northwestern Argentina and Paraguay and those that neglect either explicitly or implicitly the saturation deficit are unsatisfactory in mesopotamia. Furthermore, the question remains as to what extent the various empirical constants of the formulae (derived mostly from observations in temperate latitudes) are valid in extremely arid or humid zones of the tropics or subtropics.

No experimental proof of their validity under these quite different climatic conditions has yet been obtained.

The observational data in the climatic tables of the Appendix should therefore be regarded merely as rough approximation of the amount of evaporation in the different climatic zones. The values of Uruguay have been obtained with the Piche-evaporimeter, exposed in the meteorological shelter. In the Argentine network the American class-A pan is in use and the reduction factor of 0.7 has been applied to all monthly and annual averages; a procedure that does not make the values more reliable in all climatic zones. The highest annual potential evaporation (between 1,600 and 1,800 mm) is found in the most continental part of the subtropics (from San Juan up to the Altiplano), where a large saturation deficit is associated with intense solar radiation. Values over 1,500 mm are also recorded in the western Chaco and the pampa, where the less pronounced dryness and somewhat reduced insolation are counteracted by the higher wind speed. In the direction toward Corrientes and Misiones the annual totals decrease to about 1,200 mm while toward the Atlantic they fall to about 1,000 mm. The Uruguayan evaporation data broadly fit in with this picture. On the coast, the figures lie between 950 mm and 1,050 mm and in the northwest, at the frontier with Brazil and Argentina, they amount to 1,250 mm.

The annual *minima* occur quite uniformly in June. They decrease from approximately 100 mm on the Altiplano and the very dry longitudinal valleys of the cordillera (as far south as San Juan) to 30 mm on the Río de la Plata. The greater part of Argentina and Uruguay has values of 40–50 mm, only the Chaco in excess of 60 mm. The *maxima* are recorded mainly in December as a result of the decisive influence of radiation on evaporation in subtropical latitudes. Only in the maritime-influenced regions, in the region of tropical summer rains (Tucumán, Rivadavia) and very probably in Patagonia, is the maximum shifted to January. The regional differences are rather great in summer. Up to 240 mm per month are recorded in San Juan and the dry pampa, between 160 and 200 mm in northern and northwestern Argentina and between 140 and 160 mm in the Argentine and Uruguayan mesopotamia. The largest values registered in Punta del Este and Treinta y Tres of around 100 mm appear to be too low. As always, La Quiaca, at the Altiplano, has an annual variation of its own. The maximum is reached in October before the advection of moist air masses from the north begins, but where at the same time the radiation intensity and wind speeds are high.

All data show clearly greater evaporation in the second half of the year as compared with that in the first, e.g., in northwestern Argentina (Salta, Tucumán, Jujuy) 58% of the annual evaporation. Exceptions are found at the coast, e.g., Punta del Este and Mar del Plata, and in the southern cordillera, e.g., Bariloche, because of their special climatic conditions. In both regions the first six months account for 54% of the annual total.

Precipitation

Between the two centers of maximum rainfall in the southern part of the continent, the Sierra do Mar in the state of São Paulo (Brazil) and southern Chile with annual totals of the order of 5 m and more, lies a region of deserts and steppes (pampas) extending from northwestern Argentina to the coast of Patagonia. This region constitutes, thus, the southern extension of the coastal deserts of Peru and northern Chile. The western half of Paraguay and almost two-thirds of the area of Argentina are included in this dry region. Only Uruguay and eastern Argentina and Paraguay share the copious rains of the Atlantic and, similarly, the southern cordillera those of the south Pacific (Fig.12). Hence, north of 30°S the rainfall decreases from east to west, and south of 40°S from west to east; between 30° and 40°S rainfall decreases very rapidly from the cordillera towards the dry center of the continent and increases slowly towards the Atlantic coast. A narrow strip with intense orographic rain exists at the eastern slope of the sub-Andean sierras, extending from southern Tucumán (28°S) to the north. A coastal effect on rainfall is found only on the northern and southern coasts of Patagonia whereas the gulfs of middle Patagonia are as dry as their hinterland.

The highest registered annual totals of over 3,000 mm are found in the southern cordillera between 40° and 45°S (although situated to the lee of the rain-bearing winds), and between 1,700–1,800 mm in northern Misiones and the eastern part of Paraguay. Over a limited area in northwestern Argentina 2,000 mm may be exceeded on the eastern slope of the Sierra de Chañi by orographic effects. The driest region lies on the northern Argentinian–Chilean boundary (Atacama desert) and extends directly southwards through western Catamarca and La Rioja to San Juan where the annual rainfall is still less than 100 mm and some years are almost rainless with totals of 1–2 mm. On the main ridge of the Andes rainfall increases suddenly south of 33°S and is extending eastwards at the same time. Therefore, the driest region is displaced progressively to the east and eventually dominates the center of Patagonia with annual totals of 100–200 mm. The strongest rainfall gradients are, thus, found in the southern Patagonian Andes where annual totals increase six-fold over a horizontal distance of only about 50 km in east–west direction, and in the area of orographic rains in northwestern Argentina where the annual totals fall from 2,000 to 100 mm over the same horizontal distance and in the same direction. The 1,000-mm isohyet marks the approximate limit between the humid mesopotamia and the dry Chaco and the pampa; it almost follows the courses of the rivers Paraguay, Parana and Río de la Plata.

During the course of the year, the rainfall distribution depends primarily on the prevalence of the corresponding air masses. During the winter months the high-pressure ridge over the subtropical part of the continent produces a dominant zonal circulation originating the intense precipitation activity at the southern coast of Chile. At the same time heavy rainstorms with occasionally disastrous inundations may occur over Uruguay and the lower watershed of the Río de la Plata due to a cyclogenesis over mesopotamia and associated strong winds from the southeast, called sudestadas (SCHWERDTFEGER, 1954, 1958b). East winds and unusual rains over northern Patagonia are caused by blocking anticyclones at higher latitudes over the western South Atlantic as already mentioned (GRANDOSO and NUÑEZ, 1955). Hence, the interior is dry and monthly totals of less than 20 mm are observed in a longitudinal strip between 62° and 70°W. To the west of 70°W, at the central and southern Andes, precipitation increases very rapidly (to 200 mm and more). Towards the east the increase is slower and monthly totals of 40 mm are found along the 60th meridian, of 100–120 mm at 55°W. In *summer*, when there is a prevailing meridional circulation, the rain producing air masses have their origin in tropical zones. Monthly totals show 120–150 mm in Paraguay and northeastern Argentina; they decline slowly towards the south, to 80 mm at the Río de la Plata and in southern Córdoba.

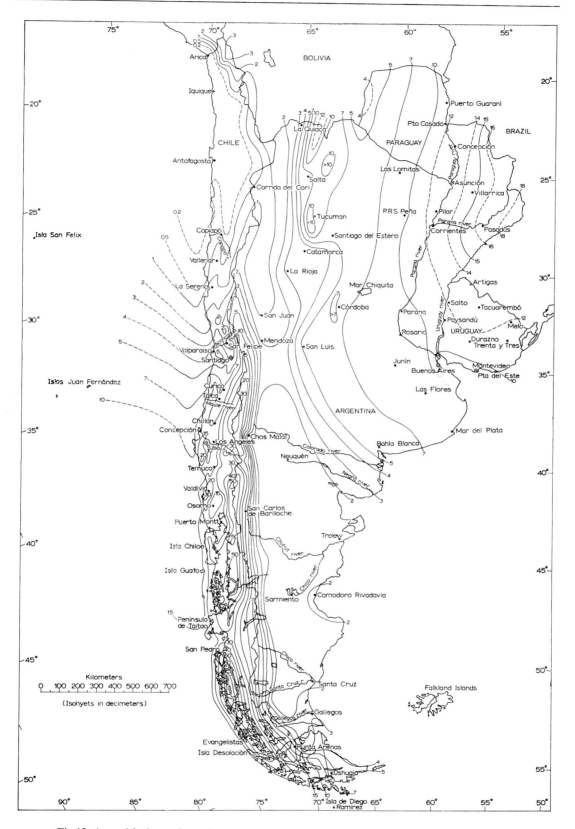

Fig.12. Annual isohyets (in mm).

Farther towards the south (Patagonia) and towards the west (Cuyo, La Rioja and Catamarca) the monthly values fall rapidly to 20–40 mm or still less. The highest monthly totals occur on the northern and eastern slopes of the sub-Andean and the pampean sierras, particularly on the former, and locally may exceed 200 mm. Maritime polar air masses bring about a considerable increase in rainfall only over Tierra del Fuego and the southernmost part of Patagonia.

Between the characteristic summer rains of tropical origin in central and northern Argentina and the typical winter precipitation of Pacific or Atlantic origin of the temperate latitudes, there lies a small transitional zone in northern Patagonia with the principal maximum in October and a secondary maximum in May. These are just the months of the beginning of the summer rainy season to the north and the winter rainy season in the south, respectively. There is another, broader transitional zone between continental and the maritime subtropical rainfall regime over the Argentine mesopotamia, with the principal maximum in March and April and a secondary maximum in November. The geographical limits and characteristic annual variations of these types of rainfall distribution in Argentina, Paraguay and Uruguay, are shown in Fig.13 (modified after PROHASKA, 1952).

Tropical regime with a single maximum

Tropical summer rains are found only in the area of the Andes in northwestern Argentina. This rainfall type is characterized by a clear division of the year into a simple summer rainy season with 2/3 to 9/10 of the annual totals and a winter dry season. The rainy season is restricted to December through March, and January always has the maximum (more than 20% of the annual total). The length of the dry season increases toward the west and south and the progressively sparser rainfall becomes more and more confined to January. A further characteristic of the tropical regime of rainfall is that the seasonal variation of cloud cover goes parallel to the rain distribution. This means that even in the areas in the neighborhood of the Tropic of Capricorn the cloudiness is less in winter than in summer. This similarity in the seasonal variation of cloudiness and rainfall is registered in the relatively rainy eastern part of the puna, but can be assumed to be also existent in the entire cordillera of these latitudes where in spite of the desertic character of the landscape occasionally very heavy precipitation occurs in association with thunderstorms. On the main range of the cordillera (frontier with Chile) it can be assumed that towards the south a direct transition to the precipitation regime with the maximum in winter takes place at about 27°S. Towards the southeast and east, however, it changes into the regime of continental summer rains on the northern boundary of San Juan (30°S) and the western border of the Chaco.

As mentioned earlier, over the eastern slopes of the pampean sierras in Jujuy, Salta and Tucumán the annual totals of more than 2,000 mm decrease on the west and southwest side of these mountain ranges abruptly to less than 400 mm and continue to decline slowly to less than 100 mm. This abrupt transition is also strikingly seen in the vegetation. The subtropical rain forest changes suddenly to the dry mountain forest and the subsequent grass and shrub steppe of the puna, which then gradually transforms into the "salares" (salt deserts) or complete deserts, or both.

Firn (névés) and glaciers are only occasionally observed in the latitudes between 20° and

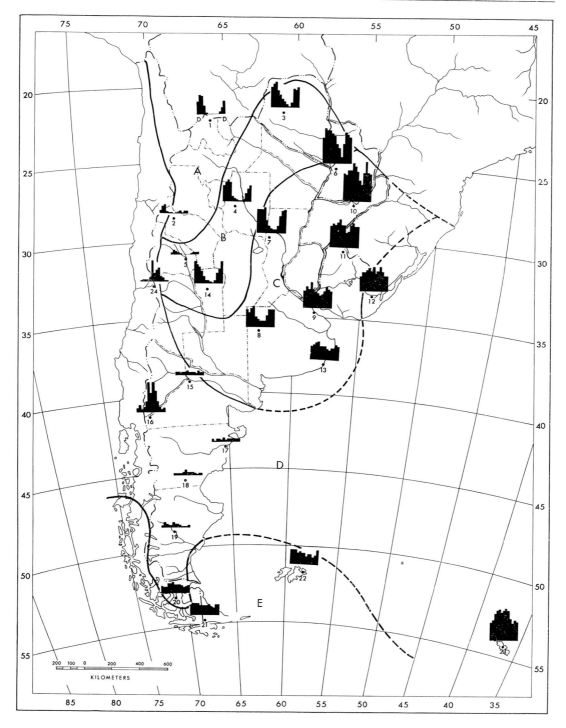

Fig.13. Precipitation regimes of southern South America according to their annual variation. *A* = tropical summer rains; *B* = subtropical continental summer rains; *C* = transition regime to maritime winter rains; *D* = maritime winter rains; *E* = all year precipitation with summer maximum: *1* = La Quiaca; *2* = Punta del Agua; *3* = Mariscal Estigarribia; *4* = Santiago del Estero; *5* = San Juan; *6* = Corrientes; *7* = Ceres; *8* = Trenque Lauquen; *9* = Buenos Aires; *10* = Posadas; *11* = Artigas; *12* = Treinta y Tres; *13* = Mar del Plata; *14* = San Luis; *15* = Neuquén; *16* = Bariloche; *17* = Trelew; *18* = Sarmiento; *19* = Cañadón León; *20* = Punta Arenas; *21* = Ushuaia; *22* = Port Stanley; *23* = Grytviken; *24* = Cristo Redentor.

27°S even though the peaks have elevations of 6,000–6,800 m and, thus, temperatures are far below the melting point. The main snow deposits are primarily caused by the advection of tropical air masses during summer. That they are not the consequence of winter precipitation which is responsible for greatly feared snowstorms and for the snow and ice fields south of 27°S, follows from the following observations. The lower limit of the snow deposits (firn line) is lower in summer than in winter and is again lower on the northern and eastern slopes, than on the southern and western slopes (ROHMEDER, 1943). The firn line rises substantially from east to west, and the frequency and extent of the snow and ice deposits decrease in the same direction. Thus, Aconquija (5,800 m, 60°W 27°S) and Chañi (6,000 m, 60°W 24°S) have extensive snow and ice fields; in the central puna, peaks above 6,000 m are only occasionally covered with snow or ice and Llullailla-co (68°30′W 24°45′S) with an elevation of 6,720 m, at the frontier with Chile, has only a seasonal snowcover. The absence of glaciers or snowfields at elevations at which even the maximum temperatures in summer are far below the melting point is not an accidental phenomenon of the present climatic conditions or more accurately of the conditions during the International Geophysical Year (1957–1958) at the time when a glaciological inventory of the Argentine cordillera was carried out (COLQUI, 1962). A joint Argentinian–Austrian archeological expedition found altars, dwellings and stables dating from Inca and even pre-Inca times on the broad volcanic cone of Llullaillaco in 1961, showing that many centuries ago even the highest elevation of the cordillera at these latitudes was at most *intermittently* covered with snow.

This raises some questions about the height of the permanent firn line in the desert climates in the subtropics, more specifically in the Andes, as they are the only mountain massif on earth which substantially exceeds 6,000 m in subtropical latitudes and, hence, with temperatures far below 0°C even in summer. In fact, one is justified in asking whether here the concept of a firn line has any meaning at all, and whether it should be represented (as is usually done) as a *continuous* line, as a function only of latitude, along a meridional cross-section of the Andes. In the first place, as we have shown, the snow limit is here not so much a function of height and latitude, i.e. temperature, as one of exposure to air masses producing precipitation, a factor which depends on the longitude in this case. Secondly, it must be assumed that in the highest and driest part of the cordillera (frontier with Chile) where there are optimum conditions for evaporation (extreme dryness of air, high wind speeds, probably the absolute maximum insolation on the earth's surface) the latter is so much greater than the total precipitation falling at very irregular intervals that permanent snow or ice cannot form, whatever the elevation. A well defined firn line starts only a few degrees of latitude farther south (approximately 27°S) with predominant winter precipitation, and also a few degrees to the north (Bolivian cordillera) where summer precipitation assumes such an intensity and frequency that glaciers develop and snowfields last throughout the dry season. Hence, it can be assumed that the firn line, whatever its definition may be, has not its highest altitude in the subtropics, as always stated and exampled at the Andes, but a gap, viz., it disappears in these latitudes.

Continental summer rains of the subtropics

The influence of the extensive plains of the Gran Chaco and its southern extension (plains and bolsons of the pampean sierras and the dry pampa) on the seasonal distribution of rainfall is so decisive that the characteristic annual variation of continental summer rains in the subtropics is observed despite the reduced width of the continent. Although the annual amounts decrease from 1,000 mm at the northern frontier of Paraguay to less than 100 mm in San Juan, the kind of the annual variation remains the same. The rainfall is much more uniformly distributed throughout the summer months than in the above-mentioned regime. The rainiest month, usually December, has more than 15% of the annual total in the western part and less in the east. Consequently, the dry season is reduced to May through September and is not so completely rainless as in the tropical regime. Like in all continental regions at extra-tropical latitudes, the cloud cover no longer decreases but increases from summer to winter or at least shows no pronounced seasonal variation, since the increased cloudiness of the west wind circulation becomes noticeable in winter. The associated disturbances, however, are too weak and the humidity is too low for frequent or intense rainfalls. On the contrary, due to the frequent advances of cold air masses and the associated thermic highs, the mean monthly isobars show the afore-mentioned high-pressure ridge over the continent linking the Atlantic and the Pacific semi-permanent high-pressure cells.

In summer, the pressure gradient between the thermal continental low and the Atlantic anticyclone produces an advection of moist air masses from the northeast, so that fronts (particularly cold fronts) as well as non-frontal convective processes are responsible for the precipitation in the area of low pressure (the ratio between the two rain producing processes is about 3:2; WöLCKEN, 1954). However, as already mentioned, the evaporative power is so great, because of the large energy supply and large moisture deficit, that even with 800 mm of rain per annum (but mainly as summer showers) only a dry forest vegetation develops without irrigation.

The limits of this rainfall regime coincide approximately with those of the influence of the thermal low and, thus, with the boundaries in the north and west of the Gran Chaco. To the southwest and south, there is a direct change into the regime of winter rains of Pacific origin observed in the cordillera of Cuyo and in northern Patagonia; to the southeast a transitional regime takes place to the subtropical winter rainfall of the Atlantic.

Transitional zone to the maritime winter rains of the subtropics

The increase in rainfall to the east of the continental summer rains exhibits a maritime character. While the summer rains are here nearly as intensive as in the interior of the continent, the rainfall in winter increases primarily in the direction to the coast. Consequently, the annual totals increase in a west–east direction and the annual range diminishes. Only to the east of the Cuchilla Grande, i.e., in eastern Uruguay, and in the adjacent part of the Brazilian coast does winter have the maximum rainfall, since the coast of the province of Buenos Aires is still too strongly under the influence of the continent to show the maximum in winter. Hence, going from west to east a shift of the main maximum can be observed from December and January (summer rain in northwestern Argentina) through March in the provinces of Santa Fé and Buenos Aires and in eastern Paraguay, April in the Argentine mesopotamia and western Uruguay, May in Misiones and June on the coast of Uruguay, thus completing the transition to the subtropical winter rains.

In this broad transitional zone, where the influence of both principal rainfall distributions is evident, a double rainy season becomes manifest with the main maximum in autumn (because of the greater supply of moisture) and the secondary maximum in spring, the main minimum in winter and a secondary minimum during the middle of the summer. The boundary between the continental- and the maritime-influenced rain regime (Fig.13) was drawn across the area where the instability produced by the heating of the continent in connection with advection of moist air masses is alternated in summer by the stabilizing influence of the Atlantic and its subtropical high-pressure system. It is, at the same time, the western limit of the maritime winter rainfall, characterized by the instability produced by the warm ocean in connection with the high degree of atmospheric moisture. The result is a relatively high rainfall frequency, at the time when the continental-influenced regime has its dry season. This so defined limit passes through the eastern part of the provinces of Formosa, Chaco, Santiago del Estero and Córdoba to the western part of La Pampa. Therefore Uruguay, eastern Paraguay and the eastern part of Argentina (25% of its area) are under maritime influence.

This seasonal distribution of rainfall in conjunction with the temperature regime of these latitudes provides almost optimum conditions for agriculture and livestock breeding, particularly in the southern part of the region called the humid pampas; in winter, rainfall is sufficient to keep meadows green and to permit cultivation with the onset of spring since evaporation is relatively low. The spring rains together with the then rapidly rising temperature permit a fast development of the crops. The decrease of the rainfall in summer (to its secondary minimum) has again a favourable effect on ripening and harvesting, while the autumn rains supply the soil with sufficient moisture to create appropriate conditions for winter sowing and enough water to last through the winter. In the north, the natural vegetation of the humid pampas changes to the subtropical rain forest in the eastern part of Paraguay and Misiones and the adjacent zones in Brazil due to the strong increase of rainfall.

The transition from the maritime to the continental rainfall regime is also visible in the change of the natural vegetation and in the type of farming and cattle raising, for it is the ecological boundary between the humid and arid pampas. But the broadly outlined location of this boundary is not stable. Its position depends on the actual atmospheric circulation and often exhibits considerable shifts in east–west direction. Due to the persistence of deviations in the atmospheric circulation there are usually several successive years in which this limit between arid and humid conditions is displaced east or west. The zone enclosed by this fluctuating boundary is generally described as "semi-arid". Unfortunately, this term has various definitions and is frequently misunderstood. It does not mean that rainfall and evaporation are in precarious equilibrium, so that semi-arid conditions actually prevail. It means, rather, that humid or arid periods occur, lasting up to several years, and that only the calculated averages reflect semi-arid conditions, which very rarely occur in an individual year. This fact presents special problems for agricultural techniques. An inappropiate soil management may lead to complete destruction of the soil by wind erosion during successive dry years and even to the formation of moving sand dunes or dust storms (dust bowls), depending on the nature of the soil (PREGO, 1961).

The winter rains of the temperate latitudes

The transition to the winter rains of Patagonia takes place over the basins of the Rio

Colorado and Rio Negro which geographically form the northern boundary of Patagonia. Here again, the transition is expressed in a double rainy season, but only a hint of it, because of the general dryness of this region. The two months of maximum rainfall are, however, very characteristic. In autumn the month is May, i.e., the month when the rainy season starts on the Pacific coast and therefore also the winter rains of Patagonia, and in spring it is October, i.e., the month when rain starts in the adjacent pampas to the north. These are thus the months when there is a rather rapid change from summer to winter circulation or vice versa. In October, in addition, a frequently occurring blocking action of the west wind circulation (high-pressure over Patagonia and the adjacent parts of the Atlantic) causes a reversal of the normal meridional pressure gradient. This gives rise to easterly winds producing rains over northern Patagonia as far as the cordillera, where they are enhanced due to orographic effects.

Except for this transitional zone, the whole of Patagonia has the same rainfall regime as southern Chile (rainy season from May to September) and will therefore be discussed in the chapter on Chile. In the Andean region, as already mentioned, this rainfall distribution extends to 35°S and on the main ranges even further north up to 27°S. Since the greater part of the water vapor is precipitated crossing the Andes, the mountains only have plentiful precipitation which reaches 3,000 mm and sometimes exceeds this figure considerably even on the Argentinian-leeward side (SCHWERDTFEGER, 1958a). Towards the east these values fall very rapidly to 200 mm, or even less. This rapid transition is reflected in an abrupt change of vegetation and the sharp limit which separates the tall Patagonian forest from the steppe. However, the causes of the extraordinary dryness of the extra-Andean part of Patagonia up to and including the coast cannot be explained merely by the lee effect of the Cordillera but must have its reasons in a differential circulation pattern over the adjacent parts of the Pacific and Atlantic oceans as well. The climatic charts of the oceans (WEATHER BUREAU, 1938) show substantially lower cloud amounts (particularly nimbus) and precipitation frequencies over the Atlantic than over the Pacific as far off shore as 1,000 miles at least, and a similar strong west–east gradient of both elements to the south of the mainland (between 50° and 60°S).

The uniform rainfall distribution throughout the year of the subpolar region

The coast and the Andes in the extreme south of Patagonia as well as the whole archipelago of Tierra del Fuego exhibit a high frequency of precipitation throughout the year, with a distinctive, but not large increase in summer and autumn, indicative of the polar character of this completely maritime precipitation regime. The annual variation is very similar to that of the sea temperature (the ocean temperature around Tierra del Fuego is 7°C in summer and 4°C in winter and spring). This rainfall variation can be ascribed not so much to higher frequencies as to the greater intensities of rain storms in summer and autumn due to the higher water vapor content of the air. This distribution is found to be the same both on the Chilean coast and the islands south of 46°S where the annual totals may be several meters[1], but also on the eastern side of the continent and the neighbouring

---

[1] Island of Guarello (50°21′S 75°21′W, 15 m), average of 9 years (discontinuous): 7,331 mm, maximum 8,496 mm (1960), minimum 6,409 mm (1953). *Pluviometría de Chile*, Fascículo II, Oficina Meteorológica de Chile (1965).

archipelago where the annual precipitation is only 200–300 mm at some places. However, the type of precipitation must of course have some effect on this annual variation since during the winter months precipitation falls predominantly as snow and owing to the high wind speeds, the measured amounts are almost certainly too low.

Similar as in Patagonia and corresponding to the large increase in precipitation from northeast to southwest, there is a clear division in Tierra del Fuego between the steppe vegetation in the relatively flat northeast and the preponderance of evergreen beech forests in the mountains of the southwest. In the western part of the island there is a direct transition to the glaciers. The snow line is at about 700 m on the windward side and at 900 m on the lee side, although the mighty glaciers of the Darwin Cordillera (with peaks between 2,000 m and 2,500 m) calve directly into the sea.

It is not by chance that the large Patagonian ice shield can persist in this precipitation regime. It is a glacierization which is extraordinary for the low latitude and elevation above sea level. The *Hielo Patagónico*, not *Hielo Continental* (according to HEINSHEIMER, 1962), stretches from 46°38′ to 51°25′S, a length of 531 km in north–south direction, with only one gap of 100 km at the latitude of the fjords Mitchell and Calen (at 48°S). It has an average width of nearly 60 km and covers an area of 17,900 km² or 7.8% of the total extra-polar glaciation of the earth (BERTONE, 1960). Roughly 60% of this surface is in the firn or névé region, i.e., in the area where accumulation exceeds ablation. This fact, as well as the large extension of the valley glaciers into the ablation zone on the Argentine side of the ice shield, corroborates the assumption of very high amounts of annual precipitation, possibly 7,000–8,000 mm at the main ridges (SCHWERDTFEGER, 1956). The longest of these glaciers cover areas as large as 595 km² in the case of Upsala, 575 km² in the case of Viedma, and 195 km² in the case of Moreno (the most thoroughly investigated of the south-Patagonian glaciers; RAFFO et al., 1953; HEINSHEIMER, 1954–1958), and their tongues extend down to 200 m above M.S.L., where they flow into the large lakes Viedma and Argentino.

Rainfall variability

In a region in which the rainfall distribution mainly depends on advective processes, in which polar, tropical, maritime and continental influences are effective for shorter or longer periods, in which sudestadas may bring heavy rainstorms lasting for several days' duration and thus bring floods and inundation to large areas, but in which there may also be several dry years in succession which alter the water budget thoroughly, the monthly and annual totals exhibit a wide variation (MARCHETTI, 1952).

The relatively smallest variability in the *annual totals* is found in the maritime-influenced region of the Argentine–Uruguayan mesopotamia, eastern Paraguay, and the coastal strips of the province of Buenos Aires, with a variation coefficient of 22–26% ($100\sigma/a$) or a relative variability of just under 20% ($100\delta/a$). In the dry pampa and the Gran Chaco these values rise to 30% and 20–25%, respectively, while in Patagonia and Cuyo they increase to more than 40 and 30%, respectively. In the extremely dry region of San Juan and the adjacent valleys of the cordillera to the north, they exceed 50 and 40%, respectively. The only exception to this general trend can be observed in the region of summer rains at the windward side (east flank) of the mountains in northwestern Argentina, where the variability is similar to that on the coast of Buenos Aires. Correspondingly, the range of the relative variability (MARCHETTI, 1951) is very high, particularly in the dry regions, and the ratio of the wettest to the dryest year may exceed 20:1 (San Juan: 202 mm and 8 mm). In general, these ratios exceed 4:1 in the entire western part of Argentina and in Patagonia, and are less in the east. In Entre Rios and the area of influence of the Río de la Plata they decrease to 2.5:1. In the rainiest region of Argentina, as in Misiones, the ratio is about 3:1 (Posadas: 2,564 mm and 888 mm). Here, also, the highest annual total was recorded at a station with an observation period of several years, namely 2,645 mm at Caa Yari. In the southern cordillera, however, annual totals, obtained by totalizers exceed 3,000 mm (Lago Quillen, 39°26'S 71°25'W, 1938–1958: 3,710 mm).

With exception of the Argentine–Uruguayan mesopotamia and the coasts of Buenos Aires, some completely rainless months are observed everywhere, particularly in winter. In western Argentina and Patagonia (except the cordillera south of 30°S) any month of the year may be free of precipitation. The absolute highest monthly amounts are found in Misiones with 750 mm in Pindapoy and in the region of orographic rainfall in Salta and Tucumán where values up to 530 mm (Pichanal) have been recorded.

Days with precipitation

The number of rainy days (≥ 1 mm) per year shows a pattern similar to that of the amount of rainfall. The subtropical region, Misiones, and the very exposed parts of the Atlantic coast such as Punta del Este and Mar del Plata have the highest values with 100 days of precipitation per year. The decrease towards the west is fast so that the transitional region to the arid areas has only 70 days of rainfall in northern Paraguay, 60 days in Santa Fé and 50 days of precipitation in Buenos Aires. In the dry parts of the Chaco, the pampa, and in Patagonia, the values are generally about 40 days and fall to 20 days per year at the foot of the central part of the Andes (San Juan). Towards the northwest the values are reduced to a few days per year approaching the Puna de Atacama. Only in the narrow strip of orographic rainfall at the eastern slope of the pre-puna (Jujuy and Salta), the frequency of rainfall increases to 90 days. In Patagonia, the increase in precipitation frequency towards the west starts appreciably sooner than the increase in precipitation (Bariloche 122 days with precipitation). This shows that the dryness of western Patagonia is not due to any lack of cyclonic activity, but to a shortage of precipitable water. The increase towards the southern tip of the continent is visible only in the southern part of Santa Cruz, where more than 100 days with precipitation are observed. A rapid increase to 150–200 days with precipitation per year is found on the island of Tierra del Fuego. Islands freely exposed to the west winds may have 300 or even more days of precipitation and thus, constitute one of the rainiest regions of the world.

Except in eastern Paraguay, Corrientes, Misiones and the southern cordillera, the rainfall is generally not very intensive. Only in the mentioned regions are there more than 30 days per year with a precipitation ≥ 10 mm. The dry western region and Patagonia actually have an average of less than 5 days with rainfall of this magnitude.

The annual variation of the days with precipitation is much more uniform than that of the amounts of precipitation. Only the region of tropical summer rain and winter dryness has a similarly extreme seasonal variation, e.g., of the 59 days of precipitation per year in La Quiaca 55 occur in summer. Summer months with more than 10 days of rain occur only in this rain regime. In winter, on the other hand, there is a similarly large precipita-

67

tion frequency on the Atlantic coast of Uruguay, in the southern cordillera and in Tierra del Fuego. Except in these widely separated and rather small regions, the monthly values of days with precipitation in the humid areas are between 6 and 8 days at the time of maximum rainfall activity and between 4 and 6 days at the time of the minimum activity. In the dry areas even the months with maximum precipitation amounts have only 4 to 6 days with rain and less than 2 days of rain in the transitional months to the dry season, where the long term average is 1 day or less. In the region of maximum rainfall in spring and autumn no predominance of the latter over the former can be found, in contrast to the rainfall totals. Consequently, the autumn rainfall maximum in the humid pampa and in mesopotamia is not due to an increased cyclonic activity but to higher water vapor content of the air at this time of the year.

The rain density (rain total/rain days) exceeds 20 mm/day during the rainiest months in northeastern Paraguay and northeastern Argentina and in the region of orographic rainfall in northwestern Argentina. Elsewhere, the figures are between 10 mm and 20 mm on a day with rainfall in the humid east, just a little over 10 mm in the central part in summer and below this value in winter, and less than 10 mm throughout the year in the dry west. The yearly maximum total in 24 h amounts to 70 mm in northeastern Argentina, 60–70 mm in mesopotamia, 50 mm in the central regions and the northwest, 30–40 mm in the west and only 10–20 mm in the northern cordillera and in Patagonia. The absolute maximum values measured during 24 h were in Argentina in the province of Entre Rios (359 mm at Gualeguaychú on April 29, 1912 and 330 mm at Concordia on May 8, 1930). In western Uruguay, the absolute maxima may be equal to those in Entre Rios, while in eastern Uruguay, and in the zone of influence of the Río de la Plata the corresponding values are of about 200 mm (Punta del Este 193 mm; Montevideo 181 mm; Buenos Aires 194 mm). In southeastern Paraguay, which is the rainiest part of this country, the absolute maxima in 24 h are probably between 200 and 250 mm and are thus similar to the values in the adjacent province of Misiones. In the dry regions, the absolute maxima are less than 100 mm (Mendoza 85 mm; San Juan 65 mm).

The synoptic situations leading to rainfall in general and to heavy rainstorms in particular have been investigated for Argentina only (WÖLCKEN, 1962), but the conclusions also apply to Paraguay and Uruguay to a certain extent. In spite of the proximity to the tropics and the marked summer rain activity in the whole of the west, about 80% of the annual totals and 70% of rainy days can be ascribed to frontal processes. It is only in the neighborhood of the mountains that the proportion of non-frontal processes (convective rain) amounts to 30% and that of the rainy days to 40%. In Buenos Aires, on the other hand, 90% of the amount of rainfall and 80% of the days with rain must be ascribed to frontal developments. Of the fronts causing precipitation, the cold fronts (in Mendoza additionally the upper cold fronts) greatly predominate (between 30 and 50% both as regards to amount and frequency) and the warm fronts (20–30%) occupy second place, except in Cuyo, where the latter have lost their activity. Here and in Tucumán the orographic rains without association of frontal processes occupy the second place (about 15% of the annual total). All other rain producing processes studied in the investigation are of minor importance if the whole region is considered.

In regard to synoptic situations which lead to *heavy rain storms*, the predominance of the frontal processes is even more marked, although in this case the warm fronts in combination with the formation of a depression over the mesopotamia (SCHWERDTFEGER, 1954,

1958b) are by far the most frequent cause (25 % of the case investigated by WÖLCKEN, 1962). Second in importance are the cold fronts and lastly the warm fronts which are not associated with the above mentioned cyclones. Regional differences are well expressed in north–south direction. In the north, where there is usually an abundant supply of water vapor, heavy rainstorms are caused mainly by cold fronts (orographically enhanced in the mountains) while farther south the warm fronts are becoming progressively important, for only these can provide the necessary water vapor.

Thunderstorms

Thunderstorm activity in this predominantly subtropical region is broadly associated with the presence of air masses of tropical origin and is particularly frequent where such masses are lifted at mountain slopes, as in the northeastern part of the puna (La Quiaca), in the eastern part of the pre-puna, and in the sierras in Córdoba. Apart from these centers, thunderstorms occur rather frequently in the humid (eastern) parts of Paraguay and Argentina and in all of Uruguay. Towards the dry zones, both westwards and south-wards, thunderstorm activity ceases abruptly. In accord with the annual variation in the occurrence of tropical air masses, the greatest frequency is in late summer and autumn, while in winter thunderstorms are registered only in the Argentine–Uruguayan mesopotamia, and in eastern Paraguay. The values given in the climatic tables in the Appendix refer to the decade 1951–1960 which had apparently a relatively small thunderstorm activity since statistics relating to the earlier although shorter period 1932–1938 (KNOCHE and BORZACOV, 1946–1947) give substantially higher figures. Thus, the values given may be regarded rather as lower limits.

The center of thunderstorm activity has more than 50 days with thunderstorms per year and a maximum of 8–10 thunderstorm days per month. In the Paraguayan–Argentine–Uruguayan mesopotamia, the values are about 40 days per year and 4 days as highest monthly averages. In the transition zone to the dry regions, the annual frequency falls to 30 days, and decreases in the latter to less than 10 days and in the very dry parts to less than 5 days per year.

No statistics are available on thunderstorm activity over the cordillera, but according to travellers and expedition reports and sferics observations as well (RÖNICKE, 1965) they must be very frequent. A well-known phenomenon, particularly in the puna, are the "relampagos de calor" or "resplandores andinos" (heat flashes or Andean lightnings) which occur on warm nights with otherwise cloudless sky. These phenomena have nothing in common with the often intense and long lasting sheet lightning of the Gran Chaco, which is observed when tropical air masses are present. Powerful lightning discharges are also the possible cause of many of the large fires in the Gran Chaco and are quite often responsible for the death of man and beast.

Another phenomenon associated with thunderstorms is hail. It is much less frequent than the former, but causes heavy damages in many regions of intensive agriculture such as the provinces of Mendoza and Buenos Aires (IRIBARNE and GRANDOSO, 1965).

Appendix – Climatic tables

Argentina (Tables XVI–XLV)

Unless otherwise stated, the period of observation is 1931–1960. However, because observations were not available for this period or could not be calculated in the time available, the following periods have been used for some elements or related statistics: maximum rainfall in a 24-h period 1921–1950; number of days with precipitation 1921–1950; number of days with thunderstorms 1951–1960; number of days with fog 1951–1960; mean sunshine hours 1941–1960; most frequent wind direction 1951–1960; mean wind speed 1951–1960; number of clear days $\leq 2/10$ 1941–1950; number of cloudy days $\geq 8/10$ 1941–1950 (as from 1951 the corresponding threshold values were changed to $\leq 2/8$ and $\geq 6/8$).

The monthly averages are obtained from observations made three times a day by means of the following formulas: $M = (08h + 14h + 20h)/3$ (the standard time of the three countries refers to 60°W). In the case of atmospheric pressure, temperature, and water vapor pressure, a correction factor was added in order to reduce these means to 24-h averages.

Precipitation totals and evaporation values were obtained from daily morning readings. Evaporation was measured with the Class-A pan. All monthly evaporation values are multiplied by the factor 0.7.

Hours of sunshine are evaluated from Campbell-Stokes recordings.

Radiation measurements have been temporarily carried out at 45 stations since 1940 with bimetal actinographs. However, continuous recordings over several years are available only for the 3 stations given in Table X. The radiation observations during the I.G.Y. (SERVICIO METEOROLÓGICO NACIONAL, BUENOS AIRES, 1962) and monthly averages from 22 stations for different periods varying in length have so far been published. The mean values in both publications were obtained from sometimes very incomplete series and the missing values were not interpolated, so that an arbitrary selection resulted. Consequently these monthly figures should not be used for climatic evaluations.

All values in the climatic tables of the Appendix and many of the data used in the text were kindly made available by the Division of Climatology (Dr. J. Hoffmann) and the Division of Hydrometeorology (J. Raffo del Campo) of the Servicio Meteorológico Nacional. A special word of thanks for gathering and controlling data are due to Miss Lic. Met. E. C. Bolognesi and collaborators at the Servicio Meteorológico Nacional, Buenos Aires.

Paraguay (Tables XLVI–XLIX)

The periods of observation are given in the respective tables. Procedures used for the observations and calculus of monthly averages in the meteorological network of Paraguay follow those for Argentina.

Evaporation data relate to the Wild evaporation balance, which is exposed in the meteorological shelter.

The author is indebted to Professor Felix Albani, then technical expert of the W.M.O. in Paraguay, and to the Director of the Weather Service, Capitán de Navío Teófilo Fariña Sanchez for making available the climatic data.

Uruguay (Tables L–LIV)

The periods of observation are given in the respective tables. Monthly averages were calculated from the daily averages which, in turn, were calculated for atmospheric pressure, temperature, and humidity from the formula:

(03+06+09+12+15+18+21+24+max.+min.)/10

The other elements were calculated from three or four daily observations at fixed hours. The atmospheric pressure is *reduced to sea level*. The evaporation data relate to the Piche evaporimeter, suspended in the meteorological shelter. Threshold values for cloudy and clear days were ⩽3.5/10 and ⩾7.1/10, respectively. All data for the climatic tables were kindly made available by the Director of the Division of Climatology of the Dirección General de Meteorología, Sr. L. A. Battione Chiarino.

References

General climatic descriptions, data, bibliographies

ALBRECHT, F., 1965. Untersuchungen des Wärme- und Wasserhaushaltes der südlichen Kontinente. *Ber. Deut. Wetterdienstes*, 14(99): 54 pp.

AMERICAN METEOROLOGICAL SOCIETY, 1952. Bibliography on climate of Paraguay, Uruguay, and Argentina. *Meteorol. Abstr. Bibliogr.*, 3(3): 243–290.

BERGEIRO, J. M., 1945. El clima del Uruguay. *Bol. Meteorol.*, Montevideo, 3(4): 7–28.

CHIOZZA, E. M. and GONZÁLEZ VAN DONSELAR, Z., 1958. Climatología. In: F. APARICIO (Editor), *La Argentina, Suma de Geografía*. Peuser, Buenos Aires, 2, pp.1–183.

CZAJKA, W., 1954. Argentinien. In: *Geografisches Taschenbuch*, 1954/55. Steiner Verlag, Wiesbaden, pp. 377–395.

DAVIS, W. G., 1910. *Climate of the Argentine Republic*. Serv. Meteorol. Argentino, Buenos Aires, 2nd ed., 111 pp.

DE LE VIN, M. M., 1921. Bibliography on the climate of South America. *Mon. Weather Rev., Suppl.*, 18: 42 pp.

DIRECCIÓN GENERAL de METEOROLOGÍA, 1963. Las lluvias en el Uruguay. *Rev. Meteorol.*, Montevideo, Supl., 3: 117 pp.

KNOCH, K., 1930. Klimakunde von Südamerika. In: W. KÖPPEN and R. GEIGER, *Handbuch der Klimatologie*. Borntraeger, Berlin, Vol.II, Pt.G, 249 pp.

KNOCHE, W., BORZACOV, V. et al., 1946–47. Clima de la República Argentina. In: *Geografía de la República Argentina*. GAEA, Sociedad Argentina de Estudios Geográficos, Coni, Buenos Aires, 5: 398 pp.; 6: 432 pp.

KUHN, F., 1937. *Argentinien*. Hirt, Breslau, 2 vols.

MARTIN DE MOUSSY, J. A., 1860. Climatologique ou climate de la confederation Argentine. In: *Description Géographique et Statistique de la Conféderation Argentine, IV*. Didot, Paris. (Spanish translation and comments by J. J. BURGOS, 1958, in: *Rev. Fac. Agron.*, La Plata, 33(1): 1–63; and *Inst. Suelos Agrotec. (INTA)*, Buenos Aires, Publ., Nr.59.

OFICINA METEOROLÓGICA, 1893–1913. *Anales de la Oficina Meteorológica Argentina*, Buenos Aires. 17 vols.

PRADA ESTRADA, R., 1946. *Climas de Bolivia*. Dirección General de Meteorología, La Paz, 52 pp.

SERVICIO METEOROLÓGICO DEL URUGUAY, 1957. *Datos Climáticos: Años 1944 al 1956. Rev. Meteorol.*, Montevideo, Supl., 3: 117 pp.

SERVICIO METEOROLÓGICO NACIONAL, BUENOS AIRES, 1958a. *Estadísticas Climatológicas 1901–1950* (44 stations).

SERVICIO METEOROLÓGICO NACIONAL, BUENOS AIRES, 1958b. *Estadísticas Climatológicas 1941–1950* (160 stations).

SERVICIO METEOROLÓGICO NACIONAL, BUENOS AIRES, 1960. *Atlas Climático de la República Argentina*.

SERVICIO METEOROLÓGICO NACIONAL, BUENOS AIRES, 1962. *Datos de radiación solar del período 1° de Julio de 1957 al 31 de Diciembre de 1958*, 135 pp.

SERVICIO METEOROLÓGICO NACIONAL, BUENOS AIRES, 1963a. *Estadísticas Climatológicas 1951–1960* (156 stations).

SERVICIO METEOROLÓGICO NACIONAL, BUENOS AIRES, 1963b. *Datos Pluviométricos, 1921–1950*, 160 pp.

THORNTHWAITE ASSOCIATES, 1965. *Average Climatic Water Balance Data of the Continents, Part VIII. South America.* Laboratory of Climatology, Elner, N. J., 18(2): 297–433.

WILHELMY, H. and ROHMEDER, W., 1963. *Die La Plata Länder: Argentinien–Paraguay–Uruguay* (2300 bibliographic entries). Westermann, Braunschweig, 584 pp.

Special regions and elements

ANTEZANA PARDO, O., 1958. Climas de Bolivia, según el sistema de Thornthwaite. *Rev. Fac. Agron., La Plata*, 34: 161–191.

ARDISSONE, R., 1937. Datos históricos acerca de las precipitaciones pluviales en la zona de Buenos Aires desde el siglo XVI hasta 1821. *GAEA, An. Soc. Argentina Estud. Geogr., Buenos Aires*, 5: 115–211.

BATTIONE CHIARINO, J. A., 1956. Resumen mensual de los valores meteorológicos correspondientes al observatorio del Prado, 1901–1950. *Rev. Meteorol., Montevideo*, 14: 29–54.

BERGEIRO, J. M., 1946. La temperatura y la lluvia estacional en el Uruguay. Delimitación regional climato-geográfica y normales respectivas. *Rev. Meteorol., Montevideo*, 5: 264–271.

BERTONE, M., 1960. *Inventario de los Glaciares Existentes en la Vertiente Argentina entre los Paralelos 47°30′S y 51°S.* Instituto Nacional del Hielo Continental Patagónico, Buenos Aires, Publ. Nr.3.

BOFFI, J. A., 1949. Effect of the Andes Mountains on general circulation over the southern part of South America. *Bull. Am. Meteorol. Soc.*, 30(7): 242–247.

BÖGEL, R., 1956. Untersuchungen zum Jahresgang des mittleren geographischen Höhengradienten der Lufttemperatur in den verschiedenen Klimagebieten der Erde. *Ber. Deut. Wetterdienstes*, 4(26): 42 pp.

BURGOS, J. J., 1963a. *Las heladas en la Argentina.* Colección Científica del INTA, Buenos Aires, 388 pp.

BURGOS, J. J. 1963b. El clima de las regiones áridas en la República Argentina. *Rev. Invest. Agríc., Buenos Aires*, 17(4): 385–405.

BURGOS, J. J. and VIDAL, A. L., 1951. The climates of the Argentine Republic according to the new Thornthwaite Classification. *Ann. Assoc. Am. Geogr.*, 41: 237–263.

BURMEISTER, H., 1864. Die Regenverhältnisse in der argentinischen Republik im allgemeinen und der ungewöhnlich starke Regenfall in Tucumán zu Anfang des Jahres 1863 im besonderen. *Petermanns Geogr. Mitt.*, pp.9–14.

CATALANO, L. R., 1927. *Datos Hidrológicos del Desierto de Atacama.* Ministerio de Agricultura, Dir. Gen. de Minas y Geología, Buenos Aires, Publ. Nr.35: 65 pp.

COLQUI, B. S., 1962. Argentine glaciology. *Antarctic Res., Geophys. Monogr.*, 7: 217–228.

CZAJKA, W. and VERVOORST, F., 1956. Die naturräumliche Gliederung Nordwest-Argentiniens. *Petermanns Geogr. Mitt.*, 100: 89–102; 196–208.

DE FINA, L. et al., 1948. *Difusión Geográfica de Cultivos Indices en la Provincias . . . (Argentinas) . . . y sus Causas.* Instituto de Suelos y Agrotecnia (INTA), Buenos Aires, Publ. Nrs.10, 22, 24, 25, 50, 58, 63, 67, 73, 80, 83, 96 (series continues).

FARIÑA SANCHEZ, T., 1946. Investigación estadística de las precipitaciones pluviométricas en Asunción del Paraguay. *Rev. Meteorol., Montevideo*, 5: 78–88.

GAIGNARD, R., 1962. La mise en valeur pionnière de la Terre de Feu (Argentine). *Cahiers d'Outre-Mer*, 15: 105–137.

GRANDOSO, H. N. and NUÑEZ, J., 1955. Análisis de una situación de bloqueo en la parte austral de América del Sur. *Meteoros*, 5: 34–35.

HEINSHEIMER, J. J., 1954–58. Zur Hydrologie und Glaziologie des Lago Argentino und Ventisquero Moreno, Argentinien. *Z. Gletscherk. Glazialgeol.*, I, 3(1): 24–32; II, 3(3): 327–333; III, 4(1–2): 61–72.

HEINSHEIMER, J. J., 1964. Die Temperaturabhängigkeit der Wasserführung schmelzwassergenährter Flüsse. *Arch. Meteorol., Geophys. Bioklimatol., Ser.B.*, 13: 404–413.

HEINSHEIMER, J. J., 1962. Hielo continental, inlandsis y el Campo de Hielo Patagónico. *An. Acad. Argent. Geogr.*, 5: 1–11.

IRIBARNE, J. V. and GRANDOSO, H. N., 1965. Experiencia de modificación artificial de granizadas en Mendoza. *Ser. Meteorol., Fac. Cienc. Exactas Nat., Univ. Buenos Aires*, 1(5): 31 pp.

LANDSBERG, H. E., 1958. *Physical Climatology.* Gray, Du Bois, Pa., 2nd. ed., 446 pp.

MARCHETTI, A. A., 1951. Oscilaciones extremas de la cantidad de precipitación en la República Argentina. *Meteoros*, 1: 46–54.

MARCHETTI, A. A., 1952. Estudio del régimen pluviométrico de la República Argentina. *Meteoros*, 2: 243–309.

MORANDI, L., 1940. Normales para el clima de Montevideo (periódo 1883–1938). *Rev. Fac. Agron., Montevideo*, 20: 9–22.

OLASCOAGA, M. J., 1950. Some aspects of Argentine rainfall. *Tellus*, 2: 312–318.

PAPADAKIS, J., 1951. *Mapa ecológico de la República Argentina*. Ministerio de Agricultura y Ganadería, Vol.1, text; Vol.2, atlas.

PREGO, A. J., 1961. La erosión eólica en la República Argentina. *Cienc. Invest.*, 17: 307–324; *Inst. Suelos Agrotec. (INTA), Buenos Aires, Publ.*, 78: 28 pp.

PROHASKA, F., 1952. Regímenes estacionales de precipitación de Sudamérica y mares vecinos. *Meteoros*, 2: 66–100.

PROHASKA, F., 1957. Über die meteorologischen Stationen der hohen Kordillere Argentiniens. *Jahresber. Sonnblick-Ver.*, 51–53: 45–55.

PROHASKA, F., 1959. El polo de calor de América del Sur. *IDIA*, 141: 27–30.

PROHASKA, F., 1961a. Las características de las precipitaciones en la región semiárida pampeana. *Rev. Invest. Agríc.*, 15: 199–232; *Inst. Suelos Agrotéc. (INTA), Buenos Aires, Publ.*, 72: 34 pp.

PROHASKA, F., 1961b. Algunos aspectos de clima de la alta cordillera y de la Puna Argentina. *Bol. Estud. Geográf. Univ. Nacl. Cuyo*, 8: 21–30; *Inst. Suelos Agrotec. (INTA), Publ.*, 79: 10 pp.

PROHASKA, F., 1962. Factores advectivos en el clima de la Puna Argentina. *Rev. Estud. Geográf. Univ. Nacl. Cuyo*, 9: 43–55.

PROHASKA, F. and DE GARCÍA, M. T. U., 1951. Gradiente vertical de temperatura en la zona central de la cordillera. *Meteoros*, 1: 342–343.

RAFFO, J. M., COLQUI, B. S. and MADEJSKI, M. E., 1953. Glaciar Moreno. *Meteoros*, 34: 293–341.

ROHMEDER, W., 1943. Observaciones meteorológicas en la región encumbrada de las sierras de Famatima y del Aconquija. *An. Soc. Cient. Argentina*, 136: 97–124.

RÖNICKE, G., 1965. Thunderstorm activity in the Andes of northern Argentina. *J. Appl. Meteorol.*, 4(2): 186–189.

SCHWERDTFEGER, W., 1951a. Ein kleiner Beitrag zur Kenntnis der aerologischen Verhältnisse über dem südlichen Patagonien. *Meteorol. Rundschau*, 4: 135–138.

SCHWERDTFEGER, W., 1951b. La depresión térmica del noreste argentino. *An. Soc. Cient. Argentina*, 151: 255–275.

SCHWERDTFEGER, W., 1952. *El Problema de la Previsión del Tiempo*. Ediciones Librería del Colegio, Buenos Aires, 142 pp.

SCHWERDTFEGER, W., 1954. Análisis sinóptico y aspecto climatológico de dos distintos tipos de depresiones báricas en el norte de la Argentina. *Meteoros*, 4: 301–323.

SCHWERDTFEGER, W., 1956. Determinación indirecta de las condiciones climáticas del hielo continental patagónico. *An. Soc. Cient. Argentina*, 161: 53–82.

SCHWERDTFEGER, W., 1958a. Ein Beitrag zur Kenntnis des Klimas im Gebiet der Patagonischen Eisfelder. *Z. Gletscherk. Glazialgeol.*, 4: 73–86.

SCHWERDTFEGER, W., 1958b. *Lectures on the Meteorology of Extratropical South America*. University of Melbourne, Melbourne, 17 pp.

SCHWERDTFEGER, W., 1961. Strömungs- und Temperaturfeld der freien Atmosphäre über den Anden. *Meteorol. Rundsch.*, 14: 1–6.

SCHWERDTFEGER, W., 1962. *Meteorología del Area del Pasaje de Drake*. Secretaría de Marina, Servicio de Hidrografía Naval, Buenos Aires, Publ. H 410, 78 pp.

SCHWERDTFEGER, W. and PROHASKA, F., 1955. Análisis de la marcha anual de la presión y sus relaciones con la circulación atmosférica en Sudamérica austral y la Antártida. *Meteoros*, 5: 223–237.

SCHWERDTFEGER, W. and VASINO, C. J., 1955. La variación secular de la continentalidad en la República Argentina. *Meteoros*, 5: 178–184.

SCHWERDTFEGER, W. et al., 1959. *Meteorología Descriptiva del Sector Antártico Sudamericano*. Instituto Antártico Argentino, Buenos Aires, 425 pp.

WEATHER BUREAU, 1938. *Atlas of Climatic Charts of the Oceans*. U.S. Department of Agriculture. W.B. No.1247, Washington, D.C.

WÖLCKEN, K., 1954. Algunos aspectos sinópticos de la lluvia en la Argentina. *Meteoros*, 4: 327–366.

WÖLCKEN, K., 1962. Regenwetterlagen in Argentinien. *Südamerika*, 12: 83–140.

TABLE XVI

CLIMATIC TABLE FOR LA QUIACA (JUJUY)
Latitude 22°06′S, longitude 65°36′W, elevation 3,459 m

Month	Mean sta. press. (mbar)	Temperature (°C)				Mean vapor press. (mbar)	Precipitation (mm)		Relat. humid. (%)
		daily mean	daily range	extreme			mean	max. in 24 h	
				max.	min.				
Jan.	672.2	12.4	14.3	27.1	−1.2	8.9	89	45	62
Feb.	672.3	12.4	14.4	27.0	−1.2	9.2	77	35	64
Mar.	672.5	12.2	15.9	27.8	−3.1	8.2	43	35	58
Apr.	672.8	10.3	19.4	25.8	−8.7	5.6	5	35	45
May	672.8	6.6	21.7	25.0	−12.7	3.5	1	5	36
June	672.8	3.9	23.3	22.0	−15.8	2.8	2	25	35
July	672.7	4.0	23.5	21.1	−15.2	2.9	1	5	36
Aug.	672.4	6.4	23.4	22.8	−14.6	3.2	0	15	38
Sept.	671.9	9.2	21.7	25.8	−12.2	4.4	2	5	38
Oct.	671.5	11.1	20.2	27.4	−10.7	5.8	9	25	44
Nov.	671.2	12.3	18.1	28.4	−4.7	7.3	31	25	51
Dec.	671.6	12.6	15.8	28.3	−1.2	8.4	63	25	57
Annual	672.3	9.5	19.1	28.4	−15.8	5.9	322	45	50

Month	Mean evap.[1] (mm)	Number of days with			Mean cloud-iness (tenths)	Mean daily sun-shine (h)	Wind		Clear days	Cloudy days
		precip. (≥1mm)	thunder-storm	fog			preval. direct.	mean speed (m/sec)		
Jan.	184	15	10	0	5.7	8.6	NE	3.6	0.7	16.7
Feb.	158	12	9	< 1	5.4	8.5	NE	3.3	0.3	13.2
Mar.	182	8	7	< 1	4.2	9.3	NE	3.1	3.2	8.0
Apr.	150	2	2	< 1	3.0	9.7	NE	2.8	8.7	4.2
May	121	< 1	< 1	< 1	2.0	9.8	S	2.8	15.2	2.9
June	102	< 1	0	< 1	1.9	9.4	S	3.3	16.7	2.5
July	99	< 1	< 1	< 1	1.6	9.6	S	3.1	16.4	1.7
Aug.	132	< 1	< 1	< 1	2.0	9.8	S	3.6	15.5	2.1
Sept.	177	1	1	< 1	2.6	9.6	S	4.2	9.9	3.5
Oct.	212	2	3	0	3.4	10.0	NE	4.4	6.5	5.8
Nov.	208	6	9	< 1	4.3	10.1	NE	4.4	2.2	6.1
Dec.	196	12	12	< 1	5.2	9.4	NE	4.2	0.5	12.5
Annual	1,921	59	54	2	3.4	9.5	NE	3.6	95.8	79.2

[1] 1955–1962 with interruptions.

TABLE XVII

CLIMATIC TABLE FOR LAS LOMITAS (FORMOSA), 1951–1960
Latitude 24°42′S, longitude 60°35′W, elevation 130 m

Month	Mean sta. press. (mbar)	Temperature (°C)				Mean vapor press. (mbar)	Precipitation (mm)		Relat. humid. (%)
		daily mean	daily range	extreme			mean	max. in 24 h[1]	
				max.	min.				
Jan.	991.9	27.1	12.7	43.4	9.4	23.5	113	105	66
Feb.	993.3	26.5	12.3	41.0	10.3	24.4	96	145	70
Mar.	995.1	24.6	13.0	37.9	9.8	22.4	112	115	72
Apr.	997.5	20.7	12.3	37.9	0.8	18.7	74	105	77
May	999.0	17.9	12.4	34.6	−2.6	16.1	40	95	79
June	999.4	17.0	11.6	32.9	−2.4	15.3	33	75	79
July	1000.3	15.7	13.6	33.7	−7.0	12.7	18	55	71
Aug.	998.7	18.4	16.2	39.7	−5.3	12.4	9	35	59
Sept.	996.3	21.2	15.1	41.2	−2.1	14.9	36	75	59
Oct.	995.4	23.3	13.3	42.6	5.5	17.9	85	115	63
Nov.	994.1	25.3	14.8	43.2	5.8	19.7	88	85	61
Dec.	992.3	26.7	14.1	42.1	8.9	22.7	109	125	65
Annual	996.1	22.0	13.4	43.4	−7.0	18.3	812	145	70

Month	Number of days with			Mean cloudiness (tenths)	Mean daily sunshine (h)	Wind	
	precip. (⩾1mm)	thunderstorm	fog			preval. direct.	mean speed (m/sec)
Jan.	8	7	1	4.9	8.4	N	3.1
Feb.	6	6	1	5.1	8.5	N	3.1
Mar.	7	4	1	4.3	8.0	S	2.8
Apr.	8	4	1	4.5	6.5	NE	2.8
May	6	2	2	4.6	5.7	NE	2.8
June	6	1	4	5.1	4.5	NE	3.1
July	5	2	2	4.2	6.1	NE	3.3
Aug.	4	2	1	3.5	7.3	NE	3.9
Sept.	7	4	1	4.0	7.0	E	3.9
Oct.	8	6	1	4.6	7.2	E	3.9
Nov.	7	5	1	4.4	8.8	E	3.6
Dec.	7	6	1	4.4	8.5	E	3.1
Annual	79	49	13	4.4	7.2	NE	3.2

[1] 1921–1950.

TABLE XVIII

CLIMATIC TABLE FOR SALTA
Latitude 24°51′S, longitude 65°29′W, elevation 1,226 m

Month	Mean sta. press. (mbar)	Temperature (°C)				Mean vapor press. (mbar)	Precipitation (mm)		Relat. humid. (%)
		daily mean	daily range	extreme			mean	max. in 24 h	
				max.	min.				
Jan.	875.7	21.4	12.7	38.4	6.1	18.9	176	95	78
Feb.	876.3	20.5	11.6	39.3	7.7	18.9	149	115	82
Mar.	877.0	19.2	11.7	34.7	2.6	17.7	94	75	80
Apr.	878.1	16.5	13.0	33.6	−1.2	14.1	25	55	75
May	878.3	13.5	15.1	33.9	−4.6	11.5	6	35	74
June	878.2	11.1	16.6	33.1	−9.5	9.8	3	15	74
July	878.6	10.6	18.3	35.0	−9.9	8.4	2	5	66
Aug.	878.5	12.4	19.4	36.3	−6.6	8.3	4	5	58
Sept.	877.7	15.9	17.7	38.0	−3.6	9.6	5	15	53
Oct.	876.6	18.4	15.6	38.8	−2.2	12.3	25	45	58
Nov.	875.7	20.7	14.4	39.0	1.8	14.8	61	45	61
Dec.	875.4	21.5	13.7	39.5	3.9	17.2	121	95	67
Annual	877.1	16.8	15.0	39.5	−9.9	13.4	671	115	69

Month	Number of days with			Mean cloud-iness (tenths)	Mean daily sun-shine (h)	Wind		Clear days	Cloudy days
	precip. (⩾1mm)	thunder-storm	fog			preval. direct.	mean speed (m/sec)		
Jan.	14	7	1	5.6	6.3	NE	1.4	0.3	15.2
Feb.	13	4	< 1	5.8	5.3	NE	1.1	0.9	14.8
Mar.	12	3	2	5.4	4.4	NE	0.8	1.2	17.0
Apr.	6	< 1	3	5.3	4.8	NE	1.1	3.0	13.9
May	3	< 1	2	4.5	5.1	N	1.1	6.1	11.6
June	1	0	2	4.1	4.6	N	1.1	7.4	9.7
July	1	0	1	3.2	6.0	N	1.4	9.3	8.1
Aug.	1	< 1	< 1	3.0	7.0	NE	1.4	12.0	7.4
Sept.	3	< 1	< 1	3.5	5.9	NE	1.4	6.4	8.8
Oct.	6	1	< 1	4.5	5.3	NE	1.7	5.1	10.2
Nov.	8	4	< 1	4.4	5.9	NE	1.7	2.1	11.0
Dec.	12	6	< 1	4.5	6.0	NE	1.7	1.1	10.6
Annual	80	26	12	4.5	5.6	NE	1.3	54.9	138.3

TABLE XIX

CLIMATIC TABLE FOR LA CASUALIDAD (SALTA), 1946–1947 AND 1954–1957
Latitude 25°03′S, longitude 68°13′W, elevation 4,000 m

Month	Temperature (°C)				Mean vapor press. (mbar)	Relat. humid. (%)	
	daily mean	daily range	extreme			mean	14h00
			max.	min.			
Jan.	8.3	19.0	25.0	−4.5	4.3	37	20
Feb.	8.9	19.5	22.5	−6.5	4.2	35	20
Mar.	6.9	19.0	21.5	−9.1	4.5	42	25
Apr.	3.4	18.0	17.5	−11.8	4.1	50	30
May	−0.8	15.5	16.0	−14.8	3.3	55	35
June	−1.4	15.0	11.0	−17.7	3.2	57	40
July	−3.4	16.0	12.0	−21.0	2.8	55	40
Aug.	−0.7	18.5	14.0	−19.8	3.3	53	35
Sept.	1.3	18.5	18.0	−17.3	3.6	48	30
Oct.	3.0	18.0	19.0	−13.0	3.6	43	25
Nov.	6.2	19.0	21.0	−9.5	3.6	35	20
Dec.	7.6	18.5	23.0	−7.5	3.4	28	15
Annual	3.3	17.9	25.0	−21.0	3.6	45	28

Month	Mean cloud-iness (tenths)	Mean daily sun-shine (h)	Wind		Clear days	Cloudy days
			preval. direct.	mean speed (m/sec)		
Jan.	1.4	9	W	5	22	< 1
Feb.	1.6	10	W	4	18	< 1
Mar.	1.0	10	W	5	23	< 1
Apr.	0.9	10	W	5	22	< 1
May	1.4	10	W	7	21	1
June	1.3	8	W	6	23	1
July	1.1	9	W	7	23	1
Aug.	0.9	9	W	6	23	1
Sept.	1.3	10	W	8	23	1
Oct.	1.1	11	W	7	23	< 1
Nov.	0.9	11	W	6	22	< 1
Dec.	1.2	9	W	6	21	< 1
Annual	1.2	10	W	6	264	6

TABLE XX

CLIMATIC TABLE FOR TUCUMÁN
Latitude 26°48′S, longitude 65°12′W, elevation 481 m

Month	Mean sta. press. (mbar)	Temperature (°C)				Mean vapor press. (mbar)	Precipitation (mm)		Relat. humid.[1] (%)
		daily mean	daily range	extreme			mean	max. in 24 h	
				max.	min.				
Jan.	953.4	24.5	12.5	44.1	9.9	22.5	183	125	73
Feb.	954.6	23.7	11.5	40.6	10.4	22.9	159	105	80
Mar.	955.8	22.0	11.0	41.0	8.9	21.2	162	206	81
Apr.	958.2	18.3	11.1	34.3	−0.2	16.5	59	65	81
May	958.6	15.4	11.2	37.5	−3.0	13.6	29	45	80
June	959.5	12.3	11.4	29.3	−6.0	11.3	19	25	81
July	959.9	12.6	14.5	37.3	−4.8	9.9	10	25	71
Aug.	959.4	14.0	15.8	39.0	−5.2	9.5	8	15	58
Sept.	958.3	17.1	15.9	41.7	−1.2	10.9	12	25	55
Oct.	956.7	20.2	14.7	43.1	2.4	14.9	77	75	65
Nov.	954.7	22.5	13.7	41.2	4.1	18.9	108	125	70
Dec.	953.5	24.6	13.2	44.2	8.0	21.3	150	95	71
Annual	956.9	19.0	13.0	44.2	−6.0	16.1	976	206	72

Month	Mean evap.[2] (mm)	Number of days with			Mean cloud-iness (tenths)	Mean daily sun-shine[2] (h)	Wind		Clear days	Cloudy days
		precip. (⩾1mm)	thunder-storm	fog			preval. direct.	mean speed (m/sec)		
Jan.	168	12	8	1	4.8	7.5	SW	2.2	2.6	9.0
Feb.	120	10	6	1	4.9	6.6	SW	1.7	2.8	9.8
Mar.	118	12	4	1	4.7	5.9	SW	1.4	3.8	12.5
Apr.	81	8	2	< 1	4.8	6.0	S	1.4	5.8	11.8
May	56	7	< 1	4	4.6	5.3	N	1.4	6.6	14.0
June	46	5	0	4	4.5	5.1	N	1.4	7.3	10.9
July	65	3	< 1	2	3.3	6.4	N	1.7	8.7	8.6
Aug.	102	3	< 1	0	3.2	6.7	N	1.7	11.4	7.4
Sept.	124	4	1	< 1	3.6	6.6	S	1.7	9.0	8.8
Oct.	140	7	4	< 1	4.4	6.6	SW	1.9	6.5	8.8
Nov.	155	9	6	< 1	4.6	7.1	SW	1.9	4.1	10.2
Dec.	161	11	9	< 1	4.7	7.5	SW	2.2	2.5	7.2
Annual	1,336	91	40	14	4.3	6.4	SW	1.6	71.1	119.0

[1] 1951–1960.

[2] 1951–1962 with interruptions.

TABLE XXI

CLIMATIC TABLE FOR PRESIDENCIA ROQUE SAENZ PEÑA (CHACO), 1931–1960
Latitude 26°49′S, longitude 60°27′W, elevation 92 m

Month	Mean sta. press. (mbar)	Temperature (°C)				Mean vapor press. (mbar)	Mean precip. (mm)	Relat. humid. (%)
		daily mean	daily range	extreme				
				max.	min.			
Jan.	997.8	27.3	14.6	43.9	10.0	22.5	136	62
Feb.	998.7	26.3	13.9	45.0	9.7	23.2	125	68
Mar.	1000.2	24.0	13.1	42.7	6.9	22.0	133	74
Apr.	1004.7	20.1	12.9	40.0	2.6	18.1	97	77
May	1004.1	17.9	12.2	37.7	−1.8	16.3	49	80
June	1005.1	15.8	11.6	38.6	−4.8	14.2	35	79
July	1005.8	15.2	13.5	36.2	−6.3	12.5	23	72
Aug.	1004.7	17.2	14.6	40.8	−4.9	12.1	23	62
Sept.	1003.0	19.4	14.9	41.7	−2.8	13.7	44	61
Oct.	1001.7	22.1	14.6	43.5	2.6	16.7	101	63
Nov.	999.5	24.2	14.6	44.0	3.7	18.9	113	62
Dec.	998.1	26.5	15.2	44.0	7.8	20.8	111	60
Annual	1001.9	21.3	13.8	45.0	−6.3	17.6	990	69

Month	Mean evap.[1] (mm)	Number of days with			Mean cloud-iness (tenths)	Mean daily sun-shine (h)	Wind		Clear days	Cloudy days
		precip. (≥1mm)	thunder-storm	fog			preval. direct.	mean speed (m/sec)		
Jan.	161	6	6	<1	3.5	8.8	NE	1.7	7.1	7.0
Feb.	131	5	3	1	3.5	8.0	S	1.7	4.3	7.3
Mar.	115	6	5	1	3.8	7.2	S	1.7	7.8	8.3
Apr.	87	6	4	3	3.6	6.5	NE	1.4	9.3	8.3
May	68	4	2	4	4.1	5.4	NE	1.9	7.2	10.0
June	61	4	1	6	4.6	4.8	NE	2.2	5.7	12.7
July	71	2	1	4	3.5	6.0	NE	2.2	8.0	8.6
Aug.	88	2	2	2	2.9	6.9	NE	2.5	11.9	5.5
Sept.	138	4	3	2	3.4	6.9	S	2.5	9.6	7.6
Oct.	135	5	7	<1	3.4	7.7	S	2.5	8.4	6.9
Nov.	154	6	6	1	3.3	8.8	NE	2.2	7.5	5.9
Dec.	162	5	7	1	3.4	9.0	S	1.7	8.4	5.3
Annual	1,371	55	47	25	3.6	7.2	NE	2.0	95.2	93.4

[1] 1959–1962.

TABLE XXII

CLIMATIC TABLE FOR POSADAS (MISIONES)
Latitude 27°25′S, longitude 55°56′W, elevation 136 m

Month	Mean sta. press. (mbar)	Temperature (°C)				Mean vapor press. (mbar)	Precipitation (mm)		Relat. humid. (%)
		daily mean	daily range	extreme			mean	max. in 24 h	
				max.	min.				
Jan.	993.2	26.2	13.0	40.7	9.8	24.2	127	115	71
Feb.	994.1	25.8	12.3	40.6	9.1	24.5	158	135	74
Mar.	995.7	24.3	12.1	39.6	7.9	23.0	149	135	76
Apr.	998.3	20.7	11.8	35.8	4.3	19.3	178	135	79
May	999.8	18.1	11.0	32.7	−4.3	17.5	167	115	84
June	1000.4	16.5	9.9	32.5	−2.2	15.9	141	145	85
July	1001.7	15.6	11.1	32.6	−4.4	14.2	104	75	80
Aug.	1000.3	17.3	12.5	36.4	−2.4	14.7	83	65	74
Sept.	998.4	18.8	11.9	37.5	0.5	16.5	133	115	76
Oct.	997.0	20.9	12.3	39.9	2.6	18.5	188	105	75
Nov.	994.7	23.3	13.5	39.5	5.5	19.7	134	85	69
Dec.	993.2	25.7	13.8	41.9	7.2	21.8	133	135	66
Annual	997.2	21.1	12.1	41.9	−4.4	19.2	1,695	145	76

Month	Mean evap.[1] (mm)	Number of days with			Mean cloud- iness (tenths)	Mean daily sun- shine[2] (h)	Wind		Clear days	Cloudy days
		precip. (≥1mm)	thunder- storm	fog			preval. direct.	mean speed (m/sec)		
Jan.	150	8	5	1	4.3	7.5	NE	2.2	4.5	8.0
Feb.	127	8	4	1	4.5	7.4	NE	2.5	3.2	7.3
Mar.	113	8	3	2	4.0	6.0	NE	2.2	5.1	9.0
Apr.	75	8	4	3	4.0	6.0	NE	2.2	8.3	8.2
May	58	8	3	7	4.3	5.2	NE	2.5	4.3	9.4
June	46	9	4	7	4.8	4.1	NE	2.5	5.1	11.9
July	52	6	3	5	4.2	5.4	NE	2.8	6.0	10.5
Aug.	61	7	4	3	3.9	5.5	NE	2.8	8.5	8.3
Sept.	90	9	5	2	4.4	4.7	SE	2.8	5.9	9.3
Oct.	113	9	6	2	4.2	6.1	SE	2.8	6.1	9.2
Nov.	129	7	3	1	3.8	7.9	S	2.5	5.1	8.1
Dec.	154	7	5	1	3.8	8.1	SE	2.5	6.7	6.6
Annual	1,168	94	49	35	4.2	6.2	NE	2.5	68.8	105.8

[1] Loreto (27°21′S 55°30′W, 163 m), 1952–1959.
[2] 1938–1944, 1953–1960.

TABLE XXIII

CLIMATIC TABLE FOR CORRIENTES
Latitude 27°28′S, longitude 58°49′W, elevation 60 m

Month	Mean sta. press. (mbar)	Temperature (°C)				Mean vapor press. (mbar)	Precipitation (mm)		Relat. humid. (%)
		daily mean	daily range	extreme			mean	max. in 24 h	
				max.	min.				
Jan.	1001.7	27.4	12.3	41.6	12.0	23.9	149	135	65
Feb.	1002.3	26.7	11.5	41.4	11.6	24.1	127	125	69
Mar.	1004.1	24.9	10.9	40.6	10.2	22.8	151	115	72
Apr.	1007.3	20.9	10.6	36.5	5.2	19.0	135	115	77
May	1008.3	18.6	9.9	34.3	2.6	17.4	86	125	81
June	1009.1	16.0	9.2	32.2	−0.6	15.5	60	65	86
July	1010.2	15.7	10.4	32.4	−1.1	13.9	47	105	78
Aug.	1009.0	17.1	11.6	38.8	0.2	13.4	42	65	69
Sept.	1007.4	19.2	11.7	40.0	0.9	15.5	75	75	70
Oct.	1005.8	21.5	11.5	41.8	4.8	17.8	139	85	69
Nov.	1003.7	24.0	12.0	40.2	8.0	19.8	139	85	66
Dec.	1002.1	26.5	12.8	42.4	10.9	21.8	119	95	63
Annual	1005.9	21.5	11.2	42.4	−1.1	18.7	1,269	135	72

Month	Mean evap.[1] (mm)	Number of days with		Mean cloud-iness (tenths)	Mean daily sun-shine (h)	Wind		Clear days	Cloudy days
		precip. (≥1mm)	thunder-storm			preval. direct.	mean speed (m/sec)		
Jan.	147	7	7	3.8	9.1	E	2.2	5.9	7.9
Feb.	127	7	6	3.8	8.6	SE	2.2	5.4	7.1
Mar.	112	8	6	3.6	7.7	SE	1.9	7.5	7.8
Apr.	84	8	6	3.7	7.0	S	2.2	9.3	8.7
May	57	7	3	4.0	6.4	NE	2.2	6.0	9.3
June	50	6	2	4.7	5.0	E	2.5	5.8	12.2
July	57	5	2	3.9	6.0	NE	2.5	6.4	9.7
Aug.	69	5	3	3.4	7.0	E	2.8	10.7	6.2
Sept.	94	7	5	4.0	6.6	E	3.1	8.2	8.6
Oct.	128	8	8	3.9	7.6	E	2.8	7.4	7.4
Nov.	149	8	6	3.7	9.1	E	2.8	6.6	6.5
Dec.	147	7	6	3.6	9.2	E	2.5	7.9	5.8
Annual	1,221	83	60	3.8	7.4	E	2.5	87.1	97.2

[1] Paso de la Patria (27°20′S 58°36′W), 1958–1962.

TABLE XXIV

CLIMATIC TABLE FOR SANTIAGO DEL ESTERO
Latitude 27°46′S, longitude 64°18′W, elevation 199 m

Month	Mean sta. press. (mbar)	Temperature (°C)				Mean vapor press. (mbar)	Precipitation (mm)		Relat. humid. (%)
		daily mean	daily range	extreme			mean	max. in 24 h	
				max.	min.				
Jan.	986.1	27.3	15.5	45.2	8.0	21.0	90	115	58
Feb.	987.4	25.8	14.3	43.2	9.7	21.0	93	135	63
Mar.	988.9	23.4	13.4	43.0	6.3	19.0	92	135	66
Apr.	991.7	19.5	13.9	38.3	0.2	15.2	27	105	67
May	992.3	16.4	13.3	35.2	−4.3	13.6	14	25	73
June	993.5	13.5	12.9	33.3	−6.7	11.6	10	35	75
July	994.1	12.9	16.1	34.4	−10.0	9.5	3	45	64
Aug.	993.3	15.2	17.3	39.0	−7.0	9.3	3	15	54
Sept.	991.9	18.6	16.8	41.8	−4.1	10.6	8	45	50
Oct.	990.2	21.9	16.0	44.4	0.2	13.7	39	105	52
Nov.	988.1	24.4	15.6	45.5	1.3	16.7	60	145	55
Dec.	985.9	26.8	15.5	44.6	8.2	19.2	79	135	54
Annual	990.3	20.5	15.0	45.5	−10.0	15.0	518	145	61

Month	Number of days with			Mean cloud-iness (tenths)	Mean daily sun-shine (h)	Wind		Clear days	Cloudy days
	precip. (⩾1mm)	thunder-storm	fog			preval. direct.	mean speed (m/sec)		
Jan.	9	5	1	4.2	8.1	NE	3.3	6.0	6.0
Feb.	8	3	<1	4.1	7.4	NE	2.8	5.7	5.9
Mar.	8	3	<1	4.2	6.1	NE	2.8	6.4	10.1
Apr.	6	1	1	4.2	5.6	NE	2.8	6.7	9.7
May	5	<1	1	4.6	5.2	NE	2.5	6.8	11.6
June	4	0	3	4.7	4.2	S	2.5	7.8	12.0
July	2	<1	3	3.5	5.8	NE	2.8	8.0	8.8
Aug.	2	<1	1	3.2	6.9	NE	3.3	11.6	7.0
Sept.	3	1	<1	3.4	6.8	NE	3.9	10.0	7.1
Oct.	6	2	<1	3.8	7.2	NE	4.2	9.2	7.0
Nov.	7	4	<1	3.7	7.7	NE	3.9	6.9	5.1
Dec.	8	5	<1	3.8	8.2	NE	3.6	7.4	4.4
Annual	68	24	11	4.0	6.6	NE	3.2	92.5	94.7

TABLE XXV

CLIMATIC TABLE FOR CATAMARCA
Latitude 28°26′S, longitude 65°46′W, elevation 547 m

Month	Mean sta. press. (mbar)	Temperature (°C)				Mean vapor press. (mbar)	Precipitation (mm)		Relat. humid. (%)
		daily mean	daily range	extreme			mean	max. in 24 h	
				max.	min.				
Jan.	945.8	27.7	13.7	44.5	9.6	17.8	66	65	48
Feb.	946.7	26.0	13.1	43.9	10.6	18.5	82	75	55
Mar.	947.9	24.0	13.4	41.3	6.7	17.3	49	45	58
Apr.	950.3	19.6	14.2	39.6	−1.3	13.5	20	55	59
May	950.9	15.4	14.6	35.2	−3.7	11.3	11	15	65
June	951.8	11.6	14.2	32.0	−5.6	9.5	6	15	69
July	952.3	11.4	16.3	37.4	−9.0	8.1	4	15	60
Aug.	951.8	14.5	16.8	40.1	−5.5	7.8	4	15	47
Sept.	950.5	18.6	15.5	42.2	−1.0	8.8	7	25	41
Oct.	948.9	22.3	14.4	42.3	2.0	11.3	29	25	42
Nov.	946.7	24.9	13.8	45.2	0.4	14.0	41	65	45
Dec.	945.5	27.2	13.9	47.2	8.9	16.4	51	25	46
Annual	949.1	20.3	14.5	47.2	−9.0	12.8	370	75	53

Month	Number of days with			Mean cloud-iness (tenths)	Wind		Clear days	Cloudy days
	precip. (≥1mm)	thunder-storm	fog		preval. direct.	mean speed (m/sec)		
Jan.	8	4	0	3.1	NE	3.9	6.7	3.8
Feb.	6	3	0	3.2	NE	3.6	7.2	4.4
Mar.	6	2	<1	2.8	NE	3.6	11.4	5.7
Apr.	4	<1	<1	2.8	NE	3.3	12.2	6.3
May	3	0	1	3.1	NE	2.2	10.8	6.5
June	2	0	1	3.0	NE	1.7	10.9	5.8
July	2	<1	<1	2.3	NE	1.7	12.1	4.7
Aug.	1	0	0	2.2	NE	2.8	15.1	3.8
Sept.	2	<1	0	2.3	NE	4.2	13.4	4.7
Oct.	4	1	0	2.5	NE	4.7	11.5	4.9
Nov.	5	3	0	2.9	NE	5.0	8.0	4.1
Dec.	6	4	0	2.9	NE	4.7	6.7	2.9
Annual	49	18	2	2.8	NE	3.4	126.0	57.6

TABLE XXVI

CLIMATIC TABLE FOR CERES (SANTA FE)
Latitude 29°53′S, longitude 61°57′W, elevation 88 m

Month	Mean sta. press. (mbar)	Temperature (°C)				Mean vapor press. (mbar)	Precipitation (mm)		Relat. humid. (%)
		daily mean	daily range	extreme			mean	max. in 24 h	
				max.	min.				
Jan.	998.8	26.1	13.8	44.8	7.9	20.2	113	225	57
Feb.	999.8	25.0	14.6	43.8	7.3	20.2	101	85	64
Mar.	1001.4	22.5	13.1	43.8	6.0	17.9	124	155	66
Apr.	1004.4	18.3	12.9	39.0	1.5	15.6	58	125	74
May	1004.9	15.8	12.5	34.7	−2.3	13.7	38	75	76
June	1006.0	13.1	11.1	32.0	−5.2	11.8	27	45	78
July	1006.8	12.3	13.0	33.0	−5.6	10.3	22	65	72
Aug.	1006.0	14.0	14.5	39.0	−4.5	10.1	20	55	63
Sept.	1005.0	16.4	14.6	39.7	−2.6	11.4	36	65	61
Oct.	1003.4	19.4	14.2	42.8	2.0	14.1	84	120	63
Nov.	1001.0	22.4	14.8	42.9	3.2	16.3	103	95	60
Dec.	999.2	24.9	14.8	42.7	8.0	18.4	110	115	58
Annual	1003.0	19.1	13.6	44.8	−5.6	15.0	836	225	66

Month	Number of days with			Mean cloud-iness (tenths)	Mean daily sun-shine[2] (h)	Wind	
	precip. (≥1mm)	thunder-storm[1]	fog[1]			preval. direct.	mean speed (m/sec)
Jan.	7	5	1	3.2	8.6	E	2.5
Feb.	7	3	1	3.0	8.8	E	2.2
Mar.	8	4	1	3.2	7.5	E	2.2
Apr.	5	0	1	3.2	6.6	NE	2.2
May	4	1	1	3.8	5.8	NE	2.2
June	3	1	4	4.3	4.6	S	2.2
July	3	1	4	3.4	5.9	NE	2.8
Aug.	3	1	1	3.0	6.6	S	3.1
Sept.	5	1	1	3.2	6.8	S	3.3
Oct.	7	4	1	3.4	7.9	E	3.1
Nov.	8	4	0	3.0	9.1	NE	2.8
Dec.	8	5	0	3.0	8.9	E	2.8
Annual	68	29	12	3.3	7.2	S	2.6

[1] 1951–1960.
[2] 1941–1950.

TABLE XXVII

CLIMATIC TABLE FOR CONCORDIA (ENTRE RIOS)
Latitude 31°23'S, longitude 58°02'W, elevation 38 m

Month	Mean sta. press. (mbar)	Temperature (°C)				Mean vapor press. (mbar)	Precipitation (mm)		Relat. humid. (%)
		daily mean	daily range	extreme			mean	max. in 24 h	
				max.	min.				
Jan.	1005.8	25.7	14.6	43.5	10.0	19.6	128	115	59
Feb.	1006.9	24.7	14.0	42.4	8.4	20.1	109	105	65
Mar.	1008.6	22.6	13.1	41.1	6.5	18.8	154	135	69
Apr.	1011.4	18.3	12.7	37.0	0.0	15.6	136	155	74
May	1012.2	15.5	12.0	33.5	−0.7	13.8	82	330	78
June	1013.1	13.2	10.6	30.5	−4.8	12.5	87	75	83
July	1014.0	12.5	11.4	30.9	−5.3	11.9	56	75	82
Aug.	1013.1	13.6	12.3	34.8	−3.2	11.7	65	85	75
Sept.	1012.2	15.1	12.2	37.2	−1.3	12.8	88	85	74
Oct.	1010.8	18.3	13.9	39.4	2.3	14.7	117	85	70
Nov.	1008.1	21.4	14.3	41.4	4.5	16.1	96	85	63
Dec.	1006.2	24.2	14.9	41.7	8.1	17.5	96	135	58
Annual	1010.2	18.8	13.0	43.5	−5.3	15.4	1,214	330	71

Month	Mean evap.[1] (mm)	Days with		Mean cloudiness (tenths)	Mean daily sunshine (h)	Wind		Clear days	Cloudy days
		thunderstorm	fog			preval. direct.	mean speed (m/sec)		
Jan.	198	5	1	3.6	9.1	E	1.9	7.8	5.6
Feb.	164	4	0	3.4	9.1	E	1.9	8.2	4.4
Mar.	151	4	1	3.4	7.8	E	1.9	8.1	6.8
Apr.	97	4	1	3.5	7.0	NE	1.9	9.1	6.1
May	70	2	3	3.9	5.8	NE	1.9	7.7	7.2
June	52	3	4	4.9	4.0	NE	1.9	5.6	11.4
July	63	3	3	4.0	5.1	NE	2.2	7.7	8.8
Aug.	78	3	2	3.7	6.1	NE	2.8	10.2	6.2
Sept.	93	5	2	4.0	6.4	E	2.5	5.9	7.7
Oct.	132	4	1	3.4	7.8	E	2.5	7.6	4.8
Nov.	174	4	<1	3.2	9.5	E	1.9	9.0	4.9
Dec.	204	4	<1	3.4	9.2	E	2.2	10.8	4.4
Annual	1,476	45	17	3.7	7.2	E	2.1	97.7	78.3

[1] Salto Grande (31°12'S 57°55'W, 37 m), 1949–1959.

TABLE XXVIII

CLIMATIC TABLE FOR CÓRDOBA
Latitude 31°24′S, longitude 64°11′W, elevation 425 m

Month	Mean sta. press. (mbar)	Temperature (°C)				Mean vapor press. (mbar)	Precipitation (mm)		Relat. humid. (%)
		daily mean	daily range	extreme			mean	max. in 24 h	
				max.	min.				
Jan.	961.3	24.2	14.9	45.0	5.7	17.2	101	95	57
Feb.	962.3	23.2	14.9	42.2	4.6	16.9	88	85	60
Mar.	963.5	20.7	13.4	41.3	1.6	16.5	93	75	67
Apr.	965.7	16.8	14.1	36.1	−1.4	13.1	39	105	68
May	965.8	13.8	13.7	35.3	−5.9	11.3	24	65	72
June	966.5	11.0	13.3	34.6	−7.7	9.4	10	35	72
July	967.1	10.6	14.7	33.2	−9.2	8.2	8	45	62
Aug.	967.0	12.3	15.5	37.3	−7.2	7.9	15	65	55
Sept.	966.5	15.1	15.1	38.5	−3.3	9.0	29	45	52
Oct.	965.1	17.9	14.7	42.0	0.1	11.6	77	135	56
Nov.	963.0	20.8	14.8	42.4	2.5	13.8	88	75	56
Dec.	961.5	23.1	15.3	44.3	4.4	15.7	108	105	56
Annual	964.6	17.4	14.5	45.0	−9.2	12.6	680	135	61

Month	Mean evap.[1] (mm)	Number of days with			Mean cloud- iness (tenths)	Mean daily sun- shine (h)	Wind		Clear days	Cloudy days
		precip. (≥1mm)	thunder- storm	fog			preval. direct.	mean speed (m/sec)		
Jan.	169	10	8	<1	4.0	9.2	NE	1.9	5.3	6.9
Feb.	138	8	6	<1	3.8	8.5	NE	1.7	6.2	6.7
Mar.	112	9	5	1	4.1	7.1	NE	1.7	7.0	10.4
Apr.	78	5	3	1	3.8	6.9	NE	1.7	7.0	9.9
May	59	4	1	2	4.6	5.7	NE	1.7	5.3	11.1
June	42	3	<1	4	4.6	5.0	NE	1.4	6.4	11.9
July	54	2	<1	4	3.9	6.0	NE	1.7	6.0	10.3
Aug.	78	3	1	2	3.6	7.0	NE	1.9	10.4	7.4
Sept.	111	4	1	<1	3.9	7.2	NE	2.2	7.1	9.1
Oct.	130	7	4	1	4.1	7.8	NE	2.2	7.8	8.1
Nov.	147	8	8	<1	3.9	8.8	NE	2.2	5.9	6.7
Dec.	169	10	8	<1	3.9	9.0	NE	1.9	6.7	6.4
Annual	1,287	73	45	16	3.7	7.4	NE	1.8	81.1	104.9

[1] 1943–1962.

TABLE XXIX

CLIMATIC TABLE FOR SAN JUAN
Latitude 31°36'S, longitude 68°33'W, elevation 630 m

Month	Mean sta. press. (mbar)	Temperature (°C)				Mean vapor press. (mbar)	Precipitation (mm)		Relat. humid. (%)
		daily mean	daily range	extreme			mean	max. in 24 h	
				max.	min.				
Jan.	937.7	26.0	16.0	44.3	4.6	15.9	18	45	47
Feb.	938.7	24.4	16.1	43.4	7.2	16.0	12	35	52
Mar.	939.8	21.4	15.7	40.0	−1.0	14.3	9	35	56
Apr.	941.9	16.2	16.0	38.0	−3.6	11.1	5	25	60
May	942.5	11.6	15.4	34.0	−5.0	8.8	1	15	65
June	943.3	8.4	15.3	33.5	−7.5	7.2	2	15	64
July	943.7	8.0	16.2	34.0	−7.3	6.4	2	15	60
Aug.	943.5	10.7	16.8	35.3	−6.0	6.4	2	15	50
Sept.	942.9	14.4	16.5	41.1	−3.0	7.7	5	25	47
Oct.	941.0	18.3	16.6	39.4	−0.6	9.8	9	25	46
Nov.	938.9	22.1	16.7	43.0	2.3	12.0	12	45	45
Dec.	937.7	25.1	16.3	43.8	6.7	14.0	10	25	44
Annual	941.0	17.2	16.1	44.3	−7.5	10.8	87	45	53

Month	Mean evap.[1] (mm)	Number of days with			Mean cloud-iness (tenths)	Mean daily sun-shine (h)	Wind		Clear days	Cloudy days
		precip. (≥1mm)	thunder-storm	fog			preval. direct.	mean speed (m/sec)		
Jan.	221	4	5	0	2.5	10.6	S	2.5	7.3	1.9
Feb.	180	3	4	0	2.2	10.2	S	2.2	11.2	1.8
Mar.	157	2	2	<1	2.2	8.9	S	1.9	14.0	3.2
Apr.	98	1	<1	0	2.5	8.1	S	1.7	12.9	2.9
May	67	1	0	<1	3.0	7.0	S	1.7	7.6	3.7
June	46	1	0	1	3.0	6.4	S	1.4	9.3	4.2
July	54	1	0	1	2.6	7.1	S	1.7	10.1	4.9
Aug.	83	1	0	0	2.5	7.9	S	2.2	15.0	3.5
Sept.	118	1	<1	0	2.5	8.6	S	2.2	11.7	3.5
Oct.	171	2	1	<1	2.3	9.6	S	2.5	14.0	2.3
Nov.	206	2	3	0	2.2	10.8	S	2.5	11.2	2.3
Dec.	242	2	5	0	2.2	10.9	S	2.8	13.4	0.9
Annual	1,643	21	21	2	2.5	8.8	S	2.1	137.7	35.1

[1] 1949–1959.

TABLE XXX

CLIMATIC TABLE FOR MENDOZA, 1944–1963
Latitude 32°53′S, longitude 68°50′W, elevation 769 m

Month	Mean sta. press. (mbar)	Temperature (°C)				Mean vapor press. (mbar)	Precipitation (mm)		Relat. humid. (%)
		daily mean	daily range	extreme			mean[1]	max. in 24 h[2]	
				max.	min.				
Jan.	916.4	23.6	11.7	37.5	9.4	15.5	28	55	53
Feb.	917.6	22.5	11.3	36.9	7.4	15.5	21	45	57
Mar.	918.5	20.2	11.0	34.9	2.9	12.8	22	45	54
Apr.	920.0	15.6	10.7	32.1	1.9	9.7	10	25	55
May	920.1	11.5	10.5	27.9	−2.5	8.0	11	45	59
June	920.6	8.1	10.2	28.1	−4.4	6.9	8	25	62
July	921.2	7.6	10.8	27.9	−6.2	6.3	7	15	60
Aug.	920.6	10.2	11.7	29.8	−2.9	5.9	10	35	47
Sept.	920.2	13.9	11.6	32.7	−1.8	6.7	14	35	42
Oct.	919.4	16.7	11.8	36.3	0.9	8.7	23	45	46
Nov.	917.6	20.4	11.8	38.2	3.8	10.9	20	45	45
Dec.	916.6	22.7	11.9	36.5	8.1	12.4	23	95	48
Annual	919.1	16.1	11.2	38.2	−6.2	9.9	197	95	52

Month	Days with		Mean cloud-iness (tenths)	Mean daily sun-shine[3] (h)	Wind	
	thunder-storm	fog			preval. direct.	mean speed (m/sec)
Jan.	5	1	3.2	9.5	S	1.7
Feb.	3	0	3.2	9.1	S	1.4
Mar.	2	1	3.3	7.8	S	1.4
Apr.	1	1	3.4	7.0	W	1.7
May	1	1	3.5	6.4	W	1.7
June	0	1	4.0	5.7	W	1.7
July	0	1	3.6	6.3	W	1.7
Aug.	0	0	3.4	7.0	W	1.9
Sept.	1	1	3.5	7.4	W	1.7
Oct.	1	1	3.4	8.1	S	1.7
Nov.	0	0	3.4	9.8	S	1.4
Dec.	4	1	3.1	9.7	S	1.7
Annual	16	4	3.4	7.8	S	1.6

[1] 1931–1960.
[2] 1921–1960.
[3] 1951–1960.

TABLE XXXI

CLIMATIC TABLE FOR ROSARIO (SANTA FE)
Latitude 32°55′S, longitude 60°47′W, elevation 27 m

Month	Mean sta. press. (mbar)	Temperature (°C)				Mean vapor press. (mbar)	Precipitation (mm)		Relat. humid. (%)
		daily mean	daily range	extreme			mean	max. in 24 h	
				max.	min.				
Jan.	1007.1	23.8	14.1	42.4	4.8	19.6	115	125	67
Feb.	1008.3	22.9	14.0	40.3	4.2	19.1	88	85	69
Mar.	1009.8	20.4	12.7	38.2	3.6	18.3	134	95	76
Apr.	1013.0	16.1	12.7	34.4	−1.5	14.7	85	75	80
May	1013.5	13.4	12.0	31.5	−4.8	13.1	55	65	85
June	1014.6	10.9	10.2	27.5	−10.6	11.4	39	55	87
July	1014.6	10.2	11.8	28.0	−9.5	10.3	33	75	83
Aug.	1014.3	11.2	13.1	32.6	−7.3	10.3	45	65	77
Sept.	1013.4	13.6	13.2	40.0	−6.0	11.5	78	75	74
Oct.	1012.5	16.5	12.9	35.4	−0.6	13.8	92	85	73
Nov.	1009.8	19.6	13.8	38.4	1.2	16.0	105	105	70
Dec.	1007.9	22.3	14.5	42.1	3.1	17.8	99	105	66
Annual	1011.6	16.7	12.9	42.4	−10.6	14.7	968	125	75

Month	Number of days with			Mean cloud-iness (tenths)	Mean daily sun-shine[1] (h)	Wind		Clear days	Cloudy days
	precip. (≥1mm)	thunder-storm	fog			preval. direct.	mean speed (m/sec)		
Jan.	7	7	1	3.6	9.8	NE	2.5	8.1	5.1
Feb.	6	6	1	3.3	9.1	NE	2.5	7.6	4.5
Mar.	8	6	3	3.7	8.0	NE	2.5	8.0	6.9
Apr.	6	4	5	3.4	7.0	NE	2.2	8.3	6.4
May	5	2	7	4.2	6.3	NE	2.8	5.3	8.7
June	5	2	8	5.0	4.8	NE	2.8	5.0	12.4
July	4	2	7	4.3	5.6	NE	2.8	5.2	10.6
Aug.	4	3	5	4.0	6.3	NE	3.3	8.5	7.9
Sept.	6	2	3	3.9	6.8	NE	3.3	7.4	8.0
Oct.	7	6	2	3.9	8.3	NE	3.3	8.6	5.3
Nov.	8	7	2	3.5	9.9	NE	3.1	8.5	4.9
Dec.	7	6	1	3.6	9.4	NE	2.8	9.8	5.2
Annual	73	53	45	3.8	7.6	NE	2.8	90.3	85.9

[1] 1921–1950.

TABLE XXXII

CLIMATIC TABLE FOR SAN LUIS
Latitude 33°16′S, longitude 66°21′W, elevation 716 m

Month	Mean sta. press. (mbar)	Temperature (°C)				Mean vapor press. (mbar)	Precipitation (mm)		Relat. humid. (%)
		daily mean	daily range	extreme			mean	max. in 24 h	
				max.	min.				
Jan.	927.8	24.0	14.9	43.4	4.0	15.3	90	125	51
Feb.	928.8	23.0	15.1	40.0	5.2	14.8	74	95	52
Mar.	929.8	20.2	13.8	38.3	2.1	14.0	60	65	59
Apr.	931.6	15.8	14.3	35.3	−2.1	10.9	33	55	61
May	931.6	12.2	13.5	32.8	−6.6	9.4	14	85	66
June	932.1	9.2	13.3	30.6	−7.0	7.9	13	35	68
July	932.6	8.8	14.2	28.6	−9.4	7.0	10	45	62
Aug.	932.5	10.7	14.6	31.9	−9.3	6.8	6	25	53
Sept.	932.1	13.9	14.3	34.0	−3.6	7.8	17	55	49
Oct.	930.9	17.1	14.6	40.8	−4.5	10.1	53	55	52
Nov.	929.1	20.5	14.8	41.6	0.0	12.0	67	65	50
Dec.	927.6	23.2	14.9	43.0	2.7	13.8	105	65	49
Annual	930.5	16.5	14.4	43.4	−9.4	10.8	542	125	57

Month	Number of days with			Mean cloud-iness (tenths)	Wind		Clear days	Cloudy days
	precip. (≥1mm)	thunder-storm	fog		preval. direct.	mean speed (m/sec)		
Jan.	8	7	<1	2.9	N	4.4	8.2	1.9
Feb.	7	4	1	2.6	N	4.4	10.0	1.4
Mar.	5	3	1	2.6	N	3.9	10.9	3.0
Apr.	3	2	1	3.0	N	3.6	10.0	4.5
May	3	<1	1	3.6	N	3.3	7.8	4.0
June	2	<1	2	3.5	E	2.8	7.1	4.6
July	2	0	2	3.1	N	3.6	8.4	4.8
Aug.	2	<1	1	2.9	N	5.3	12.0	3.1
Sept.	3	1	1	3.0	N	5.6	10.9	3.6
Oct.	5	3	1	3.0	N	5.6	10.7	3.4
Nov.	6	6	<1	2.9	N	5.0	9.6	2.4
Dec.	8	9	1	3.0	E	4.7	10.3	2.1
Annual	54	35	12	3.0	N	4.4	115.9	38.8

TABLE XXXIII

<small>CLIMATIC TABLE FOR BUENOS AIRES (OBSERVATORIO CENTRAL)</small>
Latitude 34°35′S, longitude 58°29′W, elevation 25 m

Month	Mean sta. press. (mbar)	Temperature (°C)				Mean vapor press. (mbar)	Precipitation (mm)		Relat. humid. (%)
		daily mean	daily range	extreme			mean	max. in 24 h	
				max.	min.				
Jan.	1008.3	23.7	11.5	43.3	7.8	18.8	104	125	64
Feb.	1009.4	23.0	10.9	38.7	7.2	18.9	82	195	67
Mar.	1010.9	20.7	10.3	37.9	5.5	18.2	122	95	74
Apr.	1013.5	16.6	10.1	33.1	0.9	14.3	90	125	76
May	1014.1	13.7	9.4	31.6	−2.2	12.9	79	85	82
June	1015.0	11.1	8.3	28.5	−4.7	11.1	68	75	84
July	1015.8	10.5	8.4	27.6	−5.3	10.5	61	75	83
Aug.	1015.4	11.5	9.4	30.9	−4.0	10.2	68	95	75
Sept.	1015.0	13.6	9.6	34.0	−1.1	11.5	80	85	74
Oct.	1013.4	16.5	9.9	33.5	0.6	13.8	100	105	73
Nov.	1010.6	19.5	10.7	36.8	3.2	15.4	90	75	68
Dec.	1008.9	22.1	11.5	39.3	5.0	17.0	83	95	64
Annual	1012.5	16.9	10.0	43.3	−5.3	14.4	1,027	195	74

Month	Mean evap.[1] (mm)	Number of days with			Mean cloudiness (tenths)	Mean daily sunshine (h)	Wind		Clear days	Cloudy days
		precip. (≥1mm)	thunderstorm	fog			preval. direct.	mean speed (m/sec)		
Jan.	151	7	7	<1	3.6	9.2	NE	3.2	8.5	6.8
Feb.	121	7	4	1	3.3	8.9	NE	2.8	6.5	5.5
Mar.	96	8	5	2	3.6	7.1	NE	2.5	6.8	7.8
Apr.	60	7	3	3	3.6	6.6	NE	2.5	8.7	6.5
May	40	6	3	4	4.4	5.4	NE	2.2	4.2	8.0
June	27	7	2	7	5.0	4.2	NE	2.2	4.9	11.1
July	32	6	3	4	4.7	4.8	NE	2.5	3.9	12.1
Aug.	45	7	3	3	4.4	5.7	NE	3.1	4.8	9.0
Sept.	65	7	3	2	4.4	6.2	NE	3.1	4.7	9.5
Oct.	91	8	4	1	4.2	7.3	NE	3.1	6.3	7.4
Nov.	122	8	5	1	3.8	8.8	NE	3.1	6.2	7.1
Dec.	148	8	5	0	3.7	8.8	NE	3.1	7.6	5.4
Annual	998	86	47	28	4.1	6.9	NE	2.8	73.1	96.2

[1] 1940–1962 with interruptions.

TABLE XXXIV

CLIMATIC TABLE FOR JUNIN (BUENOS AIRES), 1931–1960
Latitude 34°35′S, longitude 60°56′W, elevation 81 m

Month	Mean sta. press. (mbar)	Temperature (°C)				Mean vapor press. (mbar)	Precipitation (mm)		Relat. humid. (%)
		daily mean	daily range	extreme			mean	max. in 24 h	
				max.	min.				
Jan.	1001.4	23.4	15.1	43.9	5.2	17.9	121	105	62
Feb.	1002.6	22.2	14.9	40.5	5.2	17.6	90	75	66
Mar.	1003.9	19.7	13.3	38.5	3.2	16.8	111	115	73
Apr.	1006.7	15.3	13.2	35.6	−2.7	13.5	83	155	78
May	1006.7	12.8	11.6	33.2	−3.5	11.7	55	85	81
June	1007.9	10.0	10.0	27.5	−6.8	10.4	59	95	85
July	1008.6	9.8	10.8	27.3	−6.0	10.1	38	105	84
Aug.	1008.7	10.5	12.0	31.2	−5.3	9.5	32	55	75
Sept.	1007.9	12.7	12.7	35.5	−2.6	10.6	56	75	72
Oct.	1006.3	15.4	13.0	37.1	−1.0	12.8	99	115	73
Nov.	1003.7	19.0	14.0	39.9	0.3	14.6	96	135	66
Dec.	1001.7	21.7	14.9	40.5	0.3	16.1	101	135	62
Annual	1005.5	16.0	13.0	43.9	−6.8	13.5	941		73

Month	Number of days with			Mean cloud-iness (tenths)	Mean daily sun-shine (h)	Wind		Clear days	Cloudy days
	precip. (≥1mm)	thunder-storm[1]	fog			preval. direct.	mean speed (m/sec)		
Jan.	7	7	0.5	3.5	9.3	N	2.2	8.4	6.4
Feb.	6	4	1	3.3	8.8	N	2.2	7.5	5.7
Mar.	7	5	3	3.7	8.3	N	1.9	7.3	7.5
Apr.	6	4	4	3.5	6.8	N	1.9	8.3	6.3
May	6	2	5	4.6	5.7	N	2.2	4.1	11.2
June	5	1	9	5.0	4.1	N	1.9	4.5	12.8
July	4	1	6	4.6	6.2	N	2.2	5.5	11.1
Aug.	4	2	4	4.4	6.0	N	2.5	5.7	9.4
Sept.	6	3	3	4.5	6.6	N	2.5	5.4	10.1
Oct.	8	4	2	4.1	7.7	NE	2.5	6.7	7.3
Nov.	7	7	1	3.7	9.2	N/NE	2.5	6.8	7.1
Dec.	8	5	0.2	3.7	9.3	N	2.2	8.4	5.2
Annual	74	45	38.7	4.0	7.2	N	2.2	78.6	100.1

[1] 1951–1960.

TABLE XXXV

CLIMATIC TABLE FOR LAS FLORES (PROV. BUENOS AIRES)
Latitude 36°02′S, longitude 59°06′W, elevation 34 m

Month	Mean sta. press. (mbar)	Temperature (°C) daily mean	daily range	extreme max.	min.	Mean vapor press. (mbar)	Precipitation (mm) mean	max. in 24 h	Relat. humid. (%)
Jan.	1007.0	22.8	15.6	43.3	4.6	17.2	68	145	62
Feb.	1007.4	21.9	15.4	41.3	4.5	17.6	79	105	68
Mar.	1010.1	18.9	13.4	37.4	1.1	16.4	108	125	75
Apr.	1012.9	14.8	13.0	34.7	−0.4	13.4	91	115	80
May	1012.7	11.8	11.3	32.3	−2.9	11.7	76	175	85
June	1013.8	9.1	10.1	29.8	−7.3	9.9	68	105	86
July	1014.3	9.1	9.7	26.3	−5.5	9.8	50	85	85
Aug.	1014.5	9.7	11.2	29.0	−5.4	9.5	58	85	79
Sept.	1014.1	12.2	11.7	33.3	−2.0	11.0	82	115	77
Oct.	1012.6	15.1	12.6	31.4	−1.0	12.8	72	95	75
Nov.	1009.5	18.4	13.9	37.4	−0.1	14.5	78	75	69
Dec.	1007.4	20.9	15.1	39.4	0.8	16.1	89	105	65
Annual	1011.4	15.4	12.7	43.3	−7.3	13.3	919	175	76

Month	Days with precip.[1] (≥1mm)	Mean cloud-iness (tenths)	Wind preval. direct.	mean speed (m/sec)
Jan.	6	3.4	NE	1.9
Feb.	6	3.3	NE	1.9
Mar.	8	3.6	NE	1.4
Apr.	6	3.5	SE	1.4
May	6	4.2	NE	1.4
June	5	4.6	NE	1.4
July	5	4.3	NW	1.7
Aug.	5	4.1	NE	1.9
Sept.	6	4.2	NE	2.2
Oct.	7	4.1	NE	2.2
Nov.	7	3.7	NE	2.2
Dec.	8	3.4	NE	2.2
Annual	75	3.9	NE	1.8

[1] 1921–1950.

TABLE XXXVI

CLIMATIC TABLE FOR MACACHIN (LA PAMPA)
Latitude 37°08′S, longitude 63°41′W, elevation 142 m

Month	Mean sta. press. (mbar)	Temperature (°C)				Mean vapor press. (mbar)	Precipitation (mm)		Relat. humid. (%)
		daily mean	daily range	extreme			mean	max. in 24 h	
				max.	min.				
Jan.	993.3	23.7	17.5	43.7	4.5	14.6	59	55	50
Feb.	994.3	22.4	17.3	42.3	2.4	14.6	68	95	51
Mar.	996.1	18.9	15.5	39.1	−0.9	14.3	82	105	65
Apr.	998.3	14.5	14.9	35.4	−8.0	11.3	49	105	68
May	998.7	10.7	13.2	30.9	−6.7	9.5	39	115	74
June	999.5	7.7	11.9	25.2	−10.6	8.5	34	65	81
July	1000.3	7.4	13.3	26.0	−11.1	8.0	24	65	78
Aug.	1000.5	8.6	14.9	32.0	−9.9	7.6	16	85	68
Sept.	1000.3	11.6	15.0	34.9	−8.2	8.7	40	75	64
Oct.	998.6	15.2	15.2	37.6	−3.4	11.2	75	75	65
Nov.	995.5	19.3	16.6	39.3	−2.2	12.0	57	95	54
Dec.	993.7	22.3	17.2	43.0	2.0	12.8	65	65	48
Annual	997.4	15.2	15.2	43.7	−11.1	11.1	608	115	64

Month	Mean evap.[1] (mm)	Days with		Mean cloud-iness (tenths)	Mean daily sun-shine[2] (h)	Wind		Clear days	Cloudy days
		precip. (≥1mm)	thunder-storm			preval. direct.	mean speed (m/sec)		
Jan.	242	6	6	2.9	10.5	N	3.3	8.6	3.0
Feb.	187	6	3	2.6	9.5	N	3.3	7.3	3.8
Mar.	146	7	3	2.9	7.3	N	2.8	8.3	4.2
Apr.	87	4	2	3.3	6.5	NE	2.8	6.2	5.1
May	51	4	1	4.2	5.4	N	2.8	4.1	9.4
June	34	3	<1	4.3	4.0	N	2.5	4.4	9.2
July	47	4	1	4.2	4.8	N	2.5	4.9	9.2
Aug.	75	3	1	3.9	5.8	N	3.6	5.4	7.8
Sept.	100	5	2	4.0	6.6	N	3.6	4.8	8.3
Oct.	138	7	4	3.8	7.0	N	3.3	5.0	5.6
Nov.	183	7	5	3.4	9.8	N	3.3	4.9	3.6
Dec.	232	6	6	3.0	10.6	N	3.9	7.2	3.0
Annual	1,522	62	34	3.5	7.3	N	3.1	71.1	72.2

[1] Santa Rosa (36°34′S 64°16′W, 189 m), 1946–1958.
[2] Santa Rosa, 1941–1960.

TABLE XXXVII

CLIMATIC TABLE FOR CHOS MALAL (NEUQUÉN)
Latitude 37°23′S, longitude 70°17′W, elevation 848 m

Month	Mean sta. press. (mbar)	Temperature (°C)					Mean vapor press. (mbar)	Precipitation (mm)		Relat. humid. (%)
		daily mean	daily range	extreme				mean	max. in 24 h	
				max.	min.					
Jan.	15.4	21.2	18.5	39.5	1.3	8.4	9	35	33
Feb.	15.8	20.2	19.1	39.5	2.4	8.7	10	65	37
Mar.	16.6	17.1	19.1	37.8	−1.0	8.0	12	35	41
Apr.	17.8	12.6	17.2	33.4	−9.1	7.1	13	35	49
May	17.3	9.6	13.6	29.6	−8.3	6.3	41	55	53
June	17.6	6.7	11.9	25.3	−11.4	5.7	54	65	58
July	18.2	6.2	12.4	24.0	−11.1	5.5	32	25	58
Aug.	18.4	7.6	13.9	29.0	−9.5	5.5	29	45	53
Sept.	18.6	9.8	15.4	30.0	−6.4	5.3	13	45	44
Oct.	18.1	14.0	17.3	33.2	−3.1	5.9	11	35	37
Nov.	17.0	17.6	18.0	36.9	−0.1	7.1	7	35	35
Dec.	15.3	20.2	19.0	39.6	0.5	7.9	6	15	33
Annual	17.2	13.6	16.3	39.6	−11.3	6.8	237	65	44

Month	Days with precip. (≥1mm)	Mean cloud- iness (tenths)	Mean daily sun- shine (h)	Wind	
				preval. direct.	mean speed (m/sec)
Jan.	2	2.1	11.0	NW	3.1
Feb.	2	2.2	10.5	NW	2.2
Mar.	2	2.4	9.5	NW	2.2
Apr.	2	3.2	6.4	NW	1.9
May	5	4.4	4.8	NW	1.9
June	6	4.3	4.7	NW	1.7
July	4	4.1	4.8	NW	2.5
Aug.	5	4.0	5.8	NW	2.5
Sept.	3	3.8	7.2	NW	2.8
Oct.	2	3.4	9.1	NW	3.3
Nov.	2	3.0	10.1	NW	2.8
Dec.	2	2.4	11.0	NW	2.8
Annual	37	3.3	7.9	NW	2.5

TABLE XXXVIII

CLIMATIC TABLE FOR MAR DEL PLATA (PROV. BUENOS AIRES)
Latitude 38°08′S, longitude 57°33′W, elevation 14 m

Month	Mean sta. press. (mbar)	Temperature (°C)				Mean vapor press. (mbar)	Precipitation (mm)		Relat. humid. (%)
		daily mean	daily range	extreme			mean[1]	max. in 24 h[2]	
				max.	min.				
Jan.	1010.3	19.0	11.5	38.8	3.2	15.7	70	75	71
Feb.	1011.1	19.3	10.9	40.2	4.0	16.7	73	75	74
Mar.	1012.2	17.8	10.3	36.8	2.0	15.2	89	105	75
Apr.	1014.5	14.9	9.7	31.8	−2.0	13.2	65	75	78
May	1014.9	11.3	9.4	31.5	−3.0	11.3	66	55	84
June	1015.2	8.9	8.8	23.3	−5.5	9.7	62	95	85
July	1017.0	7.7	8.6	24.0	−6.6	9.2	51	75	87
Aug.	1015.7	8.4	9.0	23.7	−4.0	9.1	48	55	83
Sept.	1016.8	9.7	9.3	32.1	−4.0	10.0	55	95	83
Oct.	1015.2	11.6	9.7	31.7	−2.7	11.8	57	65	86
Nov.	1011.7	14.9	10.9	34.0	−0.6	13.4	68	95	79
Dec.	1010.5	17.2	11.0	37.9	0.2	14.8	64	85	75
Annual	1013.8	13.4	9.9	40.2	−6.6	12.5	768	105	82

Month	Mean evap. (mm)	Number of days with			Mean cloud- iness (tenths)	Mean daily sun- shine (h)	Wind		Clear days	Cloudy days
		precip.[3] (≥1mm)	thunder- storm[3]	fog[3]			preval. direct.	mean speed (m/sec)		
Jan.	164	8	6	2	3.8	9.0	N	5.8	4.8	5.8
Feb.	145	8	4	3	3.8	8.8	NE	6.1	4.3	4.8
Mar.	123	9	4	4	4.1	7.2	N	5.0	3.0	8.5
Apr.	83	7	2	4	4.1	6.2	N	4.7	3.7	8.4
May	60	9	1	6	5.0	5.0	NW	4.7	2.1	12.9
June	56	8	1	7	5.0	4.0	W	4.4	1.3	12.2
July	58	8	2	8	5.0	4.5	NW	5.0	2.5	12.2
Aug.	62	7	1	6	4.8	5.5	N	5.3	3.5	12.4
Sept.	78	8	2	6	4.8	5.9	S	5.3	2.6	11.6
Oct.	100	9	2	8	4.8	7.3	N	5.8	3.6	8.3
Nov.	121	8	4	4	5.0	8.8	NE	6.4	3.9	7.3
Dec.	142	8	5	2	5.0	8.6	NE	6.7	4.4	7.3
Annual	1,192	97	34	60	4.6	6.7	NE	5.4	39.7	111.7

[1] 1931–1958.
[2] 1921–1958.
[3] 1918–1940.

TABLE XXXIX

CLIMATIC TABLE FOR CIPOLLETTI (RIO NEGRO)
Latitude 38°57'S, longitude 67°59'W, elevation 265 m

Month	Mean sta. press. (mbar)	Temperature (°C)				Mean vapor press. (mbar)	Precipitation (mm)		Relat. humid. (%)
		daily mean	daily range	extreme			mean	max. in 24 h	
				max.	min.				
Jan.	979.1	21.6	18.2	41.4	−0.1	12.1	17	45	47
Feb.	979.8	20.6	18.6	41.4	0.6	12.5	9	55	52
Mar.	981.3	17.2	18.5	37.7	−3.1	11.4	14	35	58
Apr.	983.3	12.6	17.2	34.1	−5.4	9.5	13	25	65
May	983.1	8.8	13.5	29.6	−11.2	8.0	22	45	71
June	984.1	5.7	12.8	27.0	−10.0	6.8	14	25	74
July	984.7	5.8	13.9	31.9	−10.6	6.2	13	25	67
Aug.	985.1	7.6	15.6	30.2	−10.9	6.2	15	35	59
Sept.	985.3	10.9	15.5	33.7	−6.6	6.7	11	55	53
Oct.	983.7	14.4	16.5	36.4	−3.3	8.3	22	35	50
Nov.	980.9	18.7	17.2	38.7	−0.3	9.7	13	35	45
Dec.	977.9	20.8	17.4	39.6	−0.7	11.3	13	35	46
Annual	982.4	13.7	16.3	41.4	−11.2	9.0	176	55	57

Month	Mean evap.[1] (mm)	Number of days with			Mean cloud-iness (tenths)	Mean daily sun-shine (h)	Wind		Clear days	Cloudy days
		precip. (⩾1mm)	thunder-storm	fog			preval. direct.	mean speed (m/sec)		
Jan.	144	2	<1	0	2.8	11.0	W	3.3	9.4	4.1
Feb.	114	3	1	0	2.6	10.5	W	2.5	9.3	3.0
Mar.	91	3	1	0	2.8	8.6	W	1.9	8.9	6.1
Apr.	55	2	<1	1	3.4	6.5	W	1.9	8.0	8.2
May	35	5	0	1	4.9	4.8	W	1.7	4.1	12.5
June	25	4	<1	2	5.0	4.3	W	1.7	4.0	10.8
July	31	4	0	1	4.6	4.6	W	1.9	6.1	9.1
Aug.	49	4	<1	<1	4.3	6.0	W	2.2	7.0	9.3
Sept.	72	4	<1	0	4.4	6.5	W	2.2	4.6	10.2
Oct.	102	4	1	0	4.0	8.6	W	2.5	6.1	7.7
Nov.	133	2	1	0	3.8	10.0	W	3.1	5.2	6.5
Dec.	152	2	1	0	3.5	10.5	W	3.3	9.1	4.1
Annual	1,003	39	5	5	3.8	7.7	W	2.4	81.8	91.6

[1] 1941–1956.

TABLE XL

CLIMATIC TABLE FOR SAN CARLOS DE BARILOCHE, 1951–1960
Latitude 41°06′S, longitude 71°10′W, elevation 836 m

Month	Mean sta. press. (mbar)	Temperature (°C)				Mean vapor press. (mbar)	Mean precip. (mm)	Relat. humid. (%)
		daily mean	daily range	extreme				
				max.	min.			
Jan.	920.2	14.5	14.5	32.6	−5.7	9.7	37	60
Feb.	921.0	14.4	15.1	34.0	−4.0	9.9	12	62
Mar.	921.3	12.0	14.1	31.7	−8.3	9.2	28	65
Apr.	922.6	8.0	13.0	25.8	−8.6	8.1	51	72
May	921.0	5.6	9.1	18.9	−11.1	7.5	141	80
June	921.7	2.9	7.8	15.2	−15.4	6.4	89	83
July	922.1	2.3	6.7	16.7	−14.0	6.0	143	82
Aug.	921.7	2.9	9.2	19.6	−16.7	6.0	104	79
Sept.	922.6	4.7	10.5	19.5	−10.7	6.3	51	74
Oct.	923.1	7.8	11.7	24.0	−10.7	6.9	23	67
Nov.	922.1	11.2	13.3	27.6	−4.6	8.1	16	64
Dec.	919.8	13.8	14.5	32.7	−8.5	9.1	22	59
Annual	921.6	8.3	11.8	34.0	−16.7	7.8	717	71

Month	Days with		Mean cloud-iness (tenths)	Mean daily sun-shine (h)	Wind	
	precip. (⩾1mm)	thunder-storm			preval. direct.	mean speed (m/sec)
Jan.	6	<1	3.7	10.8	W	6.9
Feb.	4	<1	2.9	9.5	W	5.8
Mar.	7	<1	3.8	8.0	W	5.3
Apr.	9	<1	4.4	6.0	W	4.7
May	16	<1	5.6	4.0	W	4.7
June	16	0	5.5	3.4	W	4.4
July	17	0	5.3	4.0	W	5.0
Aug.	15	0	5.2	5.0	W	5.3
Sept.	11	0	4.7	6.2	W	5.0
Oct.	6	<1	4.2	7.8	W	5.8
Nov.	5	<1	4.0	9.9	W	6.7
Dec.	5	<1	4.1	10.6	W	6.4
Annual	117	2	4.4	7.1	W	5.6

TABLE XLI

CLIMATIC TABLE FOR TRELEW (CHUBUT), 1941–1960
Latitude 43°14′S, longitude 65°18′W, elevation 39 m

Month	Mean sta. press. (mbar)	Temperature (°C)				Mean vapor press. (mbar)	Precipitation (mm)		Relat. humid. (%)
		daily mean	daily range	extreme			mean	max. in 24 h	
				max.	min.				
Jan.	1003.8	20.6	14.8	40.0	3.2	9.0	6	25	37
Feb.	1004.2	20.0	14.7	40.4	1.7	9.5	14	45	41
Mar.	1006.2	17.3	14.0	39.5	−1.4	9.5	17	25	48
Apr.	1007.2	13.2	13.1	35.3	−2.8	8.2	11	35	54
May	1006.5	9.6	10.8	26.7	−10.7	7.4	19	35	62
June	1008.0	6.1	10.5	23.4	−9.0	6.4	11	25	68
July	1008.7	6.0	10.5	24.8	−10.8	6.1	15	25	65
Aug.	1008.4	7.6	12.2	27.4	−8.6	5.7	13	25	55
Sept.	1009.6	10.2	12.7	31.3	−8.0	6.2	15	35	50
Oct.	1007.6	14.0	14.7	34.2	−2.2	6.4	17	45	39
Nov.	1005.0	17.3	14.9	37.6	−1.0	7.5	13	25	38
Dec.	1003.6	19.3	14.5	41.2	3.3	8.4	14	55	37
Annual	1006.6	13.5	13.1	41.2	−10.8	7.5	165	55	48

Month	Number of days with			Mean cloud- iness (tenths)	Mean daily sun- shine (h)	Wind		Clear days	Cloudy days
	precip. (≥1mm)	thunder- storm[1]	fog			preval. direct.	mean speed (m/sec)		
Jan.	3	1	<1	4.3	10.4	W	6.4	1.9	8.8
Feb.	4	1	<1	4.3	9.6	W	5.8	2.3	6.5
Mar.	4	1	<1	4.0	8.0	W	5.0	3.6	8.5
Apr.	3	<1	1	4.3	6.6	W	5.0	3.8	7.4
May	5	<1	1	4.8	4.9	W	5.0	1.8	12.1
June	5	0	2	4.8	4.4	W	4.7	2.6	9.6
July	5	0	2	4.6	4.8	W	5.8	2.8	9.3
Aug.	4	<1	1	4.6	5.7	W	5.6	3.1	8.4
Sept.	5	<1	1	4.5	6.6	W	5.8	3.0	9.9
Oct.	4	1	1	4.6	8.5	W	6.1	3.2	8.5
Nov.	3	1	<1	4.7	9.8	W	6.9	1.7	10.3
Dec.	4	1	<1	4.7	10.0	W	7.2	1.6	9.3
Annual	49	6	9	4.5	7.5	W	5.8	31.4	108.6

[1] 1921–1950.

TABLE XLII

CLIMATIC TABLE FOR SARMIENTO (CHUBUT)
Latitude 45°35′S, longitude 69°08′W, elevation 266 m

Month	Mean sta. press. (mbar)	Temperature (°C)				Mean vapor press. (mbar)	Precipitation (mm)		Relat. humid. (%)
		daily mean	daily range	extreme			mean	max. in 24 h[1]	
				max.	min.				
Jan.	976.0	17.3	13.2	37.6	0.6	8.4	10	15	42
Feb.	976.3	16.9	13.4	38.3	1.5	8.2	8	45	43
Mar.	977.2	14.3	12.4	33.4	−2.0	7.6	11	25	47
Apr.	978.2	10.8	11.3	28.4	−5.8	7.2	15	35	56
May	977.3	7.0	9.2	22.0	−12.4	6.5	24	25	65
June	977.9	3.9	8.1	20.0	−18.9	5.7	16	25	71
July	978.4	4.0	8.4	19.0	−18.6	5.6	17	25	69
Aug.	979.4	5.5	9.9	20.6	−11.0	5.5	15	45	61
Sept.	980.5	8.0	11.3	25.5	−8.5	5.6	10	15	52
Oct.	979.3	11.6	12.5	30.7	−6.7	6.0	6	15	44
Nov.	977.5	14.3	12.8	35.4	−2.5	6.6	12	15	41
Dec.	975.9	16.4	13.4	35.8	0.4	7.0	9	15	38
Annual	977.8	10.8	11.3	38.3	−18.9	6.7	153	45	52

Month	Number of days with			Mean cloud-iness (tenths)	Mean daily sun-shine[1] (h)	Wind		Clear days	Cloudy days
	precip. (≥1mm)	thunder-storm	fog			preval. direct.	mean speed (m/sec)		
Jan.	8	1	0	4.7	8.7	W	6.1	2.0	6.1
Feb.	10	0	0	4.5	8.3	W	5.3	2.2	5.4
Mar.	12	0	1	4.3	6.9	W	5.3	3.3	6.3
Apr.	12	<1	1	4.4	5.7	W	4.2	3.1	6.3
May	21	0	2	4.6	4.3	W	3.9	2.9	8.1
June	20	0	2	4.7	3.3	W	2.8	2.8	8.6
July	17	0	1	4.6	3.9	W	3.9	3.0	8.4
Aug.	17	<1	1	4.4	5.5	W	5.0	3.6	7.3
Sept.	10	0	<1	4.5	6.1	W	4.7	3.1	7.2
Oct.	8	<1	0	4.4	7.3	W	6.1	3.1	6.5
Nov.	11	0	0	4.7	7.8	W	6.1	1.3	6.4
Dec.	9	<1	0	4.8	8.4	W	6.1	1.0	6.7
Annual	155	2	8	4.6	6.4	W	5.0	31.4	83.3

[1] 1916–1935.

TABLE XLIII

CLIMATIC TABLE FOR COMODORO RIVADAVIA (CHUBUT)
Latitude 45°47'S, longitude 67°30'W, elevation 60 m

Month	Mean sta. press. (mbar)	Temperature (°C)				Mean vapor press. (mbar)	Precipitation (mm)		Relat. humid. (%)
		daily mean	daily range	extreme			mean	max. in 24 h	
				max.	min.				
Jan.	999.5	18.6	12.1	37.5	5.3	8.3	16	35	39
Feb.	1000.1	18.2	12.1	37.2	4.1	8.4	11	45	40
Mar.	1001.2	16.0	11.0	35.0	1.1	7.8	21	45	43
Apr.	1002.6	12.7	9.7	28.8	−4.1	7.0	22	35	48
May	1001.8	9.4	8.3	26.2	−4.9	6.3	35	45	53
June	1002.4	7.0	7.5	22.2	−6.8	5.6	20	45	56
July	1002.9	6.9	7.5	21.0	−7.6	5.3	21	25	53
Aug.	1004.1	7.6	8.7	23.3	−8.3	5.3	18	25	51
Sept.	1005.3	9.6	9.8	29.2	−5.0	5.6	15	45	47
Oct.	1003.4	12.8	10.9	31.0	−1.6	6.1	10	45	41
Nov.	1000.6	15.4	11.7	37.0	0.5	6.9	16	55	39
Dec.	999.6	17.3	11.7	37.8	2.3	7.6	13	25	38
Annual	1002.0	12.6	10.1	37.8	−8.3	6.7	218	55	46

Month	Number of days with			Mean cloud-iness (tenths)	Mean daily sun-shine (h)	Wind	
	precip. (⩾1mm)	thunder-storm	fog			preval. direct.	mean speed (m/sec)
Jan.	8	0	<1	4.6	8.4	W	10.0
Feb.	10	<1	0	4.5	7.8	W	8.6
Mar.	12	<1	<1	4.2	6.0	W	8.3
Apr.	12	0	1	4.5	5.0	W	7.8
May	21	<1	1	4.6	3.6	W	8.1
June	20	0	1	4.6	3.4	W	7.5
July	17	<1	<1	4.5	3.7	W	9.2
Aug.	17	0	<1	4.4	4.4	W	9.4
Sept.	10	<1	<1	4.4	5.1	W	8.6
Oct.	8	<1	<1	4.6	6.7	W	9.4
Nov.	11	1	<1	4.9	8.0	W	10.8
Dec.	9	1	<1	5.0	8.0	W	9.7
Annual	155	2	4	4.6	5.8	W	9.0

TABLE XLIV

CLIMATIC TABLE FOR PUERTO SANTA CRUZ (SANTA CRUZ)
Latitude 50°01′S, longitude 68°32′W, elevation 12 m

Month	Mean sta. press. (mbar)	Temperature (°C)				Mean vapor press. (mbar)	Precipitation (mm)		Relat. humid. (%)
		daily mean	daily range	extreme			mean	max. in 24 h[1]	
				max.	min.				
Jan.	1001.7	14.3	12.0	34.1	1.4	8.9	21	45	55
Feb.	1002.1	14.0	12.4	34.6	−3.8	8.7	16	25	54
Mar.	1002.6	12.0	11.2	32.3	−3.2	8.2	20	35	58
Apr.	1003.0	8.7	10.6	26.0	−8.0	7.3	17	25	65
May	1003.7	4.6	8.4	24.0	−12.1	6.2	25	25	73
June	1003.6	1.8	7.4	17.3	−16.2	5.4	18	25	78
July	1005.3	2.2	7.2	15.3	−13.3	5.6	16	35	78
Aug.	1005.4	3.6	9.7	20.0	−11.6	5.8	15	15	73
Sept.	1006.1	6.2	11.4	23.8	−7.4	6.2	12	15	65
Oct.	1005.1	9.7	12.3	31.8	−4.9	6.7	7	15	56
Nov.	1001.6	11.9	12.3	32.0	−2.6	7.3	15	15	52
Dec.	1000.8	13.5	12.9	34.9	−0.7	8.2	18	45	53
Annual	1003.4	8.5	10.6	34.9	−16.2	7.0	200	45	63

Month	Number of days with			Mean cloud-iness (tenths)	Mean daily sun-shine (h)	Wind		Clear days	Cloudy days
	precip. (≥1mm)	thunder-storm	fog			preval. direct.	mean speed (m/sec)		
Jan.	6	1	<1	5.7	7.6	W	5.0	0.7	11.8
Feb.	5	<1	<1	5.3	7.3	W	4.7	0.1	8.9
Mar.	6	<1	<1	5.0	5.8	W	4.2	1.7	7.8
Apr.	5	<1	<1	4.8	4.8	W	3.3	2.2	7.2
May	6	0	1	4.9	3.8	W	3.3	3.2	9.9
June	5	0	2	4.6	3.2	W	3.1	4.1	6.0
July	5	0	2	4.7	3.6	W	3.6	3.2	8.9
Aug.	5	0	1	4.6	4.6	W	3.9	4.5	7.2
Sept.	4	0	1	4.7	5.3	W	4.4	2.1	9.1
Oct.	3	0	<1	5.1	6.6	W	5.0	0.6	9.4
Nov.	6	<1	<1	5.6	7.0	W	5.6	0.5	11.5
Dec.	6	<1	<1	5.6	7.2	W	4.7	0.2	11.2
Annual	62	2	9	5.0	5.6	W	4.2	23.1	108.9

[1] 1908–1933.

TABLE XLV

CLIMATIC TABLE FOR USHUAIA (TIERRA DEL FUEGO)
Latitude 54°48′S, longitude 68°19′W, elevation 6 m

Month	Mean sta. press. (mbar)	Temperature (°C)				Mean vapor press. (mbar)	Precipitation (mm)		Relat. humid. (%)
		daily mean	daily range	extreme			mean	max. in 24 h[1]	
				max.	min.				
Jan.	996.1	9.2	8.5	29.0	−0.3	8.3	58	25	71
Feb.	996.2	9.0	8.8	24.2	−4.0	8.1	50	35	70
Mar.	996.4	7.8	8.5	25.5	−2.4	7.7	57	55	73
Apr.	998.5	5.7	7.7	22.2	−6.3	6.9	46	25	75
May	998.5	3.2	6.7	19.0	−12.1	6.0	48	15	78
June	998.8	1.7	6.1	19.0	−12.6	5.5	45	65	80
July	999.0	1.6	6.3	17.5	−11.1	5.4	47	65	79
Aug.	999.9	2.2	6.6	18.0	−19.6	5.5	49	25	77
Sept.	1000.7	3.9	7.6	18.8	−7.2	5.9	38	15	73
Oct.	997.9	6.2	8.5	20.0	−5.7	6.2	37	35	66
Nov.	996.3	7.3	8.8	22.1	−3.5	6.9	50	15	68
Dec.	996.6	8.5	8.4	25.2	−3.3	7.8	49	35	70
Annual	997.9	5.5	7.7	29.0	−19.6	6.7	574	65	73

Month	Days with		Mean cloud-iness (tenths)	Mean daily sun-shine[2] (h)	Wind		Clear days	Cloudy days
	thunder-storm	fog			preval. direct.	mean speed (m/sec)		
Jan.	0	0	6.7	5.5	SW	4.7	0.2	17.0
Feb.	<1	<1	6.6	5.7	SW	4.4	0.1	15.1
Mar.	0	0	6.6	4.2	SW	3.3	0.1	15.9
Apr.	0	<1	6.2	2.8	SW	3.3	0.3	12.5
May	0	1	5.8	2.0	SW	3.1	1.3	11.6
June	0	<1	5.4	1.0	SW	2.2	2.4	9.9
July	0	<1	5.5	1.4	SW	2.8	2.5	8.6
Aug.	<1	<1	5.9	3.2	SW	3.3	1.4	13.6
Sept.	0	0	6.1	4.8	SW	4.4	0.7	13.1
Oct.	<1	0	6.1	6.2	SW	5.0	0.3	13.7
Nov.	<1	0	6.5	6.2	SW	5.6	0.3	17.3
Dec.	<1	<1	6.6	6.0	SW	5.3	0.1	16.6
Annual	1	2	6.2	4.1	SW	3.9	9.7	164.9

[1] 1928–1950.
[2] 1928–1932.

TABLE XLVI

CLIMATIC TABLE FOR MARISCAL ESTIGARRIBIA, 1941–1960
Latitude 22°01′S, longitude 60°36′W, elevation 181 m

Month	Mean sta. press. (mbar)	Temperature (°C)				Mean vapor press. (mbar)	Precipitation (mm)		Relat. humid. (%)
		daily mean	daily range	extreme			mean	max. in 24 h	
				max.	min.				
Jan.	988.2	29.7	13.0	43.1	10.6	22.4	112	125	54
Feb.	989.5	28.9	12.1	41.6	9.1	23.2	109	111	58
Mar.	990.3	27.4	12.2	41.5	7.8	22.0	80	64	60
Apr.	993.1	24.3	12.1	38.1	3.8	18.8	60	150	62
May	994.0	21.8	11.7	37.2	−3.0	17.2	42	99	66
June	995.2	20.0	10.9	35.1	−4.5	15.2	31	98	65
July	995.6	19.8	12.7	36.7	−5.1	13.2	17	42	57
Aug.	993.9	22.8	14.3	40.8	−2.0	12.9	4	12	47
Sept.	992.2	25.4	14.5	43.6	−2.0	15.1	26	80	47
Oct.	990.7	27.0	13.4	41.9	5.9	17.5	94	125	49
Nov.	989.0	28.4	13.7	42.0	6.8	19.6	84	97	51
Dec.	988.8	29.5	13.7	43.6	9.6	20.8	99	102	51
Annual	991.8	25.4	12.9	43.6	−5.1	18.2	758	150	56

Month	Mean evap. (mm)	Number of days with			Mean cloud- iness (tenths)	Wind		Clear days	Cloudy days
		precip. (≥1mm)	thunder- storm	fog		preval. direct.	mean speed (m/sec)		
Jan.	182	8	2	<1	6	N	4	2.0	5.3
Feb.	137	7	2	<1	5	N	2.5	1.1	5.3
Mar.	134	8	2	<1	5	N	2.5	2.7	6.1
Apr.	115	5	2	<1	5	NE	2.5	4.1	5.5
May	118	5	<1	1	5	NE	2.5	4.6	7.8
June	123	4	<1	1	6	NE	2.5	4.2	10.3
July	154	2	<1	2	5	NE	2.5	5.9	7.3
Aug.	222	1	<1	1	4	N	4	6.2	5.8
Sept.	226	4	1	<1	5	N	4	6.7	5.6
Oct.	207	6	2	<1	5	N	2.5	3.1	6.6
Nov.	174	7	3	<1	5	N/NE	2.5	3.7	5.1
Dec.	191	6	3	<1	5	N	2.5	3.3	5.2
Annual	1,983	63	18	7	5	NE	2.8	47.6	75.9

TABLE XLVII

CLIMATIC TABLE FOR PUERTO CASADO, 1941–1952 AND 1955–1960
Latitude 22°17′S, longitude 57°52′W, elevation 87 m

Month	Mean sta. press. (mbar)	Temperature (°C)				Mean vapor press. (mbar)	Precipitation (mm)		Relat. humid. (%)
		daily mean	daily range	extreme			mean	max. in 24 h	
				max.	min.				
Jan.	998.7	29.3	11.1	41.1	13.5	25.9	143	132	64
Feb.	999.2	28.4	10.2	39.7	11.4	27.1	128	116	69
Mar.	1000.8	27.2	10.6	38.3	9.0	25.3	145	126	70
Apr.	1004.0	24.5	11.1	37.8	6.0	22.0	106	134	72
May	1005.0	22.3	10.7	35.6	2.6	19.5	79	65	73
June	1006.0	20.4	10.0	34.7	−0.9	17.3	57	61	72
July	1006.6	20.3	11.1	34.6	0.9	16.0	47	51	67
Aug.	1004.6	22.4	12.4	37.9	1.1	15.9	28	31	59
Sept.	1002.8	24.3	12.1	40.1	0.7	18.0	71	74	59
Oct.	1001.5	26.0	11.4	38.8	6.8	21.2	125	87	63
Nov.	999.5	27.3	11.4	40.0	9.7	23.2	148	210	64
Dec.	998.4	28.8	11.2	41.7	10.2	24.7	126	100	63
Annual	1002.3	25.1	11.1	41.7	−0.9	21.3	1,203	210	66

Month	Number of days with			Mean cloud-iness (tenths)	Wind		Clear days	Cloudy days
	precip. (≥1mm)	thunder-storm	fog		preval. direct.	mean speed (m/sec)		
Jan.	9	2	<1	6	N	4	3.2	7.2
Feb.	8	2	<1	6	N	2.5	1.8	7.6
Mar.	8	2	<1	5	S	4	4.6	6.3
Apr.	6	2	1	4	S	2.5	6.6	5.7
May	6	1	1	5	S/N	2.5	7.4	7.5
June	6	<1	1	6	S	4	5.9	9.6
July	6	1	1	5	S	4	7.8	8.3
Aug.	4	1	1	4	S	4	8.4	7.3
Sept.	5	2	1	5	S	4	8.4	6.8
Oct.	8	3	<1	5	S	4	4.8	7.4
Nov.	7	2	<1	5	S	4	4.5	6.7
Dec.	7	2	<1	5	N	4	4.6	7.2
Annual	80	20	8	5	S	3.6	68.0	87.6

TABLE XLVIII

CLIMATIC TABLE FOR ASUNCIÓN, 1941–1960
Latitude 25°16′S, longitude 57°38′W, elevation 64 m

Month	Mean sta. press. (mbar)	Temperature (°C)				Mean vapor press. (mbar)	Precipitation (mm)		Relat. humid. (%)
		daily mean	daily range	extreme			mean	max. in 24 h	
				max.	min.				
Jan.	1001.1	29.3	10.5	41.8	14.8	24.4	167	107	60
Feb.	1001.6	28.8	9.9	40.8	14.0	24.8	142	202	63
Mar.	1003.4	26.9	9.7	39.9	11.5	23.1	160	95	65
Apr.	1006.6	23.6	9.2	35.8	6.4	19.7	138	128	68
May	1007.6	20.9	8.4	33.0	3.9	17.7	131	190	72
June	1008.8	18.8	7.7	32.0	2.5	16.3	87	94	75
July	1009.8	18.3	8.6	32.6	2.5	14.7	54	40	70
Aug.	1007.9	20.6	9.9	38.0	2.3	14.8	30	40	61
Sept.	1006.0	22.3	9.7	39.1	3.9	16.7	87	96	62
Oct.	1004.6	24.7	9.8	39.1	9.0	19.2	146	97	62
Nov.	1002.4	27.0	10.4	39.5	12.0	20.9	128	96	59
Dec.	1001.2	28.9	10.9	41.2	13.1	21.6	122	142	54
Annual	1005.1	24.2	9.6	41.8	2.3	19.5	1,392	202	65

Month	Mean evap. (mm)	Number of days with			Mean cloud-iness (tenths)	Mean daily sun-shine (h)	Wind		Clear days
		precip. (≥1mm)	thunder-storm	fog			preval. direct.	mean speed (m/sec)	
Jan.	120	8	2	<1	5	8.9	NE	2.5	2.9
Feb.	96	7	2	<1	5	8.8	NE	2.5	1.8
Mar.	91	7	1	<1	4	8.2	E	2.5	3.8
Apr.	69	6	1	1	4	7.6	E	2.5	6.2
May	57	6	1	1	4	6.6	E	2.5	6.9
June	46	6	1	2	5	5.5	E	2.5	4.5
July	61	6	1	2	5	6.3	E	2.5	7.0
Aug.	89	4	1	1	4	7.2	E	4	9.0
Sept.	89	6	2	1	4	6.8	E	4	7.0
Oct.	96	8	3	<1	5	7.8	E/S	2.5	4.9
Nov.	108	7	2	<1	4	9.0	E	2.5	4.7
Dec.	129	6	2	<1	4	9.5	NE	4	5.5
Annual	1,051	77	19	9	4.5	7.7	E	2.8	64.2

TABLE XLIX

CLIMATIC TABLE FOR PUERTO PRESIDENTE FRANCO, 1943–1960
Latitude 25°36′S, longitude 54°34′W, elevation 125 m

Month	Mean sta. press. (mbar)	Temperature (°C)				Mean vapor press. (mbar)	Precipitation (mm)		Relat. humid. (%)
		daily mean	daily range	extreme			mean	max. in 24 h	
				max.	min.				
Jan.	993.2	26.8	12.0	40.0	9.0	25.7	147	112	73
Feb.	993.8	26.5	11.6	39.0	9.4	24.0	120	85	69
Mar.	995.1	25.1	12.0	38.8	7.8	24.0	162	141	75
Apr.	998.4	21.6	11.8	35.5	3.2	20.3	138	102	79
May	999.6	19.0	11.1	32.3	−3.0	17.9	148	102	82
June	1000.6	17.4	10.7	30.7	−2.4	16.9	124	92	85
July	1001.5	16.7	12.2	30.9	−2.0	15.6	100	88	82
Aug.	1000.0	18.6	13.6	35.2	−2.2	16.1	72	67	75
Sept.	998.0	20.8	12.8	36.7	−1.9	18.4	156	151	75
Oct.	996.4	22.8	12.5	37.2	4.0	20.5	161	115	74
Nov.	994.4	24.2	13.6	39.7	5.2	22.0	143	96	73
Dec.	993.2	25.8	14.2	39.5	9.0	23.3	138	82	70
Annual	997.0	22.1	12.3	40.0	−3.0	20.4	1,609	151	76

Month	Number of days with			Mean cloud- iness (tenths)	Wind		Clear days	Cloudy days
	precip. (≥1mm)	thunder- storm	fog		preval. direct.	mean speed (m/sec)		
Jan.	9	1	3	6	SE, SW	2.5	2.1	7.5
Feb.	9	<1	4	5	SE	1	1.7	6.6
Mar.	8	1	5	5	SE	1	2.7	5.2
Apr.	6	<1	9	5	S	2.5	6.1	5.1
May	7	1	9	5	S	2.5	3.9	7.3
June	7	1	12	6	S	2.5	4.4	8.2
July	7	1	11	5	S	2.5	5.2	7.4
Aug.	7	1	9	5	S	4	7.4	5.9
Sept.	8	1	6	6	SE	2.5	5.2	7.9
Oct.	9	1	4	6	SE	4	3.0	9.4
Nov.	7	1	3	5	SE	4	3.9	5.8
Dec.	7	1	4	5	SE	4	5.2	4.3
Annual	91	10	79	5	SE	2.7	50.8	80.6

TABLE L

CLIMATIC TABLE FOR ARTIGAS, 1944–1960
Latitude 30°24'S, longitude 56°28'W, elevation 117 m

Month	Mean sea level press. (mbar)	Temperature (°C)				Mean vapor press. (mbar)	Precipitation (mm)		Relat. humid. (%)
		daily mean	daily range	extreme			mean	max. in 24 h	
				max.	min.				
Jan.	1010.7	26.6	13.9	40.9	8.2	20.9	115		60
Feb.	1011.9	26.1	13.6	40.4	9.7	20.9	102		62
Mar.	1013.8	23.6	12.7	40.8	6.3	20.0	122		69
Apr.	1016.9	19.1	12.2	36.0	2.0	16.5	139		74
May	1018.1	16.2	11.5	31.4	−1.0	14.9	103		81
June	1019.0	13.9	10.1	29.5	−4.5	13.6	127		86
July	1019.6	13.6	11.0	29.5	−4.0	12.5	81		80
Aug.	1018.6	15.4	11.7	33.0	−3.8	12.5	64		71
Sept.	1016.8	17.0	11.1	36.0	−1.0	14.5	123		75
Oct.	1015.6	19.2	11.4	34.0	1.2	16.1	149		72
Nov.	1013.1	22.6	14.3	39.7	5.2	17.6	81		64
Dec.	1011.2	25.2	14.8	41.4	6.3	18.5	118		58
Annual	1015.5	19.9	12.3	41.4	−4.5	16.5	1,325	193	71

Month	Mean evap. (mm)	Number of days with			Mean cloud-iness (tenths)	Wind		Clear days	Cloudy days
		precip. (⩾1mm)	thunder-storm	fog		preval. direct.	mean speed (m/sec)		
Jan.	159.9	7	2	0.3	3.5	SE	2.5	14	7
Feb.	148.1	5	1	0.2	3.5	SE	2.8	13	3
Mar.	132.3	6	2	0.8	3.3	SE	2.2	15	7
Apr.	78.9	6	2	2.0	3.3	SE	2.2	14	7
May	68.1	6	1	2.1	3.6	SE	2.2	14	8
June	53.5	7	1	3.1	4.4	SE	2.2	10	11
July	65.1	6	1	2.5	3.7	SE	2.5	13	11
Aug.	90.3	6	1	2.2	3.4	SE	2.5	15	8
Sept.	90.6	7	2	1.4	4.0	SE	2.8	12	10
Oct.	97.4	7	2	0.8	3.8	SE	2.8	13	8
Nov.	129.8	4	2	0.5	3.1	SE	2.8	15	5
Dec.	161.0	6	2	0.4	3.0	SE	2.8	16	6
Annual	1,257.0	73	19	16.3	3.5	SE	2.5	164	91

TABLE LI

CLIMATIC TABLE FOR PASO DE LOS TOROS, 1937–1960
Latitude 32°49′S, longitude 56°31′W, elevation 79 m

Month	Mean sea level press. (mbar)	Temperature (°C)				Mean vapor press. (mbar)	Precipitation (mm)		Relat. humid. (%)
		daily mean	daily range	extreme			mean	max. in 24 h	
				max.	min.				
Jan.	1009.8	24.9	13.4	42.8	8.3	19.3	112		61
Feb.	1011.0	23.9	12.2	40.4	7.5	19.7	93		66
Mar.	1012.6	21.8	11.6	38.6	5.6	18.7	128		72
Apr.	1015.7	17.5	11.2	36.0	2.0	15.5	111		78
May	1016.6	14.1	10.6	32.3	−1.7	13.9	88		87
June	1017.1	12.4	8.9	28.8	−5.0	12.7	102		88
July	1018.1	11.5	9.9	27.8	−5.0	11.7	66		86
Aug.	1017.4	12.8	10.8	32.0	−3.3	12.0	96		81
Sept.	1016.4	14.6	11.0	35.5	−1.6	13.3	107		80
Oct.	1014.8	17.4	11.4	36.8	0.4	15.2	98		76
Nov.	1012.3	20.3	12.8	39.5	3.9	16.4	81		69
Dec.	1010.7	23.2	13.6	40.9	6.3	17.7	81		62
Annual	1014.3	17.9	11.5	42.8	−5.0	15.5	1,163	244	76

Month	Mean evap. (mm)	Number of days with			Mean cloudiness (tenths)	Wind		Clear days	Cloudy days
		precip. (≥1mm)	thunderstorm	fog		preval. direct.	mean speed (m/sec)		
Jan.	191.0	6	4	0.5	3.8	E	3.6	12	8
Feb.	139.1	5	3	0.8	3.6	E	3.3	12	6
Mar.	119.8	6	3	1.2	3.8	E	3.3	13	8
Apr.	82.1	6	2	2.8	3.6	NE	3.1	13	8
May	61.8	6	2	6.0	4.0	NE	3.1	11	10
June	46.9	8	2	4.6	4.8	NE	3.3	8	12
July	55.4	6	2	4.7	4.3	NE	3.3	9	12
Aug.	78.5	6	3	3.3	4.0	NE	3.9	12	10
Sept.	83.1	7	3	2.1	4.5	NE	3.9	10	11
Oct.	104.5	6	3	1.3	4.2	E	3.6	11	10
Nov.	132.9	6	3	0.6	3.8	E	3.6	12	8
Dec.	181.6	5	3	0.3	3.6	E	3.6	13	8
Annual	1,276.7	73	33	28.2	4.0	NE	3.6	136	111

TABLE LII

CLIMATIC TABLE FOR TREINTA Y TRES, 1939–1960
Latitude 33°11'S, longitude 54°21'W, elevation 31 m

Month	Mean sea level press. (mbar)	Temperature (°C)				Mean vapor press. (mbar)	Precipitation (mm)		Relat. humid. (%)
		daily mean	daily range	extreme			mean	max. in 24 h	
				max.	min.				
Jan.	1012.2	24.4	14.2	41.1	5.0	18.4	99		60
Feb.	1013.3	23.4	13.3	40.5	6.0	18.1	92		63
Mar.	1014.9	21.5	12.9	39.5	3.0	17.2	87		67
Apr.	1017.7	17.5	12.4	36.4	0.4	14.3	131		71
May	1018.7	14.8	11.6	33.4	−4.4	12.7	79		76
June	1019.5	12.3	10.0	28.0	−4.8	11.3	117		79
July	1020.2	11.7	10.2	28.0	−4.0	10.9	78		79
Aug.	1019.6	12.4	10.9	31.0	−4.0	10.9	103		76
Sept.	1018.7	14.2	10.7	35.0	−2.2	12.3	114		76
Oct.	1017.4	16.9	11.4	36.5	−1.0	12.4	104		64
Nov.	1014.2	19.7	13.2	39.7	1.4	14.8	73		64
Dec.	1013.0	22.5	14.3	41.5	1.0	16.1	74		59
Annual	1016.6	17.6	12.0	41.5	−4.8	14.1	1,151	152	70

Month	Mean evap. (mm)	Number of days with			Mean cloud-iness (tenths)	Wind		Clear days	Cloudy days
		precip. (≥1mm)	thunder-storm	fog		preval. direct.	mean speed (m/sec)		
Jan.	96.5	7	6	2.9	4.1	E	3.3	10	10
Feb.	82.1	7	5	4.8	3.9	E	3.1	10	8
Mar.	67.1	7	5	5.5	4.0	E	3.1	12	9
Apr.	53.4	10	4	5.2	3.8	SE	3.1	13	8
May	48.5	7	4	6.6	4.2	NW	2.8	11	9
June	43.1	9	4	6.5	4.5	W	2.8	9	10
July	55.4	7	3	6.1	4.3	W	2.8	10	10
Aug.	59.7	7	6	6.4	4.1	E	3.3	12	10
Sept.	52.1	9	4	5.1	4.6	E	3.6	10	12
Oct.	64.6	7	4	4.9	4.6	E	3.9	9	11
Nov.	73.4	6	4	3.6	4.2	E	3.6	10	9
Dec.	88.7	6	6	1.9	4.1	E	3.3	11	10
Annual	784.6	89	55	59.5	4.2	E	3.3	127	116

TABLE LIII

CLIMATIC TABLE FOR MONTEVIDEO (OBSERVATORIO DEL PRADO), 1901–1950
Latitude 34°52'S, longitude 56°12'W, elevation 22 m

Month	Mean sea level press. (mbar)	Temperature (°C)				Mean vapor press. (mbar)	Precipitation (mm)		Relat. humid. (%)
		daily mean	daily range	extreme			mean	max. in 24 h	
				max.	min.				
Jan.	1011.1	22.5	11.7	42.8	7.6	17.9	76.6	104	66
Feb.	1012.1	22.2	11.2	39.5	6.8	18.1	73.4	91	68
Mar.	1013.5	20.3	10.5	38.0	4.6	17.6	99.0	135	74
Apr.	1015.5	17.0	10.1	36.7	2.0	14.7	102.8	181	76
May	1015.9	13.7	9.3	32.0	−2.0	12.9	95.2	151	82
June	1016.7	10.9	8.4	27.4	−4.2	11.3	95.4	157	87
July	1018.2	10.5	8.3	28.5	−5.0	10.7	66.5	108	84
Aug.	1017.6	11.1	9.0	30.8	−3.8	10.5	84.7	142	80
Sept.	1017.1	12.8	9.4	32.0	−1.8	11.5	89.3	79	78
Oct.	1015.9	15.1	10.0	35.8	−1.4	13.2	70.2	86	77
Nov.	1012.8	18.3	11.2	37.4	2.5	14.7	78.1	90	70
Dec.	1011.2	21.0	11.6	38.7	5.0	16.1	80.4	95	65
Annual	1014.8	16.3	10.1	42.8	−5.0	14.1	1,011.7	181	76

Month	Mean evap. (mm)	Number of days with			Mean cloud-iness (tenths)	Mean daily sun-shine (h)	Wind		Clear days	Cloudy days
		precip. (≥1mm)	thunder-storm	fog			preval. direct.	mean speed (m/sec)		
Jan.	139.5	7	5.6	0.6	3.6	10.4	ESE	4.7	13	7
Feb.	112.6	7	4.1	0.7	3.5	9.7	NNE	4.4	12	6
Mar.	95.4	9	4.3	1.1	4.0	8.4	NE	4.2	10	9
Apr.	73.2	8	3.3	2.2	4.0	7.3	NE	3.9	11	9
May	57.2	9	3.2	5.1	4.6	5.8	NE	3.9	7	12
June	43.4	9	2.8	7.3	4.9	4.9	NNE	3.9	7	13
July	50.7	8	2.9	5.1	4.8	5.1	NE	4.2	8	14
Aug.	62.5	8	3.2	4.2	4.6	6.1	NNE	4.7	9	12
Sept.	70.4	8	2.9	3.0	4.6	7.1	NE	4.7	8	12
Oct.	88.3	8	3.2	1.3	4.3	8.1	NE	4.7	9	11
Nov.	108.9	9	4.2	0.7	3.8	9.8	NE	4.7	10	8
Dec.	137.0	8	5.4	0.4	3.7	10.3	NE	4.7	12	8
Annual	1,039.1	98	45.0	31.7	4.2	7.7	NE	4.4	116	121

TABLE LIV

CLIMATIC TABLE FOR PUNTA DEL ESTE, 1940–1960
Latitude 34°58′S, longitude 54°57′W, elevation 16 m

Month	Mean sea level press. (mbar)	Temperature (°C)				Mean vapor press. (mbar)	Precipitation (mm)		Relat. humid. (%)
		daily mean	daily range	extreme			mean	max. in 24 h	
				max.	min.				
Jan.	1011.4	21.5	7.1	36.7	10.6	19.1	78.0		75
Feb.	1012.5	21.6	7.0	36.4	11.4	19.6	74.1		76
Mar.	1013.7	20.5	6.3	35.7	10.2	18.9	87.2		78
Apr.	1016.1	17.6	5.9	34.0	6.2	15.9	83.6		79
May	1016.7	14.8	5.9	29.2	2.4	14.3	81.8		85
June	1017.4	12.4	5.2	26.6	1.8	12.1	86.3		84
July	1018.1	11.4	5.7	25.8	0.4	11.2	64.9		83
Aug.	1018.0	11.5	6.0	28.2	0.2	11.1	81.7		82
Sept.	1017.6	12.7	5.8	20.0	3.8	12.0	88.0		82
Oct.	1016.2	14.7	6.2	30.0	5.3	13.5	85.8		81
Nov.	1013.4	17.2	6.8	33.6	6.3	15.2	69.2		77
Dec.	1011.9	19.8	7.4	36.5	8.3	17.1	56.6		74
Annual	1015.0	16.3	6.3	36.7	0.2	14.9	937.2	193	80

Month	Mean evap. (mm)	Number of days with			Mean cloud- iness (tenths)	Wind		Clear days	Cloudy days
		precip. (≥1mm)	thunder- storm	fog		preval. direct.	mean speed (m/sec)		
Jan.	107.8	7	2.6	0.6	4.2	NE	6.7	10	10
Feb.	98.1	7	2.4	0.5	4.0	ENE	6.7	11	8
Mar.	97.5	9	2.4	0.5	4.3	NE	6.9	10	11
Apr.	82.8	9	1.7	1.0	4.4	NNE	7.2	9	11
May	72.1	9	2.4	2.3	5.0	N	6.9	7	14
June	53.8	10	2.2	3.3	5.3	N	6.7	6	14
July	58.3	10	2.5	3.3	5.2	N	6.9	6	15
Aug.	60.3	8	2.4	4.0	5.0	NE	7.2	7	14
Sept.	58.9	9	2.3	3.5	5.0	ENE	7.2	7	14
Oct.	68.4	9	2.2	3.3	4.8	ENE	6.9	7	13
Nov.	79.7	8	2.1	2.3	4.3	E	6.9	9	9
Dec.	104.0	7	2.0	1.1	4.3	ENE	6.9	10	10
Annual	941.7	102	27.1	25.7	4.6	NE	6.9	99	143

Chapter 3

The Climate of Chile

A. MILLER

Introduction

Almost every aspect of Chile's existence, including its climate, is dominated by the massive Andean Cordillera, which occupies at least a third of the country along its eastern edge, and the vast Pacific Ocean to the west. These two geographic features rather effectively isolate the area, and the effect of the rest of the South American continent on Chile's climates is rather small. The southern half of the country is almost always bathed by the persistent westerlies of the Southern Hemisphere while the northern half, protected from easterly flow by a range that has few passages lower than 4,000 m, lies almost continuously in the slow-moving, stable air along the eastern periphery of the Pacific anticyclone.

The most striking variation within this long, narrow ribbon of land, both along its length of over 4,200 km and its average breadth of only 175 km, is that of precipitation (Fig.12 of Chapter 2)[1] (JEFFERSON, 1921; KNOCH, 1930; ALMEYDA and SAEZ, 1958; METEOROLOGICAL OFFICE OF GREAT BRITAIN, 1958; OFICINA METEOROLÓGICA DE CHILE, 1965). The northern third is extremely dry—possibly the driest in the world—while the southern third is excessively wet—probably by far the wettest extratropical region in the world. Between these two extremes, the north–south transition is pronounced: between 30° and 40°S, the average annual precipitation increases from less than 150 mm to more than 3,000 mm. The changes from west to east are, at some places, even more brusque. In the southern half of the country there are areas where the average annual precipitation changes by several meters in distances of only a few tens of kilometers.

In contrast to the pronounced variations of precipitation, the air temperature is remarkably uniform. Nowhere, even in the mountains, does the annual range of the monthly mean temperature exceed 15°C and over most of the country it is less than 10°C. The north-to-south temperature change is quite small, the mean annual temperature at sea level ranging from about 5°C in the extreme south to 19°C in the extreme north.

The rugged topography of the country is so important in determining the climatic distribution that the discussion to follow here can only be of the most general character and one must expect rather large departures at some places from the overall descriptions of each area. However, despite the latitudinal and longitudinal variations of climate, an appropriate description of the weather of almost the entire country is that it is "stable". Thunderstorms and other manifestations of intense convective activity are comparatively rare

---

[1] This map was produced by the joint efforts of A. Miller and F. Prohaska.

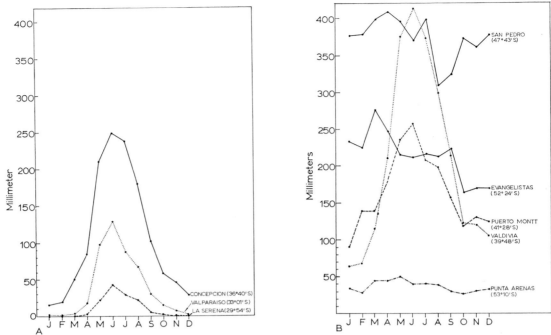

Fig.1. Mean monthly precipitation at selected stations.

over most of the country; this is as true in the wet south as it is in the arid north. Clouds everywhere tend to be predominantly of the stratiform types, even in places recording several thousand millimeters of precipitation per year. Those thunderstorms that do occur—and there are few observing sites that report as many as eight per year—are found principally over the extreme northern highlands in summer and over the mountain peaks of central Chile during the spring and summer (KNOCH, 1930; KNOCHE, 1933). Tornadoes have never been reported.

The persistence of large-scale atmospheric patterns throughout Chile is another significant characteristic (WALTER and LIETH, 1960; TREWARTHA, 1961). The latitudinal distribution of wind can be inferred from the north–south pattern of mean sea level pressure, shown in Fig.2. The highest pressure values (monthly averages) occur between 31° and

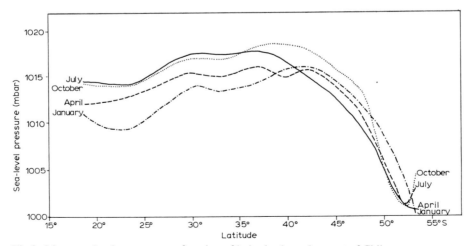

Fig.2. Mean sea level pressure as a function of latitude along the coast of Chile.

42°S, located near the southern end of this zone in summer and in the northern part in winter. North of about 31°S throughout the year, the Pacific anticyclone at low levels of the atmosphere effectively eliminates all traces of the westerlies which prevail in the upper layers. In the northernmost third of the country, at levels below 500–1,000 m, generally light southerly flow courses along the entire coast.

South of this northernmost third of the country, the westerlies play an increasingly important role in the general flow. Frontal systems from the west rarely penetrate to latitudes north of 31°S and only occasionally, during midwinter, do they reach as far north as 27°S. In the middle zone (31°–42°S), there is both a marked seasonal and (in the winter) day-to-day variation in the winds and other weather elements. Winds alternate between north-northwest and southwest, as fronts pass. The frequency of northerly winds increases with increasing latitude. In the extreme north of this central sector, southwest winds predominate during every month of the year but in the extreme south of the zone, northerly flow predominates during more than half of the year.

The southern third of the country remains under the influence of the prevailing westerlies throughout the year (LAMB, 1959). The wind prevails from the quadrant north to west, and it hardly ever lets up: except for protected interior valleys, calms seldom occur. The north–south pressure gradient, and the westerlies, increase sharply to a maximum near 50°S (TALJAARD et al., 1969).

The three latitudinal belts defined above in terms of large-scale wind flow also correspond to the annual precipitation pattern and its seasonal distribution (see Fig.1 and 12 of Chapter 2). Most of the sector north of 31°S is arid; except for a small highland area in the extreme northeast, the precipitation everywhere is less than 250 mm. Some 70% of the northern sector, mostly in the latitudes north of 27°S, receives less than 50 mm per year. The annual precipitation in the 31°–42°S belt increases rapidly toward the south to almost 2,000 mm at low elevations and much more over the cordillera. The increasing amounts of rainfall are directly related to the yearly frequency of frontal systems which also increases with latitude. At the northern end of central Chile, frontal passages never occur during the summer and more than 80% of the rainfall occurs during the four winter months of May to August. The summers are almost completely dry. But these dry summers disappear rapidly south of about 36°S and at the southern end of this central zone, there is not a single month that can be considered really dry.

Cyclones in the westerlies traverse the southern third of the country with great regularity throughout the year (ARNETT, 1958), so that no distinct dry season is possible. Frequent rain and heavy cloudiness occur at every place that is exposed to the winds from the Pacific. Extraordinary amounts of rainfall—up to 7,500 mm per year—have been recorded along the Pacific coastline near the center of this southern zone. Many places have an average of more than 300 days of precipitation per year and a mean annual cloudiness of more than 6 oktas.

An important feature of the northern two-thirds of the Chilean coastline (and all of the Peruvian coastline) is the persistent stratus and fog that extend hundreds of kilometers out over the ocean. North of Copiapó (27.3°S), southerly flow along the eastern periphery of the Pacific anticyclone is a continuous, year-round phenomenon (TREWARTHA, 1961). However, during the summer, the anticyclone is displaced several degrees of latitude southward of its winter position and southerly flow is found along the coast as far south as Puerto Montt (42°S) in the summer. Between the latitudes of 31° and 42°S, the frequency of southerly flow steadily increases toward the north. Where the relatively weak southerly flow dominates along the coast, perturbations in the circulation are rare and the cool, moist "marine" air is confined to a shallow layer above the surface. The mean vertical temperature and humidity soundings of Fig.3 illustrate the boundary between the lower marine air and the dry, warmer air aloft, and show the variation of this boundary with latitude. The average height of the base of the temperature inversion is about 900 m at the northern end of the country and about 500 m at Quintero (33°S) (see Fig.2).

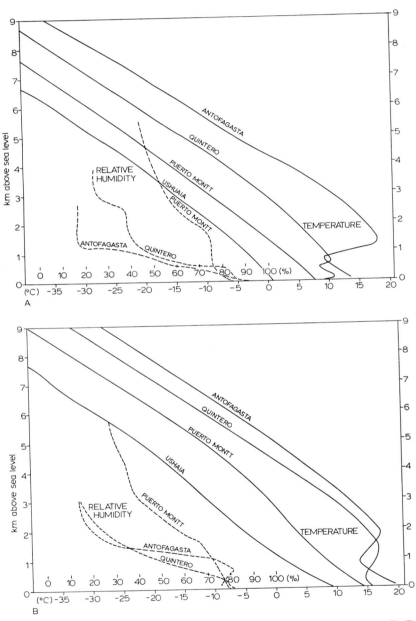

Fig.3. Mean temperature and humidity soundings. A. June, July, August; B. December, January, February. (Two years, Dec. 1957 to Aug. 1959.)

A factor that plays an important role in reducing the north–south gradient of air temperature along the northern two-thirds of the coast as well as enhancing the vertical stability is the existence of the Peru ocean current (SCHOTT, 1932; SCHWEIGER, 1959). Over the northern third of the Chilean coast, where southerly flow persists throughout the year, the mean annual temperature is 4° or 5°C cooler than it would be otherwise while over the central third, where southerlies do not remain throughout the year, the cooling by the current is only 1° or 2°C and is most evident in the summer (ANONYMOUS, 1959).

The specific direction and intensity of the southerly flow along the eastern edge of the Pacific anticyclone varies considerably along the coast. At its southern extremity, the low-level winds prevail from the southwest or west, but north of 24°S, the large-scale pressure gradient near the coast is weak, with a slight decrease in the pressure northward, giving rise to a small easterly component of the prevailing wind and a marked sea–land breeze.

The strong west to east variations in climate are due, of course, to the topography, which can be described in a highly generalized manner, as follows: north of about 27°S, the Andes are very wide, consisting of two or more parallel ranges; the western one coincides with the eastern border of Chile while the other range lies well within Argentina and Bolivia. The Andes in this sector are a formidable barrier, with few passages lower than 4,000 m and several peaks higher than 6,000 m. South of 27°S, the Andes become a single, fairly narrow range which lies approximately along the Argentine–Chilean border. South of about 34°S, the height of the Andes begins to diminish steadily from its average of about 5,000 m at the northern end to about 2,000 m in the vicinity of the Straits of Magellan. In addition to the principal Andean Cordillera, there is a lower range of mountains along much of the Chilean coast. North of Valparaiso (33.0°S), this coastal range, with an elevation between 1,500 and 2,000 m, presents a rather solid front right at the sea's edge with only a few breaks leading to the interior. South of Valparaíso, the coastal range becomes more and more broken by transverse valleys until, south of Chiloé, it appears only as a series of islands. Between the two ranges, a longitudinal valley generally exists, but its width and definition vary considerably.

The arid north (17°–31°S)

The Copiapó river (27°S) marks the southern limit of the extremely arid desert that dominates most of the region west of the Andes from the Ecuador–Peru border southward. The driest portion of the Chilean and Peruvian deserts lies in northern Chile. Not only does the polar front rarely penetrate north of the Copiapó river but the zone is also generally out of the reach of the moist, high-level flow that sometimes crosses the broad, high Andes from the east and northeast, bringing occasional high-level showers and thunderstorms over the Altiplano and western Andean slopes of Bolivia and Peru during the summer.

South of the Copiapó river, the annual precipitation begins to increase, under the influence of occasional winter incursions of the polar front. From about 25 mm per year in the town of Copiapó (27°S), the average annual amount increases southward to about 200 mm at 31°S. Although this southern sector must still be called arid, there is enough rain to support some grass and shrubs, in contrast to its northern extension, where essentially nothing exists without irrigation.

The precipitation pattern of the northern third of the country has been classified in a schematic fashion in Fig.4 in accord with the amount, source, and seasonal distribution (ALMEYDA, 1948; ANONYMOUS, 1958; OFICINA METEOROLÓGICA DE CHILE, 1965). North of about 27°S, where winter rainfall is rare, three distinct longitudinal belts can be identified: (*a*) the coast and adjacent hills and gorges that come under the constant influence of moist Pacific air; (*b*) the highlands above 2,500 m; and (*c*) the barren land between these two. The coastal range between 17° and 23°S has an average elevation of

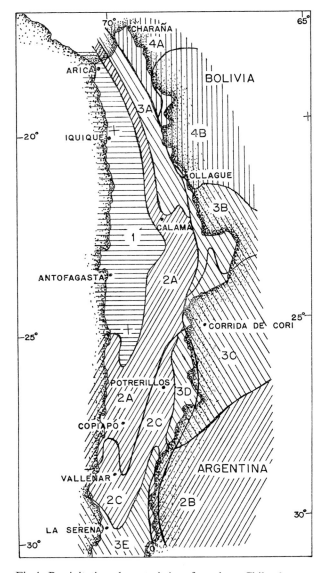

Fig.4. Precipitation characteristics of northern Chile. *1* = extremely arid; *2A* = arid, very rare winter rain; *2B* = arid, rare summer rain; *2C* = arid, occasional light winter rain; *3A* = semi-arid, occasional summer showers; *3B* = semi-arid, summer rain showers, occasional winter snow and rain (Altiplano); *3C* = semi-arid to arid, light summer rain, occasional winter snow; *3D* = semi-arid, light winter rain or snow; *3E* = semi-arid, becoming sub-humid extreme south, winter rain; *4A* = sub-humid, summer rain, occasional winter rain or snow (Altiplano); *4B* = sub-humid to semi-arid, summer rain, occasional winter rain or snow (Altiplano).

1,500 m; however, it drops sharply into the ocean, and has no wide transverse valleys, so that the marine air, which generally has a depth of about 900 m along the coast, is confined to the coastline, penetrating inland only through a few narrow gorges. Between the coastal range and the high cordillera that borders the Bolivian Altiplano, lies a narrow, irregular plain known as the Tamarugal Pampa. This area begins in the extreme north on the coast at Arica but lies inland at an elevation of between 1,000 and 1,500 m over most of the rest of its course. South of 23°S, this north-to-south central mesa tends to disappear and the coastal range and high cordillera are not clearly separated.

Coastal conditions are represented by the climatological data of Arica, Iquique, Antofagasta, and La Serena (Appendix, Tables II, III, IV, and VI; OFICINA METEOROLÓGICA DE CHILE, 1946–1964). Within the marine layer of this coast, the weather conditions are monotonously uniform. The annual range of mean monthly temperatures is only $7\frac{1}{2}$°C and the mean diurnal range only 6°C. The north–south temperature gradient is always very slight, the average difference between Arica and La Serena being only 4°C.

The winds along the coast between 24° and 31°S are almost invariably between south and west, but north of 24° they become quite variable due to the generally weak pressure gradient. Five to ten kilometers offshore the flow is from the south or southeast throughout the zone. Along the shoreline, a pronounced sea breeze produces westerly components. In the sector north of 24°S, calm or a light easterly (land breeze) wind occurs quite frequently.

The precipitation along the entire coast north of the Copiapó river is very scanty. During the past 30 years, there were only about a half-dozen days when more than 1 mm of rain fell at any of the coastal stations from Antofagasta northward. At Arica, for example, the 10 mm that occurred during one day in January, 1918, accounts for almost a third of the total precipitation during the past half century. If these rare cases were deleted from the records, the averages shown for Arica, Iquique, and Antofagasta would be reduced to half the amounts shown. In general, even north of the Copiapó river, the amount of rainfall increases slightly toward the south, reflecting the increased frequency of the rare incursions of the polar front (SCHWEIGGER, 1959; TREWARTHA, 1961).

The only reliable source of precipitation along the northern coast is the drizzle that sometimes falls from the stratus clouds. However, even this is not nearly as frequent nor as heavy as it is farther north along the central Peruvian coast or farther south in central Chile. The base of the stratus deck here is normally quite high (between 500 and 800 m M.S.L.) and usually less than 250 m thick (unlike the coastal stratus of central Peru and central Chile, where the base is normally less than 400 or 500 m above sea level). The sea level air in this sector is actually somewhat drier than it is along most of the west coast of South America. The relative humidity of the air in the sector north of 27°S is usually not more than 80%, even in the early morning, and fog is relatively infrequent. (At Antofagasta, Iquique, and Arica there are fewer than two days of fog per year.) The incidence of coastal fog is actually highest in central Chile, between 30° and 40° latitude.

In the summer, stratus occurs along the coast in the sector north of 27° on only about half the days and when it does occur, it normally breaks during the early morning, leaving the remainder of the day clear. Although stratus is almost a daily occurrence during the winter months, it usually becomes thin and frequently breaks for a few hours in the afternoon. Where the stratus is intercepted by the coastal mountains, "fog drip" is a fairly significant source of moisture, judging from the scattered seasonal vegetation that appears at these altitudes (ALMEYDA, 1957; SAA and VALDEZ, 1963). Between 24° and 27°S, this source of moisture is sufficient to support some desert shrubs and cacti. Even fog drip is considerably reduced during the summer, when the coastal stratus is thin.

The interior valleys and hills of the zone north of 27°S, protected from the coast from even the minor benefits of fog drip and drizzle, is the most barren area imaginable. Except in a few small natural or man-made oases, the area is completely devoid of life. Cultivation of any kind can only be carried on by irrigation. Evaporation is intense. In those few places where the desert has been reclaimed by irrigation, the salt had to be removed from the top soil through repeated floodings. Those few places that are under cultivation do produce very good citrus fruit.

Although the interior valleys and Andean slopes below 2,500 m north of Copiapó receive absolutely no precipitation for many years at a time, once in a great while, heavy showers and thunderstorms over the high Andes will produce a flash flood. However, except for the damage they do to the salitre pampas, their effect is negligible. The skies over this area are almost invariably clear, with an occasional thin layer of high cirrus. The dry air leads to a high diurnal range of temperatures (about 20°C in summer, 30°C in winter) and a very high potential evaporation. Even at altitudes of less than 1,000 m, the temperature often reaches the freezing point during the winter and highs of 30°–32°C in the summer. The winds in the interior zone tend to be erratic: with rather weak large-scale flow, the surface wind directions are determined largely by the local surface conditions. Dust-laden winds sometimes sweep across the valleys to the east of the coastal range on warm afternoons. From personal observation, there appears to exist a well-defined mountain–valley circulation along the slopes of the cordillera, the air generally moving from the southwest during the day and from the northeast during the night.

The climates of the northern cordillera are determined by the two large-scale circulation regimes[1] that affect it:

(*a*) During the winter, the flow in the free atmosphere is from the northwest between 2 and 5 km, shifting to west above 5 km. The only precipitation occurring in winter is that accompanying trough passages in the westerlies that occasionally (once or twice each winter) cross even the extreme northern part of this region, bringing snow to the cordillera (above about 3,700 m) and the Bolivian Altiplano (MILLER, 1954).

(*b*) During the summer, northwest winds persist to very high elevations but these are sometimes supplanted by northeast winds that bring in moisture at high levels from the Amazon basin. On these occasions, showers and thunderstorms may reach the western slopes of the Andes to elevations as low as 2,500 m. Although the rainfall amounts are usually not great, lightning and thunder accompanying these storms are frequent and intense. The summer rainfall of the extreme northeast corner of the country (Chapter 2, Fig.12) is the result of these showers. As much as 300 mm of precipitation occurs along the northwestern edge of the Bolivian Altiplano (85% during the summer months of December to March) due to this cause. However, these summer showers are a highly uncertain source of moisture for the area, since the intensity and frequency at any point varies greatly from year to year. There is evidence that an anticyclonic circulation

---

[1] There is some evidence (PROHASKA, 1962) that a monsoon type of circulation exists over the Altiplano and along the slopes of the Andes, superimposed on the general flow described here. The evidence consists of a seasonal analysis of the surface winds at two high-level stations in Argentina. There is not as yet any similar study available of high altitude Chilean and Bolivian stations.

develops intermittently over the Altiplano in summer (SCHWERDTFEGER, 1961), with northwest winds at high levels over the western slopes (northern Chile) and easterlies over the eastern slopes (Bolivia and Argentina); the appearance of these highs is strongly correlated with showery weather over the cordillera (MILLER, 1954). SCHWERDTFEGER (1965) attributes the creation of this high pressure system over the subtropical Andes to the release of latent heat during summer showers.

Although the annual temperature range over the high cordillera is small, the diurnal range can be very high in areas protected from the winds. For example, at Ollague (3,700 m), the annual range of the monthly mean is only 8°C but the diurnal range averages more than 20°C and is sometimes as much as 35°C. In the clear air of the high cordillera and Altiplano, the net fluxes of both incoming and outgoing radiation are very great.

Between 21° and 27°S even the high western range of the Andes and much of the adjacent Altiplano are very dry with few places receiving as much as 100 mm per year. Potrerillos (26°30′S 69°27′W, 2,850 m), near the southern end of this sector, gets only about 56 mm/year, almost all of which falls during an average of only one day per month. The precipitation is too scanty to support more than grazing by sheep, llama, and alpaca. However, over the easternmost portion of the Altiplano and Andean Range in Argentina, rainfall amounts increase considerably.

Considering the aridity of even the high Andes bordering Chile between 21° and 27°S, it is little wonder that the few rivers in this zone are hardly more than ravines that occasionally carry some water. Only the Loa river in northern Chile has a fairly constant flow of water throughout the year and even it contains a great deal of salt when it reaches lower elevations. Only along these river beds and where fog drip occurs along the coastal hills is there any natural vegetation below 2,500 m.

South of 27°S, troughs in the westerlies become more common in winter and precipitation in the highlands increases rapidly toward the south, from barely 50 mm to about 300 mm at 31°S. The winds in the highlands, especially at exposed points, are quite strong from the west throughout the year. (Note the winds at Corrida de Cori, Table V, p.135.)

To the west of the Andes in the 27°–31°S sector, desert conditions continue to prevail, although they are somewhat less severe than farther north. Here there is enough rain to support some grass and shrubs but what agriculture exists can only be carried on with irrigation. (Alfalfa is the principal crop.) La Serena (Table VI) is typical of the coastal conditions in this southern sector. Almost 90% of the rainfall occurs during the four winter months of May to August, associated with the occasional fronts that penetrate the area. Each month's total precipitation usually comes from a single frontal passage.

The first suggestion of the rainshadow effect to the east of the coastal range that persists throughout central Chile, appears in the vicinity of Copiapó. Here, the wet fogs or "camanchacas", as they are known in Chile, begin to have a marked effect on the coastal vegetation. Although they deposit little water, they occur so frequently (note the marked increase in fog frequency at La Serena) that they are able to support a notable amount of vegetation, including a few trees in certain areas (ALMEYDA, 1957). Also, in the 27°–31°S sector, there are more breaks in the coastal range, permitting the fogs to penetrate the valleys having an elevation of less than 800 m.

Central Chile (31°–42°S)

The steady increase in both the frequency and intensity of the westerlies with increasing latitude in the transition zone of central Chile is reflected both in heavier rainfalls toward the south and in wetter summer months. At the northern end of this zone, where only 250 mm fall per year, nine months of the year can be considered as dry: less than 1 % of the annual total is recorded during the December–February period. But at Puerto Montt, for example, which is at the southern end of central Chile, the annual precipitation is almost 2,000 mm and there is no really dry month: the three summer months account for nearly 18 % of the annual total.

The transition from dry to wet summers is quite marked. Almost everywhere north of the Bio Bio river (37°S), less than 5 % of the annual rainfall and fewer than ten days of precipitation occur during the three summer months. South of the Bio Bio river both the amount and the frequency of summer rain increase sharply toward the south: at Puerto Montt there are 40 days of rain during the December–February period. The disappearance of the dry season south of the Bio Bio is largely responsible for the increased annual amounts in the southern half of the zone. Thus, for example, the winter rainfall at Concepción is almost equal to that at Puerto Montt but the latter receives almost 700 mm more during the rest of the year than does Concepción.

The rather brusque climatic change in the vicinity of the Bio Bio river can also be seen from the types of natural and cultivated vegetation in the northern and southern halves of this zone. In the north are found mediterranean scrub forests, extensive use of irrigation in agriculture, and such fruits[1] as citrus, avocado, olives, and grapes, while to the south of the river, there are large evergreen forests and cultivation is primarily of wheat, oats, potatoes, and apples. The barrenness of the north begins to yield to the mediterranean vegetation typical of the northern half of central Chile at about the latitude of 31°S. At first, the patches of evergreen trees and bushes appear principally near the coast, but south of the Aconcagua river (33°S) they extend from the coast to the slopes of the Andes. In the sector between 31° and 33°S, where there is no interior valley, much of the hills that join the Andes with the coastal range is quite barren except for the few isolated shrubs and trees. During short periods following the irregular winter rains, grass will often appear over wide areas of previously bare soil, apparently germinated from seed on the ground.

South of the Aconcagua river, the clumps of trees and bushes become noticeably more dense. The Central Valley of Chile, located to the east of the coastal range—the agricultural heartland of the country—starts at this latitude. At its northern extremity, the Central Valley has an altitude of about 500 m, but it lowers to about 200 m south of Curicó (35°S) and finally disappears into the Gulf of Ancud to the south of Puerto Montt. South of Temuco (38°S) in the valley, beautiful lakes, fed by the high Andean rainfall, appear along its eastern edge, culminating in the largest, Lake Llanquihue, a short distance north of Puerto Montt. It is in the Central Valley and in the east–west river valleys between the Aconcagua and Bio Bio rivers that most of the fine fruits of Chile are produced. Almost all of Chile's famed vineyards are scattered among these valleys; most

---

[1] Fruit cultivation does not represent the major use of the land but it is indicative of the climate. Most of the land is devoted to cattle raising which includes growing feed such as alfalfa, both with and without irrigation.

of them are irrigated, except in the immediate vicinity of the Bio Bio river where late spring rains are adequate.

There are two reasons for the abrupt change in climate south of the Bio Bio river. First of all, it is at this latitude where the westerlies near the surface generally split, with part of the flow streaming northward around the Pacific anticyclone and part southeastward. Second, the coastal range becomes fragmented south of the Bio Bio. The coastal mountains begin to diminish both in altitude and continuity south of the Aconcagua river. Between Valparaíso and Constitución, the range is low and rolling with several broad valleys cut by westward flowing rivers. Although a rather solid barrier exists along the coast between Constitución and Concepción (35°–37°S), which partially shields the low (less than 200 m) interior valley to its east between Talca and Los Angeles from ocean air, the coastal mountains to the south of 37°S are low and discontinuous. There are only two significant remnants between 37°S and the island of Chiloé (42°–43°S): one is located to the west of Osorno between 40° and 41°S and the other lies along the northwestern edge of Chiloé island.

The reduced effectiveness of the coastal range as a barrier to flow from the west is evident from both the precipitation and temperature distributions (see Chapter 2, Fig.6, 7,12).[1] The typical rainfall pattern in central Chile is that of a maximum along the coast, a minimum along the interior some 50 km to the east, followed by a strong maximum at high elevations of the Andes. This pattern is broken to the south of 36°S: where there are gaps in the coastal range, the well-defined coastal maximum and interior minimum disappear. In fact, the prominence of the coastal range can be determined from the rainfall pattern. For example, in the valley near Osorno (40°35'S) the annual rainfall is as little as 1,200 mm per year, which is between a half and a third of that falling over the coastal mountains, while a little farther north, at 39°S, the rainfall is almost uniform along a 50-km line inland of the coast.

The temperatures over the Central Valley are also largely determined by the effectiveness of the coastal range as a barrier at each latitude. Where the coastal range is clearly defined, as in the sector north of Concepción, the Central Valley retains a relatively warm mass of air during the summer (four to six degrees warmer than the coast) and a cool mass in winter (Chapter 2, Fig.6,7). These temperature differences between the coast and the interior weaken noticeably south of 37°S and disappear south of 40°S. The protection of the valley between Talca and Los Angeles by the unbroken coastal range between 35° and 37° from marine air gives this part of the valley a much more continental character than most other sections. Summer temperatures are slightly warmer and winter temperatures considerably cooler in this part of the valley than they are farther north at, for example, Santiago. The incidence of fog and stratus in this section is also considerably less than that found farther north. Even Concepción, which is on the coast but is protected from the southwest flow by the hills of the Lebu peninsula immediately to its south, has a much more continental climate than other stations such as Valparaíso in the north or Valdivia in the south. At Concepción the average diurnal range is about 3°C greater than elsewhere along the coast, the incidence of fog is a fourth or less of that at Valparaíso or Valdivia, and the frequency of frost (20 days per year) considerably higher than at either Valparaíso or Valdivia.

---

[1] Figures produced jointly by F. Prohaska and A. Miller.

Valparaíso is illustrative of the conditions found along unprotected coastal points and westward-facing slopes of coastal hills in the northern half of central Chile. Fog or low stratus clouds are an almost daily occurrence, with an especially high incidence in the autumn months of March–May. The top of the stratus or fog deck varies between 400 and 1,000 m, with the lower extreme prevailing in summer. Although the coastal hills in north-central Chile reach as high as 2,000 m, there are several broad cuts at much lower elevations in the sector between Valparaíso and Constitución (35°S). Several of these valleys, such as those of the Mapocho and Maule river beds, permit the marine air from the coast to stream into the Central Valley between 33° and 35°S. This is especially common in the winter, when the base of the temperature inversion is high. The tempering effect of the wintertime invasion of marine air permits cultivation in the Santiago–San Fernando area of such frost-sensitive crops as avocado, lemons, and oranges, despite their distance from the coast. The deep invasion of marine air in winter is illustrated by comparing the diurnal temperature ranges at Santiago and Valparaíso. (Santiago is 80 km inland of the coast.) During June and July, the diurnal temperature range at Santiago is 11°C, about 4°C more than Valparaíso's, but during December and January, Santiago's diurnal range is 17°C, which is 8°C greater than that of Valparaíso. The absence of the marine air in this sector during summer is also evident from the almost complete absence of fog and stratus in the summer and the low relative humidity, which averages about 25% less than during the winter.

The prevailing wind in the Central Valley and along the coast north of the Bío Bío is the southwest wind throughout the year, although northwest winds also occur quite frequently during the winter. South of the Bío Bío, both along the coast and in the Central Valley, the prevailing winds in the summer come from the southwest or south and, in the winter, from the northwest. Directions vary between northwest and southwest to a far greater extent in the southern half of the zone than they do in the northern half. Wind speeds in the 31°–42°S transition zone depend largely on the direction of the wind and the location of the site. North or northwest winds are considerably stronger than those from the southwest. Along the coast and at other exposed locations, speeds as great as 25 m/sec have been observed at the northern end and greater than 35 m/sec at the southern end. Along the coast of the southern half of the zone, peak gusts of greater than 17 m/sec are observed on more than half the days of the year. Inland, at low elevations not exposed to the west, wind speeds are considerably less than along the coast; for example, the wind speed at Valparaíso is normally 40–50% higher than that at Santiago.

Over the Central Valley, a mountain breeze sometimes develops. On occasion, it may become a moderate foehn-type wind that damages crops in the valley. Known locally as "puelche", it brings in dry (as low as 15% relative humidity compared to the normal 70–80%), warm air at speeds up to 15–20 m/sec. When it occurs during the early spring, it produces extensive and rapid melting of the snow over the western slopes of the Andes. The only data available on evaporation in Chile are those for a 19-year period (1915–1933) at Frutillar (41°08'S 73°02'W, 149 m), which has an annual precipitation of 1,600 mm. The measured evaporation was less than 300 mm per year, ranging from a maximum of about 40 mm/month during January and February to a minimum of 16 mm/month during the months of April to August (OFICINA METEOROLÓGICA DE CHILE, 1965).

It is difficult to generalize about the Andes of this zone. The temperatures and precipitation at any place depend on the altitude and exposure to the west. At an exposed place like Cristo Redentor (Table VII), located in a pass to the south of the lofty Mount Aconcagua, the average diurnal temperature range (8°C) is not great. But in closed valleys within the cordillera the average diurnal temperature range may reach 20°C or more. The same criterion determines average seasonal temperature, precipitation, and winds.

Precipitation over the Andes in this zone in general increases with height, reaching a maximum close to the ridge line. The peak precipitation is usually of the order of three to four times the amount over the Central Valley at the same latitude. There is also a strong north–south gradient of precipitation amounts, especially between 30° and 35°S, where amounts increase from less than 300 mm to more than 3,000 mm per year. This increase southward is due primarily to increased frequency of precipitation both in winter and summer, toward the south. Unlike the lowlands to the west, even in periods of relatively quiet weather over the high Andes of this zone, during both winter and summer, convective-type clouds frequently build up in the late morning, produce snow or sleet showers in the early afternoon, and subside by evening. Much of the summer precipitation, especially north of 37°S where frontal activity is infrequent, comes from these cumuli. A large proportion of the summer precipitation at low elevations along the eastern slopes of the Andes results when these convective clouds drift eastward from the ridge line.

South of about 38°S over the Andes, there is little seasonal variation in the amount of precipitation. Much of the western slopes is covered by forests, with oak predominating. There is cloudiness almost every day of the year and rainfall occurs on at least one out of every two days during every month of the year. But both cloudiness and precipitation fall off rapidly on the lee side of the Andes everywhere south of 28°S. The west-to-east decrease on the lee side of the Andes is much greater than the increase on its western side. There are many points where the precipitation decreases by a factor of twenty times in a distance of less than 100 km east of the peaks. At the lower elevations (below 1,000 m) on the Argentine side, the slopes are quite arid.

Some idea of the intensity of the westerly winds that impinge on the Andes south of 27°S throughout the year can be seen from the data of Cristo Redentor (Table VII). However, the averages shown are hardly indicative of the frequency of strong winds that occur at this latitude and altitude. During the winter, speeds greater than 40 m/sec are common during storms and may exceed 65 m/sec. Many of the valleys and mountain passes become inaccessible during much of the winter because of the heavy snow that occurs along the western slopes, but remain open on the Argentine side. Clouds enshroud the high Andes during much of the winter and, south of 38°S, during much of the summer also.

Southern Chile (42°–57°S)

"The roaring forties" is an apt name for the latitudes south of 42°S, if it is understood that they extend, as well, throughout the "fifties". There is little respite at any time during the year from the buffeting that the southern Chilean coast receives from the strong westerlies (SCHWERDTFEGER, 1960, 1962). A light wind along this coast in either winter or summer is rare. Thus, for example, the average wind speed at Evangelistas (52°24′S

75°05'W, 58 m) during the year is 12 m/sec, with no month having less than 8 m/sec and every month recording maxima exceeding 30 m/sec. The average west component of the geostrophic wind between San Pedro (47°43'S 74°55'W) and Evangelistas, based on their annual mean sea level pressure difference (8.5 mbar) is 15 m/sec. The direction of the wind is also remarkably persistent along this coast. A westerly component occurs at least 75 % of the time along the entire coast. What wind variation does occur is that associated with the unending succession of cyclonic depressions that move on an eastward or southeastward course across the southern tip of the continent (ARNETT, 1958). At the northern end of the zone, the wind is generally much more nearly from the north (∼340°) while at the southern tip it is usually much more nearly from the due west (270°–290°).

With such intense westerlies blanketing the zone during the entire year, any superimposed weaker circulations, such as mountain-valley breezes, that may exist, are impossible to discern, even on the eastern side of the Andes. The last remnants of the coastal range of northern and central Chile is found on the island of Chiloé so that south of about 43°S only the Andean range remains. Its influence on the westerlies that it intercepts is by far the most important single factor to be considered in a climatic study of the zone. Although the Andes are not very high in this sector (generally less than 2,500 m, with a few volcanic peaks to 4,000 m) they are an extremely effective barrier; their height is fairly uniform, with no low passes and few valleys. They also dominate much of the land area of this southern end of the continent, since they extend all the way from the Pacific shore, where its edge is cut into a maze of islands having steep, rocky edges, to at least half way across the continent to the Atlantic Ocean.

The effect of the Andean Range in this zone is apparently twofold: air from the west at low levels is not only forced to rise abruptly near the ocean's edge but is turned southward as it approaches the coastline; the resulting convergence greatly augments the lift that the air receives at low levels. Only such a mechanism can account for the extremely high rainfall amounts recorded over the islands between 48° and 52°S at the edge of the Pacific Ocean at elevations of less than 100 m. Compare, for example, the annual rainfall of only about 2,000 mm at the west end of Tierra del Fuego (54°–56°S) to the more than 6,000 mm along the coast between 48° and 52°S (Chapter 2, Fig.12).

The persistence of the westerlies in this zone leads, of course, to an unchanging pattern of weather on the windward side of the Andes. The range of the monthly mean and the average daily range are both only about 6°C in the extreme north and 4°C or less in the extreme south. The north–south temperature gradient is small, with the mean annual sea level temperature varying from about 10°C in the north to about 7°C in the south. The average maximum temperature at San Pedro is only 11°C and at Evangelistas only 8°C; the highest maximum recorded in over 40 years of record at the latter station is only 16.6°C.

Other aspects of the climate are similar in their uniformity throughout the year. San Pedro (Table XIII), which is located near the center of the zone at the somewhat protected southern edge of the Golfo de Peñas, illustrates the conditions that persist along the lower western slopes of the Andes. Here it rains on more than 300 days every year and there is no appreciable letup throughout the year: there are between 24 and 28 days of rain in every month. Bahía Félix, at the western entrance to the Strait of Magellan, has had an average of 320 days of rain per year during the past forty years, while Evangelistas, a little to the northwest, has had an average of 300 days per year over a 40-year period.

On the average, there are no more than 6 rainless days in any month at any of these stations.

The representation of the precipitation pattern throughout this zone can at best be schematic, considering the low density of reporting stations. However, there can be little doubt that the heaviest precipitation occurs on the Pacific side of the Andes at elevations generally below 1,500 m in the extreme northern end of the zone and below 1,000 m over the rest of the area. On the island of Guarelo (50°21′S 75°21′W, 15 m) an average of 7,330 mm/year has been recorded during a 9-year period (interrupted record, 1949–1963) with a maximum of over 8,500 mm/year (Oficina Meteorológica de Chile, 1965). From available data (Almeyda and Saez, 1958; Oficina Meteorológica de Chile, 1965) it seems that precipitation decreases toward the east not only along the eastern slopes of the Andes but also along the western slopes. Unlike central Chile, south of about 46°S the peak precipitation appears to occur well below the ridge line (Fuenzalida, 1971), although Schwerdtfeger (1956) contends that the maximum falls over the crest. Lliboutry (1953) states that the precipitation over the Patagonian Ice Fields (~ 50°S) is between 3 and 5 m, which is almost certainly at least 1 m less than that near the coastline.

Above about 2,000 m in the extreme north of this zone and above 1,000 m in the extreme south, the usual precipitation type changes to snow. Because of the unvarying vertical temperature distribution, the level of snow is fairly constant at every latitude. At Evangelistas, which is at an altitude of 58 m and has a mean annual temperature of 6°C, snow occurs on only 20 of the 300 days of precipitation each year. Yet, above 1,000 m at the same latitude (~ 52°S) precipitation is almost always in the form of snow. (On the lee side, the snow line is somewhat higher and more variable.)

Glaciers straddle much of the Andean ridge line south of 42°S. The great Patagonian Ice Field (or, "Hielo Continental") that starts just south of Puerto Aisen (45½°S) stretches more than 500 km southward and reaches widths of 70–80 km. There are two more or less continuous ice fields: the northern one, which extends between 46° and 48°S, feeds the great Lake Buenos Aires; the southern one, which begins at 48½°S and reaches to at least 51°S, provides water for the lakes of San Martin, Viedma, Argentino, and Del Toro. South of these two there are several smaller fields, ending with the large one in Tierra del Fuego, just to the north of Canal Beagle (55°S). The average elevation of these glaciers is 1,500 m but they extend to 2,500 m or more and as low as 700 m (Herold, 1953; Lliboutry, 1953, 1956). At several places along the interior channels of the fjorded Pacific coast, glaciers reach sea level (for example, near Balmaceda at 51½°S). Even on the Argentine side, which is somewhat warmer, the firn line is at about 1,200 m and at places the glacier descends through narrow passes to 600 m. Lake Viedma (49½°S, 250 m) and Lake Argentino (50°S, 200 m), lie very close to the eastern edges of the snow field that feeds them. Because of the high incidence of overcast skies in this area throughout the year, melting of the snow is caused primarily by rain.

The most characteristic feature of the south Patagonian climate is the very strong and incessant wind, which often reaches 30 m/sec on the crests (Lliboutry, 1953). Precipitation falls off markedly from the crests of the Andes toward the pampas to the east: the 3–5 m annual precipitation amount over the ridges decreases generally to much less than 0.5 m within 70 or 80 km. Differences among seasons are hardly noticeable. The residents of the region say that, "all the seasons can be observed in a single day." Thunderstorms, hail, and other manifestations of strong convective activity, are practically unknown in this area.

There are no regular meteorological observations over the Patagonia glaciers but SCHWERDTFEGER (1956) has estimated, through extrapolation of San Pedro (47.7°S 74.9°W) and Evangelistas (52.4°S 75.1°W) surface data, the monthly temperatures at the mean level of the glaciers (2,000 m). He deduces that the mean monthly temperatures range from —2° to —9°C in the north (47½°S) and from —5° to —10°C in the south (51°S).

Most of the wet coastal zone and western mountain slopes is covered by dense forests, composed principally of several species of false beech and various types of conifers. A great variety of vines, shrubs, and mosses make some of the forests practically impenetrable. Thick carpets of moss cover many of the forest floors; the moss becomes especially heavy in the extreme south and most of the southern half of Tierra del Fuego is covered with it. Limited agriculture (potatoes, various cereals, apples) can be carried on in the north but south of 47°S, where rain occurs almost every day, and winds never stop blowing, agriculture is practically impossible except where there is some protection from the west, as, for instance, the area around Puerto Natales.

From the ridge line of the Andes, where the annual precipitation is generally about 3,000 mm, the rainfall decreases rapidly toward the east. Within 200 km of the ridge line, annual amounts decrease to less than 300 mm, even at elevations of more than 1,000 m. Along the slopes just to the east of the ridge line, below the level of perpetual snow there are forests and zones of mixed grassland and trees but these give way rapidly to nothing but fescue grass. The extensive, high plains immediately to the east of the Andean Ridge are generally quite arid. For example, in the vicinity of the large lakes of Buenos Aires (46½°S) and San Martin (49°S), which are fed by the melting snows of the cordillera, the rainfall is so light and evaporation so great that irrigation is required.

Much of the entire zone to the east of the Andean peaks is devoted to sheep grazing. It has excellent conditions for this purpose. At higher elevations, 1,000–1,500 m, just below the perpetual snow line, melting snow uncovers rich grassland during the summer on which herds fatten. The cool year-round temperatures permit some four million heads of sheep (1960) to prosper in Patagonia.

The temperature range along the eastern slopes is considerably higher than along the western slopes but by no means large. Even though it is dry, it is nevertheless dominated by air that originated over the Pacific Ocean. Only along the eastern slopes at the extreme southern end does very cold and dry Antarctic air invade once or twice each winter to replace the prevailing cool Pacific air that has been dried during its crossing of the Andes. Every aspect of the climate of southern Chile and Argentina is determined by the interaction of the westerlies with the Andean Range. The orographic effects are probably as "pure" and simple as can be found anywhere on earth. The range is relatively narrow and oriented perpendicular to the mean flow. The zone would undoubtedly make an excellent "laboratory" for the verification of theoretical models of air flow over mountains and resulting precipitation and cloudiness patterns. But, unfortunately, there are few surface observations and practically no measurements of the upper air. Until such observations become available, only the gross climatic features of this area can be inferred.

Appendix—Climatic tables (pp.131–145)

The tables are arranged from north to south. Mean daily temperatures were computed by adding temperature readings of 08h00, 19h00 (local time), maximum, and minimum and dividing by four.

Mean vapor pressure was computed from average relative humidity obtained by adding twice the relative humidity at 08h00, the relative humidity at 14h00, and the relative humidity at 19h00, and dividing by four.

Evaporation measurements are not available, except as noted in the text.

Cloudiness has been converted from tenths to oktas.

Radiation and sunshine hours (except for Santiago and Concepción) are not available.

Wind directions and speeds are very unreliable. Most are mere estimates by untrained observers. At other sites, anemometers and wind vanes have been moved in location and height over the years and many have rarely been serviced.

Records have been obtained from the Oficina Meteorológica de Chile in resume form. The author did not have access to the original data. For this reason, periods of summaries are not uniform and the records at many sites are discontinuous. The periods used in computing each element are shown in Table I.

References

ALMEYDA, E., 1948. *Pluviometría de las Zonas del Desierto y las Estepas Cálidas de Chile*. Editorial Universitaria, Santiago, 167 pp.

ALMEYDA, E., 1957. *Geografía Agrícola de Chile*. Escuela de Agronomía, Universidad de Chile, 213 pp.

ALMEYDA, E. and SAEZ, F., 1958. *Recopilación de Datos Climáticos de Chile*. Ministerio de Agricultura, Santiago, 195 pp.

ANONYMOUS, 1958. *Agricultura Tecnica*. Ministerio de Agricultura, Santiago, Vol.XVIII, No.2, 624 pp. (Summary of soils in Chile.)

ANONYMOUS, 1959. *Marine Climatic Atlas of the World, Vol.V. South Pacific Ocean*. U.S. Govt. Printing Office, Washington, D.C., 267 charts.

ARNETT, J. S., 1958. Principal tracks of Southern Hemisphere extratropical cyclones. *Mon. Weather Rev.*, 86: 41–44.

FUENZALIDA, H., 1971. *Climatología de Chile*. Departamento de Geofísica y Geodesia, Universidad de Chile, 73 pp.

HEROLD, L., 1953. El Hielo Continental Patagónico. *Rev. Geogr. Am., Buenos Aires*, 35: 153–161.

JEFFERSON, M., 1921. The rainfall of Chile. *Am. Geogr. Soc. Res. Ser.*, 7: 1–32.

KENDREW, W. G., 1961. *Climates of the Continents*. Oxford University Press, London, 608 pp.

KNOCH, K., 1930. Klimakunde von Südamerika. In: W. KÖPPEN and R. GEIGER (Editors), *Handbuch der Klimatologie*. Borntraeger, Berlin, Vol.2, Pt.G, 349 pp. (This volume contains an extensive bibliography on material printed before 1930.)

KNOCHE, W., 1933. *La Geografía de Chile y la Agricultura*. Editorial Nascimiento, Santiago, 39 pp.

LAMB, H. H., 1959. The southern westerlies: a preliminary survey; main characteristics and apparent associations. *Q. J. R. Meteorol. Soc.*, 85: 1–23.

LLIBOUTRY, L., 1953. More about advancing and retreating glaciers in Patagonia. *J. Glaciol., Cambridge*, Vol.2, No.13.

LLIBOUTRY, L., 1956. *Nieves y Glaciares de Chile*. Ediciones de la Universidad de Chile, Santiago, 471 pp.

METEOROLOGICAL OFFICE OF GREAT BRITAIN, 1958. *Tables of Temperature, Relative Humidity and Precipitation for the World*. H.M. Stationary Office, London, Part II, pp.19–22.

MILLER, A., 1954. *Bolivian Weather Forecasting*. M.S. Thesis, Pennsylvania State University, University Park, Pa., 82 pp.

OFICINA METEOROLÓGICA DE CHILE, 1946–1964. *Anuario Meteorológico de Chile*. Santiago. (Annual summaries, not complete for all years.)

OFICINA METEOROLÓGICA DE CHILE, 1965. *Pluviometría de Chile.* Santiago, Chile. (Tabulations by month and year of precipitation records of 137 stations in Chile.)

PROHASKA, F. J., 1962. Algunos aspectos del clima de la Alta Cordillera y de la Puna Argentina. *Bol. Estud. Geogr.*, VIII: 21–30.

SAA, G. E. and VALDEZ, M., 1963. *Captación de Agua de la Neblina.* Universidad del Norte, Instituto de Investigaciones Científicas, Antofagasta, 45 pp.

SCHMITHÜSEN, J., 1956. Die räumliche Ordnung der chilenischen Vegetation. *Bonner Geogr. Abhandl.*, 17: 1–86.

SCHOTT, G., 1932. The Humboldt Current in relation to land and sea conditions on the Peruvian coast. *Geography*, XVII: 87–98.

SCHWEIGGER, E., 1959. *Die Westküste Südamerikas im Bereich des Peru-Stroms.* Keyser Verlag, Heidelberg, 548 pp.

SCHWERDTFEGER, W., 1956. Determinación indirecta de las condiciones climáticas del hielo continental Patagónica. *An. Soc. Cient. Arg., Buenos Aires*, CLXI: 53–82.

SCHWERDTFEGER, W., 1960. The seasonal variation of the strength of the southern circumpolar vortex. *Mon. Weather Rev.*, 88(6): 203–208.

SCHWERDTFEGER, W., 1961. Strömungs- und Temperaturfeld der freien Atmosphäre über den Anden. *Meteorol. Rundsch.*, XIV(1): 1–6.

SCHWERDTFEGER, W., 1962. *Meteorología del Area del Pasaje Drake.* Servicio de Hidrografía Naval, Secretaria de Marina, Buenos Aires, 78 pp.

SCHWERDTFEGER, W., 1965. The role of latent and sensible heat for the development of a high pressure system over the subtropical Andes in the summer. *Meteorol. Rundsch.*, XVIII(3): 69–75.

TREWARTHA, G. T., 1961. *The Earth's Problem Climates.* The University of Wisconsin Press, Madison, Wisc., 334 pp.

VAN LOON, H., 1961. Charts of average 500-mb absolute topography and sea level pressures in the Southern Hemisphere in January, April, July and October. *Notos, Pretoria*, 10: 105–112.

WALTER, H. and LIETH, H., 1960. *Klimadiagramm—Weltatlas.* I. Lieferung. Fischer, Jena, unpaged.

Reference added in proof:

CAVIEDES, C., 1973. The climatic profile of the north Chilean desert at latitude 20°S. In: D. H. K. AMIRAN and A. W. WILSON (Editors), *Coastal Deserts, Their Natural and Human Environments.* University of Arizona Press, Tucson, Arizona, pp.115–121.

TABLE I

PERIODS OF RECORDS USED IN THE CLIMATIC TABLES

Table	Station	Pressure	Temp.	Humidity	Precip.[2]	Evap.	Precip.[3]	Thunder	Fog	Cloudiness	Sun	Wind
II	Arica	1912–13 1921–30 1932–50	1911–22 1922–50	1911–21 1929–50	1911–49	N.A.[1]	1911–21 1929–50	1911–21 1929–50	1911–21 1929–50	1911–21 1929–50	N.A.	1929–50
III	Iquique	1911–40	1911–28 1935–45	1911–45	1911–49	N.A.	1911–45	1911–45	1911–45	1911–45	N.A.	1929–45
IV	Antofagasta	1946–55	1946–55	1946–55	1944–55	N.A.	1944–55	1944–55	1944–55	1944–55	N.A.	1944–55
V	Corrida de Cori	1942–1960, with many breaks in record										
VI	La Serena	1911–50	1911–50	1911–50	1911–49	N.A.	1911–50	1922–50	1911–50	1911–50	N.A.	1930–50
VII	Cristo Redentor	1941–60	1934–60	1934–60	1934–60	N.A.		1951–60	1951–60	1934–60	N.A.	1941–60
VIII	Valparaíso	1899–50	1899–50	1899–50	1911–50	N.A.	1911–50	1911–50	1911–50	1911–50	N.A.	1911–50
IX	Santiago	1861–55	1861–55	1875–55	1860–50	N.A.	1937–55	1873–50	1861–50	1901–55	1930–50	1929–50
X	Concepción	1911–50	1911–50	1911–50	1860–50	N.A.	1911–50	1911–50	1911–50	1911–50	1945–53	1929–50
XI	Valdivia	1911–50	1911–50	1911–50	1911–49	N.A.	1911–50	1911–50	1911–50	1911–50	N.A.	1911–50
XII	Puerto Montt	1911–50	1911–50	1911–50	1911–59	N.A.	1911–50	1911–50	1911–50	1911–50	N.A.	1911–50
XIII	San Pedro	1932–50	1932–50	1932–50	1911–47	N.A.	1932–50	1932–50	1932–50	1932–50	N.A.	1932–50
XIV	Evangelistas	1911–50	1911–50	1911–50	1911–59	N.A.	1920–40	1911–40	1941–50	1911–50	N.A.	1911–50
XV	Punta Arenas	1911–50	1911–50	1911–50	1911–59	N.A.	1911–50	1911–50	1911–50	1911–50	N.A.	1929–52

[1] N.A. = not available.

[2] Monthly and annual averages.

[3] Maximum in 24 h.

TABLE II

Latitude 18°28′S, longitude 70°22W, elevation 29 m

Month	Mean sta. press. (mbar)	Temperature (°C)				Mean vap. press. (mbar)	Precipitation (mm)	
		daily mean	daily range	extreme			mean	max. in 24 h
				max.	min.			
Jan.	1008.1	22.1	8.3	31.5	11.0	19.2	0.3	10.0
Feb.	1008.0	22.3	8.4	30.8	10.1	20.7	tr.	0.4
Mar.	1008.1	21.3	8.7	29.5	11.5	19.7	tr.	tr.
Apr.	1009.2	19.6	8.0	30.0	10.2	17.1	0.0	0.0
May	1010.4	18.0	7.1	27.5	6.6	15.7	tr.	tr.
June	1011.4	16.7	6.3	26.5	6.8	14.6	0.1	1.0
July	1011.7	15.9	6.3	22.6	5.2	14.2	0.1	0.6
Aug.	1011.8	15.8	6.2	23.0	5.8	15.3	0.1	2.0
Sept.	1011.5	16.6	6.5	27.4	4.2	14.9	tr.	0.4
Oct.	1011.3	17.6	6.5	25.9	9.0	15.5	tr.	0.1
Nov.	1010.1	19.1	7.2	26.2	10.0	16.4	0.0	0.0
Dec.	1009.0	20.7	8.1	28.5	11.0	18.1	0.1	1.7
Annual	1010.1	18.8	7.3	31.5	4.2	16.9	0.7	10.0

Month	Number of days with				Mean cloud- iness (oktas)	Wind	
	precip.		thunder- storm	fog		preval. direct.	mean speed (m/sec)
	⩾tr.	⩾1mm					
Jan.	0.3	<0.1	0	0.0	4.2	SW	6
Feb.	0.2	0	0	0.0	3.8	SW	6
Mar.	0.1	0	0	<0.1	3.3	SW	4
Apr.	0.0	0	0	0.1	3.6	SW	4
May	0.1	0	0	0.1	4.6	SW	4
June	0.9	0	0	<0.1	5.2	SW	4
July	0.9	0	0	<0.1	5.4	SW	4
Aug.	1.4	<0.1	0	0.0	5.6	SW	4
Sept.	1.1	0	0	0.1	5.4	SW	4
Oct.	0.1	0	0	0.0	4.8	SW	4
Nov.	0.0	0	0	0.0	4.5	SW	4
Dec.	0.1	<0.1	0	0.0	3.9	SW	6
Annual	5.2	<0.1	0	0.4	4.6	SW	4

TABLE III

CLIMATIC TABLE FOR IQUIQUE
Latitude 20°22'S, longitude 70°11'W, elevation 9 m

Month	Mean sta. press. (mbar)	Temperature (°C)				Mean vap. press. (mbar)	Precipitation (mm)	
		daily mean	daily range	extreme			mean	max. in 24 h
				max.	min.			
Jan.	1011.8	20.9	7.8	30.5	13.0	19.5	0.1	2.3
Feb.	1012.0	20.8	8.2	30.8	12.0	19.4	tr.	0.8
Mar.	1012.0	19.7	8.0	31.3	9.0	19.0	0.0	0.0
Apr.	1013.0	18.3	7.4	28.6	8.4	17.0	tr.	0.3
May	1014.3	17.2	6.7	28.0	9.0	15.9	tr.	1.6
June	1015.2	16.0	5.6	22.6	9.0	14.5	0.2	4.6
July	1015.3	15.4	5.4	22.4	8.6	13.8	0.7	12.8
Aug.	1015.6	15.4	5.6	23.5	8.0	13.8	0.7	3.0
Sept.	1015.3	16.1	6.1	26.6	9.0	14.5	0.3	4.6
Oct.	1014.9	17.1	6.8	28.0	10.0	15.4	0.1	2.8
Nov.	1014.0	18.5	7.5	30.6	11.2	16.6	tr.	tr.
Dec.	1012.7	19.8	9.6	30.0	11.5	18.0	0.0	0.0
Annual	1013.8	17.9	7.0	31.3	8.0	16.4	2.1	12.8

Month	Number of days with			Mean cloudiness (oktas)	Wind	
	precip. ≥1mm	thunderstorm	fog		preval. direct	mean speed (m/sec)
Jan.	0.1	0.1	<0.1	2.6	S	2
Feb.	<0.1	0.0	0.0	2.2		
Mar.	0.0	0.0	0.0	2.0		
Apr.	0.1	<0.1	0.0	2.6		
May	0.4	0.0	0.0	3.7		
June	0.3	0.0	0.0	4.6		
July	1.1	<0.1	0.0	5.0		
Aug.	0.8	<0.1	0.1	5.0		
Sept.	1.1	0.0	0.0	5.0		
Oct.	0.7	0.0	<0.1	4.4		
Nov.	0.1	0.1	0.0-	3.5		
Dec.	0.0	0.0	0.0	2.7		
Annual	4.7	0.2	0.1	3.7	↓	↓

TABLE IV

CLIMATIC TABLE FOR ANTOFAGASTA
Latitude 23°26′S, longitude 70°28′W, elevation 119 m

Month	Mean sta. press. (mbar)	Temperature (°C)				Mean vap. press. (mbar)	Precipitation (mm)	
		daily mean	daily range	extreme			mean	max. in 24 h
				max.	min.			
Jan.	997.3	19.9	8.0	29.1	11.0	17.2	0.0	0.0
Feb.	999.2	20.1	8.5	30.1	8.4	17.6	0.0	0.0
Mar.	999.5	18.5	7.6	25.6	7.0	16.2	tr.	tr.
Apr.	1000.8	16.2	7.5	24.0	8.0	15.5	0.3	8.0
May	1001.8	15.1	7.1	24.2	5.3	13.2	0.1	2.8
June	1002.1	13.5	5.9	23.0	5.0	12.1	1.7	38.0
July	1002.0	13.1	6.7	22.0	3.2	11.0	2.4	17.0
Aug.	1002.6	13.5	6.4	21.0	3.0	11.8	1.2	28.0
Sept.	1003.3	14.5	6.4	24.0	6.0	12.5	1.0	6.8
Oct.	1002.2	15.2	6.5	22.0	6.5	12.8	0.8	9.7
Nov.	1001.0	16.8	6.5	24.0	8.2	13.7	0.2	4.0
Dec.	998.9	18.4	7.4	24.8	9.6	15.2	tr.	0.5
Annual	1001.7	16.2	7.1	30.1	3.0	14.0	7.7	38.0

Month	Number of days with				Mean cloud-iness (oktas)	Wind	
	precip.		thunder-storm	fog		preval. direct.	mean speed (m/sec)
	⩾tr.	⩾1mm					
Jan.	0.0	0.0	0	0	2.6	SW	4
Feb.	<0.1	0.0	0	0	2.1	SW	4
Mar.	0.3	0.0	0	0.6	2.3	SW	4
Apr.	0.1	0.1	0	0.1	2.8	SW	4
May	0.1	0.1	0.1	0.2	3.4	W	2
June	0.2	0.2	0	0.2	3.4	W	4
July	0.8	0.5	0	0.3	4.0	W	2
Aug.	1.3	0.4	0	0.2	4.2	W	2
Sept.	1.0	0.5	0	0.1	4.7	W/SW	2
Oct.	0.7	0.2	0	0.1	4.2	SW	4
Nov.	0.5	0.3	0	0	3.8	SW	4
Dec.	<0.1	0	0	0	3.2	SW	4
Annual	5.0	2.3	0.1	1.8	3.4	SW	4

TABLE V

CLIMATIC TABLE FOR CORRIDA DE CORI (ARGENTINA)[1]
Latitude 25°06′S, longitude 68°20′W, elevation ~5,100 m

| Month | Temperature (°C) | | | | Mean vap. press. (mbar) | Mean cloud-iness (oktas) | Wind | |
| | daily mean | daily range | extreme | | | | preval. direct. | mean speed (m/sec) |
			max.	min.				
Jan.	−1.0	11.5	14.0	−13.0	3.0	4.0	W	13
Feb.	−1.5	10.5	11.0	−12.0	3.0	3.5		13
Mar.	−2.5	10.0	9.0	−14.0	2.5	2.5		13
Apr.	−5.0	8.5	8.0	−18.0	1.5	2.5		12
May	−7.5	7.5	7.0	−18.5	1.0	2.5		13
June	−10.0	8.0	5.0	−20.0	1.0	2.5		16
July	−10.0	8.0	1.0	−23.5	1.0	2.5		14
Aug.	−9.0	8.0	2.0	−22.0	1.0	2.5		13
Sept.	−7.5	8.5	7.0	−17.0	1.0	2.5		14
Oct.	−6.5	9.5	11.0	−21.0	1.5	2.5		15
Nov.	−3.5	10.5	13.0	−15.0	2.0	2.5		15
Dec.	−1.5	11.0	14.0	−12.0	2.5	3.0		13
Annual	−5.5	9.3	14.0	−23.5	1.7	2.7		14

[1] By courtesy of the Serv. Meteorol. Nacl. Buenos Aires (see Chapter 2).

TABLE VI

CLIMATIC TABLE FOR LA SERENA
Latitude 29°54'S, longitude 71°15'W, elevation 35 m

Month	Mean sta. press. (mbar)	Temperature (°C)				Mean vap. press. (mbar)	Precipitation (mm)	
		daily mean	daily range	extreme			mean	max. in 24 h
				max.	min.			
Jan.	1010.5	18.2	7.6	27.2	9.4	16.1	0.1	1.1
Feb.	1010.7	18.4	7.8	27.8	10.0	16.5	0.8	54.3
Mar.	1010.8	16.9	7.8	29.3	6.5	15.4	0.6	3.5
Apr.	1011.9	14.9	7.2	25.8	2.9	13.9	2.6	28.0
May	1012.9	13.4	7.3	23.7	3.7	12.6	21.9	89.2
June	1013.8	12.1	7.7	24.5	2.5	11.4	43.7	80.9
July	1014.0	11.7	8.5	23.3	2.6	11.1	29.7	57.0
Aug.	1014.1	12.0	7.8	24.0	1.8	11.3	23.2	39.9
Sept.	1013.7	12.7	7.7	26.0	2.8	11.7	6.0	19.2
Oct.	1013.4	14.0	7.6	25.5	4.0	12.8	3.7	33.3
Nov.	1012.3	15.5	7.8	25.0	6.7	13.7	0.7	9.0
Dec.	1011.2	17.0	7.7	25.8	7.9	14.9	0.3	17.2
Annual	1012.4	14.7	7.7	29.3	1.8	13.4	133.3	89.2

Month	Number of days with			Mean cloud- iness (oktas)	Wind	
	precip. ≥1mm	thunder- storm	fog		preval. direct.	mean speed (m/sec)
Jan.	4.7	0	0.9	4.0	NW	2
Feb.	4.4	0	0.9	4.0		2
Mar.	5.1	0	2.2	4.0		2
Apr.	4.8	0	2.7	4.8		1
May	4.3	0	3.9	4.8		1
June	4.6	0.2	2.6	4.8		1
July	5.8	0.1	2.7	4.8		2
Aug.	5.4	0.1	3.1	4.8		2
Sept.	7.0	0	3.0	4.8		2
Oct.	6.8	<0.1	2.4	4.8		2
Nov.	5.5	<0.1	1.0	4.0		2
Dec.	4.4	<0.1	0.5	4.0		2
Annual	61.8	0.5	25.9	4.5		2

TABLE VII

CLIMATIC TABLE FOR CRISTO REDENTOR (ARGENTINA)
Latitude 32°50′S, longitude 70°05′W, elevation 3,829 m

Month	Mean sta. press. (mbar)	Temperature (°C)				Mean vap. press. (mbar)	Mean precip.[1] (mm)
		daily mean	daily range	extreme			
				max.	min.		
Jan.	643.0	4.0	10.2	20.2	−8.6	4.6	8
Feb.	643.0	3.6	10.1	17.6	−13.0	4.3	9
Mar.	642.4	1.8	8.7	18.0	−16.4	3.7	8
Apr.	641.4	−0.6	8.2	14.0	−14.2	3.2	22
May	640.2	−4.5	6.7	11.2	−20.6	2.6	96
June	638.0	−5.9	6.8	8.6	−23.5	2.2	40
July	638.1	−6.7	7.0	9.5	−30.3	2.0	56
Aug.	638.0	−6.5	7.6	9.6	−22.5	2.0	64
Sept.	638.8	−5.5	7.8	10.3	−21.7	2.4	23
Oct.	639.7	−3.5	8.4	12.8	−21.7	3.0	19
Nov.	640.6	−1.0	8.4	13.4	−19.9	3.6	7
Dec.	641.9	2.4	9.9	16.1	−13.7	4.1	7
Annual	640.4	−1.7	8.4	20.2	−30.3	3.1	357

Month	Number of days with			Mean cloudiness (oktas)	Wind		Days with	
	precip.[1] ⩾1mm	thunderstorm[2]	fog[2]		preval. direct	mean speed[3] (m/sec)	snow	wind speed >12m/sec
Jan.	5	1.0	3	2.1	SW	8.3	3	29
Feb.	4	0.5	3	2.0		8.3	3	26
Mar.	6	0.2	4	1.8		8.0	6	28
Apr.	6	0	1	2.7		7.8	6	23
May	9	0	0.6	4.2		7.5	9	18
June	10	0	1	4.6		8.0	10	17
July	10	0	1	4.1		7.2	10	14
Aug.	11	0	2	4.2		7.5	11	16
Sept.	8	0	2	3.8		7.5	8	16
Oct.	9	0	3	3.8		7.2	9	20
Nov.	7	0.3	3	3.2		8.0	6	24
Dec.	6	0.4	2	2.2		9.2	5	28
Annual	91	2.4	25.6	3.2	↓	7.8	86	259

[1] Frequencies and amounts very uncertain due to strong winds during precipitation periods.
[2] Period 1951–1960.
[3] Period 1941–1960.

TABLE VIII

CLIMATIC TABLE FOR VALPARAÍSO
Latitude 33°01′S, longitude 71°38′W, elevation 41 m

Month	Mean sta. press. (mbar)	Temperature (°C)				Mean vap. press. (mbar)	Precipitation (mm)	
		daily mean	daily range	extreme			mean	max. in 24 h
				max.	min.			
Jan.	1009.3	18.0	9.2	36.0	7.6	14.9	2.3	25.2
Feb.	1009.4	17.9	9.4	34.5	9.0	15.2	2.0	24.0
Mar.	1009.7	16.7	9.2	31.5	7.0	14.4	3.8	38.0
Apr.	1010.8	14.9	8.4	32.4	5.0	13.2	18.1	72.1
May	1011.8	13.5	7.3	31.0	3.8	12.4	96.6	171.5
June	1012.8	12.2	7.0	25.2	2.2	11.4	128.0	185.8
July	1013.1	11.8	7.3	27.2	2.0	11.1	88.0	81.7
Aug.	1013.7	12.0	7.8	27.5	3.0	11.1	66.7	103.5
Sept.	1013.3	12.9	8.0	27.6	4.0	11.6	30.4	75.5
Oct.	1012.6	14.1	8.4	34.4	4.5	12.1	16.2	41.0
Nov.	1011.2	15.7	9.3	34.0	6.5	12.7	6.8	36.1
Dec.	1009.8	17.2	9.6	33.0	6.7	13.7	2.7	25.2
Annual	1011.5	14.7	8.4	36.0	2.0	12.7	458.9	185.8

Month	Number of days with		thunder-storm	fog	Mean cloud-iness (oktas)	Wind	
	precip.					preval. direct.	mean speed (m/sec)
	⩾tr.	⩾1mm					
Jan.	2.5	0.3	0	4.1	2.9	SW	4
Feb.	1.7	0.3	<0.1	4.7	3.0	SW	4
Mar.	2.6	0.5	0.1	6.8	3.1	SW	2
Apr.	4.5	0.9	0.1	6.8	3.8	SW	2
May	8.7	5.0	0.2	6.3	4.6	SW	2
June	9.4	7.0	0.3	3.4	4.7	N	4
July	8.2	7.0	0.3	4.4	4.6	N	4
Aug.	8.0	5.0	0.1	3.4	4.2	SW	2
Sept.	5.9	2.0	<0.1	3.7	4.1	SW	4
Oct.	5.4	2.0	0.1	4.0	4.0	SW	4
Nov.	3.4	0.7	0.1	3.3	3.3	SW	4
Dec.	2.1	0.5	0	3.5	2.8	SW	4
Annual	62.4	31.2	1.3	54.4	3.8	SW	4

TABLE IX

CLIMATIC TABLE FOR SANTIAGO
Latitude 33°27′S, longitude 70°42′W, elevation 520 m

Month	Mean sta. press. (mbar)	Temperature (°C)				Mean vap. press. (mbar)	Precipitation (mm)	
		daily mean	daily range	extreme			mean	max. in 24 h
				max.	min.			
Jan.	952.9	20.0	17.2	35.6	5.9	13.8	1.6	22.4
Feb.	953.1	19.3	17.1	36.5	5.2	14.1	2.8	53.2
Mar.	953.7	17.2	16.7	34.3	2.0	13.5	4.3	24.0
Apr.	954.9	13.9	15.2	31.5	−2.5	11.6	14.2	61.7
May	955.8	10.9	12.0	31.5	−3.0	10.6	61.9	73.3
June	956.6	8.4	10.5	26.5	−4.2	9.2	84.6	103.4
July	956.8	8.1	11.2	27.2	−4.6	9.0	76.1	63.3
Aug.	957.2	9.1	11.8	29.2	−3.3	9.4	56.6	76.1
Sept.	956.7	11.6	12.4	31.4	−2.3	10.5	29.1	44.4
Oct.	955.8	13.8	14.2	33.3	−1.0	11.4	14.6	26.4
Nov.	954.7	16.5	15.9	36.0	0.9	12.0	6.3	30.0
Dec.	953.3	18.9	17.0	37.2	2.2	12.9	4.4	36.9
Annual	955.1	14.0	14.3	37.2	−4.6	11.5	356.3	103.4

Month	Number of days with				Mean cloud-iness (oktas)	Mean sun-shine (h)	Wind	
	precip.		thunder-storm	fog			preval. direct	mean speed (m/sec)
	>tr.	⩾1mm						
Jan.	0.6	<0.1	0.2	2.0	1.5	332.4	SW	2
Feb.	0.8	<0.1	0.2	2.0	1.5	276.9		2
Mar.	1.3	1	0.3	3.0	1.7	271.1		2
Apr.	4.2	1	0.2	5.0	3.0	193.9		1
May	7.8	5	0.3	8.0	4.5	116.3		1
June	9.6	6	0.2	8.0	5.0	95.5		1
July	9.5	6	0.2	7.0	4.7	113.7		1
Aug.	8.9	5	0.2	5.0	4.4	136.2		1
Sept.	8.0	3	0.4	4.0	4.4	161.4		2
Oct.	6.0	3	0.2	2.0	3.6	212.7		2
Nov.	2.8	1	0.2	2.0	2.8	263.9		2
Dec.	1.0	<0.1	0.1	1.0	1.7	327.0		2
Annual	60.5	31	2.7	49.0	3.2	2,501.0	↓	2

TABLE X

CLIMATIC TABLE FOR CONCEPCIÓN
Latitude 36°40′S, longitude 73°03′W, elevation 15 m

Month	Mean sta. press. (mbar)	Temperature (°C)				Mean vap. press. (mbar)	Precipitation (mm)	
		daily mean	daily range	extreme			mean	max. in 24 h
				max.	min.			
Jan.	1012.6	18.0	14.5	37.5	3.5	14.6	17.0	89.2
Feb.	1012.8	17.2	14.7	34.0	1.5	14.3	20.7	54.3
Mar.	1013.6	15.1	13.8	36.8	0.8	13.4	52.0	95.4
Apr.	1014.5	12.8	11.7	28.6	−0.7	12.4	85.3	92.0
May	1014.6	11.1	9.5	27.5	−0.9	11.6	211.1	137.0
June	1015.5	9.7	7.7	24.1	−2.5	10.6	249.9	126.2
July	1016.2	9.1	8.3	23.8	−4.0	10.0	238.2	116.5
Aug.	1016.9	9.1	9.5	25.2	−5.0	9.7	182.9	115.0
Sept.	1017.1	10.6	10.7	28.1	−2.5	10.6	102.8	69.0
Oct.	1016.7	12.6	11.5	34.0	−0.5	11.5	58.7	83.1
Nov.	1015.4	14.8	12.1	33.5	1.1	12.6	45.5	56.0
Dec.	1013.6	16.9	13.2	36.5	3.8	13.9	28.7	57.4
Annual	1015.0	13.0	11.5	37.5	−5.0	12.0	1,292.8	137.0

Month	Number of days with			Mean cloud-iness (oktas)	Mean sun-shine (h)	Wind	
	precip. ⩾1mm	thunder-storm	fog			preval. direct	mean speed (m/sec)
Jan.	2.3	0	0.7	2.2	257.3	SW	4
Feb.	3.5	0	0.6	2.5	223.0	SW	
Mar.	5.1	0	1.4	3.1	208.0	SW	
Apr.	9.3	0.1	1.9	4.2	153.3	SW	
May	14.5	0.5	1.2	5.5	97.8	N	
June	17.1	0.8	1.1	5.6	73.0	N	
July	16.0	0.2	1.0	5.2	99.2	N	
Aug.	13.8	0.5	0.8	4.6	130.1	N	
Sept.	11.7	0.2	0.6	4.3	156.5	SW	
Oct.	8.1	0.2	0.6	3.8	217.5	S	
Nov.	7.0	0	0.4	3.4	237.7	SW	
Dec.	4.2	0	0.2	2.7	277.0	SW	
Annual	112.6	2.5	10.5	3.5	2,130.6	SW	↓

TABLE XI

CLIMATIC TABLE FOR VALDIVIA
Latitude 39°48'S, longitude 73°14'W, elevation 9 m

Month	Mean sta. press. (mbar)	Temperature (°C)				Mean vap. press. (mbar)	Precipitation (mm)	
		daily mean	daily range	extreme			mean	max. in 24 h
				max.	min.			
Jan.	1014.7	17.0	12.4	34.8	2.7	14.1	65.4	135.4
Feb.	1014.2	16.4	12.2	36.6	1.7	14.2	68.7	77.2
Mar.	1014.8	14.5	11.3	32.5	1.0	13.2	115.3	120.5
Apr.	1013.9	11.8	9.1	28.0	—2.0	11.9	212.1	156.2
May	1014.1	9.7	6.9	21.5	—3.5	11.2	376.5	111.9
June	1015.3	8.2	5.7	17.0	—3.9	10.1	414.1	162.0
July	1015.9	7.7	6.3	19.0	—4.2	9.4	374.0	158.8
Aug.	1016.4	8.0	7.8	22.5	—3.8	9.4	301.0	174.2
Sept.	1017.5	9.3	9.2	26.0	—2.6	9.8	214.2	128.7
Oct.	1017.5	11.5	10.5	29.0	—0.5	10.8	118.9	78.0
Nov.	1015.7	13.3	10.8	32.0	0.0	11.7	121.7	131.1
Dec.	1014.8	15.3	11.3	34.0	—2.2	13.2	106.8	97.5
Annual	1015.4	11.9	9.4	36.6	—4.2	11.6	2,488.7	174.2

Month	Number of days with				Mean cloud- iness (oktas)	Wind		Days with snow
	precip.		thunder- storm	fog		preval. direct.	mean speed (m/sec)	
	⩾tr.	⩾1mm						
Jan.	10.0	7	0.2	1.1	3.1	S	2	0
Feb.	9.4	7	0.2	1.9	3.4	S	2	0
Mar.	12.6	11	0.3	4.6	4.1	S	2	0
Apr.	16.7	12	0.5	6.9	5.1	N	2	0
May	21.7	21	1.4	7.6	5.8	N	2	0
June	24.2	21	2.1	6.1	6.0	N	4	<0.1
July	22.8	20	1.1	5.2	5.8	N	2	0.1
Aug.	21.5	18	1.2	7.2	5.0	N	4	<0.1
Sept.	17.7	13	0.6	4.5	4.8	N	2	0
Oct.	14.6	13	0.2	2.0	4.2	W	1	0
Nov.	15.0	10	0.3	0.6	4.1	W	2	0
Dec.	12.6	10	0.2	0.6	3.8	W	2	0
Annual	196.9	163	8.3	48.3	4.1	N	2	0.1

TABLE XII

CLIMATIC TABLE FOR PUERTO MONTT
Latitude 41°28′S, longitude 72°57′W, elevation 13 m

Month	Mean sta. press. (mbar)	Temperature (°C)				Mean vap. press. (mbar)	Precipitation (mm)	
		daily mean	daily range	extreme			mean	max. in 24 h
				max.	min.			
Jan.	1014.5	15.2	8.8	28.5	4.5	13.7	90.1	57.4
Feb.	1014.1	14.8	8.5	29.0	4.0	14.0	138.8	133.0
Mar.	1014.4	13.2	8.0	28.6	0.5	13.0	138.6	88.2
Apr.	1014.4	11.2	7.0	24.0	−2.6	11.5	180.9	94.0
May	1013.3	9.3	6.3	19.0	−2.5	10.4	236.4	83.6
June	1012.6	8.0	5.6	17.5	−2.5	9.4	257.0	91.4
July	1014.2	7.6	5.9	20.0	−4.0	9.0	209.4	90.0
Aug.	1015.5	7.8	6.9	19.5	−3.0	8.9	197.6	116.0
Sept.	1016.5	8.8	7.6	25.5	−2.1	9.7	157.8	108.7
Oct.	1016.8	10.6	8.3	23.4	−1.0	10.5	119.1	70.0
Nov.	1015.6	12.2	8.3	26.5	−0.5	11.6	130.6	89.2
Dec.	1014.5	13.9	8.5	28.9	2.5	12.8	125.3	77.3
Annual	1014.7	11.1	7.5	29.0	−4.0	11.2	1,981.6	133.0

Month	Number of days with			Mean cloud-iness (oktas)	Wind		Days with snow
	precip. ⩾1mm	thunder-storm	fog		preval. direct.	mean speed (m/sec)	
Jan.	13.1	0.3	1.8	4.7	S	4	0
Feb.	12.2	0.4	1.3	4.8	S	2	0
Mar.	14.8	0.3	1.7	5.2	N	2	0
Apr.	19.6	0.4	2.0	5.7	N	4	0
May	24.4	0.4	1.4	6.3	N	4	0
June	23.9	0.5	0.8	6.1	N	4	0
July	22.8	0.2	0.7	6.2	N	6	0.2
Aug.	21.4	0.2	0.6	5.8	N	4	0.1
Sept.	18.8	0.2	0.5	5.3	N	4	0.1
Oct.	17.0	0.2	0.7	5.3	S	4	<0.1
Nov.	18.2	0.2	0.7	5.4	S	4	0
Dec.	16.4	0.1	0.5	5.3	S	4	0
Annual	222.6	3.4	12.7	5.5	N/S	4	0.4

TABLE XIII

CLIMATIC TABLE FOR SAN PEDRO
Latitude 47°43′S, longitude 74°55′W, elevation 22 m

Month	Mean sta. press. (mbar)	Temperature (°C)				Mean vap. press. (mbar)	Precipitation (mm)	
		daily mean	daily range	extreme			mean	max. in 24 h
				max.	min.			
Jan.	1009.7	11.2	6.5	24.0	2.0	11.8	377.9	92.0
Feb.	1008.8	11.2	6.6	26.5	0.2	12.0	385.1	113.0
Mar.	1008.5	10.0	6.1	19.7	−1.0	11.2	397.8	75.2
Apr.	1008.6	8.5	5.4	16.0	−0.5	10.2	409.2	114.0
May	1005.9	6.8	5.0	14.0	−2.0	9.1	393.9	72.2
June	1005.3	6.1	5.1	15.8	−3.0	8.6	371.9	96.4
July	1007.2	5.7	4.5	13.0	−3.0	8.4	398.0	91.2
Aug.	1009.6	5.5	4.9	13.5	−2.5	8.2	309.8	116.0
Sept.	1011.2	6.4	5.6	18.0	−4.2	8.7	326.4	79.5
Oct.	1010.9	7.8	6.1	19.5	−0.5	9.6	373.9	98.0
Nov.	1010.6	8.8	6.2	20.8	−0.1	10.3	362.8	137.5
Dec.	1008.9	10.2	6.9	26.0	−0.2	11.2	378.6	136.9
Annual	1008.8	8.2	5.7	26.5	−4.2	9.9	4,485.3	137.5

Month	Number of days with			Mean cloudiness (oktas)	Wind			Days with snow
	precip. ⩾1mm	thunderstorm	fog		preval. direct	mean speed (m/sec)		
Jan.	27.7	0.6	2.8	6.3	NW	7		0
Feb.	23.6	0.5	1.9	6.3	W	4		0
Mar.	26.6	0.8	1.8	6.6	W	4		0
Apr.	24.7	0.4	2.1	6.7	NW	9		0.2
May	25.1	0.7	2.1	6.2	NW	9		0.4
June	25.3	0.9	1.5	6.1	N	9		0.6
July	28.2	1.0	2.2	6.6	NW	9		1.0
Aug.	23.6	0.7	2.1	6.0	NW	7		1.1
Sept.	24.5	0.2	2.1	6.2	NW	9		0.5
Oct.	26.2	0.4	1.9	6.1	NW	7		0.2
Nov.	25.3	0.4	1.7	6.2	W	7		0.1
Dec.	25.4	0.4	1.7	6.1	NW	9		0
Annual	306.2	7.0	23.9	6.5	NW	7		4.1

TABLE XIV

CLIMATIC TABLE FOR EVANGELISTAS
Latitude 52°24′S, longitude 75°06′W, elevation 55 m

Month	Mean sta. press. (mbar)	Temperature (°C)				Mean vap. press. (mbar)	Precipitation (mm)	
		daily mean	daily range	extreme			mean	max. in 24 h
				max.	min.			
Jan.	997.9	8.7	4.5	16.6	1.0	9.8	233.4	73.2
Feb.	993.8	8.8	4.2	16.2	0.2	9.9	225.2	70.0
Mar.	998.0	8.3	4.6	16.5	−1.0	9.4	275.5	74.0
Apr.	995.7	7.2	4.4	14.3	−3.3	8.9	247.7	106.4
May	995.2	6.0	4.4	14.6	−2.0	8.2	215.2	73.0
June	994.3	4.8	4.1	14.2	−7.2	7.6	211.9	96.0
July	996.0	4.4	3.5	11.5	−4.5	7.3	216.4	82.8
Aug.	998.3	4.4	3.6	11.4	−3.5	7.3	214.1	136.7
Sept.	998.7	4.9	3.6	11.9	−3.5	7.6	224.5	123.0
Oct.	995.9	5.5	3.8	12.2	−1.0	7.6	164.7	72.5
Nov.	996.2	6.4	4.2	13.5	−1.9	8.3	170.2	56.0
Dec.	995.3	7.6	4.2	14.2	−1.0	9.2	170.9	50.8
Annual	996.3	6.4	4.2	16.6	−7.2	8.3	2,569.7	136.7

Month	Number of days with			Mean cloud-iness (oktas)	Wind		Days with snow
	precip. ⩾1mm	thunder-storm	fog		preval. direct	mean speed (m/sec)	
Jan.	26.0	0.2	2.6	6.8	NW	9	0
Feb.	23.6	0.2	1.7	7.0	NW	9	0
Mar.	26.0	0.2	2.3	6.8	NW	12	0.2
Apr.	25.1	0.4	2.2	7.0	NW	15	0.5
May	23.5	0.3	1.8	5.7	NW	12	2.1
June	22.5	0.2	3.3	6.6	SW	9	3.5
July	23.8	0.2	1.8	6.3	SW	9	4.7
Aug.	23.6	0.2	2.0	6.6	NW	15	3.8
Sept.	22.8	0.3	5.0	6.7	NW	12	3.5
Oct.	21.6	0.1	1.3	7.1	NW	15	1.0
Nov.	24.0	0	3.5	6 8	NW	12	0.7
Dec.	26.6	0	2.0	6.9	NW	12	0.3
Annual	289.0	2.3	29.5	6.7	NW	12	20.3

TABLE XV

CLIMATIC TABLE FOR PUNTA AREÑAS
Latitude 53°10′S, longitude 70°54′W, elevation 8 m

Month	Mean sta. press. (mbar)	Temperature (°C)				Mean vap. press. (mbar)	Precipitation (mm)	
		daily mean	daily range	extreme			mean	max. in 24 h
				max.	min.			
Jan.	999.8	11.7	8.3	29.0	0.0	9.5	32.8	26.0
Feb.	999.9	10.6	8.3	24.5	−0.1	8.8	29.0	30.0
Mar.	1000.4	8.9	7.4	22.8	−1.0	8.2	45.1	98.0
Apr.	1000.1	6.7	6.3	19.2	−3.6	7.5	46.3	65.0
May	1001.0	4.2	5.2	16.0	−7.5	6.6	50.3	46.5
June	1001.4	2.6	4.6	15.0	−7.3	6.0	40.2	44.5
July	1002.4	2.5	4.8	13.0	−9.3	5.9	41.3	45.1
Aug.	1003.4	2.9	5.2	13.8	−7.5	5.9	38.4	45.0
Sept.	1004.2	4.6	6.2	19.2	−6.0	6.4	32.9	42.0
Oct.	1003.6	7.1	7.7	20.6	−2.5	7.1	25.5	46.4
Nov.	999.6	8.5	7.9	25.0	−2.5	7.5	31.7	30.0
Dec.	999.3	10.2	8.2	25.2	−0.2	8.6	34.0	31.5
Annual	1001.2	6.7	6.7	29.0	−9.3	7.2	447.5	98.0

Month	Number of days with		thunder-storm	fog	Mean cloud-iness (oktas)	Wind preval. direct.	mean speed (m/sec)	Days with snow
	precip. ≥tr.	≥1mm						
Jan.	10.9	6	<0.1	0.1	6.5	W	4	0
Feb.	9.5	5	<0.1	0.5	5.6	NW	4	<0.1
Mar.	11.2	7	0.1	0.4	5.7	W	4	0.2
Apr.	13.0	9	0.1	0.9	5.7	NW	4	1.1
May	11.8	6	0	1.0	5.3	W	2	2.3
June	10.5	8	0	1.4	4.8	W	2	3.1
July	11.2	6	0	1.5	5.0	W	2	3.8
Aug.	10.2	5	0	1.1	5.0	W	4	3.9
Sept.	10.7	5	0	0.7	5.4	W	4	2.6
Oct.	8.1	5	<0.1	0.6	5.5	W	4	1.5
Nov.	9.7	5	0	0.2	5.7	W	4	1.0
Dec.	12.3	8	0.1	0.5	5.9	W	4	0.1
Annual	129.1	75	0.4	8.9	5.5	W	4	19.6

Chapter 4

The Climate of Peru, Bolivia and Ecuador[1]

A. M. JOHNSON

Introduction

The three states, Peru, Bolivia and Ecuador, cover a total area of 2,665,000 km² on the western side of the South American continent between 1°20'N and 22°45'S. The western edge is bounded by the Pacific Ocean from the most northern point to the southern tip of Peru at 18°21'S. The border then moves inland for some 150 km before continuing southwards for a further 5 degrees of latitude, leaving the most southern portion of the region as a landlocked area to the east of the continental divide. The most westerly point of the continental area is 81°19'W, with a small offshore group of islands, the Galápagos, lying between 89°W and 90°W around the Equator. The region's eastern boundary reaches 57°30'W, near the centre of the continent.

The whole region lies within the tropics, but its climatic conditions are dominated by the features of the land surface. The cordilleras (mountain chains) of the Andes transverse the region from north to south along the western edge. The coastal plain is about 150 km wide in northwest and western Ecuador and is of a similar width around Piura in northern Peru, but otherwise the Andes rise almost straight out of the sea, leaving a coastal lowland strip of only a few kilometers in width. At the latitude of Lima, the continental divide is only 100 km from the edge of the continent. The Andes are essentially formed by two north–south chains, joined in several places by ridges or spurs which separate intermontane basins. This can be clearly seen in Ecuador where ten such basins exist at altitudes of 2,500–3,000 m above sea level. From central Peru southwards, these basins tend to be at higher altitudes, culminating in the huge Altiplano of Bolivia which is a flat basin of some 100,000 km² at an average altitude of 3,800 m. At its northern end is the world's highest navigable lake, called Titicaca.

The western range of the Andes is an almost continuous chain, about 4,000 m high in Ecuador though generally over 5,000 m through the southern half of the region. This chain includes the probably highest peak of the region, Peru's 6,768 m Huascarán, and the highest point of Ecuador, the 6,272 m Chimborazo. Only in northern Peru is the cordillera noticeably lower. The eastern range is more dissected and breached by many tributaries of the Amazon river system. Despite the lower average elevation of this latter range, some notable peaks are included such as Cotopaxi (5,900 m), the world's highest active volcano, and the Cordillera Real containing the highest points in Bolivia with heights >6,500 m, though there is still some doubt about their exact altitudes.

---

[1] With contributions by H. Lettau (pp.188–192) and W. Schwerdtfeger (p.192–199).

To the east, the land surface descends to the lowland plains of the Amazon river system, apart from the extreme southeast where the Paraguay basin is found. The eastern slopes tend to be more extensive than those of the west, forming a less abrupt change from high sierra to lowland plain. This is particularly noticeable in Bolivia where the transitional zone is sufficiently large to be considered as a separate and distinct subregion.

The region's vegetation is very variable, being dependent principally upon rainfall and altitude. The coastline of Peru is a true desert except where mountain streams support a green strip of vegetation or crops. The coastal plain of Ecuador shows sharp contrasts in vegetation cover at short horizontal distances, thus illustrating the wide variations in rainfall. In the Guayas river basin and north of Esmeraldas, tropical rain forests dominate whilst 100 km to the west of Guayaquil a virtual desert is found, with a few low bushes and shrubs growing out of the sand. South of Esmeraldas is a narrow zone of tropical semi-deciduous forest where some of the trees drop their leaves in the dry season.

The sierra (highland) is dominated by grassland with a few eucalyptus growing up to the 3,800 m contour. Below this line reasonable grazing areas may be found particularly in well watered valleys, but above this level and in drier areas relatively poor looking bunch grass, known as ichu, is common which tends to give a brownish appearance to the landscape.

The eastern Andean slopes are covered by dense forests below about 1,800 m, this level being commonly known as the "ceja de la montaña" (the eyebrow of the mountain forest). Below this the true tropical rain forest begins which covers the Amazon basin in east Ecuador, east Peru and northeast Bolivia. In southeast Bolivia one finds tropical scrub forest interspersed with patches of grassy savanna. This is the Bolivian Chaco, a fairly low level area that varies strikingly with the seasonal climate changes. It is a veritable swamp during the rainy season and a hot semidesert during the remaining 8 or 9 months of the year. The department of Santa Cruz which lies between these two western areas is a transitional zone with a vegetation cover of approximately 40% grass and 60% forest.

The total population of the region is approximately 25.1 million of which 13.6 million are in Peru, 6.4 million in Ecuador and 5.1 million in Bolivia. The average density is, therefore, 9.4 persons per km², ranging from 22.6 in Ecuador to Bolivia's 4.7.[1] The population distribution according to physical characteristics of the region is approximately 8 million persons on the coast which includes the two largest cities of Lima and Guayaquil, 16 million in the sierra including the Ecuadorian capital of Quito and the main city of Bolivia, La Paz, and about 1 million persons in the eastern lowlands.

Climatic factors bear a great influence upon the economy of the three countries, with some 60% of the working population being engaged in agriculture. Communications are also susceptible to climatic vagaries with landslides and floods frequently cutting road and rail communications. Two other major industries in the region are mining in the highlands of Peru and Bolivia and fishing in the coastal waters of Ecuador and Peru. The fishing industry, though little influenced by the weather, is much dependent upon the Humboldt current; any fluctuations in the sea temperature can exert a great influence upon this industry as well as upon the weather of the coastal regions.

The climatological literature of the region is widely scattered through meteorological and

---

[1] U.N. estimates 1971.

148

geographical journals published in European countries and the U.S.A. during the past hundred years, special reports on studies of specific areas or places, some highly informative old travelogues, and geographical monographs (see the References, pp.200–202). Nevertheless, for a large part of the total area, the lack of reliable meteorological data has been a serious problem up to recent years, though the decade of the 1960's saw great advances by the meteorological services in all three countries. In Ecuador, for example, where there were only 34 stations in 1960 recording a wide range of data on a regular basis, by 1969 this number had increased to 119. Due to the recent nature of this output, little analysis has as yet been undertaken. Aerological information is still scarce; regular radiosonde ascents have been carried out at Guayaquil (1957–1960), Lima (1957–present), and La Paz (several years in the 1960's).

In the following description of the climatic conditions of the entire, multifarious area, many meteorological stations and other places are referred to by name. Fig.1 together with Table I will help the reader to locate them.

Fig.1. Location of climatological stations and other places mentioned in the text. The key to the numbers is shown in Table I. For the stations marked by a black circle and italicized in Table I, climatological statistics are given in the climatic tables, pp.203–218.

Precipitation

General

One of the major climatic influences common to the three countries is the wet and dry seasonal regime. This is felt throughout the whole length of the region with the exception of two areas, namely eastern Ecuador and the coasts of Peru and extreme southwestern Ecuador. The former area is wet all the year round with no marked seasons whilst the latter is opposite with only negligible precipitation. The Humboldt current and the Andes both exert strong influences so that there is no simple north–south transition from one climatic type to another.

TABLE I

NAME AND APPROXIMATE COORDINATES (LATITUDE S/LONGITUDE) OF THE STATIONS SHOWN IN FIG.1

No.	Station	°S/°W	No.	Station	°S/°W	No.	Station	°S/°W
1	Ambato	01/79	37	*La Paz*	16/68	73	Río Pita	01/78
2	*Apolo*[1]	15/69	38	La Soledad	01/90	74	Riobamba	02/79
3	*Arequipa*	16/72	39	Latacunga	01/79	75	Salinas	02/81
4	Bellavista	01/90	40	*Lima*	12/77	76	Salinas-Imbabura	00°N/78
5	Bucay	02/79	41	Lita	01°N/79	77	San Lorenzo	01°N/79
6	Cajamarca	07/79	42	Lobitos	04/81	78	San Ramón	11/75
7	Callao	12/77	43	Malacatos	04/79	79	Santa Ana	01/80
8	Caluma	02/79	44	Manta	01/81	80	*Santa Cruz*	18/63
9	Camiri	20/64	45	Matucana	12/76	81	Seymour	00/90
10	Cañar	03/79	46	Milagro	02/80	82	Sucre	19/65
11	Cañete	13/76	47	Mollendo	17/72	83	Tabiazo	01°N/80
12	Casaracra	11/77	48	Morococha	12/76	84	Tacagua	18/68
13	Cerro de Pasco	11/76	49	Moyobamba	06/77	85	*Tacna*	18/70
14	Charaña	18/69	50	Muisne	01°N/80	86	Taisha	03/77
15	*Chiclayo*	07/80	51	Oploca	21/66	87	Talara	05/81
16	Chimbote	09/79	52	Oroya	12/76	88	Talenga	10/76
17	Chiriboga	00/79	53	*Oruro*	18/67	89	Tarija	22/65
18	Chone	01/80	54	Pachamama	02/79	90	Tarma	11/76
19	Chosica	12/77	55	Pastaza	02/77	91	Tena	01/78
20	*Cuzco*	14/72	56	Pampa de Majes	15/72	92	*Tingo María*	09/76
21	El Corazón	01/79	57	*Pichilingue*	01/79	93	Tiputini	01/76
22	El Misti	16/71	58	Pisco	14/76	94	Titicaca	16/70
23	El Porvenir	12/76	59	Piura	05/81	95	Trinidad	15/65
24	El Progreso	01/90	60	Playas	03/80	96	Trujillo	08/79
25	Esmeraldas	01°N/80	61	Poopo	18/67	97	Tulcán	01°N/78
26	*Guayaquil*	02/80	62	*Portoviejo*	01/80	98	Vitor	16/72
27	*Huancayo*	12/75	63	Pucallpa	08/75	99	*Yacuíba*	22/64
28	Huánuco	10/76	64	*Puerto Baquerizo*	01/90	100	Yaupi	10/76
29	Ibarra	00N/78	65	Puerto Bolívar	03/80	101	Yauricocha	12/76
30	Ica	14/76	66	Puerto Chicama	08/79	102	Yuncan	10/76
31	Imata	16/71	67	Puerto Ila	01/79	103	Yúrac	09/76
32	*Iquitos*	04/73	68	Puno	16/70	104	Zamora	04/79
33	*Izobamba*	00/79	69	Putumayo	00N/76	105	Zaruma	04/80
34	Jama	00/80	70	*Puyo*	02/78	106	*San Juan*	15/75
35	Jauja	12/75	71	*Quito*	00/79	107	*Puerto Maldonado*	13/69
36	Juliaca	16/70	72	Riberalta	11/66			

[1] For stations in italics a climatic table is given at the end of the chapter.

The seasonal rainfall pattern is rather complex; Fig.2 gives a general idea as to which regions are strongly influenced by the wet and dry seasonal changes. The six-month period November to April has been taken as the Southern Hemisphere rainy season; an increased concentration of precipitation into this period towards the south can be seen though an east–west transition is also apparent. Eastern Ecuador and probably also extreme northeastern Peru is interesting in that despite the southern location, over half the precipitation is recorded during the northern summer. It is not until the third parallel south of the Equator is reached that the wetter half of the year occurs from November to April. To the west of the Andes a different rainfall pattern emerges, with a clear dry

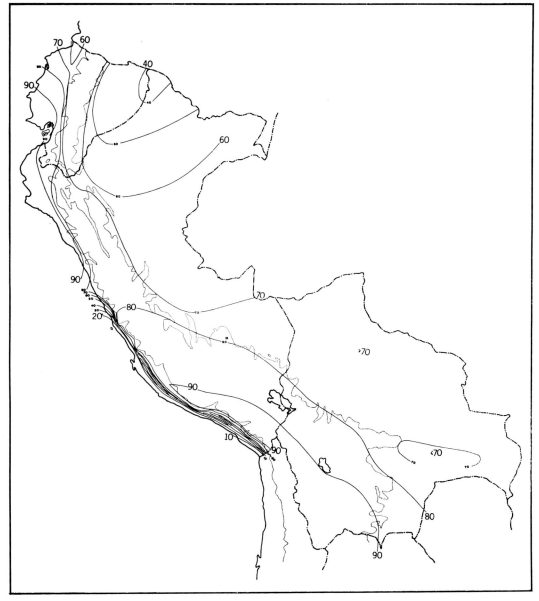

Fig.2. Percentage of annual rainfall occurring in the six months November–April. Thin dotted line: 2,000 m above sea level.

season in the middle months of the year. This Southern Hemisphere rainfall regime extends beyond the Equator, about up to the Colombian border.

The rainfall double maximum which is so common in other equatorial regions appears only in the northern sierra. The general idea is that the equatorial rains tend to follow the overhead passage of the sun, and that any location within about 10° of the Equator should experience two wet and two dry seasons. This effect is quite apparent in all sierra stations north of 3°S and to a lesser degree down to the Peruvian border at 5°S but, as

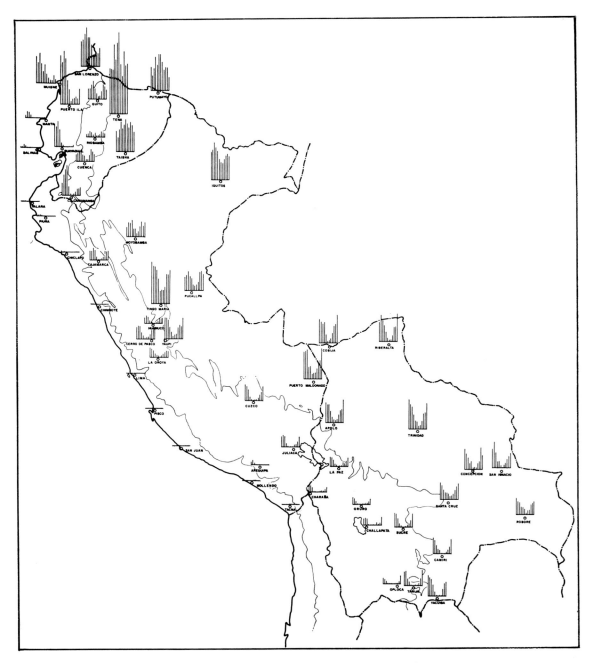

Fig.3. Seasonal distribution of rainfall. Thin line: 2,000 m above sea level.

already mentioned, the mid-year dry period is much more noticeable than the "dry" season of the months December to February. No such effect is seen on the Ecuadorian coastline, the western cordillera being the borderline.

The advance of the wet season therefore appears to be rapidly south and gradually west, from eastern Ecuador around October. Wide variations occur from year to year, the advance being more in the form of a series of pulses starting in September so that there is a gradual build up of the wet season with rainy spells becoming more and more pre-dominant. In the southern half of the region, the wet season reaches its climax in January whilst in the northern sierra a drier period takes over. In western Ecuador and the Galápagos Islands, the rains do not start until year's end with a March peak, when the offshore waters are at their annual thermal high both off the Ecuadorian coast and farther south.

With the northward passage of the sun in the early months of the year the wet season ends, normally in March in the highlands of Bolivia and in April over east Bolivia and also Peru away from the coast. In Ecuador the wet season normally lasts until May following a rainfall peak in March or April for the west and central parts of the country. The change to the dry season tends to be more abrupt than the gradual transition to the wet period. The dry season is at its most extreme in the Bolivian Altiplano, particularly in the south where virtually no precipitation occurs between May and October. Further north into Peru and on the lowlands, an odd wet spell or instability shower during the mid year period will break the dry, dusty weather.

To the west of the Andes, one finds completely different conditions. The coastal desert of Peru is part of a huge arid zone which stretches from the borders of Ecuador diagonally across the Andes to southern Argentina. In the north, the desert is restricted to the littoral where a small rise in elevation can result in a dramatic change in scenery. To the south the desert is found to move farther up the western Andean slopes until it finally crosses the Andes near southern Bolivia. Close to Lima at 12°S, vegetation becomes noticeable around the 1,800 m contour whilst the southern city of Arequipa, standing at 16°S and 2,525 m, appears as an oasis in an otherwise barren landscape due to irrigation, not to local rainfall. One notable feature of the coastal strip as seen in Table II is the complete contrast in relative seasonal rainfall distribution between the Peruvian coast north of the

TABLE II

PRECIPITATION—PERCENTAGE OF ANNUAL TOTAL RECORDED
IN THE 6 MONTHS NOVEMBER–APRIL

Latitude range	Coast (%)	Sierra (%)	East lowland (%)
North of 00°	66	63	43
00°–03°S	90	62	44
03°–05°S	90	69	56
05°–10°S	95	76	65
10°–15°S	28	80	75
15°–20°S	21	87	–
South of 20°	–	95	86

10th parallel and that to the south. South of this latitude, the period May to October is dominated by heavy clouds and drizzle. Though clouds are frequently present during this period in the north, the drizzle tends to be much less frequent whilst, in contrast, an occasional summer instability shower occurs.

Rainfall distribution in absolute terms is very variable, mainly because of the impact of the relief. The Amazon lowlands are noted for high precipitation due to a free access of the trades from the Atlantic and the large areas of swampland which, combined with the high temperatures, are favourable to evaporation along with the transpiration from the dense jungle. Iquitos with its average annual precipitation of 2,845 mm is probably fairly typical for eastern Peru. Farther south, the effect of the mid-year drier period tends to reduce annual totals and south of 15°, the wet season diminishes in importance so that the rain forest gives way to scrubland and grassland. The eastern flanks of the Andes are the first major obstacle met by the trade winds in a 4,000 km journey so that the forced uplift of this warm moist airstream with its adiabatic cooling can lead to extremely high rainfall totals. This is particularly noticeable between 2°S and the Colombian border near the Equator, reaching a peak at Tena, 1°S at an elevation of 527 m above sea level, to the east of Cotopaxi. At Tena, an average of 6,235 mm of rain has fallen in the 5 years 1965–1969, including totals of 8,380 and 8,939 mm for 1968 and 1969, respectively. The second major wet area lies on the western foothills of the Andes in northwestern Ecuador. Here again, annual totals of over 3,000 mm are recorded though a relatively dry period from July to October tends to reduce what may otherwise be values equal to those of the Amazon basin. Some high figures can still be recorded such as 7,204 mm in 1969 at Tabiazo in the Esmeraldas river basin.

Exceptionally high daily rainfall totals have been recorded in these regions and also outside. The record 24-h total in Ecuador to 1969 according to the publication *Anuario Meteorológico* from its initiation in 1959 was 310 mm at La Soledad on the San Cristobal Isle of the Galápagos on 23rd of March 1969. That date was notable throughout the islands; Puerto Baquerizo and El Progreso on the same island recorded 258 mm and 232 mm respectively, and Bellavista on the island Santa Cruz recorded 203 mm. Otherwise the exceptional wet days are found on the littoral of Ecuador such as the 287 mm at Tabiazo in February 1967, rather than in the selva which is noted for regular heavy rain instead of a few outstanding wet days. Of the eleven occasions when 24-h totals of 200 mm were recorded in Ecuador from 1959 to 1969, only one occurred east of the Andes. Peru can provide a few exceptions; 260 mm was recorded in one day in Pucallpa in October 1964, and Iquitos has seen three 24-h periods with more than 200 mm in the past 21 years, two of them being in the wet year of 1949. Tabiazo which has already received some attention has another notable record to its credit, the 3,402-mm rainfall total for January 1967 followed by 1,639 mm in February.

Rainfall in the sierra tends to be on a more moderate scale with daily totals of 100 mm being very exceptional. Though wide variations may occur at short distance due to altitude and rain shadow effects, on the broad scale, remarkable consistency of average annual totals is found over a wide range of latitude. Cuzco at 13°33'S with 750 mm of rain and even La Paz 3 degrees farther south with 488 mm receive as much rainfall as many intermontane basins of Ecuador. Quito is rather an exception here by being much wetter than most of the central highlands. It is only south of La Paz that the Andes become noticeably drier with semi-desert conditions in the Altiplano. In the extreme north, the sierra appears as a dry zone between two wet areas, whilst farther south it seems to be a transitional zone between the wet region to the east and the desert to the west.

Eastern lowlands

The eastern lowlands of Ecuador, Peru and north Bolivia form the western edge of the huge Amazon drainage basin, covered by a riot of dense tropical rain forest with rubber and cacao among its economic resources. Much of the area is still unexplored and only the steady encroachment of the oil man is helping to scratch the surface of the region. The Amazon river port of Iquitos (03°46′S 73°20′W) is the major city where rain is almost a daily feature with no dry season, despite the efforts of the tourist organisations to convince would-be visitors that June to September is the fine weather time. Heavy instability showers occur on most afternoons with no month being free. August, for example, is according to the latest 21 year average, the driest month, but at the same time August 1949 includes the wettest single day in the entire period with 247 mm. The annual average rainfall here is 2,845 mm; the peak is normally reached in March with 349 mm and the trough in August with a 157 mm average. No real evidence is available to prove a double maximum to coincide with the twice yearly overhead passage of the sun. A slight decrease in rainfall from November to December appears only to be followed by a sharp increase in January which is hardly sufficient to class December as a secondary minimum. It is often considered that great consistency in weather occurs from year to year in tropical regions, and here it will often appear to be so without any real difference being noticed between a wet month and a very wet month. Nevertheless, variances of fair magnitude do sometimes occur, with 1949 receiving 2.5 times the rainfall of 1950. The Augusts of the two years saw 656 and 34 mm of rain, respectively.

Farther west, the orographic effect of the Andes plays an important role when the moist trades, following a 4,000-km journey, finally meet a barrier. The Instituto Geográfico Militar of Ecuador produced rainfall maps for 1968 and 1969, both of which showed very similar patterns of a rainfall maximum area on the edge of the Andes between the Equator and 2°S. In both years, there was a large area with over 4,000 mm rainfall. Similarly in Peru, Pucallpa located in the Amazon basin at 151 m above sea level averages 1,463 mm per annum whilst Yurac (295 m), 130 km to the southwest near the foot of the eastern-most range, averages 4,900 mm. Tingo María at 665 m and a further 60 km west of Yurac and the Cordillera Azul, sees a slight reduction in precipitation from the rain shadow effect to average 3,072 mm. It is noteworthy that all stations of eastern Ecuador show the Northern Hemisphere rainfall regime despite the fact that nearly the whole region lies south of the Equator. North of the 3rd parallel south, as shown by data from six stations, the wettest months are June and July and February is the driest, the difference being on the order of a two to one ratio for June : February. No double maximum is apparent.

The number of days with rain indicates no real seasonable variation in eastern Ecuador. In the heavy precipitation zone, Tena has an average of 292 rain days (over 0.1 mm) and Puyo 315, both of which record 49% in the November–April period. So even though precipitation totals are lower in this season, there are just as many rainy days as for the other six months, which implies that either rainfall is heavier or is more prolonged in the latter period. As can be seen, rain is almost a daily occurrence; in the case of Puyo it is six days out of seven. Away from this area, the number of wet days tends to diminish with 198 in Taisha to the east and 218 in Zamora to the south with a slightly greater seasonal variation, 46% of such days occurring in November to April for Zamora.

More than 5 degrees from the Equator, the mid-year dry period becomes more apparent and even around Tingo María where rainfall can still be considerable in these months, the dirt roads are usually accessible, while virtually impassible by November. Heavy localised storms still regularly occur so that sometimes, early in the evening, one may sit under a starlit sky and watch an electric storm in the distance. In north Bolivia and southeast Peru, the jungle is still very dense despite the dryness of the June–August period; the annual total still ranges from 1,500 to 2,000 mm. The dry period is not consistent from year to year; a July of one year may be completely dry followed by 100 mm in the next. The wet season clearly becomes more dominated by the months December to March in the southern Selva with a fairly abrupt change to drier weather in April.

Around the 17th parallel, the dense jungle gradually gives way to more open country. Santa Cruz (17°47'S, 437 m), with 1,046 mm rain, is biologically a frontier town with dark green jungle to the north, the sand-coloured foothills of the sierra to the west, whilst southward and eastward lies a huge, monotonous grassy plain dotted with a few low trees and scrub; this is the Bolivian Chaco. The dry season becomes more and more intense towards the Paraguay–Argentina border, though the annual rainfall total still lies over 600 mm. Yacuiba (22°S) may go for several weeks without rain so that the land becomes brown and parched in the middle of the year. The wet season still receives sufficient rains to turn the low plains into swamplands. 86% of the annual total is recorded between November and April. January is normally the wettest month; at least, that was the case in five years of the decade 1960–1969.

Andes—eastern slopes

The eastern slopes of the Andes rise abruptly from the Amazon lowlands in Ecuador and Peru, dissected by deep ravines and gorges with drops of over a thousand metres. Farther south, this zone tends to broaden out; in Bolivia it is known as the Yungas, a somewhat vaguely defined term to cover the hot, humid valleys east of the Cordillera Oriental. Here one finds a complicated interplay of convection, orographic uplift and cyclonic action, particularly in the north. There is no simple relationship between altitude and rainfall though above the excessive precipitation of the lower slopes, it tends to decline to the 2,500 to 3,000 m zone. Thereafter, the orographic effect of the Andes once again becomes important, along with the effect of the high level heat source which encourages convection, leading to increasing precipitation to the higher mountains. A general rainfall decrease appears to take place south of 10°S on the lower slopes, this latitude possibly coinciding with the southern limit of the I.T.C.Z.

Aspect is an important factor for rainfall with great contrasts being common between windward (eastward facing) and leeward slopes. Huanuco (lying 80 km southwest of Tingo María) at 9°54'S, 1,900 m and to the west of the 5,000 m Carpish range, averages 420 mm rainfall, only 14% of that of Tingo María. The road between the two towns passes through two contrasting types of scenery. Tingo María is located within the rain forest from where the road rises steeply to a 3,000 m range. On its top a complete change takes place, the jungle giving way to sparse subxerophytic vegetation on the edge of the semi-arid pocket in which Huanuco lies. Above this town, the rainfall increases again to 882 mm in the sierra town of Cerro de Pasco at 4,328 m, some 100 km to the south. The Paucartambo river basin to the east of Cerro has produced a good range of rainfall data, the product of local hydroelectric installations which serve the mines in the highlands. Yaupi at 1,327 m averages 1,400 mm per annum, with the three months of December, January and February alone totalling nearly 600 mm. Yuncan, higher up the valley at 1,864 m sees a reduction in rainfall to 1,186 mm, whilst Talenga, lying at 3,995 m due east of a 5,600 m range, averages 1,968 mm. This latter total is extreme for a high altitude site but it indicates that even at this level where an outstanding cordillera blocks the path of the easterly air stream, excessive orographic rain can result. Higher altitudes tend to have more rain days but with lighter falls of shorter duration.

A change from the seasonal rainfall distribution of the eastern plains occurs at the Andean foothills. In Ecuador this is very pronounced with the mid-year heavy rain period of Amazonia changing to the mid-year dry period of the highlands. In central Peru, the east receives about 35% of its rainfall from May to October, whilst the Andean foothills see this reduced to 25% which is a very similar distribution to that of the central sierra.

The hourly distribution of rainfall tends to vary with altitude. Below 1,700 m, two thirds of all rainstorms occur during the night and early morning with the weather clearing at 9 or 10 a.m.; around 1,800 m, about half is reported at night, and above this level most occurs in the afternoon and early evening.

Humid conditions prevail on the eastern Andes to the Yungas area of Bolivia where the steep wooded slopes are deeply cut by the tributaries of the Beni river. Between here and the Argentinian border, the eastern slopes of the Cordillera Real are less steep and descend gently to the tropical plains along the Puna plateau. Here the dry season becomes very pronounced with the whole area receiving over 80% of its rainfall in the six month period November to April. Sucre (2,750 m, 19°S) totals 60% in December to February alone and Tarija (1,957 m, 21°33'S) sees 67% in these three months. The mid part of the year is usually completely dry with the odd shower in some years. The annual total for both stations is around 620 mm.

Andes—central

The highlands, through their entire 22° of latitude, are climatically dominated by the wet and dry season regimes of a Southern-Hemisphere type, with the pronounced precipitation minimum of June to August. The orographic effect of the high cordillera, the highest in the western and southern hemispheres, can result in showers in any season with the possible exception of southwest Bolivia where the dry season is normally complete. This tends to obscure the boundaries between the wet and dry seasons. In tropical lowlands one can normally state a month in which the rains are likely to start, whilst in the Andes a wide range of opinions may be given by the local people due to the wide variations from year to year and the general transition from one season to the next. Showers are very localised, particularly in the dry season when, in an otherwise dry, sunny area, large cumulonimbi may be seen a short distance away giving heavy showers or thunderstorms. Average rainfall amounts can give misleading impressions. The average July rainfall for

Cuzco in the period 1954 to 1970 was 5.3 mm, whilst July 1968 recorded 42 mm (with 38 mm in 24 h); this is almost half of the total for the entire 17-year period, thus doubling the average which would otherwise have been recorded.

The main climatic controls to the area must come from the east but there is evidence of some Pacific influence as stated by DREWES and DREWES (1957): "It seems very likely that under normal conditions the strong easterly current that brings warm and moist air from the Amazon Basin to the summits of the mountains plays an important role in the precipitation of the higher regions of the western slopes. However, the influence of the Pacific air should by no means be neglected. This air mass is especially important in summer when the height of the coastal inversion is relatively low and relatively steep lapse-rates favour convection and the break-through of the Pacific air from the lowest levels through the inversion layer. Local heating of the slopes themselves can also increase convection along the slopes independently from the prevailing lapse-rate in the atmosphere."

Only in the northern sierra a typical double rainfall maximum can be seen, with peaks in April and October, and the dry season in July more pronounced than in January. May sees a general clearance over the whole sierra.

Taking a north–south transect, three major latitudinal changes can be identified. Firstly, there is a steadily increasing concentration of rainfall into the period November–April towards the south. Secondly, rainfall decreases noticeably south of 15°S. This change towards aridity is first noticed on the western slopes of southern Peru with the desert creeping towards the continental divide. It starts to make itself felt in the high sierra close to the Bolivian border; farther south the semi-desert crosses the western ranges to give the southern Altiplano its dry appearance. One of the most obvious contrasts is between the 5,896 m volcano Cotopaxi in Ecuador and the 5,822 m cone of El Misti near Arequipa. The former peak which can sometimes be seen from Quito is heavily glaciated and appears to rise well above the snow line. El Misti, though sometimes covered white by the odd snow-storm, does not have any permanent snow as can easily be observed from Arequipa; often only the odd white patch can be identified. The average temperature and its annual variations on the two summits must be very similar so that the only major difference must be the much lower precipitation and lesser cloudiness over the more southern peak.

The third latitudinal difference is the variation in total rainfall from year to year. For the period 1960–1969 the coefficient of dispersion or variation for the annual rainfall totals of Quito was 15%; this means that a 15% variation on either side of the arithmetic mean will, on average, include about 67% of all years. In the central Peruvian sierra this fell slightly to 14% in Oroya (3,700 m), 11% in Cerro de Pasco (4,328 m) and 10% in Yauricocha (4,650 m). These latter values indicate a high degree of consistency though individual months from year to year and the commencing and termination dates or periods of the wet seasons may vary considerably. The small variability of the annual rainfall in this area compares well with that in the Amazonian lowlands where values of 11%, 15% and 15% are found for Iquitos, Pucallpa and Yaupi, respectively. Southwards greater variations tend to appear as the edge of the wet season zone is approached. It is often the climatically border areas which display the greatest variations in weather. Cuzco has a coefficient of dispersion of 27% for the 1960's, La Paz 19% and Oruro in the central Altiplano 28%.

To such a region, the quantity of precipitation is an important factor in the lives of the people. Agriculture is the dominant industry which, as anywhere else, is dependent upon climate. The main source of power is hydroelectric which relies upon water supply and gradient so that a very pronounced dry season can be a problem as occurred in Ecuador in 1956 and 1958 and in parts of central Peru in 1971. Flooding, though at times a serious problem, is reduced by the incidence of snowfall in the high levels of the river basins. Mountain storms which may produce enough precipitation to create a raging torrent leave much of their product in the form of snow which melts only gradually in the following morning, so that the full effect is spread out over a longer period. One afternoon's precipitation can therefore result in a double stream flow peak, one at the time of the rain, the second during a sunny spell in the following morning.

Snowfall first becomes felt around 3,600–3,800 m, though a little lower in Bolivia during the odd shower in the drier season. This level is also significant as being the upper limit for tree growth and extensive arable farming. By 4,300 m snow becomes as common as rain and at 4,600 m rain tends to be rare; despite the high precipitation totals at this level, snow and hail are much more common. La Paz airport (El Alto) at 4,100 m can therefore be closed because of snowstorms.

Snow does not tend to lie for much time. One factor is that, even at 4,800 m, temperatures under cloudy conditions in the afternoon are rarely below the freezing point. Heavy hail and ice pellets, common in the sierra, may persist and form a base for snow accumulation so that a cover of 6–8 cm may survive the night only to melt during some sunny spell of the following morning. Seasonal variations do play a part here. The time at which snow is likely to become more persistent is during the occasional wet spell in the mid-year period when temperatures are lower than normal. This allows the snow to settle at relatively low altitudes and to freeze during the night. Early September 1970 in the central sierra of Peru saw unusually heavy precipitation which caused snow to lie as low as 3,700 m; the higher roads around 4,500 m were blocked by snow for several hours at a time, to the detriment of the national car rally then running.

The Ecuadorian sierra contains a series of intermontane basins between the two main cordilleras, each with its own minor climatic peculiarities. North of the Equator, taking a sample of six stations ranging in altitude from 2,230 to 2,980 m (the most northerly being Tulcan at 0°49′N), the average annual rainfall for the period 1960–1969 was 830 mm. July or August are the driest months in all cases, with July receiving only 2% of the annual rainfall. The double maximum is apparent with April and November each recording 14% of the yearly total; January shows the secondary minimum with 6%. This indicates a much closer proximity of the I.T.C.Z. in January than July. The number of rain days varies considerably with Ibarra observing only 112 and Tulcan 200, this latter number being higher than that of Quito though with less total precipitation.

Quito, 0°13′S, the government seat of Ecuador, lies in one of the basins at 2,800–3,000 m above sea level. The mean annual rainfall for the ten years to 1969 was 1,109 mm with a distinct double maximum. There is no complete dry season, and the local saying is that it rains for 13 months of the year. However, July and August, the two driest months, average only 25 mm each. Wide variations can occur for these two months: 1962 saw only 3 mm of rain, whilst 125 mm was recorded two years later. A rapid build-up of the rains soon follows with October averaging 139 mm closely followed by the November total; relatively drier weather prevails in the two succeeding months. The wet season then reaches its main peak of 171 mm in April though once again this is not consistent with 59 mm in April 1967 and 285 mm in the same month 1969. The number of days with rain total 183 ranging from 7 in August to 20 in October and 22 in April. Day-to-day variations are important. A fairly typical day starts dry and sunny with a cloud build-up around noon to be followed by a short heavy shower, but dry sunny or dull drizzly days can still occur in any month. The average diurnal and annual variation of precipitation at Quito, 1948–1957, is shown in Fig.4 and 5, taken from ZIMMERSCHIED (1958).

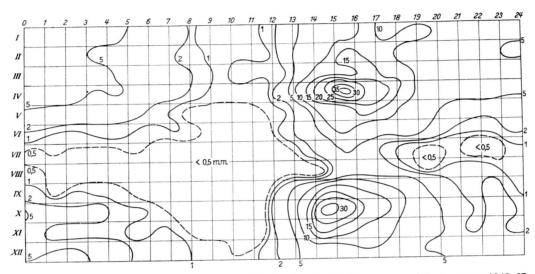

Fig.4. Diurnal and annual variation of precipitation (mm) at Quito, average of the ten years 1948–57. Abscissa: hours, local time; ordinate: months. From ZIMMERSCHIED (1958); courtesy: publisher *Meteorol. Rundschau.*

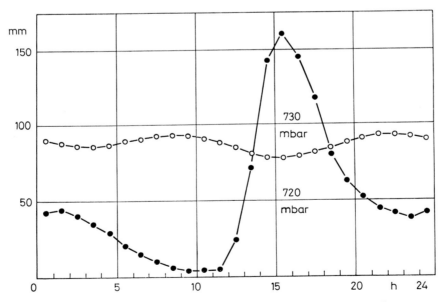

Fig.5. Annual mean of hourly values of precipitation (heavy line, dots), and pressure at station level (open circles), at Quito, 1948–1957. From ZIMMERSCHIED (1958); courtesy: publisher *Meteorol. Rundschau.*

Quito is frequently quoted as being climatically representative of the inter-Andean basins, though this is a questionable statement. The annual rainfall is unusually high being almost double the amount of many of the basins. Quito is possibly in the wettest area of any in the central Andes under 3,500 m, from the northern border of Ecuador to the southern border of Bolivia. The city lies in a relatively wet zone which crosses the Andes from the area of extremely high precipitation centered near Tena (to the southeast) to a humid area along the western Andean foothills (to the northwest). The actual peak is just to the south of the city: Izobamba at 3,058 m some 25 km away averages 1,361 mm. The surrounding countryside is distinctly different from that of the rest of Ecuador. Its rolling green hills immediately to the south with hedgerows and pastureland bear more similarity to parts of northwestern Europe or the New England states than to an equatorial region. This is very localised; immediately to the north of Quito the hills become a greyish brown with irrigated vineyards and cornfields and a noticeable lack of grass. The equatorial monument 28 km from the city stands amidst a semi-arid scrub landscape.

South of Quito, the basins tend to be drier than those of the north with an average of 577 mm annual rainfall, as taken from nine stations to the third parallel south. The same rainfall distribution prevails with wet periods from February to April and again in October and November. July and August form the centre of the dry season with a secondary minimum in January.

Wide variations in rainfall occur, with some sheltered locations becoming very dry. Latacunga, about 100 km south of Quito and at an identical altitude, averages only 475 mm of rain, being protected to the east by a high mountain chain which includes Cotopaxi. The landscape is considerably different from that of Quito with sub-xerophytic plants being much in evidence, sometimes used as field boundaries. Around 2°S, the central valley becomes a virtual desert with only tufts of rough grass and cacti to break the monotony of the sandhills. Pachamama (2°10'S) averages only 382 mm rain despite its altitude of 3,600 m. On the higher levels, the open pampa can become very bleak and inhospitable, such as on the ridges between the basins. Rainfall often rises to around 1,000 mm and with the normally dull, damp weather marshy conditions tend to prevail.

In northern Peru, the relief of the Andes is somewhat similar to that of Ecuador except that the northward flowing Marañon forms a deep valley close to the Ecuadorian border before turning east to the selva. South of 8°S, the western mountain chain presents a solid barrier of some 4,500 m with several ranges of 5,500 m. The eastern cordillera is more dissected; it becomes prominent only in southern Peru. Between the two ranges, a number of basins are again found with the Mantaro being the principal river in central Peru and the Apurimac and Urubamba the main streams of the south.

Cajamarca (7°08'S) standing at 2,621 m in the west of the Marañon basin may be taken as representative of the north Peruvian sierra.[1] Its annual rainfall is 716 mm with a pronounced dry season: June and July together bring only about 2% of the annual amount. The rains normally start in September with a rapid increase to October after which the mean monthly sums remain remarkably constant to February. A further increase is found in March, the wettest month of the year. April follows as the second

---

[1] Editor's note: A description of the meteorological conditions in the upper valley of the Chicama river, 40 km south of Cajamarca, has been given by HOWELL (1952, 1953). For a list of annual sums of rainfall on nine stations in this area, 1940–1963, see HOWELL (1965).

wettest after which the rains stop quite suddenly. This description applies to the years 1959 to 1970 whilst published statistics for 1935–1943 indicate a more gradual build up from November to the March maximum. Though the double maximum is not normal at this latitude, in some years it is not altogether absent. For the period 1964–1970 October was the wettest month, leading in 4 of the 7 years, with March lying in second place. This indicates how variations from year to year and period to period can occur so that one cannot be dogmatic about stating exactly when the wet season starts and finishes. The average October rainfall for 1964–1970 was almost double that of the previous 5 years; 105 mm against 57 mm.

The average number of days with rain (over 0.1 mm) is 115, 61 of which occur in the first four months of the year. The March maximum which generally appears is probably the result of the combined effect of the overhead passage of the I.T.C.Z. plus the high evaporation of the already saturated land surface from some five months of rain. The lowlands and the offshore waters to the west reach their thermal high around this month which may well allow a more westerly penetration of the equatorial trough from Amazonia.

Central and southern Peru lie beyond the direct effect of the I.T.C.Z. They receive their heaviest rainfall when the equatorial trough is at its nearest point and when the thermal low pressure areas to the east and southeast are at their maxima, i.e., January. Much of the rainfall is of a convective nature through the local heating of the land surface by the intense solar radiation. Clouds frequently build up around noon with outbreaks of rain in the afternoon. In Oroya (11°31'S, 3,700 m) for the year November 1968 to October 1969, it was found by JOHNSON (1970) that the rain frequency reached its maximum at 7 p.m. (local time) for the year. The frequency was 22% for this hour during the period November–April as against 4% for much of the morning. During the dry season, May–August, the peak was reached at 6 p.m. with a frequency of 6% against 2% for most hours in the morning. No data are available for rainfall amounts, but the actual rainfall maximum was probably earlier as the heavy convective showers tended to occur during the afternoon with lighter rain in the evening gradually dying away. The total number of days with rain was 151 ranging from 3 in July to 19 in January. January 1969 was rather unusual, with some interesting characteristics which indicate the inconsistency of the sierra climate: the month brought only 54% of the normal January precipitation and 32% of that for the same month in 1970; yet it had more rain days than any of the other months under consideration including both the previous and succeeding months, which in both cases recorded over twice the January rainfall amount. It was a dull, cloudy month with occasional outbreaks of rain, the type of weather which is much more associated with temperate maritime climates in the early winter than with tropical regions. Oroya's annual number of days with ⩾0.1 mm rain (102) appears to be fairly typical for the sierra. Jauja (3,400 m, some 50 km southeast of Oroya) averages 107 days and Cuzco 102. The lower number of wet days yet higher annual rainfall total of Cuzco indicates the greater intensity of rainfall experienced there, a feature typical of the more southern locations.

The localization of rainfall is statistically apparent from the very small correlation between stations only short distances apart. Thunderstorms can be quite violent in the highlands with marked diurnal and seasonal variations (Table III). In Oroya, from November 1968 to October 1969, it was found that noon to 20h00 covered all occurrences with a frequency peak between 15h00 and 16h00. 59 days with thunder were recorded,

TABLE III

AVERAGE MONTHLY NUMBER OF DAYS WITH THUNDERSTORMS AT THREE STATIONS IN THE HIGHLANDS OF THE ANDES[1]

Station :	Sucre	Huancayo	Quito
Latitude (°S):	19°	12°	0°
Altitude (m) :	2,850	3,380	2,818
Record(yrs.):	8	6	7
January	10	12	7
February	8	10	7
March	8	10	12
April	6	10	16
May	3	4	13
June	1	2	8
July	2	2	4
August	2	4	5
September	8	13	7
October	12	13	13
November	15	14	10
December	12	12	7
Year	87	106	109

[1] Sources: Sucre, REED (1928); Huancayo, COILE and CULMSEE (1953); Quito, KNOCH (1930).

28 of these in the months December to February with only three recordings from May to August. Farther south on Titicaca storms can make the lake dangerous so that daily sailings cease around noon.

Central southern Peru is inclined to be drier, particularly in the western ranges; the Arequipa–Juliaca road passes through some very arid countryside despite the high altitude. The dry season becomes more pronounced; Juliaca at 3,825 m with its annual average of 549 mm receives 88% of this from November to April. For Lake Titicaca there is a net annual excess of evaporation over precipitation and only in December to February is the reverse true.[1] The water balance has such a cycle that the lake level in April rises 0.65 m above the December minimum (CEDANO, 1971), on the average.

The Bolivian sierra is dominated by the huge Altiplano, flanked on the east and west by a high cordillera, standing at around 3,800 m above sea level with its own internal drainage system which dries up in the semi-desert of the south. In the north are tufts of coarse bunch grass and a few low shrubs with some pasture in moister spots. In the south only a few stunted bushes can be seen among the sand and stones. The climate here is generally severe which was frequently commented upon by the early Spanish writers. The *Crónicas Potosinas* recorded that in August 1557 snow fell continuously for eleven days to a depth of more than a vara (80 cm) and caused many deaths. At this latitude, the seasonal temperature range is sufficient to make a noticeable difference between summer and winter and for it frequently to be cold enough for snow in the cooler months. Fortunately for the local inhabitants, the cool months are normally very dry. Exceptions still occur; Oruro with an average August precipitation of 9 mm recorded 31 mm in that month of

[1] Editor's note: A more detailed assessment of the precipitation conditions in the Titicaca basin and of the water budget of the lake, 1957–1961, is given in Appendix III on pp.195–199.

1960 and 33 mm in 1964. The average annual rainfall decreases toward the south: La Quiaca at the Argentine–Bolivian border (22.1°S 65.6°W, 3,500 m) has a 60-year average of 320 mm. A greater variation from year to year is felt than in Peru and Ecuador: Charaña in the semi-desert border with Chile received 116 mm in 1965 and 513 mm in 1968. The wet season still, at times, can live up to its name and turn the plateau into seas of mud with only the main highways standing above and even these can submit themselves to the elements.

La Paz, the principal city of Bolivia, lies beneath the towering snow-capped peak of Illimani in a basin below the main central plateau. The city's sheltered position gives it a much more equable climate than the surroundings and with its annual rainfall of 488 mm[1] it appears to climatically relate to its more northern neighbours. The favourable geographical position was an obvious site for at least a regional capital.

Andes—western slopes

The western slopes of the Andes display a wide range of precipitation types, dependent upon altitude and latitude. The only unifying factor, apart possibly from extreme north Ecuador, is the dominance of the single rainy season in the first four or five months of the year.

In Ecuador, the western cordillera appears to form a sharp precipitation divide both to quantity and to seasonal distribution. In the north, a rapid increase in total annual rainfall is experienced west of the Andean peaks to reach a maximum near the base of the foothills. Chiriboga (1,680 m), a mere 30 km west of Quito, averages 1,994 mm against 1,109 for the capital. El Corazón at 1,500 m, 50 km to the west of Ambato, receives 2,204 mm against 420 mm for the latter town at 2,540 m in one of the highland basins. Possibly an even more surprising difference between the two areas is shown in the seasonal rainfall distribution. The central highlands ("sierra") north of 3°S are dominated by the double precipitation maximum which is completely absent in both Chiriboga and El Corazón where October, one of the year's wettest months in the highlands, records only about 4% of the year's total on the western slopes. The wet weather which spreads across the central highlands late in the year obviously only goes as far as the western ridge. The driest months are July and August with a small though steady increase to year's end. The main wet season beginning in late December leads to a climax in March or April, to be followed by a rapid change to drier conditions over the next two months. The May rainfall can vary considerably depending upon the timing of the termination of the wet season. In 1968, the rains finished in April whilst the following year saw them linger into June. The "dry" season is only relative though the more southerly station sees the odd completely rain-free month.

Despite the very humid conditions of the western slopes north of 3°S, a notable rain shadow effect can be seen in the deep valley of the river Mira. Salinas (00°30′N, 1,730 m) averaged 268 mm rain and 54 rain days per annum in its six years of records. Wide variations occurred from year to year so it is difficult to pick out any pattern though November and April appear to be the wettest months with a very dry period from July to September.

---

[1] Editor's note: This is for the ten years 1960–1969. Older series give an average between 500 and 600 mm.

The wettest single month was February 1963 with 114 mm whilst the same month in 1969 saw no rain at all.

Towards the Peruvian border and throughout Peru, the coastal plain is extremely dry so that the western Andean slopes, instead of being wetter than the highlands, become drier and form a transitional zone between the high cordilleras and the coastal desert. For short distances, the climatic belts follow the topography closely. On a broader scale, the south is drier than the north, as the desert moves further up the slope in the higher latitudes. Arequipa, the second city of Peru if one excludes Callao (the port for, and geographically part of, greater Lima) lies in the only extensive plateau west of the continental divide. It has a short season with some rainfall, covering the first three months of the year and sometimes including December. These four months cover 98% of the annual total of 104 mm which combined with the low humidity and high insolation does not encourage vegetation. The dry season is complete. In the 20 years to 1970, it never rained in either July nor August, only once in June and less than one year in five for each of April, May, and September to November. The wet season varies in intensity from year to year; the first three months of 1955 received 462 mm whilst the same period of 1966 saw only 6.5 mm. Above the city there does not appear to be any great increase in precipitation; the dull brown landscape continues up to the snows of the 6,000 m Chachani. The 5,822 m El Misti cannot support permanent snow despite its mean annual temperature of $-8.1°C$.

Coastal plain and islands

The coastline of Peru and southern Ecuador cover one of the driest regions in the world. Farther north dense tropical rain forest is found.

The Galápagos Islands, despite being crossed by the Equator, are under the continuous influence of the southeast trades with a Southern-Hemisphere rainfall pattern very similar to that of coastal Ecuador at the same latitude. Relief is an important factor so that orographic uplift on the southeast slopes can result in heavy rainfall there with very dry lee slopes. The appearance of the land surface of the islands is a good rainfall indicator. Below 200 m the bare volcanic rock often shows no signs of rainfall whilst higher up vegetation often appears, with the 1,000-m summits clad with a mantle of green which descends further down the windward slopes. The difference between the rainfall of the two sides can be seen by comparing Bellavista at 194 m on the south side of the Isla Santa Cruz with Seymour airport at sea level, 30 km to the north. The former station averages 1,100 mm rain per annum, the latter 80 mm. The wet period is universally from January to March or April. Storms can be very heavy over the islands, with some of the heaviest 24-h falls on record in the Republic of Ecuador, see p.205.

The coastal zone of Ecuador displays an interesting variety of rainfall types for such a small area not particularly marked by contrasts in relief. The northern part of the littoral is very wet with no dry season as in San Lorenzo where the period January–May is only slightly rainier than the rest of the year. Though San Lorenzo is close to the Equator (01°17′N), there is no evidence of a double rainfall maximum. A short distance to the south, even before crossing the Equator, a dry season from June to October emerges, particularly on the coast where the wet season ends suddenly in April or May. A north–south climatic border starts to appear, contrasting the relatively dry coastal strip from a "rain belt" stretching along the eastern edge of the plain to Bucay on the Andean foothills east of Guayaquil. On both sides of this belt the rainless season is sharply evidenced by the abundance of leafless trees and dried grass near year's end. Within the rain-belt, the vegetation tends to keep remarkably well with no parched appearances; this can easily be seen by passing either by road or rail from the dry sierra to the wet forests below on the way to Guayaquil. Rainfall totals are over 2,000 mm per year, particularly in the north, with towering cumulonimbi giving heavy downpours of rain.

On the coast itself, the climate and the vegetation tend to change south of Esmeraldas, despite its position still by more than one degree in the Northern Hemisphere. A pronounced dry period from May to year's end prevails with a total number of rain days of only 111, half of which occur in the first four months of the year. Between Esmeraldas and Manta, deciduous trees appear which shed their leaves in the dry season, and Manta itself by September can appear as bleak and burnt as any Peruvian port with dust flying across the pampa to the east. It is here, less than 1° from the Equator, where the first signs of the long west-coast desert begin. Its annual precipitation averaged 172 mm for the ten years to 1969 with variations from 30 mm in 1963 to 301 mm in 1965, virtually all of this in the first four months; there is almost complete aridity for the remainder of the year. It has often been stated that the great Peruvian desert ends suddenly with the border of Ecuador and that the Gulf of Guayaquil sees desert on its south side and tropical rain forests on its northern banks. This, though true, can give a misleading picture. North of this international frontier, the coastline encircles the relatively warm waters of the gulf, and the correspondingly greater evaporation leads to appreciable rainfall on its eastern and northern sides. The coast then moves west towards the cooler waters of the Pacific where the desert forthwith reappears on the immediate coastal strip. Inland from Manta, rainfall increases rapidly. Portoviejo, 30 km to the east-southeast, records over twice the rainfall of Manta, and Pichilingue in the rain belt, 140 km to the east and only 73 m above sea level, averages 1,984 mm of rain.

South of Manta, the patches of green become fewer until one reaches the Santa Elena peninsula, the driest area of the country; the town of Salinas at its tip receives only 56 mm of rain per year, some years with almost none at all. The peninsula is a level area with a few bushes and shrubs standing amidst the sand and dust. Rains that do occur tend to be in short heavy showers. In March 1964, one day saw 107 mm, almost twice the average for a normal year. Eastwards to Guayaquil, the scenery gradually changes in the 140 km journey; plantations appear half way, and close to the city woodlands and the odd swampy area become apparent.

Guayaquil, the largest city and major port of Ecuador, climatically stands on the edge of the semi-arid area to the west, the mangrove swamps of the Guayas valley to the north and the swampy areas with rice fields, banana, sugar and coffee plantations to the east. The city has a distinct wet and dry season regime, the average annual rainfall being 843 mm. The wet season starts abruptly around the turn of the year with November normally being completely dry, December usually so, yet January is one of the year's wettest months with 16 days of rain. The climax is reached in March with 27% of the annual rainfall and 19 rain days. The rains then usually die away in April but can persist into May.

South of Ecuador the main coastal desert starts which, apart from the odd irrigated valley, extends along the entire coastline of Peru and northern Chile, that is, almost 4,000 km first toward the southeast to Arica at 18.5°S, then toward the south to about 30°S, over a latitude span of nearly 28°. Only recently a theory has been developed by

LETTAU (1975) to explain this unique phenomenon, considering the vertical structure and the heat budget of the air masses over the cold coastal waters and over the land east of the Andes, and the ensuing atmospheric circulation. A summary of this theory is given as an additional note in Appendix I (pp.188–192).

The great stability of the coastal climate along with its high humidity results in a regular covering of dense stratocumulus clouds which are almost continuously present from May to October, particularly in the centre and south, with frequent cloud covers in April and November. The remaining months comprise the coastal summer with dry, sunny conditions though with the occasional cloudy day. The cloud cover of the winter half-year is the result of the continuous advection inland of maritime air in which the vertical flux of water vapor is restrained by the presence of a strong temperature inversion (see Fig.6). Cloud and precipitation conditions in this area have been described by PROHASKA (1973) as follows:

"The stratocumulus layer in the marine air is extraordinary in its persistence. During the winter of 1967, 90% of all days were overcast (that is, with a cloud cover of more than 95% in each hourly observation taken at Lima's airport, 3 km inland or 6 km downwind from the coast in the direction of the prevailing wind). Only five days had less than 6 oktas, and the clearest day during the entire period from May 16 through September 30 had an average of 3.4 oktas. The longest consecutive overcast period was 44 days, during which just 10 hours, recorded at six different days, had a cloudiness of only 5 to 7 oktas. In the above specified period (May 16–September 30) only 53 hours, or 1.6% of the time, had clear sky which generally occurred around sunset.

In spite of this stable cloudiness with ceilings generally between 150 and 300 m, fog is rather infrequent at the coast at levels below 100 m (Airport and downtown Lima). An earlier study (GRAVES, 1944) had shown that during the period 1943–1950 low ceiling frequencies below 120 m had an inverse relationship to windspeed. Therefore, it appears

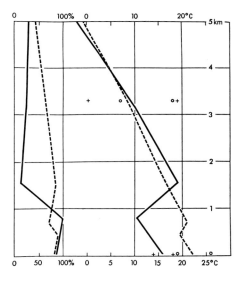

Fig.6. Average temperature and relative humidity over Lima, for February and July 1967, after PRO-HASKA (1973). The small circles near sea level and at 3,380 m indicate the mean maximum and mean minimum for Lima and Huancayo in February; the small crosses the same for July.

that nightly calms might produce fog, but the data for 1967 indicate that usually calms do not last long enough to allow the ceiling to descend to the surface. The cloud cover extends upward to the inversion base and is therefore only a few hundred meters thick. The high cooling rate in the upper part of the cloud due to the net long wave radiation outgoing from the cloud top into the extremely dry atmosphere above, increases the instability in the marine layer and leads to frequent drizzle, called *garua*. This is a typical night and early morning phenomenon all along the Peruvian coast from about 8°S latitude southward. During the 1967 winter proper (May 16–September 30), garua was recorded for 877 hours at Lima airport, that is, 27% of the total time; but this precipitation produced a total amount of only 6.1 mm. It occurred most frequently between 11 p.m. and 9 a.m.; it was least evident in the afternoons between 12 and 5 p.m. This pronounced diurnal variation applies only to coastal strip itself. On the slopes and hills ('lomas') between 100 and about 800 m, where the inland advancing marine air is lifted orographically, almost continuous drizzle is experienced in the contact zone of the cloud with the ground ('fog precipitation')[1]. Here the total amount can increase to between 100 and 200 mm during the winter, which, with the very reduced evaporation loss, is enough to produce the so-called lomas vegetation. As vegetation itself is a fog catcher, once its growth has started the soil beneath the vegetation receives amounts of water several times greater than would have been produced by garua processes on a barren slope. This in turn fosters further vegetation development. Here then is a case of vegetation acting not primarily as a water consuming factor, but as a water producing factor in a self-increasing process. All this is plainly visible on the landscape as a zone of intensely green pasture with grazing cattle between completely desertic landscapes both below and above, which respectively indicate the lower and upper limits of the clouds.

In the last two months of the year, as the stratocumulus cover and its garua precipitation fades away, the vegetation slowly disappears and the landscape once more changes back to the uniform, almost complete desert, which is so characteristic of the lower part of the western slope of the Andes."

In the summer, then, any agricultural activity in the coastal region depends exclusively upon irrigation. Not too much water is available for this purpose. Table IV gives, as an example for the Lima area, the average monthly values of the discharge of the river Rimac, to illustrate how limited this water supply really is.

In southern Peru, similar conditions prevail. The main coast road, on turning inland to Arequipa climbs steadily and passes through a zone of vegetation; though sparse and of poor quality, it is not seen on the coast itself nor on the dust and rock plateau above the coastal inversion. North of about 8°S the garua tends to die away so that the winter in the northern littoral, despite being cloudy, tends to be without precipitation.

The summers are normally dry and sunny everywhere along the coast apart from some foggy upland areas. This is the time of the sierra wet season where regular, heavy rains occur only 50–100 km from the ocean. Occasionally, during intensely wet sierra spells, showers can reach the coastal plain to give the first real rain, in perhaps, several years. This is extremely rare south of Lima but to the north the odd instability showers become

---

[1] Author's note: Some 60 km north of Lima is a permanent fog warning sign on the main Panamerican Highway after which the road starts to climb onto a high coastal hill area. Here the inevitable mist, even in the summer, envelops the road between clumps of bright green shrubs.

TABLE IV

AVERAGE MONTHLY DISCHARGE (m³/sec) OF THE RIVER RIMAC AT STATION CHOSICA (11°55'S 76°42'W, 850 m, 20 km east-northeast of Lima, years 1921–1960)
(After T. A. CORNEJO, 1970, and personal communication, 1972)

I	II	III	IV	V	VI	VII	VIII	IX	X	XI	XII
40	64	85	40	20	15	12	10	10	12	15	20

more likely until the northern border is reached where some rain is normal around the turn of the year. In Lima itself, rain is news which, when it does happen, is normally just a few heavy drops in the early evening. January 15th, 1970 was an exception when heavy rain in the early evening caused discomfort to the inhabitants who have not designed their city to cope with such phenomena. The undrained streets were awash and many poor roofs on otherwise well constructed houses let in the water. These rains are not of Pacific origin but are a spillover of the intense precipitation of the Andes. On such days, cumulonimbus clouds can be seen moving from the east at an altitude where a relatively moist layer exists; the surface inversion may then disappear. This is rare due to the intense orographic precipitation on the windward east side of the Andes and the drying of the same air by a forced downward movement of the leeward side. The seasonal rainfall distributions of the north and south littorals therefore differ as can be seen in Table II. The south receives the bulk of its annual precipitation in the winter months in the form of drizzle, the north receives what little rain there is in the form of summer showers.

The rainfall totals along this littoral are extremely low by any standard, compared with other coastal locations of similar latitudes. In the 10 years to 1970, Chimbote, 9°S, averaged 14 mm compared with 323 mm at Luanda set on the Angola coast in a somewhat similar position. Many of the averages are misleading with the occasional "wet" spell obscuring what may be zero precipitation in a normal year. Talara, less than 5° from the Equator, averaged 13 mm over a 20 year period, most of which fell in 1953 and 1958 whilst the last ten years grossed only 4 mm, half of which was recorded in March 1962 and the other 2 mm in April 1969. Probably the lowest long term averages are found in the south, beyond the range of any instability shower but favourably located away from the main foggy areas. Pisco (13°45'S) has a 16 year average of 2.2 mm. The more inland places tend to see a slight increase due, in the north, to the greater likelihood of the odd shower as in Piura (39 mm) or in the south, to the forced uplift of the coastal mist as for Tacna, 558 m above sea level (44 mm); over half of the latter amount was recorded in August and September.

The normal conditions as previously illustrated are, in some years, replaced by contrasting weather features of sufficient magnitude to warrant a separate analysis. It is the *El Niño* phenomenon which has received much discussion in the literature (SCHWEIGGER, 1959; SCHÜTTE, 1968). This name, alluding to their occurrence around Christmas time, is given to periods when the coastal waters are warmer than usual, which can drastically influence the weather of the region. It was formerly thought that the Humboldt current was replaced by a warm southward flowing stream from beyond the Equator, but now it seems unlikely that such a change could occur very far south; so now the term applies to any substantial oceanic temperature increase. There is evidence that El Niño occurs in years when the trades of both hemispheres are weak and the equatorial trough goes farther to the south in the southern summer than usual, relieving the southerly winds which maintain the upwelling. Warming of the coastal waters may occur as far south as Pisco Bay 13°45′S. BJERKNES (1966) stated: "... in weak trade wind regimes the heat transfer to the atmosphere, both in sensible and latent form, is below normal, and more than normal amounts of heat remain stored in this ocean, thus raising its surface temperature ... Weakness of the trade winds has more sudden and spectacular ocean effects in terms of temperature induced by cessation of upwelling ...". The occurrences were particularly severe in 1891, 1925, and 1953. Following 1925, the observed seven-yearly frequency brought correct forecasts for similar, though less pronounced, events in 1932 and 1939. The 1953 episode followed two phases later though since then the frequency has been more variable; 1965 saw a very mild form.[1]

TABLE V

EL NIÑO EFFECT, 1925

Location	Latitude	Main period of El Niño	Duration (days)	Average temperature of water surface	
				March normal (°C)	March 1925 (°C)
Off Lobitos	4°20′S	Jan. 20–Apr. 6	76	22	27
Off Puerto Chicama	7°40′S	Jan. 30–Apr. 2	63	21	27
Off Callao	12°00′S	Mar. 12–Mar. 27	15	19	25
Off Pisco	13°45′S	Mar. 16–Mar. 24	8	19	22

Table V shows offshore water surface temperatures for March 1925 in comparison to the March normal with the duration of the unusually warm waters. It should be noted that the sea temperature has normally just passed its peak in March. This increase in sea temperature obviously affects air temperatures but a more dramatic result are the heavy rains which then suddenly appear in the coastal desert. The change in the weather, including the rain, appears to come at a time when the temperature of the ocean begins to equal or exceed the mean temperature of the atmosphere; the exceptional rainfall continues just as long as this altered ratio is maintained. Trujillo recorded 395 mm rain in March 1925 in contrast to an average of 35 mm for the previous seven years (KNOCH, 1930). The desert was said to have become a garden in that year but with the less pleasant addition of insects and reptiles.

---

[1] Editor's note: The recurrence of the El Niño phenomenon about every seventh year appears to be old weatherlore in the coastal area of northern Peru. It already is mentioned by SEARS (1895) who also makes another interesting remark: The march of Francisco Pizarro with his troops and pack-animals through the desert southeastward from Tumbes to Cajamarca, in 1532, could not have succeeded in a climatically normal year. Anomalous rain in the preceding months, also made evident by reports on the state of two rivers, had brought about sufficient vegetation to provide food for the animals. SCHWEIGGER (1959, p.235) also writes about this idea and, after discussing several old Peruvian sources, argues against it.

TABLE VI

COMPARISON OF PRECIPITATION AMOUNTS FALLEN IN MARCH AND APRIL 1965, WITH NORMAL VALUES[1]

Station	Latitude	Total March+April 1965 (mm)	Total March+April normal (mm)	Ratio 1965/normal
Muisne	00°37′N	823	397	2.1
Jama	00°12′S	521	259	2.0
Manta	00°57′S	246	57	4.3
Salinas	02°11′S	111	32	3.5
Playas	02°39′S	487	86	5.7
P. Bolívar	03°16′S	596	79	7.5
Piura	05°11′S	76	4	19.0
Chiclayo	06°47′S	73	4	18.3
Chimbote	09°10′S	17	2	8.5
Galápagos Islands:				
Seymour	00°26′S	123	40	3.1
Bellavista	00°42′S	427	161	2.7
P. Baquerizo	00°54′S	271	103	2.6

[1] "Normal" is the average of March and April for the years 1960–1964 and 1966–1969.

The most devastating consequence is that inflicted upon the wildlife. The main food of the guano bird is anchovies which migrate away when the El Niño occurs. Hence, the fishing industry and the birds are similarly affected by the change in the sealife. The guano industry is also set back due to the heavy rains washing the guano into the sea. These spasmodic "años de abundancia" (years of abundance) can, of course, make climatic averages meaningless. Reference has already been made to the rainfall figures for Talara and the influence of the odd wet spell on the averages. April 1953 alone recorded 40% of the total 20 years rainfall.

1965 was the most recent year in which the El Niño effect was felt though only in a mild form.[1] As on previous occasions it occurred towards the end of the Southern Hemisphere summer when the Pacific waters are at their warmest. Table VI shows a range of coastal stations in Ecuador and northern Peru with a comparison of the average March and April rainfall totals for the recent decade excluding 1965, against the March and April totals for that year. The southern limit was Chimbote; Lima remained dry but with a much longer and warmer summer than usual.

Studies of the El Niño phenomenon are usually restricted to Peru with brief reference to heavy rains in southeast Ecuador.[2] Here it can be seen that the influence is felt along the entire coastal plain of Ecuador into the Northern Hemisphere. The most dramatic change is in northern Peru where any rain is a noticeable event, but a twofold increase was felt in many northwest Ecuadorian stations and even in the Galápagos Islands. In the

[1] The El Niño phenomenon has since appeared in pronounced form in March and April, 1972.
[2] Editor's note: In a recent study CAVIEDES (1973) points out that the El Niño phenomenon and the droughts in northeastern Brazil are related through anomalous displacements of the I.T.C.Z., and tend to appear in the same or in immediately preceding or following years. His list of related events contains twelve distinct appearances.

case of Bellavista the rains commenced a little later and lasted into May when the recorded precipitation was 288 mm against an average of 43 mm for the five years of records excluding 1965.

Temperature

General

Relief and the Humboldt current once again play important roles. The cold Pacific waters reduce the annual temperatures of the coast and the lower west Andean slopes of Peru and southwest Ecuador to levels not experienced in other lowland equatorial regions.

A variety of temperature regimes can be found on the Andean slopes, which has led to the popular concept of dividing the Andes into several vertical climatic regions. Boundary lines have been allocated somewhat arbitrarily, though a steady transition is normally found along the whole lengths of the valleys, and the cooling effect of the Humboldt current on the western side in Peru would have to be disregarded to make any such scheme fit. It was probably in Ecuador where the famous "tierra" (land) levels were documented. The regions of the first 1,000 m above sea level were called the "tierra caliente" (hot land), from here to 2,000 m the "tierra templada" (temperate land) and from 2,000 to 3,000 m as "tierra fría" (cold land). This latter title is something of a misnomer as the average annual temperature of this layer varies from about 13°C to 18°C, warmer than many so called temperate regions of the world. Above 3,000 m the term "páramos" (bleak uplands) is used up to the snow line. Such vertical divisions are not popularly used in Peruvian literature, as the western slopes are generally cooler than those of the east and the west is also drier, so that geographical texts have tended to treat the two sides separately. In Bolivia, the eastern slopes are more extensive and economically more important than the equivalent areas farther north. The valleys are sometimes classified into three zones, i.e., the lower up to 1,700 m, the middle around 2,500 m and the valley heads just over the 3,000 m level. Above these comes the puna (similar to paramos) around 3,600 m for the Altiplano, and higher up towards the snow line is the puna brava (fierce or rigorous bleak uplands). These latter two terms are also used in Peru. The greater attention to terms to describe the various higher levels in Bolivia than in Ecuador is indicative of the relatively greater importance of the high sierra, for in Ecuador very few towns are found over 3,000 m.

The coast of Peru and the adjacent slopes of the Andes up to the top of the inversion are thermally strongly influenced by the cold Pacific waters. Comparisons of 10 years averages for Lima with the U.S. Standard Atmosphere at 15°N show that up to 800 mbar (2 km) the troposphere is markedly cooler than the Standard Atmosphere but between 700 and 270 mbar (10 km) it becomes as much as 4°C warmer. The average temperature and moisture conditions over Lima are shown in Fig.6. In this region, there is little relation between altitude and temperature. In winter, the free atmosphere at an altitude of 1,500 m may be warmer than at sea level and daily average temperatures at 2,300 m are similar to those of the coast, but to find similar daily maxima one has to ascend to 3,500–4,000 m. On the eastern side of the Andes there is again evidence of an inversion or isothermal layer somewhere between 1,000 and 2,500 m above sea level as suggested by Table VII

TABLE VII

APPROXIMATE MEAN ANNUAL TEMPERATURE IN RELATION TO ALTITUDE[1]

Altitude (m) above sea level :	0	500	1,000	1,500	2,000	2,500	3,000	3,500	4,000	4,500
Temperature (°C):	27	24	21	20	17	16	13	9	6	3

[1] Applicable to Ecuador and east and central Peru, excluding the coastal regions.

though not as pronounced as that above the coast. By regression analysis of some 50 stations in Peru and Ecuador a clear relationship emerged between temperature and altitude from an annual mean of 15.8°C at 2,500 m to 2.9°C at 4,500 m, giving a lapse rate of 6.5°/1,000 m. The lowest thousand metres produced a similar relationship for stations north of 15°S, a rate of 6.8°C/1,000 m. The sea-level figure of 27.4°C is an extrapolation, but it is reasonable in view of the 26.4°C average for Iquitos (104 m) and 27.1°C for Riberalta (172 m).

The average lapse rate from 1,000 to 2,500 m is 3.2°/1,000 m. Unfortunately, the available records are insufficient to make a thorough study of this altitude range, but there is a suggestion of two stable layers. Very little change appears between 2,500 and 2,000 m whilst from here down to 1,500 m a more normal lapse rate reappears. In the layer 1,000–1,500 m almost no temperature change is apparent; indeed, an inversion may exist. For the daily maxima an increase with altitude clearly takes place, Malacatos (4°S, 1,600 m) having a 3°C higher value than the stations some 600 m lower.

Annual temperature variations are dependent upon latitude, proximity of the Humboldt current and seasonal differences in cloud cover and humidity. The latitudinal effect is basically that the farther from the Equator one goes, the greater is the annual variation. The Humboldt current, though varying little in itself, has a profound and seasonally varying influence on the lower troposphere and the resultant cloud cover so that a large seasonal temperature variation (for such a low latitude) is found close to the Pacific seaboard; this is another unusual characteristic of the area. Cloud cover variations have a wider implication still. The rains which move south late in the year tend to follow the overhead passage of the sun; in most of Peru and Bolivia the rainy season is not fully effective until December. The warmest weather, therefore, is frequently found in October and November over the sierra and selva when the noon sun is nearly overhead and when there are still frequent sunny spells. This affects particularly the daily maxima and the year's highest temperatures. From December to March the heavy cloud cover lowers day time temperatures though the reduced loss of heat by long wave radiation inevitably leads to the highest mean minimum temperatures in this period. The occasional sunny day in the wet season will result in warm weather as the hours of daylight are relatively long from December to February when the midday sun is making its second overhead passage of the southern summer. In April and May, the clearer days and increased sunshine will tend to counterbalance the effect of the reduced daylight hours and lower position of the sun and can sometimes result in a slight increase in daily maximum temperatures, but usually they remain fairly constant for the first five months of the year. Night minima, in contrast, normally show a sharp decline following the departure of the rains. The months June to

August are the driest for all of Bolivia and Peru away from the coast, and invariably have the coolest nights. The nights are at their longest in this period and, combined with the clear skies and low humidity, heat is rapidly lost after dusk through outgoing radiation. In Bolivia the days are also relatively cool, though the high insolation leads to maximum temperatures only 2°–4°C below those of the early part of the year. The odd cloudy day can result in much lower temperatures. Similarly, in the southern districts of Peru the daily maxima tend to be at their lowest in the middle months of the year, but farther north the smaller reduction in daylight hours leads to a lesser reduction in temperatures. On the western slopes in the mid-year months, the combined effect of high insolation with subsiding air from the subtropical high, at its maximum at this time of the year, results in many areas having their warmest days when the coast is said to be having its winter.

In Ecuador virtually no change in day length is experienced through the year and the precipitation regimes of the sierra and selva do not show such pronounced changes as those farther south. Temperature variations are therefore small. Over the eastern lowlands the annual range between the coolest month (still normally July) and the warmest month (around year's end) is rarely more than 2°C. The sierra is similar, but there is very little consistency between stations. June–August again tends to be the cooler period at several stations; nevertheless at Quito August happens to be the year's second warmest month, September is first. There is a tendency for a double thermal maximum to coincide with the double overhead passage of the sun though the differences are too small to be of any importance.

The coastal lowlands of Ecuador show very little difference between the months apart from the extreme southwest where the Humboldt starts to make itself felt so that from here southwards there is a marked difference between the cool cloudy mid-year months and the warm sunny period from December to March.

Maximum temperatures for the region tend to be found in the summer months of the drier lowland areas. The tropical rain forests are not noted for outstanding temperatures and 40°C is extremely rare despite the regular occurrence of days with 35°C. The highest published temperature for Ecuador in the eleven years to 1969 was 39.8°C in April 1963 in the central coastal plain; 38.6°C was the highest for the same period in the east, at Putumayo in January 1968. Trinidad on the southern edge of the jungle recorded 40.3°C in January 1960 and Pucallpa in Peru has recorded 42°C, an exceedingly high value for a rain forest area though the annual rainfall here tends to be low for an Amazonian location. Southwards onto the plains of the Chaco the summers are exceedingly hot with 40°C frequently exceeded in the last four months of the year and into January and February during years when the wet season is not so pronounced. During the 1960's, Santa Cruz recorded 40.5°C in December 1967, Yacuiba on the Argentinian border 41.5°C in November 1963 and Camiri (20°S) 43.0°C in October 1965. In southern Bolivia, high temperatures are not necessarily confined to lowland stations. Oploca (21°19′S) at 3,120 m recorded 39.0°C in January 1961. This could be a world record temperature for such a high altitude station caused by a favourable set of circumstances. In the preceding year there was no rainfall from April to October and only 42 mm in the last two months; the month concerned had 15 mm, less than 25% of the January normal. The ground must have been very dry, the cloudiness less than normal. The mean maximum temperature resulted to be 33.9°C, 6.2°C above the January normal.

Extremely low temperatures are very much a function of altitude and latitude. The lowest published temperature for Ecuador to 1969 was —6.8°C at Rio Pita (3,860 m) in December 1968. As there is no pronounced dry or cool season in Ecuador, temperatures do not fall to the levels experienced in Bolivia. It is in the Titicaca basin and farther southward where in mid-year severe night frosts are experienced. Peru's record low is on the edge of this basin at Imata (4,405 m on the Arequipa–Puno railway) where —25°C was recorded in July 1961. It is in the frost hollows of the zone 4,000–4,500 m that the lowest temperatures appear to occur, for on the summit of the nearby El Misti, some 1,400 m higher, the lowest observed in 27 months of records was —17.8°C. On the Bolivian Altiplano —20°C is not unusual. Charaña on the border with Chile has recorded —23.5°C and Oruro —22.4°C.

Eastern lowlands

The eastern lowlands between 10°S and the Colombian border are hot and humid with almost no seasonal temperature variations though with moderate diurnal ranges. The lower plains average 25°–26°C ranging from around 32°C for the day maximum to about 21°C at night. Monthly extreme highs are usually in the range 34°–37°C in any month though the extreme lows are more variable. The occasional mid-year cold spell from the south can push the thermometer down to below 10°C, but usually 13°–15°C are the lower limit. The lowest record in eight years in Tiputini (00°45′S) at 220 m was 14°C.

Seasonal variations are too small to indicate any real trend. In Iquitos the range between the warmest and coolest months is 1.7°C and in Tingo María at 665 m it is 0.9°C. This latter figure at first sight is surprisingly low being from a station 9° from the Equator but in fact this hides a slight trend towards cooler nights in the dry season from June to September, balanced by the rather warmer days in this period.

Southwards to the edge of the tropical rain forest, the first signs of significant changes appear. The temperatures for the first four months of the year are similar to those further north, but a slight decrease occurs in the period May–July when maximum temperatures frequently stay below 30°C and night minima drop to 16°C, and as far as 6° or 8°C in the odd cool spell. Coinciding with the southward track of the sun, ahead of the rainy season, day temperatures show an increase in August, leading to a September or October peak. This change is around 3°C between July and September. The increase in the nightly minimum tends to occur later along with rains so that the highest mean daily temperature tends to occur in November, i.e., when the night-time temperatures are approaching their peak and the day-time values have only just passed theirs. Riberalta (12°S, 172 m) appears to be the warmest place in the entire three countries. Its mean annual temperature is 27.1°C. The warmest month is September with a mean of 28.7°C and a mean daily maximum of 34.1°C. In 1963, both September and October averaged 30.2°C; this happened after an exceptionally dry mid-year period, the September being completely without rain and October receiving only 57 mm against a normal 133 mm. The September of the following year recorded 146 mm rain and was 3° cooler.

South of the selva onto the plains of southeast Bolivia the middle months of the year show a noticeable thermal decrease with the annual averages below 25°C. The annual range increases to the south. The first lowland frosts appear around 20°S in the period May–September. Yacuíba recorded —7.0°C in July 1969. The mean daily temperature ranges are a little greater than farther north being about 12°C in the most months, but the monthly extremes are very much more. This area, during the mid-year dry season, frequently sees a 40°C range between the absolute monthly extremes; at Yacuíba, it was as much as 42.7°C in September 1966, from 38.7°C down to —4.0°C.

Andes—eastern slopes

Above the 1,000-m contour of the eastern Andean front the steady decrease of temperature with altitude tends to be modified. It is around this level where the deep narrow valleys begin; farther down they are wider and less confined, with a tendency for the rivers to be braided due to an overload of sediment.

The lower slopes can record some very high temperatures in favourable conditions. San Ramón (11°S) is only 830 m above sea level but is within the Andean valleys and outside the rain forests with a sunny climate and only moderate rains. Its record high temperature is 42.0°C. Tingo María, 2° farther north and 160 m lower, is situated well into the dense jungle with almost double the annual rainfall and a record high of only 37°C. This indicates the effect of the different rainfall and site factors. South into Bolivia, similar conditions prevail: the lower valleys of the Yungas region average around 21°C, the middle valleys (2,500 m) about 17°C, and the upper reaches of the valleys become cool with temperatures around 14°–15°C. Annual variations are not large. The high temperatures of November prevail at this altitude as well as for the plains, coinciding with the overhead position of the noon sun and the period immediately prior to the main rains. The temperatures of November vary inversely to the amount of rain and cloudiness.

Towards southern Bolivia diurnal and seasonal variations become pronounced. There appears to be quite a sudden change between the 17th and 20th parallels from even year-round temperatures to a seasonal variation of 7°C to 10°C for both lowlands and highlands.

Andes—central

The temperatures of the sierra depend primarily upon altitude but towards the south the greater seasonal variations in day length become important, thus bringing in the influence of latitude. Table VII shows the approximate average temperatures in relation to altitude, with a uniform lapse rate between 2,500 and 4,500 m. By extrapolation, this line continued to sea level would reach a temperature of 31.9°C which (as annual mean value) probably does not exist anywhere. Thus it appears that regardless of the effects of heavy clouds on temperature at the lower levels, the sierra is anomalously warm. The warming of the subsiding air on the edge of the subtropical high above the inversion layer is one reason. There is also the effect of the intense insolation which must heat the land surface to temperatures above that of surrounding free air. Above 4,500 m data are scarce, but by continuing the trend in Table VII the freezing point is reached at 4,950 m which is close to the snow line in humid areas. Extreme temperatures show a less clear pattern because site features become important, though generally the decrease of extreme and average maxima with altitude is less than that for the average temperatures. This indicates a greater diurnal range in the high sierra up to about 4,300 m. The same month– July 1961—that Imata recorded Peru's lowest temperature of —25°C, a daily maximum of 20.3°C was also recorded, giving a range of 45.3°C. A diurnal range of 25°C is common in the southern sierra and the Bolivian Altiplano during the months June to August. Charaña at 4,057 m had an average diurnal range of 31.9°C for August 1969 and an extreme range of 42.5°C.

The relationship between altitude and night minima is very indefinite but steep lapse rates can occur as shown by a study carried out near the head of the Cañete valley just west of the continental divide at 12°S, for the period July 15th to August 27th, 1970. Two valley floor locations were chosen, one at 4,100 and the other at 3,600 m, some 10 km downstream by road. The mean minimum for the higher station was −7°C with a range from −12° to −1°C to be compared with an average of −2°C and a range of −6° to +3°C for the lower. The period was generally clear and dry, and the deep valley sites of both stations were ideal for high nocturnal cooling. The average difference of 5°C is large considering the altitudinal range of only 500 m. In Ecuador, night minimum temperatures tend to follow the trend observed elsewhere in the sierra but on a reduced scale. Following a peak in February–April they steadily drop to a July–September low.

In Peru, the mid-year dry season becomes more pronounced and increases to the south. This has two major effects on temperature. Firstly, the increased number of sunshine hours tends to offset the consequence of shorter days and the lower height angle of the sun to keep day temperatures up to a relatively high level. Secondly, the clear skies combine with the longer nights to give lower minima and therefore greater diurnal ranges.

The night minima display considerable seasonal variations in all locations. The mildest nights occur around the start of the year. During mid-year, nights are relatively long, dry and clear, favoring sharp temperature drops. Night frost is common in the sierra from May to August to be followed by a rapid rise in temperature after sun-rise. It is not unusual to be quite comfortable in the strong mid-morning sun and yet still see ice on the ground.

It is in the highlands of Bolivia where the excessive winter cold and extreme day and night temperature differences are felt. In the months May–September, the air is clear and thin in the high altitude, the water vapor content is small. Strong daytime insolation burns through the frigid air to heat up the dry land surface. Night frost is normal from the Titicaca basin southwards though the lake itself tends to ameliorate the temperatures along its edges; the Peruvian lake port of Puno does not record the hard frosts which are experienced in Juliaca, 40 km to the northwest. Mean night minima are regularly around −10°C in much of the Altiplano, with extremes to below −20°C. The warmest nights are during the wet season when the mean minimum is normally above freezing point though the occasional frost can still be expected. Oruro, which occupies a fairly central position in the Altiplano, sees an average minimum of 4°C in January and February, which falls below freezing point in April, reaches a low of −11.1°C in July and rises above freezing point again in November. The influence of Lake Poopo on the immediate surroundings is also quite pronounced. The lakeside station of Tacagua has a mean annual night minimum of 5.2°C higher than that of Oruro some 120 km away at the same altitude of 3,700 m; the difference reaches 8.4°C in July when the unfrozen waters of the lake prevent extremely cold nights. Day-time temperatures show a much smaller annual variation.

La Paz, despite its position between 3,500 and 3,800 m on the edge of the Altiplano, has a much more equable climate than any true Altiplano location. It lies at the bottom of a deep valley on a perennial stream in a position sheltered from the fierce plateau winds.

The mean daily maximum ranges from 16.4°C in July to 19.1°C in November which is about normal for Altiplano conditions, but the nights are much milder than anywhere else in highland Bolivia. Frost is rare with some years completely missing it.

The volcanic cone of El Misti (16°S, 5,822 m), west of the Titicaca basin and overlooking Arequipa, provides some interesting statistics. The astronomical observatory of Harvard University established a meteorological station on the summit, in operation from October 1893 to December 1895, with a second station located 1,100 m below the summit, known as Mont Blanc, for 1894 and 1895 (BAILEY, 1899, and later, HANN, 1909). Table VIII shows a summary of the hourly thermograph evaluations for the period. The full daily range data are not available so that the average of the coldest and warmest hours of each month have been substituted.

A number of interesting facts emerge from these values. The annual variation of the mean monthly temperatures for Mont Blanc (2.3°C) is similar to Arequipa to the west (2.7°C) but less than that of Juliaca at a closer altitude to the east (5.2°C). The summit has a greater annual range (4.4°C). Compared with other sierra stations the seasonal range for the day maxima is high whilst that of the night minima is exceptionally low.

TABLE VIII

EL MISTI—TEMPERATURE DATA (October 1893–December 1895)
(After HANN, 1909)

Elevation (m)	Temp. (°C)	Jan.	Feb.	Mar.	Apr.	May	June	July	Aug.	Sept.	Oct.	Nov.	Dec.	Year
5,835	max.[1]	−2.7	−3.2	−3.6	−4.3	−5.4	−5.6	−4.9	−4.4	−4.0	−2.7	−1.8	−2.2	−3.7
5,835	min.[1]	−8.3	−8.4	−9.1	−10.2	−11.6	−12.2	−12.4	−12.6	−12.2	−10.7	−9.3	−8.6	−10.5
5,835	mean	−5.9	−6.1	−6.8	−7.8	−9.1	−10.0	−10.3	−9.7	−8.7	−7.4	−6.5	−6.1	−7.9
4,750	mean	1.7	1.8	1.6	1.0	0.2	−0.3	−0.5	−0.2	0.4	0.9	1.1	1.4	0.8

[1] Maximum and minimum are averages of the warmest and coldest hour of each day, respectively.
(Recent measurements show the summit of El Misti to be 5,822 m).

Andes—western slopes

The western Andean slopes of Ecuador are climatically similar to those of the east by lying between hot humid lowlands and warm, relatively dry highlands. An inversion layer again apparently exists, with some variation in the course of the day, as shown in Table IX. Seasonal variations are very small (< 1°C).

In Peru, the coastal inversion is much more extreme and visibly identifiable by a layer of stratus clouds for over half the year. Above the level of the inversion (see Fig.6), the sinking motion is strongest and most persistent in the mid-year months. The corresponding warming effect and the more favourable insolation conditions lead to warmer weather in the Southern Hemisphere "winter" than in the "summer". Moving inland from the coast, the annual range of temperature decreases and becomes nil around the 1,500-m contour which is the same altitude where the highest annual temperature is

TABLE IX

COMPARATIVE TEMPERATURE DATA—ECUADORIAN ANDES

Location	Temperature (°C)					
	mean 07h00	mean 13h00	mean 19h00	mean day	extremes	
					max.	min.
Quito[1]	9.4	19.2	13.0	13.0	27.6	2.3
Salinas[2]	16.6	25.8	16.7	19.7	32.8	5.0
Chiriboga[3]	11.6	20.0	16.9	16.1	28.4	4.0

[1] 00°13′S 78°30′W; 2,818 m.
[2] 00°30′N 78°08′W; 1,730 m.
[3] 00°16′S 78°44′W; 1,680 m.

observed. Above this level, the annual variation becomes slightly inverted with the dry season being warmer than the rainy season. Monthly temperatures at intermediate levels on western slopes depend more on the annual variation of the mean maxima than the mean minima because of good air drainage. According to PROHASKA (1969) the mean daily maximum during the period January through March 1967 in Matucana (2,380 m) was 17.8°C and during July–September 20.8°C. During the same periods, the minima were 11.1° and 9.5°C respectively, giving mean temperatures of 14.4°C for the southern summer and 15.2°C for the southern winter.

In southern Peru, the seasonal variations in day length and angle of the sun's rays tend to affect the opposing thermal influence of the seasonal subsidence changes. In Arequipa July is the coolest month with an average temperature 2.7°C below that of November, the warmest. The onset of the rains in December, little though they may be, then depresses temperatures slightly. The effect of the rains on temperature is shown by the 1965–1966 wet season which did not start until February; January 1966 was the warmest of any month in the 21 years to 1970, with an average of 18.9°C, 3.1°C above the January norm. The mid-year dip in temperature is mainly due to the cooler nights in this period. Frosts occasionally occur from June to October with a record low of −4.0°C in July 1953. The average diurnal range is 16°C, reaching 17.6°C in August, so that hourly temperature changes are more important than seasonal variations. Nights are cool to be followed by rapidly rising morning temperatures with the afternoons being considered as "hot", particularly under the intense rays of the sun. The extremes of the Arequipa weather sometimes receive attention by the Peruvian press. It was reported in the dry season of 1971 that one day at noon the thermometer at Arequipa reached 28.8°C with a relative humidity of 16%, followed by a minimum of 5.8°C at 05h00. The range of 23°C was regarded by the Lima press as a health hazard.

Coastal plain and islands

The coastal plain of Ecuador is dominated by year-round hot weather with almost no seasonal variation apart from the extreme southwest where the mid part of the year is noticeably cooler. The mean daily maximum is usually around 30°C and the night minimum around 20°C; most inland lowland areas remain within 2°C of each of these. Coastal sites show a reduced diurnal range as is normal.

Around 1°S, the annual range becomes at least 2°C. Portoviejo, a short distance from the coast, sees a range of 2.3°C for both maximum and minimum temperatures. For the daytime, the coolest month (June) is only two months after the warmest, to coincide with the change from the wet to the dry seasons. The relatively cool southeast trades tend to become more dominant in May and June, replacing the northerly winds of the rainy season.

Guayaquil on the southern edge of the plain is a hot, humid city with daily maximum temperatures normally over 30°C from December to May, coinciding with the rains. December tends to see the year's highest temperature; that occurred in six of the ten years to 1969. In this month, relative humidity normally is at its lowest for the rains have not yet seriously begun. The warmest nights occur in March which is, on the average, the month with the greatest number of rain days. July records the coolest days and nights; then the southerly winds come from the nearby cool Peruvian coast which is normally under a thick cloud layer at this time.

To the west of Guayaquil lies the Santa Elena peninsula which has a thermal regime more akin to that of Peru than to the Ecuadorian littoral in general. Salinas, at the tip, is the driest and thermally the most seasonally variable place of this lowland area. A decrease of 7.3°C in the daily maxima occurs from March to August and 5.1°C for the night minima from February to August. The mean daily range for this latter month is only 4.6°C. This is the first real sign of the tropical anomaly of small diurnal and large seasonal temperature variations which typifies the entire Peruvian coastal strip.

The Galápagos Islands bear some similarity to the southwestern coast of Ecuador despite the distance of about 1,000 km. The seasonal temperature changes (5° at Puerto Baquerizo and 6° at Bellavista on Santa Cruz) are unusually high for island stations lying almost on the Equator and must be primarily the result of ocean temperature changes. The climate is what one might expect some 2,000 km from the Equator.

The principal single feature of the temperature of the Peruvian coastline is its anomalous coolness, particularly in the winter. PROHASKA (1973) stated that: "In Lima the winter temperature is 9°C below the latitudinal average, a temperature observed in Salvador (Bahía) at the Atlantic coast of South America at the same latitude. In the Northern Hemisphere the same latitude and exposure (Pacific coast of Nicaragua) has 10° higher temperatures than Lima. To find temperatures similar to those in Lima, one has to go to 24°S along the west coast of Africa and to 28°S along the west coast of Australia."

The coolness of the Humboldt current and variations in the overhead inversion are mainly responsible for the thermal peculiarities of this littoral. During the winter months the inversion with its thick low clouds is almost always present. The clouds are sufficiently dense to prevent much insolation from reaching the land surface so that temperatures are virtually a function of those of the coastal waters. The diurnal range is correspondingly small; in the month of July, it is only 4°C on the average for Callao and rarely over 6°C for any station farther south. In the summer the cloud layer is not normally present, so that the sun which is at a near overhead position at midday can raise the day-time temperatures quite considerably.

The area affected by these climatic features is a very narrow strip of land. Inland of Lima, between 12 and 24 km on the central west–east highway are fog warning signs beyond which the winter fog and cloud cease, indicating that here the width of the coastal climatic zone amounts to no more than 30 km. Of course, even within this belt some east–west variances are found. Even in mid-summer Callao tends to experience frequent mist and low clouds.

The change from winter to summer tends to be gradual with periods of winter and summer types alternating until the latter becomes more frequent and extended, and eventually prevails around year's end. In autumn, the transition may take place in a few days with a sudden change in the inversion. In 1967 this change in Lima occurred in one day, May 16, when the warm, sunny weather was suddenly replaced by cool damp weather. A composite graph designed by PROHASKA (1973) has been reproduced in Fig.7; it eloquently describes the different character of the two halves of the month. The time of this change varies from year to year so that average temperatures do not indicate such a sudden transition.

Variations in temperature occur along the coast with a fairly steady gradient from an annual mean of 21°C at Talara (4°34′S) to 19°C at Callao (12°C). Farther south, less variation is found. The Lima–Callao conurbation is located in a particularly cool and humid part of the coast. Though the sun passes over the entire coastline twice yearly, no indication of a double temperature maximum is apparent. This is due to the ocean being relatively cold in October and November following a cool cloudy winter, and to the fact that the cloud-free days do not become the norm until January. February is usually the warmest month though a March maximum is found at some northerly stations, indicative of the later overhead passage of the sun. The hot weather sometimes continues into April; the record temperature of Talara (39°C) occurred in that month. The mean daily summer maximum is usually around 26°–28°C, the night minimum is around 20°C in the north, and 18°C in the south. The seasonal range shows a slight tendency to be greater in the south, particularly for day-time temperatures, due to the more persistent cloud cover

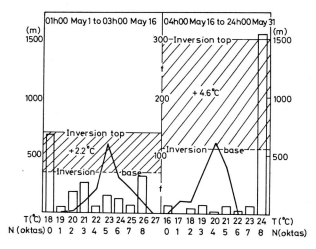

Fig.7. Mean height and strength of the inversion-layer over Lima-Callao in the first and second half of May, 1967. In the lower part of the graph, the two heavy lines indicate the frequency of the daily maximum temperature listed at the bottom; the top value is 6 (days) in both halves. The vertical columns refer to the cloudiness scales from 0 to 8 oktas at the bottom, and represent the frequency of hourly observations (number of hours on the ordinate in the center of the graph) with the respective amount of clouds.

there in the winter. The mid-winter daily temperatures normally range from 15° to 24°C north of Chiclayo and from 13° to 19°C in the southern districts.

The El Niño phenomenon has been discussed at the end of the "Precipitation" section as it is the rainfall that brings the most dramatic change. However, as the coastal temperatures are greatly dependent upon the temperature of the Pacific waters, any warming of these waters is likely to raise the coastal temperatures. In 1925 when there was a particularly severe effect, the mean monthly temperatures of Trujillo for March through August were 3°C above those of the previous year. In 1965, temperatures tended to be a little above normal from March to June for much of the coastline from Chiclayo north through Ecuador and out to the Galápagos. Though the rains did not reach Lima, the higher temperatures did and the year was noted for its exceptionally long summer with sunny weather until June. Table X shows the temperature differences for this year against the average for the remaining years of 1961–1970 for Lima. The increased positive deviations of the day temperatures over the night particularly for June to August illustrate how clearer skies can increase the diurnal temperature range. The inclusion of the pressure deviations which were negative for all months in 1965 indicate that there is probably some correlation between the El Niño effect and changes of the pressure field. Of the first eight months of this year, five recorded the lowest pressure for those months of the ten years to 1970.

TABLE X

ATMOSPHERIC TEMPERATURE AND PRESSURE COMPARISON BETWEEN 1965 AND NORMAL (1961–1964 and 1966–1970) AT LIMA

| | Difference from normal: | | | | | | | | | March– |
	Jan.	Feb.	Mar.	Apr.	May	June	July	Aug.	Sept.	August av.
Mean min. temp. (°C)	−0.3	−0.9	+2.1	+1.6	+2.0	+2.7	+2.4	+1.9	+0.7	+2.1
Mean max. temp. (°C)	+0.4	−0.9	+2.1	+2.2	+1.9	+4.1	+2.9	+2.6	−0.1	+2.6
Atm. pressure (mbar)	−0.5	−1.7	−0.8	−0.6	−1.6	−1.7	−1.9	−1.7	−0.7	−1.4

Radiation, sunshine and cloudiness

References to sunshine and cloudiness have already been made particularly in connection with the discussion of the temperature regime. Records of sunshine hours and radiation data tend to be scarce but Table XI shows a fair coverage of the former for Ecuador and Table XII gives a few series of the latter for Ecuador and Peru.

Over both sierra and selva, the dry season becomes more pronounced to the south which in turn generally means less cloud and therefore more hours of sun. A comparison between the two sierra towns of Izobamba in Ecuador and Huancayo in Peru indicates that the latter, nearly twelve degrees farther south, receives 64% more global radiation. The value for Izobamba is probably quite representative for the Ecuadorian highlands as suggested by Table XII. Huancayo, in contrast, occupies a favourable site in the centre of a broad basin so that cloud development over the surrounding mountains stays some distance away whilst many other valley areas tend to be overshadowed by large

TABLE XI

MEAN MONTHLY AND ANNUAL SUMS OF SUNSHINE HOURS, ECUADOR

	Jan.	Feb.	Mar.	Apr.	May	June	July	Aug.	Sept.	Oct.	Nov.	Dec.	Year
Coast													
San Lorenzo	106	110	123	133	95	78	84	82	80	91	78	78	1,138
Portoviejo	82	78	109	145	115	73	100	130	125	118	95	101	1,271
Pichilingue	78	82	102	112	88	55	47	65	62	52	50	64	857
Guayaquil	106	112	141	149	165	123	130	146	161	126	115	140	1,614
Sierra													
Tulcán	127	111	109	93	123	122	128	127	129	114	113	105	1,401
Ibarra	168	138	136	125	161	148	186	183	164	150	146	157	1,862
Quito	180	144	141	131	175	170	212	199	189	166	156	186	2,049
Izobamba	168	139	130	118	163	150	193	193	166	155	152	168	1,895
Ambato	174	144	139	144	146	122	124	124	138	165	169	179	1,768
Riobamba	149	139	112	137	152	150	157	175	140	149	124	160	1,744
Cañar	148	118	124	127	160	149	190	174	148	141	153	157	1,789

TABLE XII

GLOBAL RADIATION DATA: MONTHLY AVERAGES (cal./cm²/day) AND ANNUAL SUM (kcal./cm²)

	Izobamba 00°22'S 3,058 m 1964–67	Portoviejo 01°04'S 44 m 1964–67	Pichilingue 01°06'S 73 m 1964–67	Puyo 01°35'S 950 m 1965–66	Huancayo 12°07'S 3,380 m 1964–66	Arequipa 16°19'S 2,525 m 1964–66
Jan.	364	280	314	244	634	720
Feb.	354	297	333	243	577	637
March	349	330	330	224	563	634
April	322	345	339	248	563	608
May	353	304	289	259	531	529
June	348	253	238	255	540	505
July	362	276	228	252	542	520
Aug.	383	304	247	280	580	583
Sept.	385	329	267	296	617	650
Oct.	358	305	257	292	642	703
Nov.	359	283	248	286	667	726
Dec.	362	289	264	245	620	685
Average	358	300	280	260	590	625
Annual sum	131	110	102	95	215	228

cumuli over the nearby ranges. The most sunny area is probably the western Andean slope of southern Peru, above the coastal inversion but below the rainy regions of the high sierra. Arequipa is in this optimum area where very high solar radiation values are recorded.

The cloudiest regions are probably on the lower slopes of the eastern Andes where the first orographic uplift causes intense condensation. Puyo is an example of this. The lower western slopes in Ecuador are similar particularly in the "rain belt" in which Pichilingue is located. Further sunless places are on some coastal uplands of Peru where the forced uplift of the moist marine layer causes regular dull conditions with low cloud and fog in the summer months. In the winter the entire coastline is normally cloud covered.

The Amazonian lowlands generally have a cloud cover of 6–7 oktas so that sunshine totals are low. The middle part of the morning is often the clearest following the dispersion of early mist and low cloud, but around mid-day clouds usually build up again prior to the afternoon showers. Towards the south, the mid-year dry period tends to see sunnier weather particularly in Bolivia. The eastern foothills of the Andes cause the first uplift to the westward moving warm, humid airstream, resulting in condensation and heavy cloud over locations around 1,000 m altitude. Daily sunshine totals may be under 3h/day as for Puyo, though wide daily variations can occur with 9h in one day following one completely overcast.

The upper eastern slopes are much sunnier and the level between 2,000 and 3,000 m in central Peru has some popular sites for tourists. The valley below Tarma (3,000 m) lying some 4 to 5 hours by road from Lima is notable for its warm sunny climate with little rainfall. Above 3,000 m cloudiness tends to increase again with the higher peaks under the frequent storm clouds.

In the highland basins of Ecuador, annual sunshine is around 1800 h, with minor seasonal variations. There is, of course, a strong correlation between sunshine hours and daily temperature maxima and inversely with rainfall.

In Peru, seasonal variations are more marked. For the period November 1968 to October 1969 at Oroya, August was the sunniest month with 75 % of possible sunshine and March with the lowest percentage of 30. The dry season of May to August averaged 67 % whilst November to April saw 43 %. The early mornings are affected by low stratus which soon dissipates to give the sunniest period from 10h00 to 13h00 but thereafter the build-up of cumulus clouds reduces insolation.

The north and central parts of the Ecuadorian coastal plain have climatic conditions, including sunshine, similar to that of the selva. Annual totals are around 1,000 h or 3 h/day but tend to be higher on the drier coast and less in the "rain belt" of the eastern part of the plain. Global radiation values are low for tropical regions and rather comparable to places some 40° to 50° from the Equator. Seasonal variations are much more marked than for the lowlands to the east of the Andes. Maximum values for both sunshine hours and global radiation occur in April. Cloudiness then increases rapidly to give an insolation minimum in June or July. In the case of Portoviejo, the daily sunshine halves from 4.8 h in April to 2.4 h in June. Guayaquil, on the southern edge of this plain, is generally sunnier in all months with a peak in May, the start of the dry season, a secondary peak in September, and a minimum in January which is one of the wettest months.

Sunshine and cloudiness are very variable on the Galápagos Islands and are dependent upon aspect. The windward, i.e., southeastern slopes are damp and cloudy with average cloud covers of 6–7 oktas. The leeward slopes in contrast average 3–5 oktas. Puerto Baquerizo on the western side of Isla San Cristóbal is the only station which provides the sunshine data: 6.4 h/day is the annual average; 7.7 h/day are recorded from March to May.

The coast of Peru has the greatest seasonal cloud and insolation variations. The winter is normally overcast and the summer generally clear, but some variations along the coastline are found. The Lima area is notoriously damp. Some 250 km to the south, in the vicinity of Pisco, brighter weather often occurs; a nearby hotel is boasting year-round sunshine, a somewhat extravagant claim. Southwards, a line of hills rises straight out of the sea so the plain lying between this coastal range and the Andean Massif is cut off from a direct marine influence and records much more sunshine hours than does Lima. In this city the first sunny days become noticeable in October and steadily increase in frequency to year's end. The first three months of the year regularly see days with 6 to 9 h of sunshine along the entire coastal strip. The mornings are usually clear though a bank of fog may advance along the coast just after sunrise, only to be transformed into very low fractostratus or fractocumulus which are dissolving inland. This is the season of the sierra rains and the nearby development of the huge cumuli can often be seen from the coast around mid-day. The clouds then descend down the western slopes and dissolve, frequently forming altocumulus lenticularis which may cover the coastal plain in the late afternoon.

At the time of the coastal winter the highlands are enjoying the dry season and though some cloud development is normal over the higher peaks during the afternoon very little overshadows the zone from 1,000 to 2,000 m. Chosica, some 35 km from Lima averages 2 oktas of cloud cover from June to August. This area at around 800–1,000 m of altitude has seen the development of the city's "winter" country clubs, and each weekend there is a mass exodus of the Limeños (citizens of Lima) to the middle reaches of the Rimac valley, with the central highway lined with cars for almost 20 km. At year's end cloud amounts increase to reach 6–7 oktas by January; the annual average for Chosica is 4 oktas. The cloudiest part of the day is normally around 19h00 which coincides with the hour in which precipitation is most frequently observed in Oroya, directly inland. The clearest time is late morning following the early clearance of low clouds.

Southwards with the decline of the rains in the western cordillera, the cloudiness of the western slopes also decreases. Arequipa is undoubtedly in the sunniest area of the three countries. Its annual sum of global radiation, 228 kcal/cm², is equal to that of much of the central Sahara of north Africa. The peak is reached in November prior to the short rainy season, when the sun is making its overhead southern passage.

Surface winds

Relief has a twofold impact upon surface air flow. Firstly there are the valley winds themselves which are upslope (anabatic) in the daytime and downslope (katabatic) at night; these become important in most Andean areas. Secondly there is the channeling effect of the regional wind into certain directions; thus an east–west valley may show a dominant easterly wind all year and obscure a seasonal change from northeast to southeast.

In the northern part of the eastern lowlands, Iquitos is well away from the influence of the highlands and should be representative of this region. It mainly experiences southerly winds from May to August and is dominated by northeasterlies for the rest of the year. The predominance of the northeast wind indicates that the equatorial trough in this region tends to spend much of the year well to the south of the Equator particularly as northeasterlies still frequently occur in the mid-months of the year. Northwards into Ecuador, this airstream becomes more common still; Puyo has only one month with dominant southeast winds whilst Tena has none. This wind regime which resembles that of more northerly regions, ties in with the slight seasonal rainfall changes which show a mid-year wet rather than a mid-year dry period. Winds are normally very light, particularly in Ecuador where calms are frequent and the average speeds around 2 knots.

Southwards, the southeast trades become more important. In Pucallpa they form the main airstream from April to September and sometimes to year's end, with northerlies for the rest of the time. Relief may play a role so that Moyobamba (6°S), with highland to the southeast, sees easterlies for the mid-part of the year. Tingo María is even more dominated by relief and rarely experiences anything but upvalley northerlies during the day. The southeast trades are most noticeable in the latitudinal zone from 4°S to about 10°–12°S where average speeds increase to 3–6 knots. Towards the southern edge of the selva northwesterlies become dominant. The southeasterlies are restricted to April and May and strong northwest winds follow, averaging 13 knots from July to September at Santa Cruz (17°47'S). The occasional mid-year month in which the south wind predominates will bring cooler weather. In July 1962, for instance, this flow pattern affected all lowland Bolivia; the average temperature of Santa Cruz was 3.1°C below normal and 4.1°C below at Yacuíba. In the extreme south, southerly winds are frequent, particularly in the early months of the year with the thermal lows of Paraguay to the east.

In the highlands most population centres are located on valley floors which are influenced by the anabatic/katabatic wind regimes. Strong nocturnal cooling is frequently experienced with a light downdrift until early morning. Around 11h00 valley winds take over which can become rather strong so that maximum temperatures are frequently recorded before midday; the wind dies away just after sunset. Local instability through powerful insolation heating of the land surface can cause updrafts of air and small whirlwinds or twisters. These may often be seen moving across some of the more southern highland basins in the dry season along with other clouds of dust. There is a tendency for winds to be stronger in the dry season. Dust storms can appear, mostly in the early afternoons. These are very noticeable in the Altiplano where strong gusts make driving difficult and prevent sailing on Lake Titicaca after noon. High winds often precede thunderstorms which can be heralded by an advancing cloud of dust, often a reddish brown in the limestone areas.

The winds in the sierras of Ecuador are so influenced by relief that in most areas no clear pattern emerges. Calms are normal in many locations particularly during the night and early morning, with little seasonal variation. In some cases, even annual totals can indicate little; for Quito Observatory the most frequently recorded wind directions for the years 1965, 1966 and 1967 were respectively north, west and east. Izobamba, 25 km to the south and in a higher location, shows a clearer pattern. The dry months of May–September record mainly southwest winds with easterlies starting to appear in the wet month of October. From then until April east and northeast winds are the most common.

In the Peruvian highlands similar variations between locations appear though wind speeds are greater than in Ecuador. There also is a more noticeable seasonal change, the mean wind speed being greater during the dry season. Radiosonde data from Lima indicate for the mid-troposphere a predominant flow from east or southeast in the wet season and a northwest airstream in the dry. Cuzco shows some tendency of this, but Cajamarca in the north with its predominant easterlies from June to October and southerlies for the rest of the year obviously has different controls.

In the Bolivian highlands there is some evidence of the same seasonal change as indicated by the Lima radiosonde data. La Paz, despite its sheltered position, sees a predominant east or southeast airstream of 3–4 knots for the seven months of November–May and a mainly westerly flow of 4–5 knots for the remaining five months. The city's airport on the plateau some 500 m higher has a similar regime except that the west-wind dominance is restricted to the four months May–August with an average speed of 1 knot higher. Oruro in the plateau centre has a seasonal change but of a slightly different nature. The wet season wind direction has easterly and northerly components whilst the dry season largely sees southerlies, somewhat similar to many lowland stations to the east. Most valley locations are little influenced by regional airstreams; Tarija in a northwest–southeast valley rarely experiences anything other than a southeast wind.

Over the coastal lowlands of Ecuador winds tend to be light and variable with frequent calms inland and stronger breezes on the coast. The extreme north receives frequent northerlies, but southwards southwest winds tend to predominate. The south to southwest wind direction is particularly noticeable in the dry season from May to near year's end. Portoviejo has a prevailing northwest flow for the first four months of the year bringing moist air from the coastal waters. Guayaquil has in these months regular light east to northeast winds but still with the odd southwesterly. Pichilingue in the "rain belt" has a prevailing southwest wind all year apart from October (southeast), but the wetter months see occasional east to northeast airflows.

The coast of Peru is dominated by the south to southeast winds for the entire year. At Lima at 07h00 62% of all observed surface winds are from the south and have, therefore, a small onshore component. The 24-h frequency distribution is basically the same except for a higher frequency of the direction parallel to the coast (south-southeast) at night-time and in winter, and from the west in the morning hours of the summer. Farther north, due to a change in the direction of the pressure gradient, wind directions show a more easterly component. The 07h00 readings at Puerto Chicama (7°40′S) show southeast 37%, south-southeast 43%, and south 15%. The afternoon winds are similar but with an increase in those from south-southwest. There still prevails a slight onshore movement but without any real land–sea breeze circulation.

Average wind speeds are 4–8 knots around Lima but become stronger, reaching an average of 13 knots at Chimbote (09°10′S) and around 10 knots in many other locations. They are particularly strong around dusk. There is also a tendency for the windspeed to be higher in the latter half of the year than in the former.

A notable exception regarding the general direction of the surface winds is found in the Pisco Bay area. Pisco (13°45′S) has a prevalent southwest wind which means a strong onshore component; it is the result of the local relief. South of Pisco the lowlands lie between the Andes to the east and a coastal range to the west. This broad valley, being cut-off from the cold Pacific waters, is relatively warm and draws in the cooler marine air from Pisco Bay. The sand dunes of this very dry area indicate a prevailing northwest wind between Pisco and Ica in the centre of these lowlands. A description of wind and weather in this area was given by SCHWEIGGER (1959, p.188).

Radiosonde data indicate that the prevailing southerly flow has a vertical extent of no more than 1,000–1,500 m with westerly or northwesterly winds above this level. Arequipa at 2,525 m, with Mt. Chachani (6,100 m) 20 km to the north, and El Misti (5,822 m) 16 km to the northeast, has all year long westerly winds during the warmer part of the day,

TABLE XIII

DIURNAL VARIATION OF THE ANNUAL AVERAGE FREQUENCY (%) OF WIND DIRECTIONS AT AREQUIPA, AT THE PLACE OF THE OLD HARVARD OBSERVATORY, ABOUT 3 KM NORTH OF, AND 100 M ABOVE, THE MAIN PLAZA OF THE CITY
(After HANN, 1909)

Local time	N	NE	E	SE	S	SW	W	NW
01 and 02h	47	26	2	2	2	4	6	11
03 and 04h	46	35	0	0	2	2	6	9
05 and 06h	46	38	2	0	2	2	2	8
07 and 08h	27	35	8	4	8	6	6	6
09 and 10h	2	6	4	8	27	39	12	2
11 and 12h	0	0	0	0	2	43	55	0
13 and 14h	0	0	0	0	0	16	84	0
15 and 16h	0	0	0	0	0	10	88	2
17 and 18h	2	0	0	0	0	4	84	10
19 and 20h	30	4	0	0	2	4	32	28
21 and 22h	48	12	2	2	2	6	13	15
23 and 24h	51	19	2	2	4	2	7	13

from 10h00 to 19h00, but northerly and northeasterly winds during the night and early morning hours, see Table XIII. Notwithstanding, the alignment of the dunes some 40 km towards the coast suggests a prevailing flow from south-southeast.

The Galápagos Islands lie within the trade-wind circulation regime of the southern Pacific Ocean. All south and east facing points show a great predominance of southeast winds in all seasons which may exceed 80% of all direction frequencies. Wind speeds are regularly around 10 knots on the coast but tend to be less inland.

Acknowledgements

I wish to thank the following persons and organizations for information provided: Ing. Mario Canedo Daza, Servicio Nacional de Meteorología e Hidrología, Bolivia; Mayor General Jaime Cayo Murillo, Servicio Nacional de Meteorología e Hidrología, Peru; Ing. Antonio Garcia S., Servicio Nacional de Meteorología e Hidrología, Ecuador; and B. H. Wadia, Centro Minería del Perú (formerly Cerro de Pasco Corporation), Peru.

Appendix I—Dynamic and energetic factors which cause and limit aridity along South America's Pacific coast[1]

A climate will be arid if either the regional atmosphere lacks precipitable water (as for example in polar deserts) or tends to be dominated by descending air (subsidence). In the framework of the planetary circulation of the atmosphere descending air will prevail either in regions of quasi-permanent anticyclones (free subsidence) or on the lee-side of

---

[1] By H. Lettau, University of Wisconsin, Madison, Wisconsin, U.S.A.

mountain barriers (dynamically forced subsidence). Air will descend along terrain slopes if cooled by terrestrial radiation (thermally forced subsidence or katabatic wind).

The atmosphere in the arid climate along South America's Pacific coast does not lack precipitable water. The region is extremely narrow (hardly wider than 100 km) in contrast to its meridional extent of more than 3,000 km (between about 30°S and 1°–2°S). The fact that subsidence prevails here is convincingly documented by a narrow band in which clouds appear to be erased, as shown on practically every picture of "planet earth" taken by cameras on orbiting or geostationary satellites.

This subsidence is conventionally explained by a combination of regional and planetary factors. A regional cause is seen in the suppression of updrafts due to inversional stratification above the abnormally cold water of the Peru current along this coast. A factor of the planetary circulation could be free subsidence due to the quasi-permanent South Pacific anticyclone normally centered between 25°–30°S and 90°–105°W. Also, it has been claimed that the regional topographical barrier of the Andes Cordillera produces forced subsidence in the planetary trade wind regime of the Southern Hemisphere.

For brevity we shall refer to the general flow pattern of the lowest 1–2 km layer above the ground in the arid zone as the desert wind. A thorough examination of its structure (in terms of annual average states as well as seasonal and diurnal variations) reveals control by regional energetics and dynamics to such a degree that planetary factors are eliminated or even reversed. For example, in their seasonal variations, the planetary winds (including the trade winds) reach maximum strength during the hemispheric winter months when the Equator-to-Pole gradient of insolation is at peak value, while the desert winds are strongest during the summer months when the local insolation is highest. This is documented not only by direct climatic records but also indirectly, for example, by the significant annual course of the monthly rates of dune migration in the Pampa de La Joya. Near the inland border of the arid zone (approximately described by the isohyets of 200 to 300 mm/year or, very crudely, by terrain contour lines between 2 and 3 km above sea level) the observed surface winds are often from west and occasionally from north, the opposite of the trade wind components, which eliminates forced subsidence. While the trade winds show no significant lateral (cross-wind) gradient and shear, they produce significant advection (downstream) gradients of heat and moisture. In spite of its long meridional fetch the desert wind appears to have only insignificant or even reversed downstream gradients of heat or moisture. Roughly parallel to the coast and a few hundred meters above it, there appears to be a well-developed core in the form of a velocity maximum with pronounced speed-decrease not only upwards and downwards but also landwards as well as seawards. Significant lateral–horizontal shear is consistent with a steady helical vortex circulation in the vertical plane perpendicular to the jet axis.

The two main factors which determine the steady-state dynamics of the desert winds are of strictly regional nature and can be described by the following: (*1*) a thermal wind primarily caused by horizontal large-scale density gradients and Coriolis effects in the warm-air envelope of the beaches and the rising terrain; (*2*) the Ekman-drift effect in the ocean littoral and the associated upwelling of cold water which, by natural feedback, enforces and stabilizes the thermal slope-wind effect, thus fixing the core position of the low-level jet above the average shoreline.

The cause of the warm air envelope is relatively high insolation due to lack of clouds on the arid land above the maritime layer (inversion). The resulting large-scale thermal slope wind tends to blow parallel to the general contour lines of the Andes Cordillera; it can be compared with the circulation pattern of a warm low-pressure cell. This represents the counterpart to the circulation pattern similar to a cold low-pressure cell as exists in the cold-air envelope of the slopes of the Antarctic continent; see LETTAU (1968), and LETTAU and SCHWERDTFEGER (1967). The common characteristic in both cases is the overruling of direct circulations (katabatic–anabatic winds) by Coriolis forces (tendency to geostrophic adjustment) and a modification by surface stress and internal friction (geotriptic balance).

Dynamic consequences of the Ekman-drift effect in the region have been quantitatively formulated and discussed by LETTAU (1975), employing a mathematical solution of the problem of boundary layer velocity spirals for height-independent eddy diffusivity and exponential thermal wind, originally discussed by MAHRT and SCHWERDTFEGER (1970). The Ekman-drift effect changes sign at the Equator which serves to explain the northern limit of the arid zone near the border between Peru and Ecuador, and the rather abrupt transition from aridity associated with cold-water upwelling to humid coastal climates in Colombia with the warm ocean littoral; see also FLOHN (1969).

Essential for the steady-state energetics of the desert wind is that daily insolation in the arid zone is high because unimpeded by cloudiness, while terrestrial radiation is moderate because of relatively cool surface temperatures. In comparison with inner-continental "hot" deserts this results in relatively intense net radiation; for example, about 200 ly/day on the annual average for the Pampa de La Joya. Latent heat transfer is nil and conduction into the soil insignificant. Therefore, the entire 200 ly/day are loaded on the lower atmosphere. In cases of classical boundary layer flow such surface heating is laterally uniform and the calories convected into the flow are transported downstream with resulting increase in air temperature so that horizontal advection (of relatively cool air) balances the vertical (eddy) convection. LETTAU (1975) has shown that the average structure of the desert wind is significantly different. Its most important and unique feature is that surface heating on the land-side of the jet-core changes to surface cooling on the ocean-side. Advection by the component parallel to the coastline is insignificant. The entire 200 ly/day convected into the flow system on the land-side are transported laterally by the steady and continuous helical vortex circulation to the ocean littoral where the heat is absorbed by the water which compensates the cold-source of the upwelling. Finally, ocean currents export the heat to distant regions on the deep sea.

Along the coastline the low-level jet component of the desert wind represents the energy-dissipating phase and the helical vortex circulation the work-producing cycle of a huge thermodynamic engine, fueled by insolation on land, with the ocean littoral serving as the condenser. LETTAU (1975) discussed data which show that the efficiency of this engine is about 2%, comparable to that of other atmospheric systems.

In summary, regional dynamics and energetics serve to explain the annual mean desert winds and their constancy, together with the northern limit at the Equator. The cause of subsidence remains to be clarified. LETTAU (1975) suggests that the coastline centered low-level jet will have an anti-parallel counterpart along the continental border of the arid zone, approximately 100 km to the east. The rough topography of the Andes Cordillera will prohibit there the development of an organized flow pattern comparable with the pronounced core of the coastal jet. However, tendencies to low-level west and north wind components instead of south and east components should be expected and are, in fact, observed at these altitudes. If the height-averaged east component of boundary layer flow decreases toward the east (even without sign change to west components) while the main flow components do not change in the downstream direction, a divergence results which requires subsidence at the top of the regional boundary layer. According to data discussed by LETTAU (1975) this divergence undergoes a significant diurnal variation which results in a "thermotidal surge". Theoretically its amplitude and phase is controlled by the difference between the solar day (which measures the frequency of the forcing pulses) and the pendulum day (which measures the inertia of atmospheric dynamic response and is defined as one sidereal day divided by $2\sin\varphi$, where φ is geographic latitude). This difference changes sign at $\varphi = 30°$. When the theory is applied to the large-scale conditions along the Andes Cordillera it follows for the Pacific slope that day-time divergence prevails equatorwards of 30°S but changes to day-time convergence polewards of this latitude, with opposite conditions and reversed changes on the eastward facing slope. Consequently, the "thermotidal surge effect" serves not only to explain the southern limit of the coastal desert near 30°S but also the shift of aridity to the Argentine side of the Andes Cordillera farther polewards.

Finally, some remarks on the El Niño phenomenon may be added. The regional energetics of the desert wind require a fairly intense net radiation to be balanced over land solely by dry convection which is well organized in the form of steady helical vortex flow superimposed on the coastal low-level jet which, in turn, maintains the Ekman-drift and the cold source of upwelling. The balance between the interconnected system of winds and currents would be disturbed if either net radiation over land were reduced or moist convection would compete with dry convection for a significant share of net radiation.

Consistent with the above the following sequence of events can be logically constructed. Suppose that an occasional migrating disturbance which originates in the winter circulation of the Northern Hemisphere, causes a transient spell of precipitation activity in the region between 4° and 5°S which normally has little or no rain at all. The disturbance may move out within a couple of days. However, soil moisture supplied by the rain remains until it is evaporated. This process requires 600 ly per cm of water so that the normal rate of net radiation (200 ly/day in the region) can dispose of only 3 mm/day, or less if anomalous cloudiness prevails.

As long as evaporation goes on, the portion of net radiation normally available for dry convection is diminished, and with it the diurnal amplitude of air temperature. Thus, the thermal forcing for the boundary layer jet is reduced which weakens wind stress and the upwelling in the ocean littoral. This may signal the onset of a destructive phase in the feedback chain. With the reduced efficiency of the drift current for heat export the ocean surface warms up which reduces the thermal forcing of the desert wind even more. Net radiation received by the warmer water stimulates maritime evaporation whereupon the lower atmosphere along the coast will become moister as well as warmer. As is normal for the tropics moist convection will be involved in disposing of net radiation.

Another phase of the chain reaction begins if the reversal of the cross-shore gradient of surface temperature lasts more than about one-half pendulum day, which can be considered the minimum period for geostrophic adjustment. For instance, at 3° latitude one-half pendulum day equals ten sidereal days, but only five at 6°. Thus, the farther south the disturbance reaches and the longer the duration of a reversed temperature gradient, the more likely is the establishment of anomalous boundary layer flow with north wind components. The resulting stress reversal on the ocean causes an Ekman-drift southward and on-shore, whereupon the ultimate phase of the destructive feedback begins. With the disappearance of upwelling, heat export to the deep sea is cut off and in addition to local utilization of net radiation, relatively warm water from farther north moves in. Evaporation increases and local showers may become heavy. Dynamic factors enter also because the anomalous northerly flow generates a confluence zone with the normal jet flow that still prevails in the dry region farther south. Consequently, with a relatively limited initial investment of moisture from foreign sources acting merely as a trigger, rain showers sustained by local sources develop and can continue long after the original disturbance has moved out. The end of the El Niño period may come only after the sun returns to the Equator in March, when seasonally increased insolation re-establishes normal conditions of net radiation.

In conclusion, it should be emphasized that our discussion of strictly regional factors which cause and limit aridity along South America's Pacific coast are of necessity restricted to the broadest outlines. The need for further detailed climatic studies as well as a refinement of theory is obvious. However, our concepts of regional energetics and dynamics are essentially based on first principles of physics. This implies that a potential exists for the explanation of arid zones along other tropical or subtropical coastlines, as for example in Africa.

Appendix II—High thunderstorm frequency over the subtropical Andes during the summer; cause and effects[1]

A notable feature of the atmospheric circulation over the subtropical Andes is the existence of an upper-tropospheric high pressure system in the region of the high plateau of western Bolivia and southern Peru during the summer months, and at the same time the regular appearance of intense convective activity, most pronounced over the surrounding mountain chains.

The violent thunderstorms over the mountains, forming almost every summer-afternoon or evening, have impressed the scientists exploring the area in years long passed much more than the meteorological observers of recent times. This discrepancy is of more than local climatological interest. It is suggested that the contribution of western South America to the total, worldwide thunderstorm activity has been significantly underestimated by the thunderstorm frequency statistics used in studies regarding the maintenance of the earth's electric field. Nearly every old travelogue, e.g., VON TSCHUDI (1846), RECK (1865), CONWAY (1901), BOWMAN (1916), and many others, gives eloquent descriptions of the weather conditions in the mountain regions proper. To quote one of them (Reck) briefly: "... almost always, the rainshowers are accompanied by violent thunderstorms, hail, and snow. The thunderstorms seldom appear before noon, and much more frequently in the afternoon, evening, and night; their intensity surpasses all imagination ...". The few, and sometimes poorly attended, meteorological stations of the

---

[1] By W. Schwerdtfeger.

area are located in the less affected, inhabited places, some of them at considerable distance from the higher mountain ranges, which may explain that their records are not fully representative as far as the thunderstorm frequency in the wider area is concerned. Close to the higher ranges of the western cordillera a meteorological station was maintained, May 1913–March 1914, at Santa Lucia (15.8°S 70.7°W) at about 4,700 m in a little "mining center" at the head of the valley running northeast toward Lake Titicaca. BOWMAN (1916, p.170) summarizes the relevant weather conditions of the five summer months as follows:

Nov.: "tempest" 11 times, distant thunder and lightning 9 times;

Dec.: beginning of rain and strong wind frequently observed 5–6 p.m., "tempest" 19 times;

Jan.: rain, actual or threatening, characteristic of afternoons; "tempest" generally 5–6 p.m., 7 times;

Feb.: frequent afternoon or evening rains; "tempest" 16 times, generally 4 p.m. or later;

Mar.: "tempest", generally in the late afternoon and accompanied by hail, 19 times.

Measured precipitation during these five months accounted to 550 mm. Of course, as most of this precipitation fell with violent winds, more credit should be given to the storm frequency than to this number.

One more passage in BOWMAN's book (1916, p.169–170) is revealing in several aspects: "Of special interest are the strong contrasts between the comparatively high temperatures of mid-day and the sudden 'tempests' accompanied by rain or hail that follow the strong convectional movements dependent upon rapid and unequal heating. The furious winds drive the particles of hail like shot ... We encountered one after another of these 'tempestades' on the divide (about 4,900 m) between Lambrama (13.8°S 72.8°W) and Antabamba (14.3°S 72.9°W) in 1911. They are among the most impetuous little storms I have ever experienced. The longest of them raged on the divide from two-o'clock until dark, *though in the valleys the sun was shining*." Still, older statistics (KNOCH, 1930) for the city of Sucre (19°S 65°W, 2,850 m on the eastern slopes) report an average of 80 days with rain in the summer half-year, 65 of them with thunderstorms. Farther to the north, at Huancayo (12°S), eight years of reliable observations (COILE and CULMSEE, 1953) indicate an average number of 106 days with thunderstorms per year, only 13 of them in the four winter months May–August.

The meteorological conditions over the Altiplano and surrounding mountains, leading to the high thunderstorm frequency, have been discussed by SCHWERDTFEGER (1961) and analysed in more detail, including estimates of the heat budget of the air over the Altiplano in the summer, by GUTMAN and SCHWERDTFEGER (1965). Since then, a radiosonde station has been in operation temporarily at La Paz (LABORATORIO DE FÍSICA CÓSMICA, 1968). Its results clearly confirm the earlier conclusions, which can be summarized as follows.

(a) In the summer there exists a thermal (warm) anticyclone or anticyclonic ridge in the upper troposphere over the subtropical highland and mountains. This feature modifies the general flow pattern of the southern circumpolar vortex with its prevailing westerly winds which extend northward at least up to 12°S. At the Pacific coast, the wind turns between 500 and 200 mbar from west to northwest, indicating cold-advection in this layer. In the winter, the average wind pattern of the upper troposphere—straight westerly flow, stronger than in the summer—does not respond to the hypothesis of a dynamically produced and maintained anticyclonic ridge over the Andes. See Fig.6 of Chapter 1, and TALJAARD et al. (1969).

(*b*) In the summer, the ground of the high plateau and hence the lower layers of the atmosphere are heated by the intense solar radiation, incoming particularly during the morning hours when there is little cloudiness. The lapse rate of temperature in the lowest, say, 500–1,000 m of the troposphere becomes superadiabatic. Strong convection and then condensation and precipitation take place, with the corresponding upward flux of sensible heat and, amounting to about twice as much, the release of latent heat. This is the main heat source which counteracts the above mentioned cold-advection and the radiative cooling so that a balance can be maintained, as it is evidenced by the insignificant temperature changes from day to day. Table XIV shows a comparison of the mean daily temperatures near the surface over the Altiplano with those in the free atmosphere over the coast, at the same pressure level. It may be noted in passing that at some stations in the northernmost part of the Altiplano and in the mountains farther to the northwest the mean daily maximum temperatures are about as high in the winter as they are in the summer. In winter, however, the air is very much drier, so that convective clouds do normally not develop.

TABLE XIV

COMPARISON OF TEMPERATURES: 12H GMT, FREE ATMOSPHERE OVER THE PACIFIC COAST (F), AND DAILY MEAN SURFACE VALUES OVER THE HIGH PLATEAU (S)*

Station	Lat. (°S)	Long. (°W)	Pressure level (mbar)	Summer (°C)	Winter (°C)
Lima (F)	12.1	77.0	625	5.2	5.2
El Alto (S)	16.5	68.5	625	8.2	5.8
Antofagasta (F)	23.5	70.4	672	8.5	5.2
La Quiaca (S)	22.1	65.5	672	12.5	4.5

* Three month averages, December–February = summer; June–August = winter.

(*c*) This difference in the moisture supply is related to a striking change of the wind regime in the lower layers over the Altiplano. In the summer, relatively moist air masses are advected from northeast or east to feed the convective activity; while dry air masses from the south to west move into the area of the plateau in the winter. (See Table XV for some corroborating data.)

(*d*) It is to be expected that the pronounced diurnal variation of the upper tropospheric temperature field caused by the periodic appearance of high-reaching convection over the subtropical mountain area produces noticeable diurnal variations in the wind field in the free atmosphere, in particular of the jetstream south of the plateau. However, as most of the upper-air soundings in South America are launched in the early morning hours (12h00 G.M.T.), there are not sufficient data available to prove this point which, of course, would be interesting for aviation-meteorology.

TABLE XV

CLIMATOLOGICAL DATA CHARACTERIZING SUMMER AND WINTER CONDITIONS ON THE ALTIPLANO FOR (a) LA PAZ (EL ALTO)[1] AND (b) LA QUIACA[2]

1. Mean temperature ($\bar{T}$), relative ($\bar{U}$) and specific humidity ($\bar{q}$) near the surface

		$\bar{T}$ (°C)	$\bar{U}$ (%)	$\bar{q}$ (g/kg)
Summer (XII–II)	(a)	8.2	68	7.6
	(b)	12.4	69	9.3
Winter (VI–VIII)	(a)	5.8	36	3.4
	(b)	5.0	39	3.2

2. Surface-wind directions (%)

		N	NE	E	SE	S	SW	W	NW	Calm
Summer (XI–II)	(a)	1	9	*74*[3]	1	2	3	9	1	0
	(b)	18	*35*	8	7	11	6	5	8	2
Winter (VI–VIII)	(a)	1	5	23	2	1	4	*62*	2	0
	(b)	8	8	6	12	*24*	10	15	12	5

3. Resultant wind at 6 km (m.s.l.) over La Paz[4]

Summer (XII–II)	125 soundings	091°	2.4 m/sec	0.48
Winter (VI–VIII)	126 soundings	281°	5.3 m/sec	0.66

[1] 16.5°S 68.2°W, 4,103 m, 5-years record.
[2] 22.1°S 65.6°W, 3,458 m, 9-years record.
[3] The most frequent directions are given in italics.
[4] Direction, vector mean speed, and constancy, computed from soundings made at La Paz in 1965, 1967 and 1968 (LAB. DE FÍSICA CÓSMICA, 1968).

Appendix III—The water-budget of Lake Titicaca[1]

This often debated problem has recently been given a thorough analysis by KESSLER and MONHEIM (1968) who could make use of the data of numerous meteorological stations installed in the Titicaca basin since 1957, supplemented by their own observations in the area. The following description is essentially based on Kessler and Monheim's study.

One of the assets of their work is a new map of the mean annual precipitation for the five year period 1957–1961 (Fig.8). The map indicates greater values than had previously been assumed or derived from sparse observations, and in particular shows a pronounced maximum over the central part of the lake. This maximum area, manifested by the measurements on the two islands Soto and Taquili, is an interesting phenomenon by itself: the prevailing winds and hence the moisture-transport toward the Titicaca basin are from the east sector. As the map shows (elevations of more than 5,000 m stipled), there is a stretch of about 100 km length where the chain of the eastern cordillera reaches considerably lesser height (around 4,500 m) than farther north and south. One might

[1] By W. Schwerdtfeger.

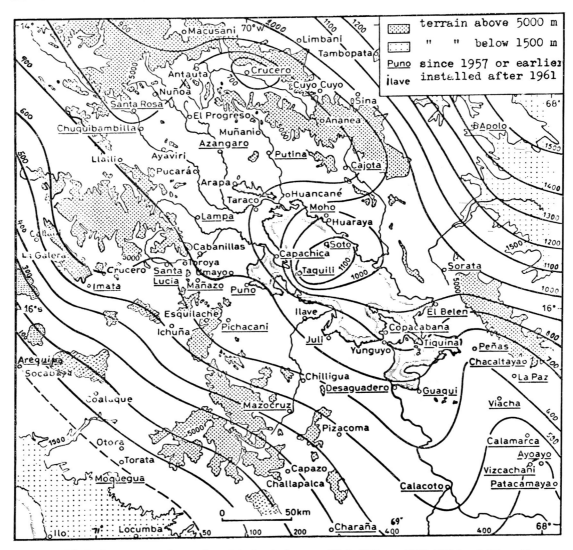

Fig.8. Average annual sum of precipitation in the area of Lake Titicaca, 1957–1961 (mm). (From KESSLER and MONHEIM, 1968, with the permission of the editor of *Erdkunde*.)

infer that the air masses advected from central South America through the gap can carry a correspondingly greater amount of moisture which then is available for condensation and precipitation over the basin, downwind of the gap. Another circumstance helping to produce the precipitation-maximum over the lake can be seen in the evaporation itself which must be greater over the large water surface than over the surrounding land. Finally, it appears possible, though not proven, that a convergence of the flow in the lower layers of the atmosphere over the lake, due to a nightly land-breeze from all sides, leads to periodic enhancement of rising motion and eventually of precipitation.

The at-surface inflow into the lake through the five main rivers of the basin, in the order of average discharge the Ramis, Coata, Ilave, Huancané, and Suchez, has been measured. The 5-year mean amounts to 162 m³/sec; 5% of this value have been added to account for minor tributaries and a probable subterranean inflow (though the latter might be more than cancelled by subterranean discharge).

The only surface discharge from the lake is through the Rio Desaguadero which in the five years of Kessler's and Monheim's analysis, 1957–1961, carried water at the rate of 8 m³/sec, varying between 5 m³/sec in November and 14 in April. In these five years, the level of the lake was first decreasing, then rising, and on the average slightly below normal, which explains that the long-year average discharge is greater (20 m³/sec). Since the evaporation from the lake surface turns out to be 20 to 50 times more copious than the surface discharge, the exact number of the latter matters little for the overall water-budget.

Changes of the level of a lake can be observed, of course, with comparatively greater exactness than the other three measurable budget-components. For the Titicaca, however, a long and continuous series of gauge-data exists only from one station, Puno, at the extreme western end of the lake. CEDANO (1971) shows graphically the variations between 1918 and 1967, and gives some interesting comments: the average height (above mean sea level) of the water surface has been determined to 3,808 m, and the area of the lake at this level to 8,170 km². The area varies by 250 km² per meter of change of the level. In the 49 years on record, a maximum of about 2 m above the average has been observed in 1933 and 1963. The level was more than 1 m below average from 1940 through 1948, reaching its minimum, almost 3 m below normal, in 1943; in that year, the Desaguadero river remained dry.

The fact that the only gauge-station is located in the Bay of Puno (Bahía de Puno) can introduce a slight bias into the monthly water-budget values computed by KESSLER and MONHEIM (1968), by means of a wind effect on the lake. The mean annual budget-values, though, should not be significantly affected. Table XVI shows the mean monthly and

TABLE XVI

THE WATER-BUDGET OF LAKE TITICACA, AVERAGE MONTHLY AND ANNUAL COMPONENTS, 1957–1961 (in mm)
(After KESSLER and MONHEIM, 1968)

Month	Inflow (I)	Precipitation (P)	Discharge (D)	I+P−D	Change of level (C)	Evaporation (E)
January	116	181	2	295	+199	96
February	151	157	3	305	+251	54
March	153	116	4	265	+128	137
April	75	81	4	152	+6	146
May	36	43	4	75	−56	131
June	20	9	3	26	−118	144
July	14	2	3	13	−117	130
August	11	15	2	24	−116	140
September	10	57	2	65	−53	118
October	11	52	2	61	−68	129
November	16	86	1	101	−39	140
December	51	132	1	182	+67	115
Year	664*	931	31**	1,564	+84	1,480

* ≈ 171 m³/sec.

** ≈ 8 m³/sec.

annual components of the water-budget. The value of almost 1,500 mm for the average annual evaporation finds a certain confirmation in the fact that the respective values of the five individual years of the analysis vary only between 1,440 and 1,547 mm, this range being no more than 7% of the average. In clear contrast, the range of the annual precipitation sums amounts to 43%, the range of the annual inflow to 85% of the respective averages. A further check of the derived evaporation value has been made by considering separately those individual four winter months of the period in which *no* rainfall on the lake was observed. As inflow and discharge in June and July are very small in comparison with the change of the lake-level, this latter value should be, and has been found, close to the average evaporation of the months in question. Such a test is important also because it strongly suggests that the possible subterranean infiltration has not been underestimated by the authors of the study.

Besides the mean annual evaporation value itself, it is interesting to note that precipitation on the lake itself contributes more (58%) to the water-budget than the run-off from the 33,900 km² drainage area (42%), and that the discharge of water through the Rio Desaguadero and by seepage is insignificant (2%) in comparison to the evaporation (98%).

The annual march of the various budget components shows several characteristic details. Precipitation has the largest annual amplitude, 179 mm for January minus July. It is followed, with a delay of two months, by the inflow, 143 mm for March minus September. Evaporation, in contrast, is lowest in February when there is still an ample amount of precipitation and the relative humidity of the air is high, and strongest in April when the water of the lake is still warm, but the air already relatively dry. This April maximum, however, established by a series of measurements of only five years, cannot be considered statistically significant; in fact, all nine non-summer months show large evaporation, between 4 and 5 mm/day.

The question of the most probable annual variation of the evaporation from Lake Titicaca has been further investigated by KESSLER (1970) who compares the above results with measurements made at various stations with a Piche evaporimeter, and with computations using the well-known Penman formula and a mass-exchange formula. He concludes that the maximum evaporation should be expected in the months of October and November. The minimum certainly belongs to the summer months, January or February, but it appears doubtful whether the real February-value is as low as the 5-year mean water balance analysis suggests.

The analysis given by KESSLER and MONHEIM (1968) includes the water-budget of the land area of the Titicaca basin or, more exactly, of the drainage areas of the five major streams flowing into the lake. For these areas, the 5-year average of the annual precipitation is 726 mm. With the discharge of the five rivers into the lake, 151 mm, the discharge-coefficient becomes 0.21. The average evaporation from the land area, 575 mm per year, amounts to 39% of the evaporation of the lake.

Though Lake Titicaca is rather unique by virtue of its elevation above sea level, it may be of interest to compare the mean annual evaporation as determined by Kessler and Monheim, with estimates of this important parameter for various large lakes in central Africa, at elevations of a bit more than 1,000 m (see Table XVII).

TABLE XVII

ESTIMATES OF THE AVERAGE ANNUAL EVAPORATION FROM VARIOUS LAKES AT HIGH ELEVATION IN TROPICAL REGIONS[1]

Name	Elevation (m)	Latitude (°S)	Av. ann. evapor. (mm)
Titicaca	3,808	15–16	1,480
Kivu	1,460	2	1,400
Victoria	1,134	0–2	1,580
Bangweulu	1,067	11	1,650

[1] WALKER, 1956, for Lake Victoria; BULTOT and GRIFFITHS, 1972, for the other African lakes.

Appendix IV—Station pressure values vs. elevation of places in the high mountains[1]

There are significant discrepancies between the elevation above sea level and the average pressure values, as given in the climatic tables. Reliable information regarding the average height (in geopotential meters, abbreviated g.p.m.) of the standard pressure levels (850, 700, 500 mbar) over South America is available in a new upper-air atlas (TALJAARD et al., 1969). From these data one can compute the pressure to be expected at the indicated elevation of any given station, and vice versa the elevation which would justify the reported pressure values, with an estimated error of $\pm$ 1 mbar or $\pm$ 10 m, respectively. The conversion of geopotential to geometric meters has been made according to standard procedure (LIST, 1958, pp.218–219). The results are summarized in Table XVIII; they leave no doubt that *either* the height values *or* the barometers used at the stations in the Peruvian mountains, or perhaps both, are in need of a thorough re-examination.

TABLE XVIII

PRESSURE – HEIGHT DISCREPANCIES AT SOME MOUNTAIN STATIONS[1]

Name of station	Latitude	Values given in climat. tables		Pressure if elevation is correct (mbar)	Elevation if pressure is correct (m)
		(mbar)	(m)		
Izobamba	0°22′S	704	3,058	707	3,110
Cajamarca	07°08′S	778	2,621	746	2,270
Huancayo	12°07′S	688	3,380	681	3,300
Cuzco	13°33′S	707	3,312	686	3,070
Arequipa	16°19′S	788	2,525	756	2,160

[1] For the two important stations, Quito (Ecuador), and El Alto, the airport above La Paz (Bolivia), full agreement between station pressure values and indicated elevation has been found.

Appendix V—Climatic tables

The climatic tables XIX–XLII are shown on pp.203–218 and are arranged from north to south.

[1] By W. Schwerdtfeger.

References[1]

ALPERT, L., 1946. Note on the weather and climate of Seymour Island, Galápagos Archipelago. *Bull. Am. Meteorol. Soc.*, 27: 200–209.

BAILEY, S. I., 1899. Peruvian Meteorology, 1888–1890. *Ann. Astron. Obs. Harv. Univ.*, 39(1): 1–153.
1906. Idem: 1892–1895. 39(2): 154–292.
1907. Idem: Observations made at Arequipa, 49(1): 1–103.
1923. Idem: Observations made at auxiliary stations, 49(2): 104–232; and 86(3): 123–194.

BJERKNES, J., 1966. Survey of El Niño 1957/58 in its relation to tropical Pacific meteorology. *Inter. Am. Trop. Tuna Comm., Bull.*, 12(2): 25–86.

BOWMAN, I., 1916. *The Andes of Southern Peru*. Holt, New York, N.Y., 336 pp.

BRANDT, B., 1920. Über Luftspiegelungen und Tromben im Hochland von Bolivien. *Naturwissenschaften*, 15: 111–112.

BULTOT, F. and GRIFFITHS, J. F., 1972. The equatorial wet zone. In: H. E. Landsberg (Ed.-in-chief), *World Survey of Climatology, Vol.10. Climates of Africa* (J. F. GRIFFITHS, Vol.-Ed.). Elsevier, Amsterdam, pp.259–311.

CAVIEDES, C. L., 1975. Secas and El Niño: two simultaneous climatical hazards in South America. *Proc. Assoc. Am. Geogr.*, 7: 44–49.

CEDANO, L. V., 1971. Estudio de niveles del lago Titicaca. Resumen del trabajo presentado por el Perú, en el 5 seminario latinoamericano de irrigación, Caracas, Venezuela, 1968. *Rev. SENHAMI, Lima*, I(1): 6 pp.

COILE, R. C. and CULMSEE, W., 1953. Note on thunderstorms at high altitudes in the tropics. *J. Geophys. Res.*, 58: 280–281.

CONWAY, W. M., 1901. *The Bolivian Andes. A Record of Climbing and Exploration in the years 1898 and 1900*. Harper, London, 403 pp.

CORNEJO, T. A., 1970. Resources of arid South America. In: H. E. DREGNE (Editor), *Arid Lands in Transition. AAAS, Washington, D.C., Publ.*, 90: 345–380.

DREWES, W. V. and DREWES, A. T., 1957. *Climate and Related Phenomena of the Eastern Andean Slopes of Central Peru*. Syracuse Univ. Research Institute, Syracuse, N.Y., 51 pp.

ELLENBERG, H., 1959. Über den Wassergehalt tropischer Nebel-Oasen in der Küstenwüste Perus. *Geobot. Forsch., Inst. Ruebel, Berlin*, 12: 47–74.

ESCOBAR VALLEJO, I., 1948. *Régimen Pluviométrico de Bolivia*. Publicación del Ministerio de Agricultura, Ganadería y Colonización, La Paz, 57 pp.

FLOHN, H., 1969. Local wind systems. In: H. E. LANDSBERG (Ed.-in-chief), *World Survey of Climatology, Vol. 2. General Climatology* (H. FLOHN, Vol. ed.). Elsevier, Amsterdam, pp.139–171.

GRAVES, M. E., 1944. An objective method for forecasting ceiling heights. Panagra publication, Lima, 22 pp.

GUNTHER, E. R., 1936. Variations in behaviour of the Peru coastal current—with an historical introduction. *Geogr. J., Lond.*, 88: 37–65.

GUTMAN, G. J. and SCHWERDTFEGER, W., 1965. The role of latent and sensible heat for the development of a high pressure system over the subtropical Andes, in the summer. *Meteorol. Rundsch.*, 18(3): 69–75.

HANN, J., 1909. Zur Meteorologie von Peru. *Sitzungsber. Akad. Wiss., Wien, Math. Nat. Kl., II, 118*: 1283–1372.

HANN, J., 1915. Temperaturverhaeltnisse von Quito. *Meteorol. Z.*, 32(11): 488–500.

HASTENRATH, St., 1971. Beobachtungen zur Klimamorphologischen Höhenstufung der Cordillera Real (Bolivien). *Erdkunde*, XXV(2): 102–108.

HAUTHAL, R., 1911. *Reisen in Bolivien und Peru*. Duncker und Humblot Verlag, Leipzig, 247 pp.

HOWELL, W. E., 1952. El tiempo en el valle del Chicama. *Nimbus*, 3(10/11): 11–23.

HOWELL, W. E., 1953. Local weather of the Chicama valley (Peru). *Arch. Meteorol., Geophys., Bioklimatol., Ser. B*, 5: 41–51.

HOWELL, W. E., 1965. Twelve years of cloud seeding in the Andes of northern Peru. *J. Appl. Meteorol.*, 4(6): 693–700.

---

[1] Editor's note: Mr. A. M. Johnson, the author of the main part of this chapter, wrote his contribution while working for the Cerro de Pasco Mining Corporation, at La Oroya, high up in the Peruvian Andes. Inevitably, his access to meteorological literature was severely limited. Therefore, the editor of this volume has taken the liberty to add some bibliographic notes to Mr. Johnson's text, and to compile the following list of references which might be helpful to the reader.

JOHNSON, A. M., 1970. One year in the Peruvian Andes. *Weather*, 25(11): 487–494.

KESSLER, A., 1970. Über den Jahresgang der potentiellen Verdunstung im Titicaca Becken. *Arch. Meteorol., Geophys., Bioklimatol., Ser. B*, 18: 239–252.

KESSLER, A. and MONHEIM, F., 1968. Der Wasserhaushalt des Titicaca Sees nach neueren Messergebnissen. *Erdkunde*, XXII: 275–283.

KING, C. A. M., 1962. *Oceanography for Geographers*. Arnold, London, 225 pp.

KINZL, H. and WAGNER, A., 1939. Pilotaufstiege in den Peruanischen Anden. *Gerlands Beitr. Geophys.*, 54: 29–55.

KNOCH, K., 1930. Klimakunde von Südamerika. In: W. KOEPPEN and R. GEIGER (Editors), *Handbuch der Klimatologie*, Vol.II, Part G. Borntraeger, Berlin, 349 pp.

LABORATORIO DE FISICA COSMICA, UNIVERSIDAD MAYOR DE SAN ANDRES, 1968. *Final Technical Report on the Study of the Cosmic Ray Intensity Variations Observed at Mt. Chacaltaya, Part III*. La Paz, Bolivia. Two issues: data of 1965–67: 45 pp., data of 1968: 55 pp.

LETTAU, H., 1968. Small to large-scale features of boundary-layer structure over mountain slopes. *Colo. State Univ., Dept. Atmos. Sci., Publ.*, 122: 1–74.

LETTAU, H., 1975. Thermal forcing of the wind regime as common cause of regional aridity, dune migration, and cold-water upwelling. Unpublished manuscript.

LETTAU, H. and SCHWERDTFEGER, W., 1967. Dynamics of the surface-wind regime over the interior of Antarctica. *Antarct. J. U.S.*, 2(5): 155–158.

LIST, R. J. 1958. *Smithsonian Meteorological Tables* (6th revised ed.). Smithsonian Inst., Washington, D.C., 527 pp.

MAHRT, L. J. and SCHWERDTFEGER, W., 1970. Ekman spirals for exponential thermal wind. *Boundary-Layer Meteorol.*, 1: 137–145.

MARKSON, R. and NELSON, R., 1970. Mountain-peak potential-gradient measurements and the Andesglow. *Weather*, 25(8): 350–360.

MOSTAJO, B., 1928. La lluvia en Lima en el período 1909–1927. *Co. Guano, Bol.*, 4(12): 585–599.

MURPHY, R. C., 1926. Oceanic and climatic phenomena along the west-coast of South America during 1925. *Geogr. Rev.*, 16: 26–54.

ORTOLANI, M., 1965. Osservazioni sul clima delle Ande Centrali. *Riv. Geogr. Ital.*, 72(3): 217–235.

PROHASKA, F., 1970. Distinctive bioclimatic parameters of the subtropical–tropical Andes. *Int. J. Biometeorol.*, 14(1): 1–12.

PROHASKA, F., 1973. New evidence on the climatic controls along the Peruvian coast. In: D. H. K. AMIRAN and A. W. WILSON (Editors), *Coastal Deserts, Their Natural and Human Environments*. The University of Arizona Press, Tucson, Ariz., pp.91–107.

RECK, H., 1865. Geographie und Statistik der Republik Bolivia. *Petermanns Geogr. Mitt.*, *1865*, 281 pp.

REED, W. W., 1928. Climatological data for northern and western tropical South America. *Mon. Weather Rev. Suppl.*, 31: 21 pp.

RICO NEGRETTI, J. J., 1949. Variaciones de las precipitaciones annuales. *Nimbus*, 1(1): 7–20.

ROESSEL, H., 1963. *Informe Sobre las Lomas de Lachay*. Publicación del Servicio Forestal y de Caza, Ministerio de Agricultura, Lima, 17 pp.

SCHÜTTE, K., 1968. Untersuchungen zur Meteorologie und Klimatologie des El-Niño-Phaenomens in Ecuador und Nord-Peru. *Bonner Meteorol. Abhandl.*, 9: 152 pp.

SCHWEIGGER, E., 1959. *Die Westküste Südamerikas im Bereich des Peru-Stromes*. Keyser Verlag, Heidelberg, 548 pp.

SCHWERDTFEGER, W., 1961. Strömungs- und Temperaturfeld der freien Atmosphäre über den Anden. *Meteorol. Rundsch.*, 14(1): 1–6.

SEARS, A. E., 1895. The coast desert of Peru. *Bull. Am. Geogr. Soc.*, 28: 256–271.

STENZ, E., 1950. Condiciones climatológicas del Altiplano de Bolivia. *Nimbus*, 2(6–7): 3–23.

TALJAARD, J. J., VAN LOON, H., CRUTCHER, H. L. and JENNE, R. L., 1969. *Climate of the Upper Air, Part I. Southern Hemisphere, Vol.I. Temperatures, Dew-Points and Heights at Selected Pressure Levels*. NCAR, ESSA, and D.D., 6 pp. and 134 tables.

TREWARTHA, J. T., 1966. *The Earth's Problem Climates*. University of Wisconsin Press, Madison, Wisc., 334 pp.

TROLL, C., 1930. Die tropischen Andenländer. In: F. KLUTE (Editor), *Handbuch der Geographischen Wissenschaft*, *Vol.IV. Süd-Amerika*. Borntraeger, Berlin, 400 pp.

TROLL, C., 1952. Die Lokalwinde der Tropengebirge und ihr Einfluss auf Niederschlag und Vegetation. *Bonner Geogr. Abhandl.*, 9: 124–182.

VAN LOON, H., TALJAARD, J. J., JENNE, R. L. and CRUTCHER, H. L., 1971. *Climate of the Upper Air: Southern Hemisphere, Vol.II. Zonal Geostrophic Winds*. Naviar 50-1C-56 and NCAR TN/STR-57, 40 pp.

VON TSCHUDI, J. J., 1846. *Peru, Reiseskizzen 1838–1842*. St. Gallen, Switzerland, 252 pp.

WALKER, H. O., 1956. Evaporation from Lake Victoria. *Weather*, 11: 382–384.

WEISCHET, W. and HAVLIK, D., 1966. La diversa distribución vertical de la precipitación pluvial en las zonas tropicales y extratropicales; sus razones y efectos geográficos. *Union Geográfica Internacional, Conferencia Regional Latinoamericana, Mexico 1966, Vol.III*, pp.457–478.

ZIMMERSCHIED, W., 1958. Vorläufige Mitteilung über die Niederschlagsverhältnisse in Ecuador. *Meteorol. Rundsch.*, 14(5): 156–162.

TABLE XIX

CLIMATIC TABLE FOR QUITO OBSERVATORY, ECUADOR
Latitude 00°13'S, longitude 78°30'W, elevation 2,818 m

Month	Mean sta. press. (mbar)	Temperature (°C)					Mean relat. humid. (%)	Precipitation							Mean thunderstorm freq.[3] (days)	Wind				Mean daily cloudiness (oktas)	Mean monthly sunshine (h)
		mean			extreme			mean[2] (mm)			max. (mm)	min. (mm)	>0.1 mm (days)	max. in 24h (mm)		av. speed (m/sec)		preval. direct.	calm (%)		
		max.	min.	daily[1]	max.	min.										07h	13h				
Jan.	727.1	22	8	13.0	26	2	81	94	106	124	149	19	15	44	7	0.3	0.7	E	45	6	180
Feb.	727.4	21	8	13.0	25	3	81	99	109	135	219	35	16	47	7	0.3	0.8	E	48	6	144
Mar.	727.7	21	8	12.9	25	4	82	134	132	159	220	44	18	40	12	0.3	0.8	E	45	6	141
Apr.	728.2	21	9	13.0	25	4	84	171	188	180	285	59	22	54	16	0.4	0.7	W	49	6	131
May	728.2	22	8	13.1	26	4	80	99	127	130	171	47	19	32	13	0.6	0.8	W	34	6	175
June	728.2	21	8	13.0	26	2	76	52	39	49	93	8	13	24	8	0.7	1.0	W	29	5	170
July	727.8	22	7	12.9	26	3	70	24	22	18	50	1	8	18	4	0.8	1.3	S	23	5	212
Aug.	727.8	23	7	13.1	26	3	69	25	38	22	75	2	7	27	5	0.8	1.4	E	23	5	199
Sept.	727.7	23	7	13.2	28	2	72	69	76	83	131	37	11	28	7	0.6	1.1	W	30	5	189
Oct.	727.7	22	8	12.9	27	3	81	139	93	133	223	65	20	54	13	0.5	0.7	W	41	6	166
Nov.	727.5	21	8	12.8	26	3	82	117	96	110	254	22	18	35	10	0.4	0.7	W	43	6	156
Dec.	727.3	21	8	13.0	27	3	81	86	97	107	122	29	16	34	7	0.4	0.7	E	43	5	186
Annual	727.7	22	8	13.0	28	2	78	1,109	1,123	1,250	1,366	890	183	54	109	0.5	0.9	W	38	6	2,049
Rec.[4] (yrs.)	10	10	10	30	10	10	10	10	16	65	10	10	10	10	7	8	8	8	8	10	8

[1] Data from KNOCH (1930) for Quito (0°14'S 78°32'W, 2,850 m), 1895–1907 are 0.4° lower.

[2] Second column data from Knoch (see footnote 1), period not specified; third column data from ZIMMERSCHIED (1958) for Quito (0°13'S 78°30'W, 2,818 m, 1890–1898 and 1901–1957.

[3] Data from Knoch (see footnote 1), period not specified.

[4] 10 = 1960–1969; 8 = 1962–1969; 30 = 1931–1960.

TABLE XX

CLIMATIC TABLE FOR IZOBAMBA, ECUADOR
Latitude 00°22'S, longitude 78°34'W, elevation 3,058 m

Month	Mean sta. press. (mbar)	Temperature (°C)				Mean relat. humid. (%)	Precipitation					Wind				Mean daily cloudiness (oktas)	Mean monthly sunshine (h)
		mean		extreme			mean (mm)	max. (mm)	min. (mm)	≥0.1 mm (days)	max. in 24h (mm)	av.speed(m/sec)		preval. direct.	calm (%)		
		max.	min.	max.	min.							07h	13h				
Jan.	703.5	18	6	24	1	82	122	200	61	19	36	2.4	3.3	NE	5	6	168
Feb.	703.9	18	6	23	2	83	152	253	59	18	39	2.1	3.4	NE	6	6	139
Mar.	704.3	17	6	22	1	83	170	236	45	22	40	2.4	3.4	NE	4	7	130
Apr.	704.7	17	6	21	2	86	193	315	73	23	35	2.2	3.2	NE	4	7	118
May	704.7	18	6	22	1	82	123	193	28	20	57	2.7	3.2	SW	4	6	163
June	704.7	18	5	21	1	79	75	112	6	15	24	3.0	3.7	SW	2	6	150
July	704.3	18	5	22	1	74	25	60	1	9	14	3.6	4.8	SW	1	5	193
Aug.	704.3	19	4	23	1	72	29	68	10	10	23	3.4	4.6	SW	2	5	193
Sept.	704.2	19	5	22	0	75	78	158	35	15	39	3.3	4.3	SW	2	6	166
Oct.	704.3	18	5	22	1	82	145	206	74	21	41	2.3	3.3	S	9	7	155
Nov.	704.1	18	5	21	1	84	145	239	85	19	33	1.7	3.2	E	5	6	152
Dec.	703.8	18	6	23	1	82	104	216	39	19	38	2.0	3.3	E	5	6	168
Annual	704.2	18	5	24	0	80	1,361	1,718	1,221	210	57	2.6	3.6	SW	4	6	1,895
Rec.[1] (yrs.)	8	8	8	8	8	8	8	8	8	8	8	8	8	8	8	8	8

[1] 1962–1969.

TABLE XXI

CLIMATIC TABLE FOR PUERTO BAQUERIZO, ISLA SAN CRISTOBAL, GALÁPAGOS
Latitude 00°54'S, longitude 89°37'W, elevation 6 m

Month	Temperature (°C)				Mean relat. humid. (%)	Precipitation					Wind				Mean daily cloudiness (oktas)	Mean monthly sunshine (h)	Number of thunderstorms
	mean		extreme			mean (mm)	max. (mm)	min. (mm)	>0.1 mm (days)	max. in 24h (mm)	av. speed (m/sec)		preval. direct.	calm (%)			
	max.	min.	max.	min.							07h	13h					
Jan.	29	23	31	17	80	48	135	10	12	34	2.8	4.3	SE	17	5	178	–
Feb.	30	23	33	20	80	67	205	5	8	70	1.7	3.4	SE	15	4	216	–
Mar.	30	23	32	19	81	85	488	–	7	258	1.3	3.1	SE	24	3	234	0.1
Apr.	30	23	32	18	80	35	217	–	5	67	2.7	4.7	SE	14	3	233	–
May	29	22	31	16	78	16	84	1	4	35	3.4	5.3	SE	4	3	238	–
June	27	21	30	16	77	2	3	1	4	2	3.8	6.3	SE	2	5	227	–
July	25	19	29	17	79	4	12	–	12	2	3.5	5.3	SE	–	6	193	–
Aug.	24	19	28	17	79	5	12	1	11	3	3.4	5.2	SE	1	6	174	–
Sept.	24	18	28	16	77	7	14	–	13	10	3.3	5.2	SE	1	6	144	–
Oct.	25	19	29	16	75	7	14	1	11	5	3.3	4.9	SE	1	6	162	–
Nov.	26	20	29	17	78	5	10	2	9	5	3.7	5.4	S	2	5	165	–
Dec.	27	21	30	18	78	7	12	1	11	4	3.7	5.5	SE	1	5	179	0.1
Annual	27	21	33	16	79	288	765	93	107	258	3.1	4.9	SE	7	5	2,343	0.1
Rec.[1] (yrs.)	4*	4*	10	10	10	10	10	10	10	10	8	8	8	8	10	6	8

[1] 1960–1969.
* 1959–1962.

205

TABLE XXII

CLIMATIC TABLE FOR PORTOVIEJO, ECUADOR
Latitude 01°04'S, longitude 80°26'W, elevation 44 m

Month	Mean sta. press. (mbar)	Temperature (°C) mean max.	mean min.	extreme max.	extreme min.	Mean relat. humid. (%)	Precipitation mean (mm)	max. (mm)	min. (mm)	≥0.1 mm (days)	max. in 24h (mm)	Wind av.speed (m/sec) 07h	13h	preval. direct.	calm (%)	Mean daily cloudiness (oktas)	Mean monthly sunshine (h)	Number of thunderstorms
Jan.	1005.8	31	22	35	19	76	75	185	18	16	55	0.7	2.3	NW	10	7	82	–
Feb.	1006.0	31	22	35	19	78	84	231	17	15	63	0.9	1.9	NW	12	7	78	2
Mar.	1006.6	32	22	36	18	80	104	205	22	18	80	0.9	2.0	NW	14	7	109	3
Apr.	1007.0	32	22	35	17	78	52	96	2	13	47	0.8	2.0	NW	13	7	145	3
May	1007.2	31	21	35	16	77	21	83	–	8	24	0.9	2.0	SE	9	7	115	1
June	1008.1	30	20	34	17	78	12	50	–	5	37	1.1	1.7	S	10	7	73	–
July	1007.7	30	20	34	15	77	4	17	–	2	11	1.1	1.8	S	9	7	100	–
Aug.	1008.0	30	19	35	15	76	1	8	–	1	5	0.9	2.4	S	10	6	130	–
Sept.	1007.8	31	20	35	16	76	1	8	–	1	8	1.0	2.6	S	11	7	125	–
Oct.	1008.0	31	20	35	15	76	1	4	–	1	4	1.0	2.5	S	8	7	118	–
Nov.	1007.7	31	20	34	15	76	1	4	–	2	2	1.1	2.5	S	7	7	95	–
Dec.	1007.0	31	21	35	15	73	6	15	–	3	14	1.1	2.6	S	8	7	101	–
Annual	1007.2	31	21	36	15	77	362	503	186	85	80	1.0	2.2	S	10	7	1,271	9
Rec.[1] (yrs.)	6	10	10	10	10	10	10	10	10	10	10	8	8	8	8	10	7	7

[1] 1960–1969, with interruptions for various elements.

206

TABLE XXIII

CLIMATIC TABLE FOR PICHILINGUE, ECUADOR
Latitude 01°06'S, longitude 79°29'W, elevation 73 m

Month	Mean sta. press. (mbar)	Temperature (°C) mean max.	mean min.	extreme max.	extreme min.	Mean relat. humid. (%)	Precipitation mean (mm)	max. (mm)	min. (mm)	≥0.1 mm (days)	max. in 24h (mm)	Wind av.speed (m/sec) 07h	13h	preval. direct.	calm (%)	Mean daily cloudiness (oktas)	Mean monthly sunshine (h)	Number of thunderstorms
Jan.	1002.2	30	21	34	18	86	468	629	247	28	139	0.8	1.4	SW	20	7	78	0
Feb.	1002.0	30	22	33	17	86	438	636	195	25	144	0.9	1.5	SW	22	7	82	1
Mar.	1002.4	30	22	34	19	87	389	622	150	27	117	0.8	1.5	SW	19	7	102	1
Apr.	1002.9	30	22	37	17	87	358	712	101	22	135	0.8	1.4	SW	17	7	112	2
May	1003.2	29	22	34	18	87	116	378	12	15	74	1.0	1.5	SW	17	7	88	0
June	1004.1	28	20	33	16	88	41	149	1	11	39	0.9	1.5	SW	16	7	55	0
July	1003.8	27	19	31	15	87	5	20	–	3	4	1.0	1.4	SW	14	7	47	0
Aug.	1003.8	28	19	35	16	84	9	40	–	5	40	0.9	1.4	SW	19	7	65	0
Sept.	1003.7	29	20	34	17	82	7	38	–	4	21	0.9	1.4	SW	20	7	62	0
Oct.	1003.8	29	20	35	17	81	19	107	–	7	102	1.0	1.4	SE	18	7	52	0
Nov.	1003.7	29	20	35	15	80	14	47	1	5	19	1.1	1.4	SW	12	7	50	0
Dec.	1002.8	30	21	36	16	80	120	300	27	15	68	0.9	1.5	SW	14	7	64	0
Annual	1003.2	29	21	37	15	85	1,984	3,005	1,236	167	144	0.9	1.4	SW	17	7	857	4
Rec.[1] (yrs.)	7	10	10	10	10	8	10	10	10	9	9	7	7	7	7	7	7	8

[1] 1960–1969; first years missing for various elements.

TABLE XXIV

CLIMATIC TABLE FOR PUYO, ECUADOR
Latitude 01°35'S, longitude 77°54'W, elevation 950 m

Month	Temperature (°C) mean max.	mean min.	extreme max.	extreme min.	Mean relat. humid. (%)	Precipitation mean (mm)	max. (mm)	min. (mm)	≥0.1 mm (days)	max. in 24h (mm)	Wind av. speed (m/sec) 07h	13h	preval. direct.	calm (%)	Mean daily cloudiness (oktas)	Mean monthly sunshine (h)	Number of thunderstorms
Jan.	26	17	30	11	89	299	411	203	26	67	0.1	1.4	NE	62	7	74	2
Feb.	26	17	31	11	89	294	369	206	23	112	0.1	1.4	NE	61	7	75	5
Mar.	26	17	31	13	89	391	513	287	27	75	0.1	1.4	NE	56	7	50	6
Apr.	26	17	30	11	89	453	556	380	27	95	0.1	1.5	E	60	7	61	6
May	26	17	30	12	89	324	422	204	27	74	0.2	1.4	NE	60	6	74	2
June	25	16	29	9	89	391	453	276	26	102	0.2	1.2	NE	62	7	79	2
July	25	16	29	12	89	339	593	236	27	83	0.1	1.3	SE	64	7	80	1
Aug.	26	16	30	12	87	345	470	227	26	84	0.2	1.4	NE	62	7	101	5
Sept.	27	16	30	11	87	354	495	231	28	74	0.1	1.4	E	60	6	98	7
Oct.	27	16	30	13	88	360	429	304	26	84	0.1	1.4	E	60	6	98	8
Nov.	27	17	30	13	88	367	528	295	26	79	0.2	1.6	E	55	7	105	9
Dec.	27	17	31	13	88	377	564	212	26	130	0.1	1.4	NE	64	7	90	5
Annual	26	17	31	9	88	4,294	4,688	4,000	315	130	0.1	1.4	NE	60	7	985	58
Rec.[1] (yrs.)	5	5	5	5	5	5	5	5	5	5	5	5	5	5	5	5	5

[1] 1965–1969.

TABLE XXV

CLIMATIC TABLE FOR GUAYAQUIL AIRPORT, ECUADOR
Latitude 02°12'S, longitude 79°53'W, elevation 6 m

Month	Mean sta. press. (mbar)	Temperature (°C) mean max.	mean min.	extreme max.	extreme min.	daily[2]	Mean relat. humid. (%)	Precipitation mean (mm)	max. (mm)	min. (mm)	≥0.1 mm (days)	max. in 24h (mm)	Wind av. speed (m/sec) 07h	13h	preval. direct.	calm (%)	Mean daily cloudiness (oktas)	Mean monthly sunshine (h)	Number of thunderstorms
Jan.	1009.7	31	23	35	19	25.5	78	217	519	52	16	155	1.3	2.8	NE	18	7	106	–
Feb.	1009.7	30	23	35	20	26.0	81	189	365	95	16	93	1.1	2.4	NE	22	7	112	1
Mar.	1009.9	31	23	35	20	26.4	81	231	434	111	19	98	1.2	2.5	NE	21	7	141	3
Apr.	1010.1	31	23	35	19	26.3	78	133	385	11	13	53	1.3	3.0	NE	20	6	149	2
May	1010.5	30	22	34	19	25.6	78	38	207	–	5	41	1.7	2.9	SW	15	6	165	1
June	1011.3	29	21	34	18	24.4	79	15	127	–	2	125	2.2	2.7	SW	9	7	123	–
July	1011.3	28	20	32	17	23.5	78	–	2	–	–	2	2.9	3.2	SW	3	6	130	–
Aug.	1011.4	28	20	33	17	23.2	76	–	–	–	–	–	3.2	4.0	SW	2	6	146	–
Sept.	1011.2	29	20	34	17	23.8	75	–	2	–	–	2	2.7	3.8	SW	4	6	161	–
Oct.	1011.2	29	21	35	18	24.0	74	4	18	–	1	17	2.9	3.8	SW	3	6	126	–
Nov.	1011.1	29	21	34	17	24.6	74	1	6	–	1	5	2.4	3.3	SW	6	7	115	–
Dec.	1010.1	31	22	35	19	25.4	72	15	60	–	4	28	2.3	3.6	SW	7	6	140	–
Annual	1010.6	30	22	35	17	24.9	77	843	1,332	397	77	155	2.1	3.2	SW	11	6	1,614	7
Rec.[1] (yrs.)	9	10	10	10	10	30	10	10	10	10	10	10	8	8	8	8	10	6	5

[1] 1959; 1961–1969.
[2] 1931–1960.

TABLE XXVI

CLIMATIC TABLE FOR IQUITOS, PERU
Latitude 03°46'S, longitude 73°20'W, elevation 104 m

Month	Mean sta. press. (mbar)	Temperature (°C)				Mean relat. humid. (%)	Precipitation				Wind	
		mean		extreme			mean (mm)	max. (mm)	min. (mm)	max. in 24h (mm)	aver. speed (knots)	preval. direct.
		max.	min.	max.	min.							
Jan.	1010.4	32	22	36	17	80	256	361	109	160	2	NE
Feb.	1010.6	31	22	36	15	81	276	719	63	144	2	NE
Mar.	1011.0	31	22	36	17	82	349	1,069	90	201	2	NE
Apr.	1011.8	31	22	36	18	84	306	553	176	135	2	NE
May	1012.7	31	21	36	16	83	271	620	153	192	2	S
June	1013.8	31	21	36	15	82	199	838	61	198	2	S
July	1014.4	30	20	35	14	81	165	345	25	90	2	S
Aug.	1013.3	32	21	37	11	79	157	656	34	247	2	S
Sept.	1012.3	32	21	37	15	78	191	336	51	114	2	NE
Oct.	1011.2	32	21	37	18	79	214	410	99	105	2	NE
Nov.	1010.0	32	22	37	18	80	244	620	75	201	3	NE
Dec.	1009.9	32	22	36	16	80	217	439	86	150	2	N
Annual	1011.8	31	21	37	11	81	2,845	5,621	2,259	247	2	NE
Rec.[1] (yrs.)	22	22*	21**	22*	22**	22	22	22	22	22	22	22

[1] 1949–1970.
* Excluding January 1949.
** Excluding year 1949.

TABLE XXVII

CLIMATIC TABLE FOR CHICLAYO, PERU
Latitude 06°47'S, longitude 79°50'W, elevation 31 m

Month	Mean sta. press. (mbar)	Temperature (°C)				Mean relat. humid. (%)	Precipitation				Wind	
		mean		extreme			mean (mm)	max. (mm)	min. (mm)	max. in 24h (mm)	aver. speed (knots)	preval. direct.
		max.	min.	max.	min.							
Jan.	1010.5	28	19	34	13	73	10.3	105.4	–	105.4	11	S
Feb.	1010.2	30	20	35	15	73	2.3	17.0	–	5.4	10	S
Mar.	1010.2	30	20	34	15	73	7.8	70.0	–	70.0	10	S
Apr.	1010.9	28	19	33	14	75	2.8	11.9	–	4.2	11	S
May	1011.8	27	17	32	13	75	0.2	2.0	–	1.6	11	S
June	1013.0	25	16	31	13	76	–	–	–	–	10	S
July	1013.1	24	15	29	12	80	0.1	1.2	–	1.2	10	S
Aug.	1013.2	23	15	29	10	80	–	–	–	–	11	S
Sept.	1013.0	24	15	29	12	79	0.4	4.0	–	1.0	11	S
Oct.	1012.9	24	15	33	10	79	1.1	6.0	–	3.0	12	S
Nov.	1012.6	25	16	29	13	77	4.0	26.0	–	23.0	11	S
Dec.	1011.7	27	17	33	14	76	1.5	18.0	–	18.0	12	S
Annual	1011.9	26	17	35	10	76	30.5	115.9	0.4	105.4	11	S
Rec.[1] (yrs.)	10*	16	16	16	16	16	16	16	16	10*	16	16

[1] 1954–1969.
* 1960–1969.

TABLE XXVIII

CLIMATIC TABLE FOR CAJAMARCA, PERU
Latitude 07°08′S, longitude 78°28′W, elevation 2,621 m

Month	Mean sta. press. (mbar)	Temperature (°C)				Mean relat. humid. (%)	Precipitation				Wind	
		mean		extreme			mean (mm)	max. (mm)	min. (mm)	max. in 24h (mm)	aver. speed (knots)	preval. direct.
		max.	min.	max.	min.							
Jan.	777.6	22	8	26	0	67	89	165	33	31	3	S
Feb.	777.7	21	7	27	0	67	89	183	4	51	3	S
Mar.	778.0	21	7	25	0	72	114	235	16	41	3	S
Apr.	778.5	21	7	25	0	69	96	212	15	38	3	S
May	778.6	22	5	25	0	64	32	72	1	26	3	S
June	778.9	22	3	26	0	58	8	25	–	12	5	N
July	778.9	22	3	25	0	55	5	26	–	13	6	NE
Aug.	778.6	22	4	26	0	55	11	44	–	10	6	E
Sept.	778.2	22	5	26	0	57	25	65	3	18	5	E
Oct.	778.0	22	7	29	1	64	85	126	23	29	4	SE
Nov.	777.8	22	6	25	0	64	79	143	20	40	3	S
Dec.	777.6	22	6	25	0	64	83	176	22	44	3	S
Annual	778.2	22	6	29	0	63	716	924	441	51	4	S
Rec.[1] (yrs.)	8*	12	12	12	12	12	12	12	12	12	12	12

[1] 1959–1970.
* Dec. '62–Oct. '70.

TABLE XXIX

CLIMATIC TABLE FOR TINGO MARIA, PERU
Latitude 09°08′S, longitude 75°57′W, elevation 665 m

Month	Temperature (°C)				Mean relat. humid. (%)	Precipitation			Wind	
	mean		extreme			mean (mm)	max. (mm)	min. (mm)	aver. speed (knots)	preval. direct.
	max.	min.	max.	min.						
Jan.	30	19	36	15	81	394	796	229	1	N
Feb.	30	19	35	13	81	359	585	126	1	N
Mar.	30	19	35	13	81	353	618	47	2	N
Apr.	31	19	36	13	80	319	579	63	1	N
May	31	19	35	14	78	238	419	26	2	N
June	31	18	34	12	79	119	252	5	2	N
July	31	18	34	10	78	132	240	–	2	N
Aug.	31	18	36	11	76	125	278	47	2	N
Sept.	31	18	36	12	77	158	308	71	2	N
Oct.	31	18	37	10	79	280	455	130	2	N
Nov.	31	19	36	13	78	278	422	169	2	N
Dec.	31	19	35	12	80	317	550	137	2	N
Annual	31	19	37	10	79	3,072	3,860	1,994	2	N
Rec.[1] (yrs.)	14*	16	14*	16	16	16	16	16 16		16

[1] 1954–1969.
* July '54–June '59; 1961–1969.

TABLE XXX

CLIMATIC TABLE FOR LIMA AIRPORT, PERU
Latitude 12°00′S, longitude 77°07′W, elevation 11 m

Month	Mean sta. press. (mbar)	Temperature (°C)					Mean relat. humid. (%)	Precipitation				Wind	
		mean		daily²	extreme			mean (mm)	max. (mm)	min. (mm)	max. in 24h (mm)	aver. speed (knots)	preval. direct.
		max.	min.		max.	min.							
Jan.	1011.5	26	19	21.5	31	15	83	1.2	10.5	–	8.2	7	S
Feb.	1011.0	26	19	22.3	30	15	83	0.4	2.4	–	1.2	6	S
Mar.	1011.2	26	19	21.9	29	15	84	0.6	3.4	–	1.9	6	S
Apr.	1012.1	24	17	20.1	28	11	85	0.1	0.5	–	0.5	6	S
May	1013.2	22	16	17.8	27	10	86	0.5	1.7	–	0.9	5	S
June	1014.4	19	15	16.0	25	8	85	0.8	3.0	–	1.3	4	S
July	1014.5	18	14	15.3	23	9	85	2.0	8.1	–	5.0	5	S
Aug.	1014.5	18	14	15.1	23	10	87	2.3	9.6	0.9	3.3	6	S
Sept.	1014.2	19	14	15.4	22	12	87	1.2	3.7	–	0.7	6	S
Oct.	1014.0	20	15	16.3	23	13	85	0.4	1.4	–	0.9	6	S
Nov.	1013.8	22	16	17.7	27	10	83	0.1	0.3	–	0.1	7	S
Dec.	1012.5	24	17	19.4	28	13	83	0.4	1.5	–	1.5	7	S
Annual	1013.1	22	16	18.2	31	8	85	10.0	20.4	3.0	8.2	6	S
Rec.¹ (yrs.)	10	10	10	10	30	10	10	10	10	10	10	10	10

¹ 1961–1970.
² 1931–1960.

TABLE XXXI

CLIMATIC TABLE FOR HUANCAYO, PERU
Latitude 12°07′S, longitude 75°20′W, elevation 3,380 m

Month	Mean sta. press. (mbar)	Temperature (°C)				Mean relat. humid. (%)	Precipitation					Mean daily cloud-iness (oktas)	Mean monthly sun-shine (h)
		mean		extreme			mean (mm)	max. (mm)	min. (mm)	≥0.1 mm (days)	max. in 24 h (mm)		
		max.	min.	max.	min.								
Jan.	687.2	18	7	24	0	73	119	200	52	22	43	6	177
Feb.	687.6	18	7	24	0	78	123	203	43	22	46	7	152
Mar.	688.0	18	6	24	−1	77	107	185	39	21	54	7	173
Apr.	688.5	19	5	23	−4	72	55	94	23	13	29	6	195
May	689.0	19	3	23	−6	65	25	63	1	7	34	4	228
June	689.2	19	0	23	−9	57	8	35	–	3	19	3	254
July	689.0	19	0	22	−10	56	8	23	–	3	19	3	258
Aug.	688.9	20	2	24	−6	55	14	68	–	5	22	4	238
Sept.	688.5	20	5	24	−2	61	40	95	19	11	37	5	197
Oct.	687.8	20	6	25	−3	64	69	125	26	14	38	5	205
Nov.	687.3	20	6	26	−3	62	67	115	33	14	37	6	207
Dec.	687.0	19	6	25	−3	68	89	144	19	18	39	6	204
Annual	688.2	19	4	26	−10	66	724	902	575	153	54	5	2,488
Rec.¹ (yrs.)	20	20	20	38*	38*	20	20	20	20	20	36**	20	16***

¹ 1948–1951 and 1955–1970.
* 1930–1951 and 1955–1970.
** 1931–1933, 1935–1951 and 1955–1970.
*** 1955–1970.

TABLE XXXII

CLIMATIC TABLE FOR PUERTO MALDONADO, PERU
Latitude 12°38′S, longitude 69°12′W, elevation 256 m

Month	Temperature (°C)				Mean relat. humid. (%)	Precipitation			Wind	
	mean		extreme			mean (mm)	max. (mm)	min. (mm)	aver. speed (knots)	preval. direct.
	max.	min.	max.	min.						
Jan.	31	21	36	15	80	262	372	160	5	N
Feb.	31	21	36	14	81	271	488	184	5	N
Mar.	31	20	36	15	80	289	494	177	4	N
Apr.	31	20	36	12	77	118	395	33	5	E
May	31	19	37	7	76	119	420	20	4	E
June	29	17	39	10	76	54	187	4	4	E
July	30	16	38	10	72	55	99	6	4	E
Aug.	32	18	38	10	67	53	131	4	5	E
Sept.	33	19	38	9	63	97	177	25	5	N
Oct.	32	20	38	14	72	140	201	74	5	N
Nov.	32	21	38	14	74	173	267	89	5	N
Dec.	31	21	35	14	78	296	491	105	5	N
Annual	31	19	39	7	75	1,927	2,541	1,433	5	N
Rec.[1] (yrs.)	9	9	12*	12*	9	9	9	9	9	9

[1] 1961–1969.
* 1958–1969.

TABLE XXXIII

CLIMATIC TABLE FOR CUZCO, PERU
Latitude 13°33′S, longitude 71°59′W, elevation 3,312 m

Month	Mean sta. press. (mbar)	Temperature (°C)				Mean relat. humid. (%)	Precipitation				Wind	
		mean		extreme			mean (mm)	max. (mm)	min. (mm)	max. in 24h (mm)	aver. speed (knots)	preval. direct.
		max.	min.	max.	min.							
Jan.	706.8	19	7	25	1	64	151	253	57	38	5	NW
Feb.	706.9	19	7	24	2	66	139	196	69	43	5	E
Mar.	707.3	19	6	24	1	65	106	165	37	47	5	NE
Apr.	707.7	19	5	24	1	61	39	82	10	20	4	NW
May	708.5	19	3	23	−3	55	12	27	−	14	4	W
June	708.6	19	1	25	−4	48	2	15	−	8	5	NW
July	708.6	19	1	23	−4	47	5	42	−	38	6	NW
Aug.	708.4	20	2	25	−4	46	6	24	−	11	7	W
Sept.	707.8	20	5	26	−3	51	27	48	10	22	7	NE
Oct.	706.9	21	6	26	0	51	52	84	22	26	7	NE
Nov.	706.5	21	6	27	−1	52	77	183	32	37	6	NE
Dec.	706.5	20	7	27	3	59	134	254	40	47	6	W
Annual	707.5	20	5	27	−4	55	750	982	390	47	6	NE
Rec.[1] (yrs.)	17	17	17	17	17	17	17	17	17	17	17	17

[1] 1954–1970.

TABLE XXXIV

CLIMATIC TABLE FOR APOLO, BOLIVIA
Latitude 14°43′S, longitude 68°30′W, elevation 1,382 m

Month	Temperature (°C)				Precipitation			Wind	
	mean		extreme		mean (mm)	max. (mm)	min. (mm)	aver. speed (knots)	preval. direct.
	max.	min.	max.	min.					
Jan.	25	17	31	13	194	367	75	7	N
Feb.	25	17	30	14	176	267	86	6	N
Mar.	25	17	29	11	177	255	75	5	N
Apr.	24	16	29	8	113	162	54	5	N
May	23	15	27	5	55	109	16	5	N
June	22	14	26	6	32	65	7	6	N
July	22	14	26	7	25	55	–	6	N
Aug.	24	15	29	6	32	65	1	6	N
Sept.	26	16	31	11	69	137	5	6	N
Oct.	26	17	31	10	111	274	46	7	N
Nov.	26	17	32	11	120	267	26	6	N
Dec.	26	17	32	13	220	292	189	6	N
Annual	25	16	32	5	1,324	1,650	968	6	N
Rec.[1] (yrs.)	10	10	10	10	10	10	10	10	10

[1] 1960–1969.

TABLE XXXV

CLIMATIC TABLE FOR SAN JUAN, PERU
Latitude 15°22′S, longitude 75°12′W, elevation 30 m

Month	Mean sta. press. (mbar)	Temperature (°C)				Mean relat. humid. (%)	Precipitation			Wind	
		mean		extreme			mean (mm)	max. (mm)	min. (mm)	aver. speed (knots)	preval. direct.
		max.	min.	max.	min.						
Jan.	1010.9	26	18	29	13	77	0.3	1.9	–	10	S
Feb.	1010.3	27	19	29	15	76	0.3	4.0	–	10	S
Mar.	1010.6	26	18	29	15	75	0.1	0.8	–	10	S
Apr.	1011.6	25	17	28	13	75	0.3	4.2	–	10	S
May	1012.8	22	16	26	12	76	0.1	1.5	–	11	S
June	1014.2	20	14	23	11	78	1.3	11.3	–	11	S
July	1014.5	18	13	22	10	78	0.3	1.9	–	11	S
Aug.	1014.5	18	13	21	8	79	0.7	6.5	–	12	S
Sept.	1014.0	19	13	22	11	79	1.6	6.8	–	12	S
Oct.	1013.7	20	14	23	11	79	1.5	10.1	–	12	S
Nov.	1013.2	22	15	25	10	78	0.9	4.5	–	11	S
Dec.	1011.8	24	17	27	12	78	0.4	1.3	–	10	S
Annual	1012.7	22	16	29	8	77	7.8	31.6	0.1	11	S
Rec.[1] (yrs.)	12*	14	14	14	12*	14	14	14	14	14	14

[1] 1957–1970.
* 1959–1970.

TABLE XXXVI

CLIMATIC TABLE FOR AREQUIPA, PERU

Latitude 16°19′S, longitude 71°33′W, elevation 2,525 m

Month	Mean sta. press.[1] (mbar)	Temperature (°C)					Relat. humid. (%)	Precipitation (mm)					Wind	
		mean			extreme			mean[3]		max.	min.	max. in 24h	aver. speed (knots)	preval. direct.
		daily[2]	max.	min.	max.	min.								
Jan.	788.2 (761.3)	13.9	22	8	27	2	57	31	(30)	150	–	27	8	W
Feb.	788.0 (762.1)	13.9	22	8	27	1	63	48	(46)	173	1	59	7	W
Mar.	788.2 (762.1)	13.5	22	8	28	1	59	19	(15)	184	–	44	7	W
Apr.	788.5 (762.5)	14.1	22	7	27	1	48	–	(4)	6	–	5	7	W
May	788.9 (762.9)	13.8	22	6	26	0	37	–	(1)	3	–	3	8	W
June	788.8 (762.9)	13.2	22	5	27	−3	29	–	(0)	1	–	1	9	W
July	788.6 (762.9)	13.1	22	4	27	−4	27	–	(1)	–	–	–	9	W
Aug.	788.6 (762.9)	13.8	22	5	28	−3	26	–	(0)	–	–	–	9	W
Sept.	788.7 (762.6)	14.4	23	6	27	−2	29	1	(0)	9	–	6	8	W
Oct.	788.6 (762.1)	13.6	23	6	27	−1	30	–	(1)	3	–	2	8	W
Nov.	788.4 (761.9)	13.9	23	6	28	0	34	1	(1)	8	–	5	7	W
Dec.	788.3 (761.8)	14.1	23	7	29	0	44	4	(9)	19	–	13	8	W
Annual	788.5 (762.3)	13.8	22	6	29	−4	40	104	(108)	466	13	59	8	W
Rec.[4] (yrs.)	21 (4)	33	21	21	21	21	21	21	(37)	21	21	21	21	21

[1] Values between brackets given by KNOCH (1930) for Arequipa (16°22′S 71°36′W, elevation 2,451 m).Period 1892–1895.

[2] Given by KNOCH, 1930 (see footnote 1). Period 1888–1920.

[3] See footnote 1. Period 1888–1924.

[4] 1950–1970.

TABLE XXXVII

CLIMATIC TABLE FOR EL ALTO, LA PAZ, BOLIVIA

Latitude 16°30′S, longitude 68°12′W, elevation 4,105 m

Month	Mean sta. press. (mbar)	Temperature (°C)				Mean relat. humid.* (%)	Precipitation					Wind	
		mean		extreme			mean (mm)	max. (mm)	min. (mm)	⩾0.1 mm (days)	max. in 24h (mm)	aver. speed (knots)	preval. direct.
		max.	min.	max.	min.								
Jan.	624.6	13	3	21	−1	71	116	209	48	20	24	4	E
Feb.	624.9	13	3	22	0	73	110	211	45	16	19	4	E
Mar.	625.1	14	3	18	−2	70	66	153	24	11	12	4	E
Apr.	625.6	14	2	19	−3	66	26	40	4	6	9	4	E
May	625.7	13	0	17	−4	52	14	55	1	3	9	4	W
June	625.6	13	−2	17	−6	45	3	10	–	2	2	5	W
July	625.3	13	−2	17	−6	52	6	18	–	2	4	5	W
Aug.	625.7	14	−1	18	−5	48	12	34	–	3	9	5	W
Sept.	625.6	14	0	20	−4	56	42	79	5	8	12	5	E
Oct.	624.8	16	2	23	−3	61	29	51	12	8	11	5	E
Nov.	624.3	16	3	22	−1	60	44	80	12	9	12	5	E
Dec.	624.3	14	3	20	−1	64	96	160	41	16	17	5	E
Annual	625.1	14	1	23	−6	60	564	684	401	104	24	5	E
Rec.[1] (yrs.)	15	6	9	13	26	16	9	9	9	17	17	10	10

* Another set of data for the same station gives values 5% less, throughout the year.

[1] 1942–1969; data missing for different years for various elements.

TABLE XXXVIII

CLIMATIC TABLE FOR LA PAZ, BOLIVIA
Latitude 16°30′S, longitude 68°08′W, elevation 3,632 m.

Month	Temperature (°C)				Mean relat. humid. (%)	Precipitation			Wind	
	mean		extreme			mean (mm)	max. (mm)	min. (mm)	aver. speed (knots)	preval. direct.
	max.	min.	max.	min.						
Jan.	18	7	25	3	68	92	144	38	4	E
Feb.	18	7	24	3	71	89	130	38	3	SE
Mar.	18	6	24	1	65	62	124	24	4	E
Apr.	18	5	23	−1	56	26	50	7	4	SE
May	17	4	23	−1	49	11	38	2	3	SE
June	17	3	22	−3	42	2	5	−	4	W
July	16	2	21	−2	40	4	15	−	5	W
Aug.	17	3	22	−1	47	7	19	−	4	W
Sept.	17	4	24	1	56	34	76	3	4	W
Oct.	19	6	25	2	53	28	48	12	4	W
Nov.	19	6	24	1	55	48	113	10	4	E
Dec.	18	6	25	2	65	85	113	46	3	SE
Annual	18	5	25	−3	56	488	664	369	4	W
Rec.[1] (yrs.)	10	10	10	10	5*	10	10	10	10	10

[1] 1960–1969.
* 1960–1964.

TABLE XXXIX

CLIMATIC TABLE FOR SANTA CRUZ, BOLIVIA
Latitude 17°47′S, longitude 63°11′W, elevation 437 m

Month	Temperature (°C)				Precipitation			Wind	
	mean		extreme		mean (mm)	max. (mm)	min. (mm)	aver. speed (knots)	preval. direct.
	max.	min.	max.	min.					
Jan.	31	21	38	13	161	377	88	10	NW
Feb.	30	21	37	13	136	279	46	9	NW
Mar.	30	20	37	10	94	219	8	8	NW
Apr.	29	19	36	9	68	156	12	8	NW
May	26	17	35	4	63	104	17	9	S
June	24	16	31	3	53	107	16	12	NW
July	25	15	32	4	34	123	3	13	NW
Aug.	28	17	35	5	40	134	2	13	NW
Sept.	30	19	37	6	52	111	2	13	NW
Oct.	31	20	38	11	88	235	11	12	NW
Nov.	31	21	40	13	116	248	28	10	NW
Dec.	31	21	40	13	141	265	50	10	NW
Annual	29	19	40	3	1,046	1,235	820	11	NW
Rec.[1] (yrs.)	10	10	10	10	10	10	10	10	

[1] 1960–1969.

TABLE XL

Latitude 17°58′S, longitude 67°07′W, elevation 3,708 m

Month	Temperature (°C)				Precipitation			Wind	
	mean		extreme		mean (mm)	max. (mm)	min. (mm)	aver. speed (knots)	preval. direct.
	max.	min.	max.	min.					
Jan.	19	4	25	−2	62	143	12	6	E
Feb.	18	4	23	−3	60	115	20	5	E
Mar.	18	2	23	−8	32	90	2	5	N
Apr.	18	−1	22	−10	11	26	–	4	N
May	16	−6	20	−16	5	19	–	4	S
June	14	−11	18	−22	2	14	–	4	S
July	14	−11	19	−20	1	6	–	4	S
Aug.	16	−8	19	−20	9	33	–	5	N
Sept.	17	−3	24	−15	13	33	1	5	N
Oct.	19	−1	25	−13	11	35	–	5	N
Nov.	20	1	25	−8	24	56	4	5	S
Dec.	19	3	25	−4	52	90	26	6	E
Annual	17	−2	25	−22	282	415	152	5	N
Rec.[1] (yrs.)	10	10	10	10	10	10	10	10	10

[1] 1960–1969.

TABLE XLI

Latitude 18°04′S, longitude 70°18′W, elevation 558 m.

Month	Mean sta. press. (mbar)	Temperature (°C)				Mean relat. humid. (%)	Precipitation				Wind	
		mean		extreme			mean (mm)	max. (mm)	min. (mm)	max. in 24h (mm)	aver. speed (knots)	preval. direct.
		max.	min.	max.	min.							
Jan.	1013.3	27	16	31	9	65	0.5	4.0	–	4.0	5	S
Feb.	1012.9	28	16	31	7	65	0.8	16.0	–	6.0	5	S
Mar.	1013.2	27	15	33	6	66	0.3	7.0	–	5.0	5	SW
Apr.	1014.5	25	12	30	4	70	0.8	12.3	–	3.0	5	SW
May	1015.7	22	10	29	3	76	2.8	27.0	–	8.0	5	S
June	1016.9	20	9	28	2	79	3.6	30.0	–	5.0	4	SW
July	1017.3	19	8	25	2	78	5.3	32.0	–	6.0	4	SW
Aug.	1017.3	20	9	28	4	78	8.5	41.0	–	8.0	4	SW
Sept.	1016.8	21	10	29	1	77	14.0	40.0	–	11.0	4	SW
Oct.	1016.1	22	11	27	5	74	5.2	39.5	–	6.5	5	SW
Nov.	1015.3	24	13	32	6	70	0.9	12.6	–	5.0	5	S
Dec.	1014.1	26	14	30	7	67	1.0	7.8	–	5.0	5	S
Annual	1015.3	23	12	33	1	72	43.7	114.1	–	11.0	5	SW
Rec.[1] (yrs.)	14*	19	20	20	20	20	21	21	21	21	20	20

[1] 1950–1970.
* 1957–1970.

217

TABLE XLII

CLIMATIC TABLE FOR YACUÍBA, BOLIVIA
Latitude 22°01′S, longitude 63°43′W, elevation 580 m

Month	Temperature (°C)				Precipitation			Wind	
	mean		extreme		mean (mm)	max. (mm)	min. (mm)	aver. speed (knots)	preval. direct.
	max.	min.	max.	min.					
Jan.	31	18	39	10	196	424	94	4	S
Feb.	30	19	39	11	175	300	96	4	S
Mar.	28	17	39	7	153	386	18	4	S
Apr.	26	15	35	1	105	311	20	4	S
May	24	13	34	−2	40	115	–	4	N
June	21	10	31	−3	11	40	–	4	N
July	23	8	36	−7	6	24	–	5	N
Aug.	26	9	39	−6	6	56	–	6	S
Sept.	28	12	40	−4	8	25	–	7	S
Oct.	29	16	41	5	62	152	3	6	S
Nov.	30	17	41	5	118	208	53	6	S
Dec.	31	18	38	8	129	289	77	5	N
Annual	27	14	41	−7	1,009	1,661	653	5	S
Rec.[1] (yrs.)	7	7	7	7	10	10	10	10	10

[1] 7 = 1963–1969; 10 = 1960–1969.

Chapter 5

The Climate of Brazil

L. R. RATISBONA

Introduction

A survey of the climate of Brazil is a difficult task. In an area of more than 8.5 million km², the data from only 515 climatological stations were available for the elaboration of this chapter. In addition to this, the stations are not uniformly distributed. Though in certain areas of the more highly developed part of the country they are satisfactorily numerous, in vast regions like the Amazon basin, the basin of the rivers Parana and Paraguay, and in other regions their number is much smaller. Also the inadequacy of upper-air observations, especially by radiosondes, made at only four stations, renders difficult the comprehension of weather conditions necessary for an adequate interpretation of the climate.

Furthermore, although valuable works on the climate of Brazil have been published by Brazilian climatologists, they are in general regional studies. On the country as a whole we can cite only *Météorologie du Bresil* and *Climatologie du Bresil* by Delgado de Carvalho, *Contribuição ao Estudo do Clima do Brasil* by H. Morize, *Notas sobre o Clima do Brasil* by S. Serebrenick, and some articles in geographic reviews mainly by L. C. Bernardes. The first three were published in the beginning of the century, when there were much less data available than now. The third one is of interest principally because the author introduces a useful climatic classification. In view of these limitations, we will only be able to achieve a general view of the climate of Brazil.

Before giving a description of the various aspects of *climatic elements*, we will discuss the *climatic factors* which determine them.

The understanding of these factors will be instrumental for the explanation of the geographical distribution and annual variations of the climatic elements. For selected stations, the most important elements are shown in the climatological tables in the Appendix (see p.267). For some regions, special climatic features will be described in the section on regional characteristics.

We thought it interesting, finally, to include a short exposition of the most important *climatic zones* of Brazil and of some problems in Brazilian climatology.

Climatic factors

As a first step, it is appropriate to review the principal factors which determine the distribution of climatic elements in Brazil and their evolution through the seasons; they are: latitude, relief, nature of the land surface, pressure centres and air masses.

219

Latitude

The amount of solar energy which a particular point of the surface of the earth receives, depends upon the angle at which the solar rays strike the surface and upon the duration of time during which it undergoes the action of the radiation. These two factors depend, in the first place, upon the latitude.

Brazil extends from 5°10′N to 33°47′S. The sun passes through the zenith on the first of these latitudes on September 8 in its north–south movement, and reaches the latitude of 23°27′, that is, the Tropic of Capricorn, on December 21. On its return northward it passes again over the northernmost parallel of the country on April 6. Thus the area between the Tropics is crossed twice a year by the sun, those at the Tropic just once, and those between the Tropic and latitude 33°47′S are not crossed.

On the southernmost latitude of the country the length of the longest day of the year is 14 h 23 min on December 21, and of the shortest, 9 h 37 min on June 21, the difference being 4 h 46 min. At the lower latitudes the difference varies between this figure and less than 3 min at the Equator.

Relief

The relief of Brazil is characterized by plateaus surrounded by plains, the first occupying 5/8 of its area and the second 3/8. It is not a country of high mountains; no point reaches an altitude of 3,000 m, and only 3% of its territory exceeds 900 m, while the lowlands with an altitude lower than 200 m account for less than 40% of the total area. In spite of this, relief is very important as a climatic factor in Brazil. We assume that the principal aspects of this relief are known. However, to facilitate the description of the distribution of meteorological elements and regional characteristics, we will indicate here the names of the principal regions into which, based on relief, we will divide the country, using the names of the plateaus, plains, and basins of the large rivers, although they are not in strict accord with the designations given them by Brazilian geographers, which are not universally agreed upon anyway.

The Planicie Amazonica (Amazon plains), which have as their axis the Amazon river, can be divided into the upper Amazon, the area west of 60°W, and the lower Amazon, the area east of this line as far as the mouth of the river. The area bounded on the east by the Serra de Ibiapaba range and on the south by the Piaui Mountains, including the basins of the rivers Grajau and Parnaiba, is referred to as the Meio Norte (middle north) by Brazilian geographers.

The northeast is comprised of two parts, the western part which consists of the crystalline peneplain between the mountain range Serra de Ibiapaba, the river São Francisco, and the Borborema escarpment, and the eastern part, that is, the coastal plain, and the eastern slopes of the Borborema.

The area to the west of the Vale do São Francisco (valley of the São Francisco) from the Amazon plains in the north to the slopes of the valley of the rivers Parana and Paraguay in the south, is called the Planalto Central (central plateau). We shall make a distinction between a western part, less elevated, situated beteeen the Tapajos and the Tocantins rivers, the altitude of which ranges from 200 to 500 m, and an eastern part, with altitudes from 500 to 1,000 m.

We call the east the region which, bounded toward the west by the valley of the São Francisco and in the south by the valley of the Paraiba, extends along the Chapada Diamantina (Diamantina plateau), the mountain ranges Serra do Espinhaço and Serra da Mantiqueira and the sea coast.

The Brazilian portion of the middle-upper Paraguay is called the Pantanal. The plateau to the south of the Planalto Central and the upper São Francisco, including in the south and east the mountain range Serra da Mantiqueira, is known as the Planalto Centro-Oriental (central-eastern plateau). To the south of this plateau as far as the Planicie Riograndense (Rio Grande plains), extends the Planalto Meridional (Meridional plateau), bordered on the west by the Vale do Parana (valley of the Parana). The less elevated portion to the north is called the Planalto Paulista (Paulista plateau). The coastal area to the east of these plateaus, to the south of the mouth of the Paraiba, is called the Litoral Sudeste (southeast sea board).

Nature of the surface

The atmosphere absorbs only a small quantity of direct solar radiation; it is warmed principally by terrestrial radiation and convection, from the surface. Therefore reference must be made to the nature of the types of surface which might influence the climate of Brazil. The principal distinction to be made initially is between liquid and solid surfaces. Brazil, with its long coastline, bordered in the north and east by the Atlantic Ocean from which the winds usually blow, undergoes a great deal of influence from this liquid surface. Isotherms over the southern part of this ocean do not have, however, a zonal direction. Marine currents make its western side warmer than its eastern side. The equatorial current, which has an east–west direction, branches at the Cape of São Roque, one of the branches following the north coast of Brazil, the other, named the Brazil current, the length of the east coast, carrying the isotherms to the south because it is a warm current. In the continental area of the country it is necessary to distinguish two principal characteristic types of surface according to its vegetation which has influence on, and is at the same time a consequence of, the climate: the Amazon forest and the caatinga.

The first is the equatorial rain-forest in the Amazon basin which covers approximately 40% of Brazil's territory. Here it is possible to distinguish three types: the forest on solid ground above the flood level is the most common type, with trees whose tops reach various heights, rich in lianas and epiphytes; the varzea, subject to periodic floods by the rivers and temporarily flooded; and finally the igapo, the permanently flooded forest. It is important to note that the sun's rays never penetrate directly the canopies of dense plant masses.

The caatinga in the northeast region and in the basin of the São Francisco has vegetation made up of small trees of the xerophytic type the small leaves of which are deciduous during drought periods, with open areas and absence of grass. All these characteristics allow the sun to reach the soil.

We will refer occasionally also to the cerrado, consisting of sparse trees and underbrush with a cover of grasses over the soil, and to the tropical, or atlantic, forest with fewer species and smaller trees than the equatorial forest.

In the equatorial rain-forest, most of the solar energy is used for the evaporation and transpiration of water, so that the land surface acts as a regulator of temperature and as a supplier of moisture. In the caatinga, the atmosphere receives a great quantity of sensible heat and less humidity from the surface.

Pressure centres

Since pressure gradients are the determining factor of atmospheric motion, acquaintance with the various pressure centres which affect the country is necessary for an understanding of the circulation and the weather conditions throughout Brazil. This is true although atmospheric pressure does not reach very high values, as it happens in the middle latitudes, nor very low values, the South Atlantic being free of the violent hurricanes characteristic of the other oceans.

It is difficult to trace the mean isobars for the South American continent, especially for Brazil. The South Atlantic except next to the coast, is a little-navigated ocean, hence the lack of data sufficient for an analysis of the pressure field. Furthermore, uncertainty of the exact altitudes of stations in the interior of the country makes the reduction of pressure values to a standard level a doubtful procedure, in particular over the plateaus.

Considering the average annual distribution of pressure over the globe, Brazil due to its latitudinal position is preponderantly under the influence of the subtropical high pressure belt of the Southern Hemisphere and of the equatorial pressure trough. However, the subtropical anticyclone over the North Atlantic also has some influence.

The profile of average annual pressure values along a meridian shows that the pressure increases from the Equator to 30°S and from there decreases to the low-pressure band at 60°S. The thermal effect of the South American continent produces, however, a break in the high-pressure belt, separating one anticyclone in the Atlantic from another in the Pacific, placing the continent between these two anticyclones. Thus, some of the isobars, principally in the summer[1], do not follow the parallels across the continent. In fact, the average annual chart shows that the pressure (reduced to sea level) is lower on the continent with values of 1,010–1,012 mbar for the Amazon basin, an isobar of 1,016 mbar being located along the east coast.

In this way, the country is situated in the western part of the quasi-stationary anticyclone of the Atlantic, between the equatorial and the subpolar low-pressure belts. Because the thermal effect of the continent is more accentuated in the summer than in the winter, this isobaric pattern varies according to the season of the year.

In winter, all of the country lying south of the equator is dominated by the anticyclone of the South Atlantic, with a low pressure area only in the equatorial zone. All stations, even stations situated north of the Equator, record their highest monthly mean pressure in this season. The anticyclone of the South Atlantic is then at its maximum intensity, with 1,022.5 mbar as the central isobar. On the average, its axis extends to the latitude of 27.5°S, where the 1,020-mbar isobar is found in the eastern part of Brazil, the Amazon having a pressure of 1,013 mbar. Starting in July, the pressure begins to fall all over the country. In October, the continental heat low begins to form with its centre approximately over the Tropic. In summer, this feature is well established with an isobar of 1,008 mbar near the centre. The 1,014-mbar isobar is tangent to the east coast of Brazil, facing

---

[1] Here and in the following, Southern Hemisphere seasons are meant (December–February = summer).

a value of 1,010 mbar over the Amazon basin. With the return of the anticyclone beginning in March, the heat lows decline until they disappear in the winter.

The part of Brazil situated in the equatorial zone undergoes the influence of the North Atlantic semi-stationary anticyclone. While the anticyclone of the south grows and spreads over the country from east to west with a certain regularity, beginning in late summer, attaining its maximum in July, then declining until January, the anticyclone of the North Atlantic behaves more irregularly. Like its counterpart of the South Atlantic, it is strongest in July with more than 1,025 mbar in the centre, situated approximately at 37°N, extending its influence over the part of Brazil north of the Equator. It then begins to weaken until November. However, while the latter remains over the ocean as the Vitoria curve shows (Fig.1), the anticyclone of the north re-intensifies until February, then decreases through April and increases again until July. This rhythm can be seen in the diagram from the station at Oiapoque.

In the equatorial zone, between the two anticyclones, lies the equatorial pressure trough. Throughout the year it undergoes a process of deepening and filling. The Brazilian stations situated therein, Oiapoque on the seacoast and Boa Vista in the interior, show that their annual pressure curves respond to the simultaneous expansion of the anticyclones with higher values in the winter than in the summer of the Southern Hemisphere, indicating a filling of the trough in the former season, and a deepening during the latter. There are not sufficient data to determine exactly the change of the location of the trough's axis, either over the ocean or the continent. Over the ocean it ought to follow the

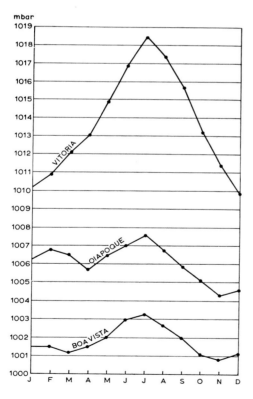

Fig.1. Annual variation of average monthly pressure at station level. Vitoria: 20°19′S 40°21′W; Oiapoque: 3°50′N 51°50′W; Boa Vista: 2°49′N 60°40′W.

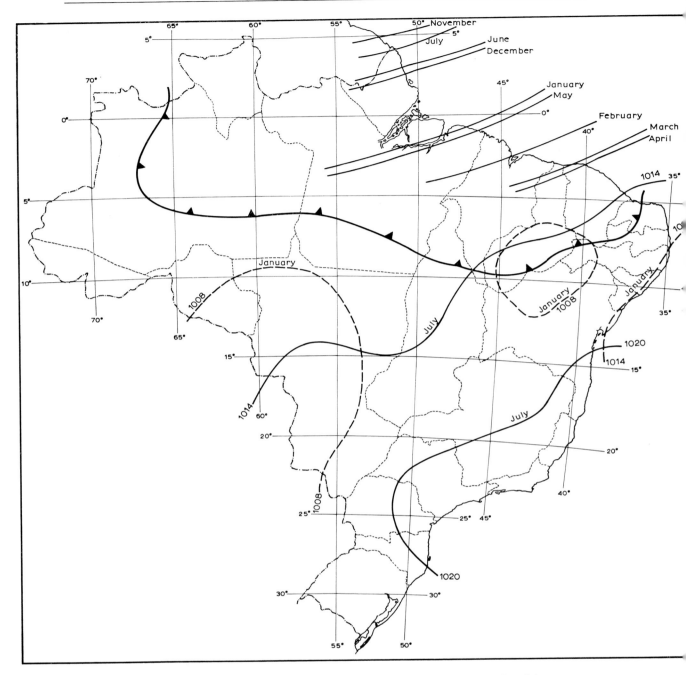

Fig.2. Average position of the Intertropical Convergence in the various months of the year; extreme position reached by the cold fronts in normal years; selected average isobars at sea level in January and July.

displacement of the convergence of the wind field and have its extreme southerly position in March–April (Fig.2), its extreme northerly position in September–October.

The anticyclone of the Pacific has no influence on the climate of Brazil because of its great distance from Brazilian territory and the barrier formed by the Andes.

Considering secondary circulations, the country is periodically invaded by cold anticyclones of polar origin which traverse it following two trajectories, one continental, reaching the Amazon basin, the other maritime, reaching the northeast coast. Frequently, southern Brazil is also crossed from northwest to southeast by dynamic cyclones. In winter and spring, intense cold anticyclones can extend themselves over the whole country and join with the Atlantic high pressure system. In summer and autumn there are only weak cold air invasions along the continental trajectory, and invasions following the maritime trajectory often do not pass beyond the tropic.

The facts presented show that the isobaric systems which have the most importance in the study of Brazil's climate are the following: the semi-stationary anticyclones of the South and North Atlantic; the cold anticyclones and the dynamic low pressure systems; the continental heat low, and the equatorial trough.

As for the seasonal variations of these systems, we must keep in mind the following: (*1*) advance over the country, of the anticyclone of the South Atlantic south of the Equator, or rather, the progressive increase in pressure from summer until July; (*2*) the subsequent general decline in pressure beginning in July, or the recession of the anticyclone over the ocean and the formation in summer of the continental heat low; (*3*) the advance over the country to the north of the Equator of the anticyclone of the Azores, its withdrawal until November, followed by a new advance until February and recession until April; (*4*) the increasing influence of the equatorial trough at the end of summer and in the fall over the northern coastal area of the country, that is, its southerly movement until April and return to its northernmost position in October; (*5*) periodic invasions of cold polar anticyclones, most intense in winter and spring, and formation of dynamic cyclones which move from northwest to southeast over southern Brazil.

Air masses

On the basis of these facts about the nature of the surface and pressure systems, it is possible to determine the principal air masses which act on the climate of Brazil.

The tropical maritime mass (mT) forms over the South Atlantic and, because of its contact with the ocean surface and the subsidence in the semi-stationary anticyclone, it consists of two layers, the lower one being cooler and more humid, the higher one, warmer and dryer. Moving in the same direction from east to west, the two layers are separated by a strong temperature inversion and a discontinuity in humidity. In this east–west motion the air rises and the altitude of the temperature inversion increases until it reaches the coast of Brazil at 1,800–2,000 m. The lower layer acquires a great quantity of moisture and a lapse rate near to the adiabatic, due to the warming over the warm Brazil current and irradiation from the top of the clouds. The upper layer, because of its subsidence, is dry, warm, and stable. The inversion disappears toward the south letting the moisture penetrate the upper levels. This air mass goes on to feed the warm sector of the Atlantic polar front. In winter, this air mass dominates the major part of the country. Soundings reveal the frequent presence of an inversion between 3,000 and 4,000 m. The lower layer, however, also contains relatively little moisture, due to the losses through precipitation in the regions of the east coast.

The mass formed in the equatorial trough where calms, also called doldrums, prevail, is the equatorial maritime mass (mE). The inversion of the trade winds disappears, moisture penetrates great heights, instability is great and any local convergence produces convective clouds with great vertical development. This air mass affects the north of Brazil at the end of summer and in autumn.

Over the area of dense forest in the Amazon basin, where calms and weak winds prevail, the equatorial continental mass (*cE*) forms. Solar energy is used up largely in the evapotranspiration of plants which reaches values of 1,300–1,600 mm/year. For this reason it is a very humid and convectively unstable air mass. In winter it covers only the upper Amazon because the rest of the basin is dominated by the less humid *mT* air mass. In summer, however, with the recession of the anticyclone to the ocean and the formation of the continental heat low, the latter draws in the air from the Amazon region which dominates the whole interior of the country, even in the south. This phenomenon may be considered as analogous to a monsoon.

The tropical continental mass (*cT*) can only be distinguished during the summer over a narrow, low, hot and dry zone centered over the Tropic. It is there that the highest maximum temperatures are observed and the continental heat low forms.

From time to time, the country is invaded by cold masses of polar origin (*mP*) reflected in the isobaric field by the cold anticyclones described above which, in the region where they originate, have low temperatures and therefore low moisture content, and stable stratification. However, when the mass moves northward it warms up, becomes more unstable, and in its oceanic trajectory acquires more moisture, especially in the summer. It is less humid in its continental trajectory. Fig.2 shows its northernmost position during normal winters.

Fig.3 indicates a typical succession of the positions of a cold front, a warm front, and a tropical discontinuity over Brazil, linked in a cyclonic circulation which occurred between November 7 and 13, 1963. Numbers indicate the dates of their positions at 12h00 G.M.T. On November 8 we see that a cyclone is formed over a stationary front, which moves northwest–southeast, from the continent to the ocean. Between the 8th and the 11th the cold front follows a continental trajectory and disintegrates over the upper Amazon and the Planalto Central. Near the coast, however, it continues to move northeastward until November 13, when it begins to dissolve north of the Baia de Todos os Santos. Behind the cold front the mass *mP* advances, preceded by the masses *cE* and *mT*.

Frequently, within the trade winds system there appear undulatory disturbances, as yet little studied, which affect the plains of the eastern northeast and the east (VUORELA, 1950).

Climatic elements

Winds[1]

The principal isobaric systems which determine the surface circulation over Brazil have been described above. These factors, in their annual variation, combine to determine the circulation over the various regions of the country. Predominance of one or the other of them determines the prevailing direction and speed of winds. These are the general factors to which we must add the local ones which modify them in each locale, that is, relief and nature of the surface, the most important being the land–sea difference. The

---

[1] Observations of 07h00, 14h00, and 21h00 local time have been used for this discussion.

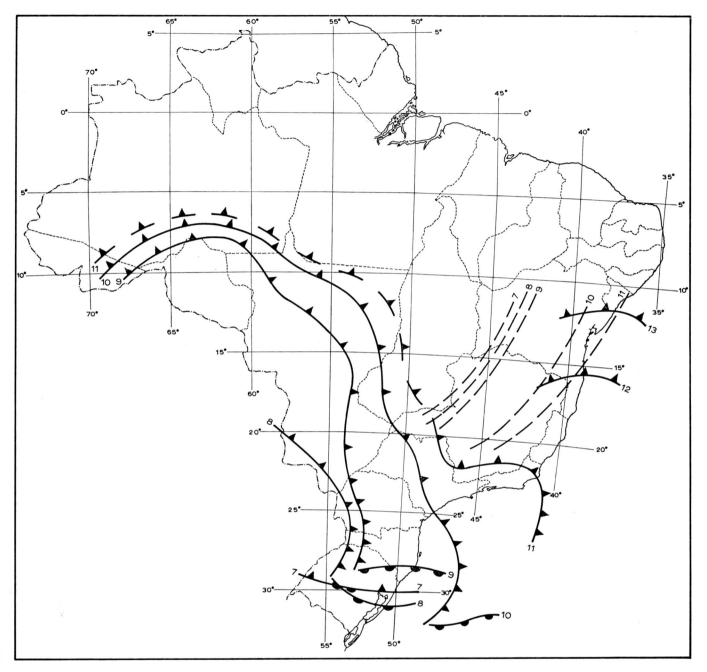

Fig.3. Successive positions of the fronts and of the tropical discontinuity between November 7 and 13, 1963, at 12h00 G.M.T.

first, relief, channels the air currents, obstructs their movement and produces mountain and valley breezes. The second, the nature of the surface, brings about land and sea breezes along the vast extension of the Brazilian coast.

Nevertheless, the most important determinant of the circulation over the country is the semi-stationary anticyclone of the South Atlantic. It is a centre of action subject to continuous subsidence. From there, the winds diverge in all directions. Those reaching the coast of Brazil have a direction between southeast and northeast. If this were the only influence the winds would blow from southeast to east in the north of the country (south of the Equator), that is, they would be trade winds, shifting to northeast and north in southern Brazil, following the isobars.

Relief, the formation of the heat low in summer, and cold air invasions in winter somewhat modify this aspect on the surface. This last phenomenon favours a southerly component of the winds, principally in the south of the country, where it is more intense, although its effect is felt in the upper Amazon and on the east coast. The heat low, on the other hand, introduces a northerly component in the circulation of northern Brazil. In effect, this formation of the continental heat low causes winds north of the tropic to rotate counter-clockwise from winter to summer giving a northerly component to the wind south of 6°S and west of the mountains.

On the north coast of the country, the influence of the equatorial trough in its southerly movement is felt. This plays an important role in the diminution of speed and change in direction of winds and in the increase in the number of calms, when it is in its extreme southerly position.

The annual wind chart indicates that the southern limit of the southeast trade winds, near the east coast of Brazil, is found at about 15°S. However, the position of the line of change of the wind direction from southeast to northeast varies in latitude during the year. In the period from April to August it stays south of that latitude, moving north in September, and from October to March it remains over the east coast of the Brazilian northeast, north of 15°S, related to the displacement of the points where the ridge of the anticyclone reaches the coast.

We now will consider the wind regimes in winter and summer. In the winter, when the anticyclone of the Atlantic dominates the continent, the southeast trade winds blow, well defined, over all the littoral plains and the slopes of the east region, and over the Planalto da Borborema as far as the Serra do Espinhaço, the Chapada Diamantina, and the Serra da Ibiapada.

The average speed of these winds, which is 5–10 m/sec over the ocean, falls sharply to 1–2 m/sec as they penetrate the continent in the coastal lowlands due to the friction effect. However, over the Planalto Central and the Chapada, the speed increases to 5 m/sec because of the altitude. Calms, which are practically non-existent over the ocean, are very frequent in the eastern coastal lowlands (40–60%), but rare at higher altitudes.

To the west of the relief barrier, in the valley of the São Francisco river, the dominant winds are from the east and northeast. There are many calms. To the west of the São Francisco over the Planalto Central winds blow with directions east and northeast. In the valley of the Parana–Paraguay northeast winds predominate in the valleys as well as over the Serra de Maracaju which separates the two rivers.

On the north coast to the west of the Serra da Ibiapaba and in the valley of the lower Amazon, a northeast current dominates with 2–3 m/sec. This current shifts to the southeast and increases up to an altitude of 1,500 m, with about 9 m/sec. Beginning at this altitude the wind rotates to east and diminishes in speed. Thus we see that the northeast current is very shallow, comprising a layer of several hundred metres only. Above it in the low troposphere the general current of southeast to east of the trades predominates. This is purely a continental phenomenon, not found over the ocean. It must be attributed to the progressive diminution of influence of the anticyclone of the South Atlantic and to the decrease of pressure over the continent produced by continental warming.

As we near the upper Amazon and go upstream along the tributaries to the north, weak winds blow in all directions with a slight predominance of east winds and very frequent (40–60%) calms. However, over the Rio Branco region the prevailing winds are easterlies. In the southern part of the country, especially over the Planicie Riograndense, successive passing of dynamic cyclones and anticyclones tend to equalize the frequencies of various directions. Gales occur from 6 to 20 days/year, most commonly at the coast, where gusts of 30–34 m/sec have been recorded at some stations.

The winds over the whole country tend to turn counter-clockwise with altitude. The southeast current of the trade winds tends toward east, northeast winds take north and northwest directions, and above 5,000 m, from the extreme south of the country to 10°S, west winds strongly predominate; this reflects the effect of the southern polar vortex. Comparing the prevailing wind directions in different seasons, one sees a counter-clockwise rotation from winter to summer, due to the formation of the heat low.

In summer dominant winds over the coast of the eastern part of the Brazilian northeast are now east, followed by southeast. South of the Baia de Todos os Santos (in the state of Bahia) dominance passes to northeast, because the ridge line of the anticyclone appears over the northeast. Predominance of southeast continues over the plateaus and plains. From the sea to land, the winds' speed diminishes less in summer than in winter.

On the coast and in the interior of the central part of the northeast, spring and summer wind patterns change completely in relation to those of winter (southeast). Due to the heating of the caatinga region, the prevailing winds are from the northeast, 2–4 m/sec, and all over the north coast the northeast dominates.

West of the mountains, over the São Francisco basin and the Amazon basin to the south of the river, in the interior of the Meio Norte, on the Planalto Central and in the valleys of the Parana and Paraguay, calms prevail and winds from north and northeast, caused by the heat low, show only a low frequency maximum.

In southern Brazil, with diminished effect of secondary circulations, dominant winds are from the east quadrant, due not only to the flow toward the heat low, but also to the maritime trajectory of the cold anticyclones.

In altitude, the southeast winds behave as in winter. However, the west winds of the polar vortex appear at 5,000 m not far beyond the tropic.

Considering the intermediate seasons, spring is characterized by a general increase of wind speed. In autumn equatorial calms reach the equatorial area of the country, reducing the speed of northeast winds at the north coast.

Cloudiness

Distribution

Only in the south of the country, when and where depressions move from northwest to southeast along the polar front, the typical succession of middle latitude cloud formations is found: cirrostratus, altostratus, nimbostratus, followed by cumulonimbus with the invasion of the cold air.

In the entire tropical continental region of the country, principally in the summer, there prevail the huge cumulus and cumulonimbus clouds with their typical diurnal life cycle.

At the north coast, the development of these clouds is intensified, at the end of summer and in fall, by the action of the Equatorial Convergence Zone. Contrarily, in the eastern part of the northeast region the presence of the trade inversion hinders the vertical development of tall, convective clouds.

In a general way, the annual distribution of cloudiness reveals a vast area with a cloud cover greater than 5/10, comprised of the Planalto da Borborema, the seacoast and the mountain slopes of the Região Leste, the Planalto da Mantiqueira, the Planalto Meridional, the southeast coast, and the Planicie Riograndense. A second area with a cloud cover greater than 5/10 includes the Amazon basin with the exception of the valley of the Rio Branco, the basins of the Meio Norte, the Chapada dos Parecis, and the Maracaju, São Jeronimo and Caiapos Mountains.

Between these two areas extends a belt with a south-southwest–north-northeast direction, in which average cloudiness is less than 5/10, including the Parana and Uruguay basins, the Pantanal (low-lands of Mato Grosso state), the basin of the São Francisco and the central part of the northeast.

The upper Amazon reveals the highest mean annual amount of clouds in the country, where in the eastern part of the river's upper course it rises to over 8/10 of the sky covered. On the southeast coast, in the valley of Itajai, cloudiness rises above 7/10.

The smallest cloud cover values are observed on the seacoast in the central part of the northeast and in the middle São Francisco, where they are under four tenths, and over the Serra da Canastra which is the area of least cloud cover, with less than three tenths.

Duration of sunshine

The chart of Fig.4 represents the distribution of insolation over Brazil; it is accompanied by graphs showing the annual variation. For the points where sunshine records were not available, estimates based on cloudiness have been used to facilitate the drawing of the isolines.

Distribution

A comparison of the annual insolation and cloud cover charts, for Brazil, shows the intimate relationship in the distribution of these two elements.

On the insolation chart we see the same pattern as described for average cloud amounts, except that, due to their complementary nature, where cloudiness is least, duration of sunshine is longest. Thus the great belt running southwest–northeast from the Parana–Paraguay basin through the basin of the São Francisco to the western part of the northeast, where the weakest cloudiness is found (less than 5/10), has the highest insolation with values in excess of 2,400 h/year. The maximum values are found in the middle São Francisco valley with 3,400 h and immediately to the west of the Borborema with 3,000 h. On the island of Fernando Noronha, northeast of Natal, the average duration of sunshine amounts to 3,215 h, which indicates that this oceanic area has rather high insolation. This is because for a large part of the year it is under the influence of the subsidence of the anticyclone of the Atlantic.

In the large area of cloudy skies which extends over the plateaus and the slopes of the

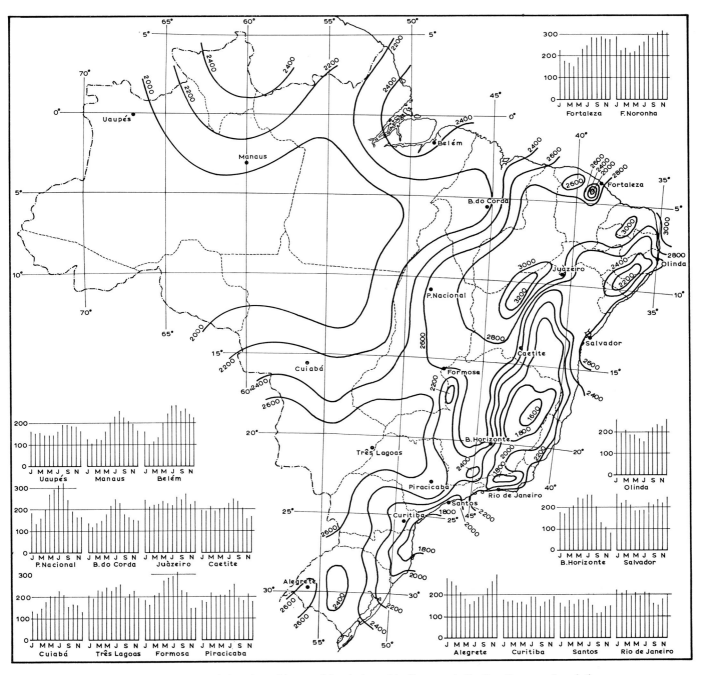

Fig.4. Average annual duration of hours of insolation with diagrams indicating the annual variation.

east region, insolation decreases to values lower than 2,200 h. The lowest value of the entire country is found over the middle Rio Doce with 1,600 h.

For the Amazon basin, data were available only for some points over the lower course of the river, over its mouth, and over the Rio Negro and the Araguaia–Tocantins; therefore, over the southern part of the basin isolines of insolation were drawn on the basis of values deduced from cloudiness.

Over the Planalto Meridional and the southeast coast, insolation varies from 2,400 to less than 1,800 h/year. This last figure is recorded in the valleys of the rivers Itajai and Iguape, where the average cloud cover is over 7/10.

Annual variation

Excepting the south of the country, where the difference in the length of day in winter and in summer is appreciable, the principal cause of variation in insolation is cloud cover. The small annual variation of cloudiness in Uaupés, a typical station of the Rio Negro region, causes a slight corresponding variation in insolation. Greatest insolation occurs in August and September when cloudiness is least; minimum insolation is in April and May when cloudiness is greatest. Manaus has maximum cloud cover at the end of summer and in autumn, and the minimum in winter. Hence the least insolation in February and the greatest in winter.

In Fortaleza and on Fernando Noronha where the tropical convergence appears in April with maximum cloudiness, minimum insolation occurs in this month and the maximum in spring. The graph of Belém shows high insolation in winter and spring. It decreases rapidly from December to February, when it reaches its minimum because of the rapid increase of cloudiness with the appearance of the tropical convergence.

Maximum duration of sunshine occurs in Porto Nacional and Barra do Corda in winter and the minimum at the end of summer. The graph for Joazeiro shows high insolation throughout the year, due to little cloudiness and its slight variation during the year.

In the northeast area on the slopes of the Borborema and along the whole east coast, insolation is greatest in summer and least in winter due to the inverse variation with cloudiness, as the diagrams for Olinda and Salvador show.

All the diagrams for the central area of the country where cloud cover is greatest in summer and least in winter as in Belo Horizonte and Formosa, show minimum insolation in summer and maximum in winter.

This effect is less well defined in Tres Lagoas, farther south, where distribution of insolation is more homogeneous.

Over the Planalto Meridional the homogeneous distribution of cloudiness during the year causes a corresponding homogeneous distribution of insolation, as the Curitiba diagram indicates. The slight decline of cloud cover in winter is counterbalanced by the shortness of the days.

The increase in cloudiness in spring reduces insolation during that season of the year along the southeast coast, as the diagrams for Rio de Janeiro and Santos indicate.

The Alegrete graph shows a great variation in insolation between summer and winter, the minimum occurring in the latter season. In this case, however, it cannot be attributed only to the small increase in cloud cover in winter, but also to the difference in possible insolation, almost five hours more at the summer solstice than at the winter solstice, as we have already seen.

Temperature

The various climatic factors already described account for the distribution of temperature over the country. Temperature observations were made at the Brazilian climatological

232

stations in the period of the normals used, at 07h00, 14h00, and 21h00 local time. The average temperature values have been corrected to represent a 24 h average.

Average temperature

Charts of the average annual temperature over the Atlantic Ocean indicate that north of 30°S the isotherms of the air do not follow the parallels. This must be attributed to the influence of the surface temperature of the sea, which over the Atlantic Ocean is higher in the west than in the east, because of the action of the warm Brazil current and the cold Benguela current, the first moving toward the Pole and the second toward the Equator. The annual average varies from 28°C in the Planicie Amazonica and in the interior of the northeast region, to 16°C in the Planicie Riograndense. Due to the relatively high elevation, the lowest values appear in the Planalto Meridional, reaching 14°C in the Serra Geral.

Summer

Fig.5 and 6 contain isotherm maps for the middle months of summer and winter. In the southern part of the country, this is an important factor for the annual temperature variation. It is less important in the part of the country that lies between the Tropics, where the noon-sun always remains high above the horizon. In January the sun remains south of 17°S. To the south of this latitude temperatures are all higher than the annual mean temperature, and the difference increases with latitude. To the north, temperatures are almost the same as the annual average.

The map (Fig.5) reveals that the most important factor in temperature distribution, however, is altitude. Isotherms follow the contour lines of the plateaus. In southern Brazil, at the same latitude, between the interior plains and the highest areas of the plateaus the difference in temperature reaches 6°C, the same as that existing horizontally between the Planicie Riograndense and the plains of the northeast more than 20° of latitude apart.

West of the Planicie Riograndense and the Pantanal (swamplands of Mato Grosso), the isotherms of 26 and 28°C indicate the existence of a large nucleus of high temperatures centered over the Tropic, the obvious cause of the formation of heat lows and the corresponding change of wind patterns in the summer.

The southern part of the Amazon basin is cooler than the northern part. This is due to much more intense cloudiness in the first than in the second. On the average, the highest values are found in the semi-arid caatinga region of the northeast, with values higher than 28°C.

Winter

In July (Fig.6) the sun is in the Northern Hemisphere. Over the east coast of the northeast and over the Borborema plateau, as well as south of 10°S in the rest of the country, the chart indicates temperatures lower than the annual mean temperature. The temperature difference between the Planicie Riograndense and the Planicie do Nordeste in July is 14°C.

The most important factor in this cooling of southern Brazil and consequent increase in

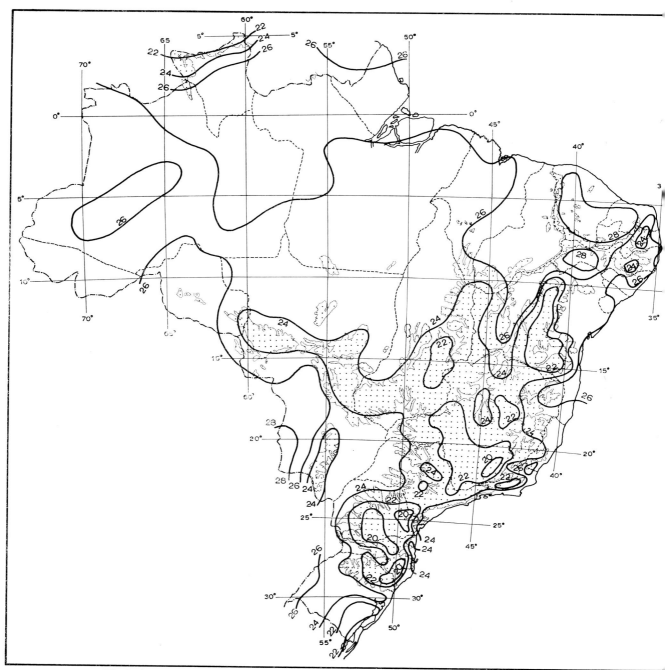

Fig.5. Average temperature in January (°C).

the north–south horizontal temperature gradient should be sought not only in the inclination of the sun's rays, but also in the polar invasions during that time of year. In their continental trajectory they turn the isotherms for 24 and 22°C northward, in the southern part of the Amazon basin. In their maritime trajectory they lower the average temperatures in the eastern part of the northeast. Greater cloudiness also contributes to this, however.

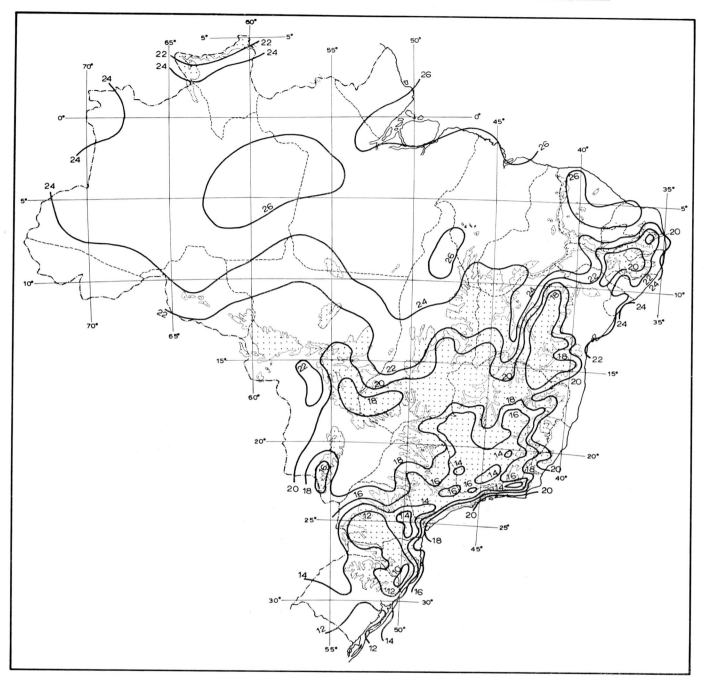

Fig.6. Average temperature in July (°C).

The "crowding of isotherms" on the south coast should be attributed not only to differences in altitude, but also to descendent northwest winds preceding the cold invasions and heating the lowlands on the coast.

The curve for 18° adopted by Köppen to separate the temperate and tropical zones follows the contours of the Planaltos Centro-Oriental and Meridional. Over the Chapada Diamantina temperatures are also lower than 18°C. The lowest means of the country are recorded in the extreme south of the Serra Geral, with temperatures less than 10°C.

Spring and autumn

The sun crosses the Equator at the equinoxes when inclination of the sun's rays is at the minimum. However, isotherm charts for the vicinity of the Equator reveal that at the vernal equinox for the Southern Hemisphere, temperatures for the equatorial zone of Brazil are about 2° higher than at the autumnal equinox. At the same time of year, in the semi-arid region of the caatinga the highest monthly mean temperatures of the country are recorded. This fact is influenced by the difference in cloud cover. In autumn, due to the presence of the trough and the equatorial calms, there are more clouds to act as an obstacle to the penetration of solar radiation.

Southern Brazil and the eastern part of the northeast are reached from time to time by invasions of cold air. As these are much more intense in spring than in autumn, temperatures are lower in the former than in the latter season. This causes the temperature gradient between the south and the north, in the interior of the country, to be greater in spring. In addition, while in March the temperature on the coast of the eastern part of the northeast is the same as in the central part, in September there is a difference of 4°C.

Maximum temperatures

The semi-arid area is the region where the mean annual maximum temperatures are highest, 32–34°C. Temperatures higher than 30°C are also found over the Pantanal, the valley of the Parana and in some parts of the Amazon basin. The lowest average maxima are recorded over the plateaus with 20–24°C. The highest monthly mean maximum temperatures occur in the semi-arid region and in the southern part of the Amazon basin with a reduction in cloudiness and an increase in insolation beginning in August. Temperatures above 36°C are recorded. They remain so until January, when increased cloudiness and precipitation lower temperatures by 2°C in the semi-arid region and by 4°C in the southern part of the Amazon basin. The lowest mean maximum temperatures are recorded in July over the Planalto Meridional, with 16–18°C.

Absolute maximum temperatures rise above 40°C at several points in the country, and in different times of the year, depending on the location.

In the semi-arid region west of the mid-São Francisco river, temperatures rise even higher than 42°C in August and September, when insolation is most intense with more than 300 h/month. During the same months, and also due to high insolation, maximum temperatures in excess of 40°C occur, along the mid- and upper Tocantins river.

In the Amazon valley and in the basin of the Meio Norte absolute maximum temperatures vary between 35°C and values a little below 40°C, never reaching 40°C. The lower temperatures occur on the lower Amazon and on the coast. This must be due to dominance of northeast winds.

The most important factor lowering the maxima in the Amazon valley is the cloud cover and the fact that the greater part of the solar energy is consumed in evapotranspiration by the plant cover.

Temperatures of 42°C also occur in the Pantanal Matogrossense (swamp-lands of Mato Grosso) and of 40°C, in the valley of the Parana. This takes place in October and November with insolation of 220 h/month.

TABLE I

TIME OF MAXIMUM TEMPERATURE AT VARIOUS STATIONS

Station	Local time	Station	Local time
Uaupés	14h00–14h30	Corumbá	14h00–14h20
Manaus	14h00–15h00	Três Lagoas	14h30–15h10
Recife	14h00	Curitiba	14h00–14h30
Salvador	14h10–14h30	Florianópolis	13h15–14h00
Belo Horizonte	14h30–15h00	Alegrete	14h10–15h00

tion occurs on the island of Fernando Noronha, where temperature is highest between noon and 13h00.

The minimum temperature generally occurs between 05h00 and 06h00. The longest retardation is found in the semi-arid zone, where it occurs between 06h00 and 07h00.

The average daily range, that is, the difference between the mean maximum and the mean minimum, varies between 8°C in the equatorial zone of the country and 16°C in the semi-arid region, over the mid-São Francisco. The large range in the latter region is due to the difference between the strong daytime radiation and nocturnal radiation permitted by the low cloudiness and the low humidity. The same phenomenon occurs in the valley of the Parana, where the range reaches 13°C.

The regulating action of water makes the temperature range relatively small at the coast, where it is only 6–8°C. The range increases toward the interior of the country: it is smaller over the plateaus (less than 12°C) than over the plains.

Day to day variation

From the bioclimatic point of view, an important element is the temperature variation from day to day. In southern Brazil, strong variations are due to penetrations of polar air, i.e., rapid prefrontal rise in temperature and its abrupt fall as the cold air arrives.

Mean variation and maximum variation here described are those recorded for a 24 h period, between 09h00 of one day and 09h00 of the next, in 5 years of observations computed by SERRA (1939).

The data ought to show greater values in the interiors of continents and lesser values near oceans, although this is not always true of Brazil. They should also indicate greater values in winter than in summer.

The influence of cold air invasions is very clear in winter (July) when, in the regions of the continental trajectory over the Planicie Riograndense and over the Pantanal Matogrossense, mean values of 2.4–3°C and maximum values of 12–14°C are found. On the Planicie Riograndense the lowest values occur on the coast due to the influence of the ocean. The upper Amazon, although it is reached by invasions of cold air in this part of the year, shows lower mean and maximum variation, 1–2°C and 8–12°C respectively. The air masses which reach the area have undergone great changes in their continental trajectory. With the decrease in latitude in southern Brazil's interior, the mean variation is between 1° and 2°C, the maximum variation between 6° and 12°C, with values decreasing toward the east coast. In the northeast, with the cold air invasions in their maritime trajectory reaching principally the east coast, the mean variation falls from 1.4°C on the coast to 0.8°C in the interior, and the maximum variation from 6 to 4°C, thus altering their normal distribution.

In summer, in January, when the cold invasions are limited to southern Brazil, mean and maximum variation decrease normally from south to north and from the ocean to the interior.

In spring and fall the general pattern is the same as in January, but it is interesting to observe that in April on the north coast, under the influence of the equatorial convergence and of possible polar air masses from the Northern Hemisphere, mean variation falls abnormally from 1.6°C on the coast to 1°C in the interior and the maximum from 8 to 4°C.

On the Planicie Riograndense and in the southern part of Mato Grosso, which are more influenced by the cold air invasions, in some cases exceptional temperature variations somewhat exceed the indicated values. In intense invasions temperatures may fall 25°C in 24 h on the Planicie Riograndense.

Warm days and nights

The term *warm* applies to those days when temperatures exceed 25°C, and to those nights when the temperatures exceed 20°C. Tropical days are those with temperatures above 30°C.

In the Amazon basin and in the semi-arid region almost all days are warm, as they are on the coast of the eastern part of the northeast. Southern Brazil and the northern Brazilian plateaus have fewer than 300 warm days annually. In southern Brazil there are fewer at high elevations, under 120 days. On the extreme southern coast of the country there are fewer than 100 warm days annually.

Only near the Equator are all the nights of the year warm. There are fewer than 50 warm nights on the plains of the extreme south and the plateaus of northern Brazil, including the Planalto da Borborema, which lies at a northerly latitude. Warm nights do not occur over the Planalto Meridional or the Planalto Centro-Oriental.

The Amazon basin, the Meio Norte, the semi-arid zone and the Pantanal have 250–300 tropical days and the valley of the Parana from 200 to 250 warm days. The whole east coast, the plateaus, and the Planicie Riograndense have fewer than 100 tropical days per year.

The number of warm days, tropical days, and warm nights is greater in summer than in winter. In the latter season, warm nights are limited to the equatorial zone of the country.

Moisture[1]

Vapour pressure

The quantity of water vapour contained in the atmosphere is determined by evaporation

---

[1] Humidity measurements are made at the Brazilian climatological stations at 07h00, 14h00 and 21h00 local time, with non-ventilated psychrometers. The average values have been corrected to represent a 24-h mean.

from oceans, lakes and rivers. Vapour pressure, which expresses this quantity, ought to be higher near these sources of moisture. In Brazil we must take into account, besides the North and South Atlantic, another important source of water vapour, the Amazon basin, covered with dense forests, wide rivers, and flooded areas where real evapotranspiration is high. This varies between 1,000 and 1,500 mm, thus being higher than evaporation obtained from the oceans, which is less than 1,000 mm/year.

The air's capacity to hold moisture is in direct relation to the temperature, so that vapour pressure varies with latitude as latitude affects temperature. In addition, vapour pressure is greatest at the lowest layers of the atmosphere, changing rapidly with altitude. For this reason and because of their lower temperatures, the higher regions of the country show lower values of vapour pressure.

An examination of the geographical distribution of vapour pressure justifies these affirmations.

On the part of the east coast subject to southeast–northeast maritime winds, vapour pressure varies from 24 to 27 mbar over the lowlands. Humidity decreases toward the interior reaching values lower than 18 mbar in the Planalto da Borborema. This is due partly to increasing distance from the ocean and partly to altitude. The humidity originated in the ocean penetrates into the valleys of big rivers such as the São Francisco, the Doce, the Jequitinhonha, and the Paraiba.

On the southeast coast vapour pressure also decreases toward the interior over the Serra do Mar range. However, through the lowering effect of latitude on temperature, the values are from 18 to 21 mbar at low altitudes.

Over the southern plateaus, vapour pressure falls to 14–18 mbar, over the Planalto Central 18–22 mbar, and over the southern part of the Serra Geral range to less than 14 mbar.

In the Amazon basin and the Meio Norte basin, vapour pressure values are found between 24 and 28 mbar, the highest for the country.

In the zone along the mid-São Francisco, vapour pressure falls sharply to less than 17 mbar. This region is not reached by moisture originating in the Atlantic to the east, or in the Amazon to the north.

The influence of altitude is clearly visible over the Maracaju range, with values of less than 20 mbar, separating the Parana basin with 20–22 mbar, from the Paraguay basin, with 22–24 mbar.

This description indicates that pressure varies with latitude from 28 mbar in the Amazon valley to 15 mbar in the extreme southern part of the country, that is, a difference of 13 mbar.

As for variation with altitude, on the northeast coast, there is a variation from 26 mbar on the coast to 22 mbar over the Planalto da Borborema; in the southern part of the country, from 18 mbar at the coast to 14 mbar over the Serra Geral on the Planalto Meridional, at about 900 m.

The amplitude of the annual variation of the vapour pressure increases with latitude, being less than 2 mbar in the equatorial region and 10 mbar in the extreme southern part of the country. The maximum occurs in summer when temperatures are higher and consequently the air can contain more moisture. The region with the least variation is the Amazon valley, from 28 mbar in the summer to 26 mbar in the winter.

Over the Planalto Central, the Planalto Centro-Oriental and the São Francisco valley, with the substitution of the humid continental mass in summer for the dry subsiding maritime mass in winter, the vapour pressure varies from 6 to 9 mbar. For the same reasons, the variation reaches its maximum for the country, 10–11 mbar, over the Pantanal and the Parana valley.

The variation is small over the eastern part of the northeast and the semi-arid region, about 3–5 mbar.

Relative humidity

Relative humidity depends upon the quantity of water vapour contained in the air *and* its temperature. Thus the distribution of relative humidity is somewhat different from the distribution of the amount of water vapour over the country.

The coastal areas of Brazil, due to their proximity to large sources of moisture from the ocean, and the Amazon basin have the highest relative humidity, above 80%, as well as the highest vapour pressure. Between these regions, extending in a northeast–southwest direction, is a wide band of lower relative humidity, from the interior of the northeast to the Parana–Paraguay basin.

The lowest average annual values for the country are found in this band, over the mid-São Francisco, where humidity goes down to about 48%. Temperature contributes in part to the low humidity of this area, because vapour pressure is higher here than over the Planalto Meridional, where the relative humidity is above 80% in the southern part. On the east coast, south of Baia de Todas os Santos, an area of heavy precipitation, humidity rises above 85%. The same occurs over the upper Amazon and the Rio Negro. On the Planalto Centro-Oriental and in the São Paulo part of the Planalto Meridional, the relative humidity is under 80%, varying between 75% and 80% over the former and between 65% and 80% over the latter. Over the southern part of the Planalto Meridional it is above 80%, and 85% on the coast and in the eastern part of the plateau.

In the Amazon basin the relative humidity corresponds to the rhythm of the rains. Over the upper Amazon and the Rio Negro, where rains are more or less constant, it varies little, remaining between 85% in the spring, when precipitation is lightest, and 90% in autumn, when it is heaviest.

In April, the wettest month in the region of the mid-Amazon, the humidity is 85%, August, the driest month, has 75%.

The range of the annual variation over the lower Amazon is 10–15%. Autumn is the rainy season and has a humidity of over 90%, while in the spring it falls below 80%.

The east coast, constantly under the influence of maritime winds and small annual temperature variation, has a humidity range of only 4–8%. Lowest relative humidity here is 80% and the highest is 88%.

The largest annual variation in relative humidity occurs over the Planalto Central, especially in the eastern part, where it reaches more than 30%.

Precipitation

Generally, precipitation in Brazil falls in the form of rain. In some years, however, snow reaches the ground in the Planicie Riograndense and in the southern part of the Planalto Meridional.

The widely extended cloud layers which appear with cold air invasions, can produce continuous rainfall, occasionally persistent through several days. Such a situation is preceded and followed, in the rain-season of northern Brazil, by the development of convective cloudforms, with afternoon or evening showers, or both.

In the coastal areas of the Meio Norte and in the northeast region, a maritime regime prevails, with rainfall most frequent in the early morning hours. No pronounced daily variation appears in the Amazon region.

Hail is not a frequent phenomenon in Brazil. Only in some years 1–3 days with hail are reported by the majority of the meteorological stations; in other years, none. However, a region with a somewhat greater frequency of hail is the Planicie Riograndense and the Planalto Meridional, where up to 10 days with hail have occurred in one year.

Distribution

The latitudinal position of Brazil, as stated in the discussion of pressure and winds, places the country between a zone of convergence and rising motion of air in the equatorial area and the zone containing the path of dynamic cyclones formed in the polar front between 40 and 50°S. These two belts of precipitation are separated over the oceans and in winter also over the continent by the zone of high pressure, divergence and absence of precipitation. Both belts show some meridional displacement, the former reaching the north of the country in autumn. The latter affects the south of Brazil more intensely during winter. To these two zonal bands, however, we must add the meridional band which affects the eastern part of the country, and the precipitation area of the western Amazon region. The yearly variation of both is related to the expansion and contraction of the anticyclone of the Atlantic. The meridional band becomes more intense in winter and the precipitation area of the Amazon spreads over the country in summer.

The principal areas of precipitation, as indicated by the isohyets in Fig.8, are: the lower Amazon and Amapa; the upper Rio Negro and upper Amazon; the central area of the country extending southeastwards to the Planalto Centro-Oriental and Serra da Mantiqueira; the eastern part of the northeast; the eastern region; the southeast coast; and the southwest of the Planalto Meridional.

Orographic effects become clearly visible over the northern slopes of the Chapada do Araripe, the east of the Serra da Canastra, in the Mantiqueira region and the southwest of the Planalto Meridional.

Number of days

Generally, the regions with the greatest number of days with rain (Fig.9) coincide with those which have the most precipitation. On the Planalto Meridional, however, the opposite occurs. On the western part of the plateau, where there is the heaviest precipitation, the number of rainy days varies between 90 and 150, while in the eastern part, with the lightest precipitation, there are 150 to 180.

Average intensity

By calculating the quotient of the total rainfall and the number of rainy days we can get an idea of the average intensity of daily rainfall.

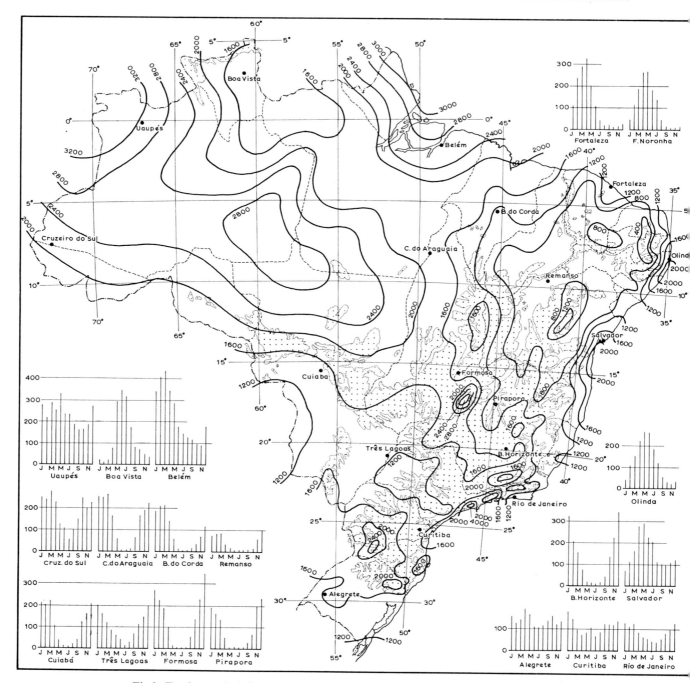

Fig.8. Total annual rainfall with diagrams indicating annual variation.

The concordance of the regions of isohyets with those of isolines of the number of rainy days in most of the country causes the average intensity to have the same pattern as the isohyets. The region in which this does not occur and which shows the highest average daily rainfall is southwest of the Planalto Meridional. There, in autumn and winter, the average intensity reaches 24 mm/day, in contrast with the eastern part, where it is only 8 mm/day. This is because the rains are of frontal origin, produced by

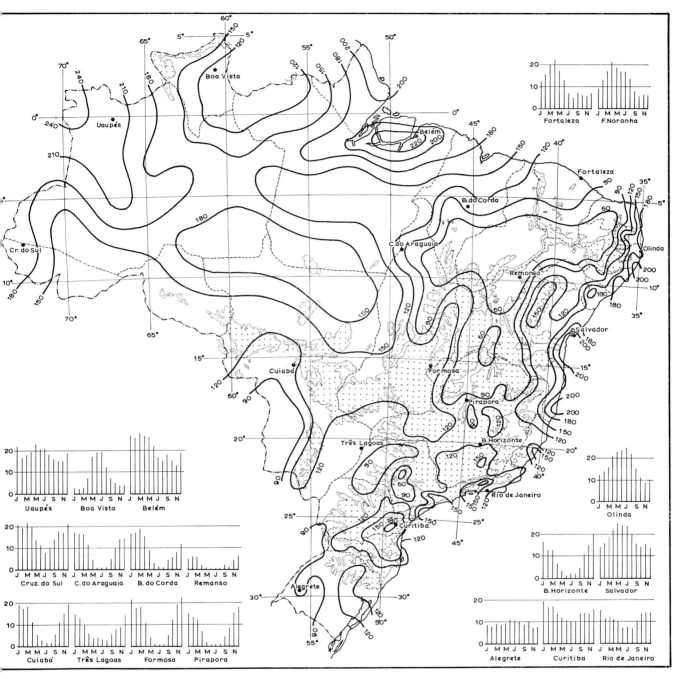

Fig.9. Number of days with precipitation (>0.1 mm), with diagrams indicating annual variation.

cold fronts from the cyclones of the polar front, moving west–east and intensified by the relief (Fig.3). Thus the rains are heavier but not as lasting on the windward side. On the leeward side, however, the quantity of precipitation diminishes and the influence of the ocean causes the rains to be more continuous, making for a lower average intensity.

The interior of the northeast and the São Francisco valley, because of lighter rainfall in winter and spring, shows the lowest average daily intensity in the country, that is, only

4 mm/day during winter and spring, rising to 16 mm in summer and fall. It is also the region in which the average intensity of precipitation undergoes the greatest variation during the year: over the upper Amazon and Rio Negro, 20 mm in summer, 16 mm in winter; over the upper Rio Branco 16 mm in summer, 8 mm in winter; in the central area of the country 16–20 mm in summer, 8–17 mm in winter.

Maximum intensity

Generally the heaviest precipitation in 24 h varies between 60 and 300 mm. Its distribution over the country does not follow a determined pattern. Local conditions, particularly mountains, determine exceptional amounts of rain in certain isolated spots.

Over the upper Amazon and the Rio Negro it is under 160 mm, at most of the stations in this area under 100 mm. In the lower Amazon region it varies between 100 and 150 mm and at some points of the coast of the Meio Norte it reaches 250 mm.

The heaviest rainfall occurs on the east coast, on the southeast coast, and on the eastern slopes of the Planalto Central. In some points of the east coast, maximum 24 h rainfall rises above 270 mm, and on the southeast coast as well as in the Espigao Mestre several stations have recorded values of about 370 mm.

Regimes

Let us now classify the annual variation of rainfall, that is, the precipitation regimes, and enter into a discussion of some details of their behaviour (Fig.8). The rainy and dry seasons which characterize these different regimes are determined by the prevailing synoptic situations and the general circulation in the atmosphere.

Brazil has three distinguishable types of rainfall regimes:

(*1*) The *equatorial regime*, related to the equatorial trough. One may distinguish a continental type over the upper Amazon and the Rio Negro from a maritime type over the mid- and lower Amazon, extending over the north coast of the country. It shows a well defined high in fall and a low in spring; at some points a second high, poorly defined, is distinguishable.

(*2*) The *tropical regime*, related to the expansion and shrinking of the semi-stationary anticyclones of the North and South Atlantic. This regime also shows two types: a continental one, subdivided into north and south tropical, which dominates the whole interior of northern and central Brazil north and south of the Equator; and a maritime type, which dominates the east coast. The continental type has a high in summer and a low in winter according to the hemisphere. The maritime one shows the high at the end of autumn and the beginning of winter and the low in summer.

(*3*) The *subtropical regime*, related to migratory anticyclones which follow the invasions of cold polar air moving south to north, and the cyclones of the polar front moving northwest–southeast.

The continental or maritime trajectories of invasions of cold polar air allow two types to be distinguished below the tropic: one over the continent and another on the coast. In the continental type, precipitation is evenly distributed throughout the year. Over the coast, rainfall increases a little in summer.

Maritime equatorial type

It is called maritime because the rains are produced by penetration of maritime air masses into the continent.

The whole area of the mouth of the Amazon and the Amapa zone is subject to this type, as well as the mid-Amazon and, on the north coast, the central part of the northeast. Rains are heaviest in autumn and lightest in spring. This rhythm is conditioned by the movement of the equatorial convergence. In this movement it first reaches the Amapa zone, where rains begin in November. From the mouth of the Amazon eastward they begin in December (see Fig.8, Belém graph), in January (Fortaleza graph) or in February (Fernando Noronha graph). Maximum rainfall occurs in March and April in Fortaleza, when the equatorial convergence is in its extreme southern position. As the convergence moves northward, precipitation falls off, beginning to decrease sharply in August and reaching its low in October. The rains begin and reach their maximum later and later, because the convergence reaches these stations later as they lie farther south and east.

The dry and rainy seasons become more clearly defined as we move toward the east, as the Belém and Fortaleza charts indicate. Fernando Noronha also undergoes the influence of the maritime tropical type.

Continental equatorial type

In the area of the Rio Negro and north of the upper Amazon precipitation is well distributed throughout the year. There is no real dry season: the heaviest rains occur at the end of autumn and the beginning of winter (May), and the lightest in spring.

The typical graph of Uaupés shows a maximum in May and a minimum in September and October, indicating that this type is related to the equatorial convergence.

Northern continental tropical type

In the continental region north of the Equator there is only one station, Boa Vista do Rio Branco. It has a tropical pattern of the Northern Hemisphere, with the maximum in the summer and the minimum in the winter of this hemisphere. Dry and rainy seasons are clearly defined. It is the symmetrical pattern of north-central Brazil, that is, corresponding to the high and low positions of the sun in the Northern Hemisphere. Heaviest rains occur in July and the lightest in February, as the Boa Vista graph indicates.

Southern continental tropical type

It predominates in the whole central part of the country. It extends as far as the mid and upper São Francisco valley, the Planalto Central, the Planalto Centro-Oriental, the Depressão Paulista, the Rio Doce valley, the Pantanal, and the Parana valley. The graphs of Cruzeiro do Sul, Barra do Corda, Remanso, Cuiabá, Formosa, Belo Horizonte, Três Lagoas and Pirapora show clearly this type of annual precipitation distribution, with heaviest rains in southern summer and lightest in winter.

It is a tropical type. Rains occur when the sun is at its high point, and the dry season when it is at its low point. In the first stage a depression pattern prevails, with rising air and prevailing winds from the north to northwest. Daily charts show that rains are produced by periodic air currents from northwest and west carrying the Amazon air mass. In the second stage, in winter (June–August), the anticyclone of the Atlantic dominates with dry air and east winds, and the rains disappear almost entirely.

From west to east, that is, from Cruzeiro do Sul in the humid tropical zone, to Remanso in the dry tropical zone, the drought period, with rainfall of less than 60 mm/month, increases from one to three, five, and eight months, and begins later and later. A similar variation occurs from north to south, toward the dry subtropical zones. Non-existent in Uaupés, the dry season lasts only a month in Cruzeiro do Sul, and five months in Cuiabá. The rains spread southeast and south over the country, so that while in Conceicão do Araguaia they begin in August, they begin in October in Formosa and Belo Horizonte in the southeast, and in Cuiabá and Três Lagoas in the south.

This rhythm of precipitation is then conditioned by the movement of the anticyclone of the South Atlantic, which in turn is related to the annual motion of the sun.

Maritime tropical type

It dominates the whole east coast of the country and is characterized by the Olinda (Recife) and Salvador graphs.

These graphs indicate differences in rain distribution throughout the year which permit a distinction between two separate sub-types of this maritime pattern. One is north of the mouth of the São Francisco river, extending to the cape of São Rogue, where the rainfall maximum over the eastern part of the northeast is located (see the Olinda graph). The other sub-type is found south of the mouth of this river extending to 20°S (Salvador graph), where the large precipitation maximum of the east region coast predominates. In the winter the coast of the eastern part of the northeast is under the influence of the convergent air currents and disturbances of the trade winds. In addition, it is in the southern winter that heavy penetrations of cold air reach these latitudes on their maritime trajectory.

In the region characterized by the second type, the greatest rains occur in autumn and winter. As already seen, the divergence of trade winds moves southward in April and the entire coast north to about 20°S is subject to southeast winds. With their increased intensity they bring about the autumn and winter rains.

When the divergence moves northward, beginning in September, part of the south coast is affected by *cE* air masses, which cause the summer rains with a second maximum in November or December (Salvador graph).

Thus, the two tropical types, the maritime and the continental, are superimposed.

Sub-tropical regimes

South of the tropics the influence of summer rains of the continental tropical type diminishes, as the Curitiba graph indicates. Precipitation distribution here is more homogeneous throughout the year, with slightly more rainfall in fall and spring on the Planicie Rio-grandense as the Alegrete graph shows.

Cold and warm fronts from the cyclones of the Atlantic polar front, principally responsible for precipitation, pass over this region throughout the year, although less frequently in summer.

The tendency toward maximum in summer and minimum in winter which is found on the southeast coast (Rio de Janeiro graph) must be attributed to the fact that penetrations of cold air masses follow a more maritime than continental trajectory, carrying a greater quantity of water vapour. In addition to this, they are more stable in winter.

Variability

The distribution of the average annual precipitation over the country and its yearly variation have already been described. This distribution varies, however, from year to year.

The average annual variability of precipitation at a given station is defined as the ratio:

mean deviation of the yearly values
———————————————————
over-all average of the yearly values

expressed as a percentage .The chart of Fig.10 was drawn using these values.

With the chart it is seen that the areas of least variability, 10% of normal, are the areas of heavy rains of the upper Rio Negro and the mouth of the Amazon. The coastal rainy areas like those of central Brazil and the plateaus have a variability of less than 20%. It is a little higher in the driest area of the Planalto Paulista.

The areas of great variability are within the region with rains of less than 1,200 mm annually, and the highest values are found where there is least precipitation, that is, to the west of the Planalto da Borborema in the central part of the northeast, where the variability exceeds 50% at some places. In the northeast of the country, the variability increases from northwest to southeast.

This indicates that there is an appreciable constancy in annual precipitation in the continental equatorial, the tropical patterns, and the subtropical regimes. The same does not occur with the maritime equatorial type, however. As stated above, its precipitation is produced by the air mass *mE* formed within the equatorial pressure trough in its southerly movement. In some years it reaches farther south than in others producing more rain than normal. Thus the areas farther southeast are more strongly affected by these variations because they lie at the end of the path of the equatorial trough and are rather irregularly reached by the maritime air masses.

Thunderstorms

The annual chart for the distribution of thunderstorms for South America and the Atlantic Ocean shows a belt of high frequency over the continent and extending to the South Atlantic in a northwest–southeast direction. Between this belt of thunderstorms and that which exists over Africa, a large area of smaller frequency of thunderstorms extends over the Atlantic in the same direction, touching the northeast coast of Brazil.

As thunderstorms are phenomena produced by rising currents of air, this distribution may be explained by the pressure field. In the tropical zone, in the hot period of the year, the continents are warmer and pressure over them is lower. Thus the strong convection currents necessary for thunderstorms are formed. In the area of the ocean occupied by the semi-permanent anticyclone, on the other hand, air is descending, impeding the formation of thunderstorms.

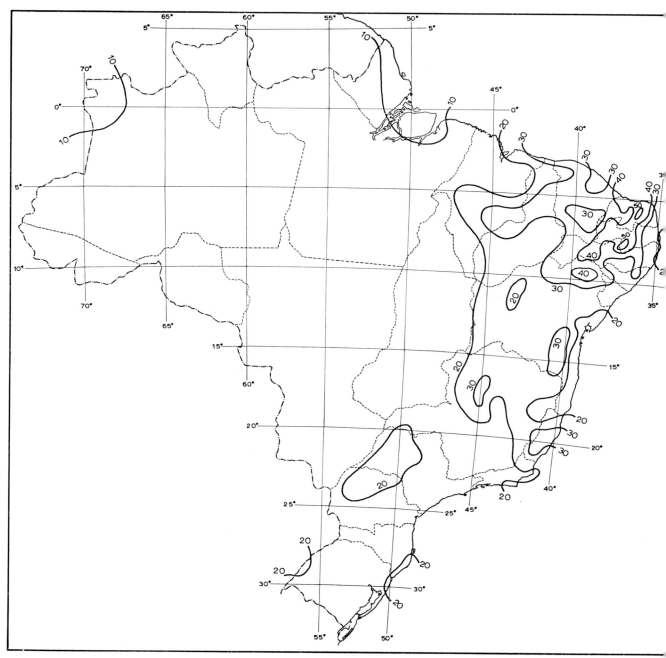

Fig.10. Average annual variation in amount of rainfall in %, as defined in the text.

To this cause we must add for southern Brazil the thunderstorms of frontal origin, and for the north coast those produced by the equatorial convergence.

From this we may understand the distribution of thunderstorms over Brazil, namely their infrequent occurrence in the eastern part of the country and their increasing frequency toward the interior.

The number of days when thunderstorms occur during the year is low in that part of the country east of the Serra de Ibiapaba, the Chapada Diamantina and the Serras de Canastra and Mantiqueira. This is the region which with fewer than 20 days of thunderstorms, undergoes the strongest influence of the anticyclone of the Atlantic.

On the east coast of the northeast and over the Planalto da Borborema there are fewer than 10 days, as the dry layer of air above the inversion of the trade winds impedes the development of large cumulus clouds with rapid rise of large masses of water in liquid and in frozen form to great heights, with temperatures much below freezing, conditions necessary for the formation of lightning and thunder, which define the thunderstorm.

Their frequency increases sharply toward the west, beginning in the São Francisco valley and the Serra de Ibiapaba, with the intensified and/or more frequent upward motion of air at larger distance from the influence of the anticyclone. The highest thunderstorm frequency is found over the Planalto Central with 60–120 days, and over the Serras dos Parecis and Caiapos with 80–130 days.

Except for the Amazon basin, where thunderstorms are frequent, they form principally over the mountainous areas of the country; there are fewer thunderstorms in the river valleys.

During the winter, thunderstorms are rare in Brazil, and do not occur at all in vast areas of the country. This is the time of year when the sun is in the Northern Hemisphere and the country is under the influence of the anticyclone of the Atlantic. At that time, the Amazon valley registers only from 3 to 12 thunderstorms per month.

Southern Brazil has from 3 to 6 thunderstorms per winter month, and they are frontal in character, formed in the cold fronts of the polar front.

When the anticyclone withdraws in the spring and low pressure prevails over the country, the frequency of thunderstorms increases. In the summer, there are over 18 days/month in the Amazon basin and on the Planalto Central, giving the summer distribution an appearance similar to the annual distribution. Conditions for the formation of thunderstorms in the summer are exceptionally favourable, with the high humidity of the Amazon air mass occupying central and eastern Brazil, with the change of winds caused by the formation of the heat low, and the warming due to the altitude of the sun, now in the Southern Hemisphere.

There are fewer thunderstorms again in fall over this whole continental region, although there are slightly more on the north coast because of the equatorial convergence. The scanty data for the oceanic area off the north coast show that within the convergence there are only slightly more thunderstorms, as they are principally a continental phenomenon.

This description reveals three types of thunderstorms in the country:

(*1*) Storms caused by daily warming within the continental air mass in the hot season of the year, reinforced by the effect of relief over the plateaus and mountain ranges.

(*2*) Frontal and prefrontal thunderstorms in southern Brazil intensified by the effect of relief.

(*3*) Thunderstorms formed in the equatorial trough when it reaches the northern part of the country in the fall.

Fog and haze

Fog occurs most frequently over the mountain ranges along the east and southeast coast of the country. In the Serra do Mar and Serra Geral ranges there are 60–120 days/year with fog. In the interior of the Planalto Meridional and on the southeast coast, the frequency falls to fewer than 30 days/year. On the Planalto da Mantiqueira, the Espinhaço range, and the Chapada Diamantina, fog occurs on 30 –90 days annually. The same frequency is found on the Planicie Riograndense.

Fog is very rare on the east coast, in the eastern part of the Planalto Central and in the São Francisco valley, and almost non-existent in the semi-arid region. In some parts of the Amazon basin fog occurs as often as 90 days in a year.

On the coastal slopes of the mountains fogs are formed by the rise of stable air. Seen from the coast they are classified as stratiform clouds. Over the plateaus they are radiation fogs, formed at dawn, occurring more frequently in the winter. Over the forests in the Amazon basin there is also this type of fog, due to the high moisture content of the air.

When the big cold anticyclones are centered north of the tropic, retrogressing cold air carried by the northeast winds causes thick advection fogs in the southern part of the country. Nocturnal radiation also contributes to the formation of these fogs.

On the southeast coast, principally in the spring, after some days of prefrontal northwest winds causing upwelling and resurgence of deep cold water, the arrival of the polar air mass on its oceanic trajectory forms fogs and very low stratus clouds, which penetrate inland from the coast.

Haze[1] is a very common phenomenon in almost all regions of Brazil. Unlike thunderstorms, which are a consequence of unstable stratification of the atmosphere in the summer within the mass cE, haze appears in late winter and spring due to the dominance over the continent of the mass mT with its stable stratification. The top of the haze layers varies between 3,000 and 4,000 m.

From January to April, haze can be observed mainly in the dry tropical zone and in the southwestern part of the country, with a frequency of 3–9 days/month, on the average. In the month of May, with the advance of the quasi-stationary South Atlantic anticyclone over the continent, haze frequently appears in the region of the Planalto Central. In July, it is found also in the Amazon basin and the Planalto Centro-Oriental, again with a frequency of 3–9 days/month. In August and September, haze is the dominant phenomenon over large parts of the country. In the Planalto Central, the average frequency reaches 15, in the Planalto Centro-Oriental even 21 days/month.

In October, when the Atlantic anticyclone recedes and the stable mT air masses over the continent are transformed into unstable air masses of the type cE, the frequency of the occurrence of haze diminishes first over the Planalto Central and then over most of the country, so that at the end of the year only the above mentioned dry tropical zone remains affected. Parallel to this development goes a gradual increase of the frequency of thunderstorms, and during the transition from the stable to the unstable regime, huge cumulus and cumulonimbus clouds can be observed in the still hazy air.

The intensity and extension of the haze varies considerably from year to year, inversely depending upon the frequency of the occurrence of rain. As far as specific weather situations are concerned, haze becomes more intense with the prefrontal northwest winds and disappears with the arrival of cold air, as it can often be observed along the southeast coast.

---

[1] The Portuguese word is "nevoa seca"; it may be understood as dry haze and/or dust.

The nature of the particles which make it up has not yet been studied, but they must be particles of dust and products of combustion. Winter (June–August) is the dry season in the whole interior of the country, when dry vegetation of the cerrado (areas covered with dense scrub growth) is burned, either spontaneously or intentionally for the cultivation of the land. This explains the increased occurrence of haze or dust in the centre of the country. Spring is the season with the strongest winds, the turbulence of which contributes to the dissemination of particles of combustion. At this time of year, due to its intensity, the phenomenon occasionally causes serious problems to air traffic. Regarding the haze occurring in the summer, one can assume that it consists mainly of dust particles originating in the semi-arid zones and lifted to considerable height in turbulent air currents.

On the Planicie Riograndense where haze is very frequent although not very intense, especially in the eastern part, throughout the year, it seems to be made up of dust from the subtropical arid region. This must be the same type of haze found over the tropical semi-arid region of the northeast.

Climatic types

Several types of climate classification have been applied to Brazil. Among the Brazilian classifications most often used, those of DELGADO DE CARVALHO (1916), MORIZE (1927) will be cited, along with that of SEREBRENICK (1945).

Delgado de Carvalho and Morize recognize three principal types of climate in Brazil: (*1*) the equatorial climate in the north, within an annual isotherm of 25°C; (*2*) the subtropical climate between this isotherm and that of 18°C in the coldest month; and (*3*) the temperate climate south of the latter isotherm. Each of these types of climates has several subdivisions.

SEREBRENICK (1945) adopts the 18°C line in the coldest month to separate the tropical zone from the temperate zone. Then he takes into account the existence or non-existence of the dry season, amount of yearly rainfall, and the time of year in which most precipitation falls.

Several Brazilian geographers and climatologists have applied to the entire country, or to the various states, the classifications of Köppen and Thornthwaite. BERNARDES (1951a,b) and SERRA (1960) prepared detailed maps of the Köppen classification for the entire country, and SCHMIDT (1942) for the Amazon valley. PAES DE CAMARGO (1960) applied Thornthwaite's classification to the state of São Paulo. SETZER (1946b) drew a climate map of São Paulo based on his humidity index.

In the following, attention will be drawn only to certain well characterized climatic types of Brazil.

Of the two principal climatic elements, temperature and precipitation, the more important in the temperate zone is temperature. In the tropical zone, however, more importance belongs to precipitation. In fact, the seasonal rhythm of humidity in the tropics regulates the seasonal activities of organisms the same way that temperature regulates them in the temperate zone.

As for seasonal variation of temperature and precipitation, in the south there is an appreciable constancy of precipitation, while temperature varies considerably during the year. In the north, on the other hand, the temperature is more constant, and the annual

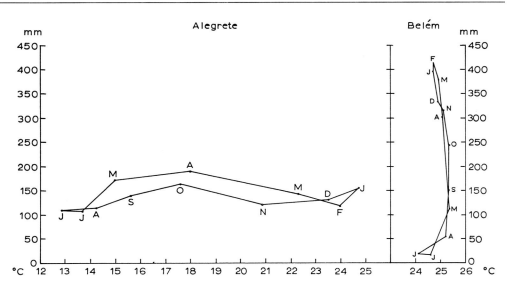

Fig.11. Climate diagrams; mean monthly temperature vs. precipitation, for Alegrete (29°46′S 55°47′W 104 m) and Belém (1°28′S 48°27′W, 24 m).

variation of precipitation is great. This is clearly visible in the climatic diagrams for Belém and Alegrete (Fig.11).

It is interesting to observe that this difference in the behaviour of meteorological elements determines the use, by the inhabitants of the region, of two distinct criteria for classification of seasons. When the temperature varies more, the year is divided into winter and summer according to the criterion of the temperate zones, that is, considering temperature exclusively. Spring and fall are poorly defined in Brazil. When, however, the principal variation is in precipitation, the rainy season is called winter and the dry season is called summer. This is so because the rainy season is the one with lower temperatures. As the temperature varies little, the concepts of summer and winter are linked to the presence or absence of precipitation.

It is difficult to draw an exact line of separation between the zones in which variation of the one or the other of these elements is the more important. Instead, this separation is constituted by a band along the Tropic of Capricorn. The limit which separates the climates *Aw* and *Cw* of Köppen, in the north, from the climates *Af* and *Cf* in the south is found within this band and could be considered as the border between the two zones. In fact, the letter *f* indicates a more homogeneous distribution of rain throughout the year, as it indicates that the difference between the wettest and driest months is less than in *w* and *s*.

This limit follows approximately the Tropic of Capricorn. Because of the lack of data for the Parana basin it is difficult to draw it with precision. Approximately, therefore, it might be indicated by the tropic in Fig.12.

The area to the north of this line is the tropical zone of the country. For the area to the south, as in summer and fall tropical conditions tend to predominate and in the rest of the year temperate conditions, the designation subtropical zone is preferable to temperate zone. During the year, this zone is subject to continually changing air masses with the advances and withdrawals of the polar mass. In winter and spring the masses *mT* and

Fig.12. Climatic zones. Solid lines = Thornthwaite's moisture index; dotted lines = limits of the "humid tropical zone", see text; dash-dotted lines = temperatures of 18°C in the coldest month and temperatures of 22°C in the warmest month.

mP predominate. In the summer and fall the masses *cT* and *cE* also intervene. The homogeneity of precipitation distribution must be attributed to the fact that in winter rains are formed mostly by invasions of cold air. In summer, when these invasions are less intense, principally in the northern part of the zone, the rains originate mostly in the mass *cE*. In the tropical zone a gradual modification and/or exchange of air masses occurs throughout the year, certain types of masses prevailing at certain times of year. The frequent abrupt exchange of air masses like that of the subtropical zone, does not exist here, although the tropical zone occasionally is subject to invasions of cold air, principally in winter.

The above mentioned band along the Tropic is the region where these invasions of air masses and cold fronts lose the characteristics shown at higher latitudes. North of this region, that is, in the tropical zone, due to the diminution in thickness of the polar mass and to the decrease in the slope of the fronts, the accompanying cloudiness ceases to have a convective nature. For this reason, in the interior of northern Brazil, the cold penetrations have vast layers of stratiform clouds, producing continuous precipitation and not sudden downpours of rain.

In the summer the band along the Tropic is the zone where the cold fronts become stationary, especially in the years when secondary circulation is weak.

Köppen separates the tropical from the temperate climates by the annual isotherm of $18°C$ in the coldest month. This isotherm coincides approximately, over the southern part of the Pantanal and the Parana basin, with the band of separation referred to above. Beginning at $50°W$, however, due to relief, it turns northward, moving away from the tropic, and follows the contour lines of the Planalto Centro-Oriental. As a name for the climate of the area enclosed by the isotherm of $18°C$ BERNARDES (1962) suggests "tropical de altitude" (elevated tropical), which to us does appear to be quite appropriate.

By taking humidity into account, it is possible to separate two well defined zones in tropical Brazil: the dry tropical zone and the humid tropical zone.

To determine them the moisture index of Thornthwaite (1948) was used:

$$I_m = \frac{100s - 60d}{n}$$

in which I_m is the moisture index, s the water surplus, d the water deficiency, and n the water need (the annual potential evapotranspiration).

With these values on the map of Fig.12, curves (solid lines) were drawn to delimit the following zones; the humid zone, with positive values of I_m, and the dry zone, with negative values. The dry zone was divided into semi-arid, between the curves for —40 and —20, and arid, below —40. In the humid zone, values larger than 100 define the "perhumid" province.

The zero-line delimits approximately the caatinga, with its xerophytic vegetation conditioned primarily by climatic factors. This curve is within the annual isohyets of 800 and 1,200 mm. A large part of the Pantanal has to be included in the dry zone; actually, it is linked to the dry subtropical zone. A small area in the extreme southern part of the east coast also belongs to the dry zone.

The curves for index —20 limits the semi-arid tropical zone.

SEREBRENICK (1945) calls semi-arid an area in which precipitation remains between 250 mm and 600 mm. Other climatologists use the limits of 250 mm to 500 mm to define this climate. Compared with this area, the one defined by the humidity index of Thornthwaite is considerably larger. It has, however, an appreciable coincidence with the boundaries of Köppen's dry steppe climate (*ES*). In the northern part of the mid-São Francisco area and in the Seridó region west of Planalto da Borborema, the curve for index —40 delimits the two arid zones. According to Köppen's classification some parts of the Seridó have the climate *Bw*, a desert climate.

The dry zone lies between three major precipitation regimes, the maritime equatorial, the continental tropical and the maritime tropical.

This dry zone which covers part of the northeast, of the São Francisco basin, of the Meio Norte and of the east region is caused by the fact that little moisture reaches here from the ocean and the Amazon basin. Prevailing winds over the dry zone are from the east. The moisture contained in the layer of the trade winds is precipitated over the eastern part of the northeast and the east. On the other hand, the continental mass *cE* only extends over the country during the hot season, with the withdrawal of the anticyclone of the Atlantic, which in summer tends to maintain a ridge over the northeast. Thus the continental mass can only move toward the east, reaching the coast south of 12°S latitude, as the precipitation maps show. The semi-arid zone is mostly free of its influence. In its turn, the intertropical convergence line only very rarely enters the interior of the northeast, carrying considerably southward the rains which it causes, over the dry zone. The cold masses, blocked by the Mantiqueira range, only produce rains over the upper São Francisco.

The remaining portion of Brazil's tropical zone is the humid tropical zone. In the major part of this zone, in the interior of the country, rains occur in the summer; they belong to the continental tropical regime. This climate ought to be classified *Am–Aw*. On the north coast the north equatorial pattern prevails, with fall rains, and the climate tends to be *Aw*. In the extreme northwestern part of the country, in the area of the Solimões river and the Rio Negro, rain is rather homogeneously distributed throughout the year. This area includes Thornthwaite's perhumid province, and its climate would be *Af* according to Köppen.

In the eastern part of the northeast, under the tropical maritime regime, the heaviest rains occur in winter, giving the region a climate *As*. Due to the superposition of the two tropical regimes, maritime and continental, over part of the coast of the east region, the rains are well distributed throughout the year; the climate would be *Af*.

B. J. Garnier and A. W. Küchler (in FOSBERG et al., 1961) proposed criteria for the delimitation of the humid tropical zones, the former a climatic criterion, and the latter a vegetational criterion.

Applying Garnier's criterion to Brazil with the data used in this study, the humid tropical zone is found to have two areas, on either side of the dotted line on Fig.12. There is considerable coincidence between this line and those of index *O* and temperature of 18°C in the coldest month.

In this way, the criterion excludes from the humid tropical zone the "zona tropical de altitude", which becomes part of the subtropical zone.

To characterize the dry zone there are several dryness indices, although Thornthwaite's hydric index is preferable. The choice of the appropriate index will depend on its capacity to determine limits for the dry zone which coincide with the caatinga. Unfortunately, the limits of the caatinga in many places have not been determined exactly either.

The humid tropical zone has three types of vegetation: the equatorial forest, the tropical or atlantic forest, and the "cerrado" (more or less dense scrub growth). According to the vegetation criterion of A. W. Küchler (FOSBERG et al., 1961) the area occupied by the cerrado is not included in the humid tropical zone. It ought to be observed, however, that its climate is the same as that of the Atlantic forest area. The difference between their respective vegetation must be sought in the difference between their types of soil, that is, the depth of the water table, deeper in the cerrado, and deficiencies in nutritive elements for this type of vegetation (RIZZINI and PINTO, 1964). The destruction of the vegetation by man-made fire appears to date back to pre-Columbian times. Besides natural fires, it must be considered one of the causes of the formation of the cerrado.

The difference between the equatorial and tropical forests is due only in part to climate, that is, higher humidity and temperature in the former than in the latter. Other factors, like age, soil and elevation also have some influence. The tropical forest is much older than the equatorial forest. The soil in the former is crystalline, sedimentary in the latter. Finally, the tropical forest stands on higher ground, while the equatorial forest is in the Amazon basin.

The whole subtropical zone of the country belongs to the climate *Cf*, except for a small area in the northern part of the southeast coast which is included in the climate *Af*, although it is different from the climate which has this classification to the northwest of the Amazon region. On the map of Fig.12 the curve for 22°C in the hottest month separates the climate *Cfb*, in which temperatures are below this figure, from the climate *Cfa*, in which they are above it. The line for index 100 surrounds the perhumid province. The area included within these curves is occupied by the pinheiral (pine forest).

In conclusion, it may be said that the behaviour of air masses and of precipitation patterns permits a delimitation of Brazil's climatic types.

Regional characteristics

The climatological tables (see Appendix p.267) contain the normal values of the principal meteorological elements for some stations which are typical for the climate of the more important geographical regions of the country. The delimitation of these regions has been discussed together with the description of the relief of the country. The regions and stations are shown in Table II.

The normal conditions of some of these regions are altered by phenomena or processes which produce characteristic deviations from the norm. The most outstanding are: the brief periods of low temperatures in the upper Amazon; snowfall and frost in the southern part of the country caused by strong cold air invasions; droughts in the northeast, linked to alterations of the normal seasonal displacement of the equatorial trough.

The Amazonic region is characterized by the high values of temperature, cloudiness, precipitation and humidity, and by the constancy of these parameters during the year. In the upper Amazon area, however, this prevailing situation is interrupted by a phenomenon which can persist through several days, popularly known under the name "friagem"; it is a pronounced decrease of temperatures in the southern winter. As in these low latitudes the normal temperature level is rather high, the sudden "cool change" accompanied by variations of other meteorological elements impresses the inhabitants of the region not only as an exceptional phenomenon *per se*, but also because it directly affects their usual way of life.

A weakening of the subtropical high pressure belt makes it possible that polar air masses advance over the continent to low latitudes. Due to the orographic conditions, that is, the

TABLE II

REGIONAL POSITION OF THE CLIMATOLOGICAL STATIONS

Region	Station	Table
Upper Amazon	Uaupés	III
Upper Amazon	Sena Madureira	IV
Intermediate region	Manaus	V
Lower Amazon	Belém	VI
Lower Amazon	Santarem	VII
Meio Norte	Barra do Corda	VIII
Central part of the northeast	Quixeramobim	IX
East part of the northeast	Recife	X
Region between Planicie Amazonica and		
Planalto Central	Alto Tapajós	XI
Planalto Central	Porto Nacional	XII
Planalto Central	Formosa	XIII
São Francisco valley	Remanso	XIV
São Francisco valley	Caetite	XV
Eastern region	Salvador	XVI
Eastern region	Vitoria	XVII
Pantanal	Cuiabá	XVIII
Pantanal	Corumbá	XIX
Parana valley	Três Lagoas	XX
Planalto Central (east)	Belo Horizonte	XXI
Planalto Meridional	Campinas	XXII
Planalto Paulista	Curitiba	XXIII
Southeast coast	Rio de Janeiro	XXIV
Planicie Riograndense	Alegrete	XXV
Planicie Riograndense	Porto Alegre	XXVI

Andes in the west and the Brazilian sierras in the east, in the winter the cold air can move northward between these two elevations across the Pantanal to the Amazon basin, and with a deviation to the left, to the upper Amazon region. When the drop of temperature is very pronounced, we have the "friagem".

During a "friagem", the following absolute minimum temperatures have been recorded in the normal period used in the present text: Cuiabá, 1.2°C; Corumbá 0.8°C; and Sena Madureira at a latitude of 9°08′S and 135 m elevation 7.3°C, which is less than at Rio de Janeiro.

With the passage of the cold front, thunderstorms occur in the region south of the Pantanal. When the front advances farther north, the thickness of the cold air mass decreases and vast layers of stratiform clouds can be observed. During the night, these cloud decks become more intense due to the loss of heat by radiation, and produce precipitation in form of drizzle (garoa), followed by a clearing of the sky. Generally, the "friagem" affects the upper Amazon region. Exceptionally strong cold air invasions, however, can reach Manaus and Santarem. If the cold air in the upper Amazon region passes the equator, it experiences a deviation to the east and penetrates the Orinoco plains between the Andes and the Guianas, and eventually can reach the Atlantic Ocean in the region of the estuary of the Orinoco. Fig.13 shows the extreme northern position of a cold front, in the occasion of the great "friagem" of June, 1933.

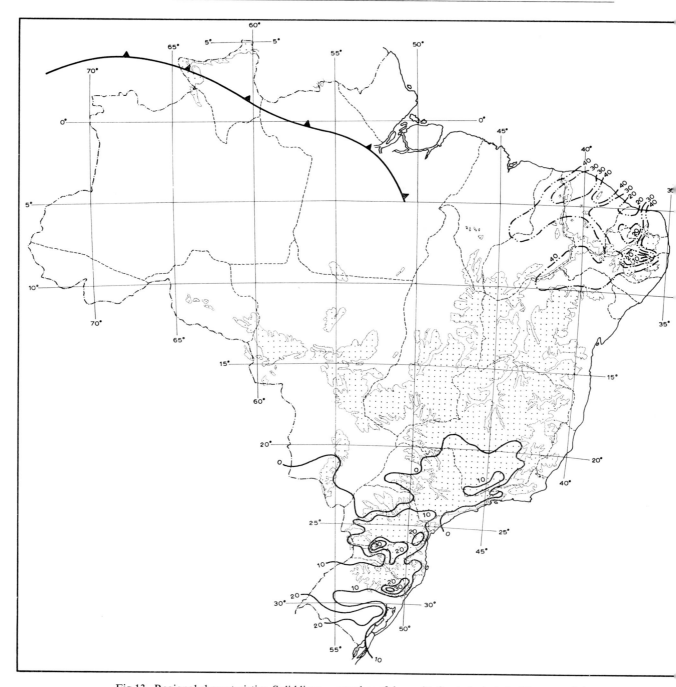

Fig.13. Regional characteristics. Solid lines = number of days with frost; dash-dotted lines = minimum percentage of normal precipitation; upper left = extreme northern position of a well defined cold front, June 1933.

In the upper Amazon region, "friagems" appear between May and August. In some years, up to five intense cold invasions have been observed, but in other years they practically do not occur. Their duration is between 3 and 5 days, but there are reports on exceptional "friagems" in May 1878 and in August 1882, which lasted 15 and 10 days, respectively.

Snowfall cannot be considered an exceptional phenomenon in the southern part of the Planalto Meridional and the Planicie Riograndense. In the period 1912–1942, no snowfall was observed at any station only in the three years 1914, 1919 and 1938. However, the Depressão Central, the valley of the Uruguay river and the coastal region always remain free of snow.

Snow can appear as early as May, and as late as September; July is the month with the highest frequency. Maximum duration of snow at the ground is about three days. Accumulations between 5 and 10 cm can be observed, and in the Planalto as much as 20–30 cm has been recorded.

Nevertheless, the more important deviations from the normal climatic pattern, with impact on the economy of the country, are the occurrence of frost, and the drought situations of the northeast.

Frost during a winter night can afflict the entire area of the country south of 20°S (Fig. 13). The highest frequency of frost is found in the Planalto Meridional. In our reference period (1912–1942), remarkable frosts occurred in 1917, 1924, 1927, 1933, 1937 and 1941. "Cold nights", that is, nights with a minimum temperature below freezing, are linked to invasions of polar air, between May and September. Near the centre of a cold anticyclone, with clear sky and weak winds and correspondingly a pronounced radiative loss of heat, so-called white frosts are observed; their appearance depends upon the intensity of the cold air invasion, and upon height and latitude of the affected place. Occasionally, so-called black frost also occurs. In the southernmost part of the region, in the Serra Geral, the earliest frosts can appear in April, the latest in October; July is the month of most intense frosts. In the Planalto Paulista, less high and at lower latitude, the occurrence of frost is a rare phenomenon.

Droughts can occur in various regions of the country. Nevertheless, those which due to their intensity and serious consequences constitute a characteristic of an entire region, are the droughts of northeast Brazil. As Fig.10 shows, this is the area with the greatest variability of precipitation, that is, with an intermittent appearance of periods with an accentuated lack or a pronounced excess of precipitation. The former ones are most important because of their effects upon the economy of the region, but also the periods of floods can bring considerable damages and losses.

These droughts and floods are well known throughout the history of the country. In our reference period, the years 1915, 1919, 1932–1933, 1936, 1942–1943 were dry, 1915 and 1932–1933 extremely dry; 1912–1913, 1917, 1921–1922, 1924 and 1935 were years of excess, with the most serious floods in 1917 and 1924*.

For the severe drought of 1915 and 1932, the map in Fig.13 shows the deviations from normal precipitation amounts, in the northeast region. One can see that the area most strongly affected lies west of Borborema.

In normal years, the rains in the northeast occur between January and April, and it is the lack of sufficient precipitation in this time interval which determines drought conditions. Of course, there are years with earlier and years with later start of the rains, and it has been shown that a dry year is bound to occur if intense rains have not begun to fall until

---

* SAMPAIO FERRAZ (1950) tried to establish a correlation between sunspot numbers and the occurrence of drought in the northeast and to find a possible cycle of recurrence.

mid-March, though there are some years in which the rainy season extends into the month of June.

How severe a drought can become depends not only upon its intensity in one year, but rather upon the occurrence of successive dry years, which leads to the so-called "great droughts", with disastrous effects. The most notable "great droughts" have been reported for the years 1790 through 1794 and 1877 to 1879 when dry conditions extended all over the northern belt of the country, and the same happened in the years 1932–1933. In years of drought, comparatively little cloudiness and correspondingly increased incoming solar radiation also bring relatively high maximum temperatures.

As far as the other meteorological elements in years of drought and in years of flood are concerned, the most interesting are the variations of the wind field. While in times of drought the trade winds (from the southeast) predominate with great constancy and more than normal speed, in times of copious rainfall the central part of the northeast region experiences the frequent occurrence of calms, a general weakening of the trades and the appearance of weak winds from the north.

Altogether, then, drought conditions represent a persistence of the type of atmospheric circulation which normally characterizes only the dry season from May to December; and the occurrence of floods is linked to an accentuation of conditions which are typical of the rainy season, i.e., frequent calms and a penetration of maritime air masses from the north.

Some problems

Difficulties connected with lack of data due to deficiencies in the climatological network were indicated in the Introduction (p.1). Here attention must be called to some problems which appear in the interpretation of certain aspects of Brazilian climatology.

The droughts of the northeast

The intermittent droughts of the northeast area are certainly the most serious meteorological problems of Brazil, principally because of the disastrous consequences for the country's economy. Several attempts have been made to explain the phenomenon.

SAMPAIO FERRAZ (1931) attributes the modification of the northeast's precipitation regime to the trajectory of the cold anticyclones. If during the rain season these systems reach abnormally low latitudes, so that the cold air penetrates in altitude more intensely into the northeast causing instability and convection, abundant rains occur. If they remain in the southern part of the continent, there are droughts.

SERRA and RATISBONA (1942) showed that the precipitation regime of the central part of the northeast is conditioned by the movement of the intertropical convergence, and suggested that the intermittent droughts which occur in the region must be attributed to the fact that in drought years the intertropical convergence remains north of the Equator and does not reach the northeast.

SERRA (1946), in a certain way, links these two explanations. He sees a certain symmetry between the activity of the cold invasions in the Northern Hemisphere and in the Southern Hemisphere. In rainy years the intense cold invasions over the Gulf of Mexico and the

Caribbean Sea force the intertropical convergence toward the south. At the same time the symmetrical cold invasions in the Southern Hemisphere, rapidly succeeding one another, push the anticyclone of the South Atlantic out to sea, weakening the trade winds and permitting the convergence and its precipitation to move into the semi-arid region.

On synoptic weather maps, he observes that in the southern summer and autumn when a cold invasion from the Southern Hemisphere is weak and only reaches the tropic, with the cold front oriented southwest–northeast in eastern Brazil, the anticyclone of the Atlantic invades the northeast, and east winds and fair weather prevail. If, however, the polar air is very active, the cold anticyclone goes beyond the tropic, the cold front is oriented northwest–southeast, and pressure falls in the semi-arid region. That is, the warm anticyclone withdraws over the sea and the convergence reaches the northeast with precipitation and north to northwest winds.

Thus the principal cause of the rains must be sought in the southward movement of the convergence over the northeast, determined by the behaviour of cold air penetrations in eastern and southern Brazil. If and when this southward movement does not take place, the drought is due to occur.

Determination of the convergence's position and of its monthly movement might possibly permit prediction of its southerly extension in March and April somewhat in advance. Ship observations are not sufficient to accomplish this task, but the recent photographs from artificial satellites will be very helpful, as the cloud patterns of the convergence show up quite clearly on most of the photographs.

Rains in the interior of tropical Brazil

Frequently in the meteorological literature the rains of the southern tropical regime are attributed to the penetration of maritime masses from the Northern Hemisphere into the interior of Brazil, carrying the intertropical convergence to the tropic.

The intertropical convergence, however, does not in our opinion penetrate deep into the continent, and is limited to the equatorial part of the country, as Fig. 2 shows.

As stated above, it is necessary to separate circulation changes produced by the intensification of the heat low in the summer from those produced by movements of the intertropical convergence. The first causes rains of the southern continental tropical, the second those of the maritime equatorial regime.

Average monthly rainfall charts show the formation and progressive growth, beginning in September, of the continental precipitation area, with the withdrawal of the anticyclone of the Atlantic and the formation of the heat low, without much change of the precipitation over the north coast of the continent. Only in December, when this situation has established itself over the interior of the country, do rains begin in Oiapoque and Belém with the appearance of the intertropical convergence, which passes the Oiapoque station in that month.

The prevailing moist northeast current in the whole lower Amazon region does not reach above about 1,000 m. The dry trade winds blow above that altitude. Such conditions do not favour any precipitation. Rains then are produced by the mass *mE*, originating in the Northern Hemisphere, and they only penetrate the continent beginning in January, accompanying the convergence. Its influence lasts through May. During this period, however, the continental precipitation area weakens.

Thus it seems that the rains of the southern tropical regime ought to be attributed to the unstable Amazon mass *cE*, carried by the northwest to west currents of the heat low in the summer, and not to a current of maritime air penetrating the country from the north with a type of monsoon. It would be at most a monsoon originating not over the ocean, but in the upper Amazon basin.

The absence of high altitude observations in the central part of the country makes an interpretation of air mass movements in the region difficult. During the winter the stable east current prevails completely. When it prevails also in the summer, weather remains dry. A change to north or northwest winds brings on the appearance of large convective clouds and precipitation which are most intense in the afternoon and evening.

The precipitation regime of the continental area of Brazil north of the Equator, which is called the northern tropical continental regime, has maximum rainfalls in the northern summer, in July, and lightest in February. Rains begin in May and end in September. Here, however, precipitation occurs in the opposite way to what happens in the southern regime, in the period of highest atmospheric pressure.

In the description of pressure centres it was pointed out that in the northern summer the anticyclones of the North *and* South Atlantic are most intense. At that time of year, pressure also rises over Brazil north of the Equator, that is, the equatorial trough appears to "fill".

On the other hand, at the Boa Vista station, at about 3°N, beginning in April, winds from the east and northeast become less frequent, and south winds and calms appear more often. Between May and June the intertropical convergence crosses the area of Brazil north of the Equator from south to north (Fig.2).

All this seems to indicate that the dry period, between November and April, is due to the dry subsident air mass of the northern anticyclone, and that with the northward movement of the equatorial convergence from May to July the Amazon mass invades the region, causing precipitation. Finally, we should mention the large area of more or less constant rains in the upper Amazon region where, as it appears to us, the two precipitation regimes we just described originate. A part of the rains in this area comes from local evaporation. In this way, a closed cycle is formed. The rainfall is not due only to moisture advection from outside. It is true that there are winds from the east quadrant in the troposphere above the friction layer, but these winds bring, as we have seen, little moisture; they reach the upper Amazon region only after a long way across the continent. There must be more moisture advection into this region during the months December to May with the southward displacement of the equatorial trough.

The rains of the east coast

In the description of the climate of the eastern part of the northeast and of the east region, an attempt was made to explain the precipitation of the maritime tropical regime. In spite of some categorical affirmations, it must be admitted that the problem is still unresolved. Some comments on the probable causes of this precipitation follow.

Charts of the resultant winds over the tropical Atlantic (RIEHL, 1950; MINTZ and DEAN, 1952) indicate that in January there is convergence over the southern part of the east coast, divergence over the northern part and vice versa in July. This notion is confirmed by the rainfall patterns.

The rapid decrease in rainfall from the coast toward the interior must be attributed to the decrease in wind velocity due to friction as the wind goes from the ocean to the continent, causing convergence and rising of the air near the coast.

The orographic effect is felt on the eastern slopes of the Planalto da Borborema, as the transition from the wet area on the east and the dry area on the west is clear.

In the winter the cold air invasion passes beyond the line at which the winds diverge, at that time of year located on the south coast, and are incorporated into the trade winds. This is a probable cause of precipitation. In the summer, however, with the divergence over the coast of the northeast, the cold air invasions do not go beyond the Baia de Todos os Santos. At this time of year, the rains south of the bay decrease from the interior toward the coast and are brought about within the mass *cE* by cold fronts and tropical discontinuities (Fig.3).

On the east coast of the northeast, rains are probably also produced by disturbances in the trade winds. These disturbances have not yet been sufficiently studied. Finally there is a possibility that the equatorial trough exerts some influence on the east coast of the northeast region.

Acknowledgements

The author expresses his gratitude to Adalberto Serra who made available the material he elaborated for an extensive study of the climatology of Brazil. The help of meteorologist Nysia Sobreira who typed the manuscript, and of the auxiliaries Sonia Losso, Lucia Bastos, Helena Muller Lopes and Edna Bittencourt is also acknowledged, as well as the collaboration of the geographers Maria do Carmo Galvão, E. Nimer and Lysia Bernardes and the statistician Jacques Ellis.

The translation of the original Portuguese text was made by Mrs. Mary Lillywhite, graduate student at the University of Wisconsin.

Appendix—Climatic tables (pp.270–293; see also Table II, p.261)

Concerning the climatic tables the following notes must be made.

(*a*) Pressures are given at station level corrected for gravity.

(*b*) Average temperatures were corrected for 24 h averages.

(*c*) Evaporation data were obtained with a Piche evaporimeter.

(*d*) Wind data are for an altitude of 7 m.

(*e*) The basic period for the climatic tables is 1912–1942, but the record of several stations is incomplete.

References

ALMEIDA SIMÕES, R. M., 1954. Notas sôbre o clima do Paraná. *Rev. Brasil. Geogr.*, 19(2): 235–241.

ANGOT, A., 1928. *Traité élémentaire de météorologie*. Gauthier-Villars, Paris, 4me éd.rev., 417 pp.

BARROS, L. F., 1957. Contribuição ao estudo das massas de ar na Bacia do São Francisco. *Rev. Brasil. Geogr.*, 19(3): 301–340.

BERNARDES, L. M. C., 1951a. Clima do Brasil. *Bol. Geogr.*, 9(103): 727–739.

BERNARDES, L. M. C., 1951b. Os tipos de clima do Brasil. *Bol. Geogr.*, 9(105): 988–997.

BERNARDES, L. M. C., 1962. Aplicação de classificações climáticas ao Brasil. *Bol. Geogr., Rio de Janeiro*, 20(160): 437–442.

BEZERRA DOS SANTOS, L., 1962. Clima. In: M. G. C. HEREDA (Editor), *Grandes Regiões Meio-Norte e Nordeste*. Conselho Nacional de Geografia, Rio de Janeiro, pp.113–134.

CONRAD, V. and POLLAK, L. W., 1950. *Methods in Climatology*. Harvard University Press, Harvard, 459 pp.

COUSSIRAT ARAÚJO, C., 1930. *Memória sôbre o Clima do Rio Grande do Sul*. Serviço de Informação do Ministério da Agricultura, Rio de Janeiro, 100 pp.

DELGADO DE CARVALHO, C. M., 1916. *Climatologie du Brésil*. Bale, London, 91 pp.

DELGADO DE CARVALHO, C. M., 1917. *Meteorologie du Brésil*. Bale, London, 527 pp.

FOSBERG, F. R., GARNIER, B. J. and KÜCHLER, A. W., 1961. Delimitation of the humid tropics. *Geogr. Rev.*, 51(3): 333–347.

GALVÃO, M. C., 1959. Aspectos da geomorfologia do Brasil. *Bol. Geogr.*, 148: 54–63.

GALVÃO, M. V., 1959. Clima da Amazônia. In: A. T. GUERRA (Editor), *Grande Região Norte*. Conselho Nacional de Geografia, Rio de Janeiro, 1(A): 61–112.

GALVÃO, M. V., 1960. Clima. In: M. V. GALVÃO (Editor), *Grande Região Centro-Oeste*. Conselho Nacional de Geografia, Rio de Janeiro, 2(A): 71–117.

GUERRA, I. A. L., 1955. Tipos de clima do nordeste. *Rev. Brasil. Geogr.*, 17(4): 449–491.

GUIMARÃES, F. M. S., 1943. Relêvo do Brasil. *Bol. Geogr.*, 4: 63–73.

JOHANNSON, O. V., 1931. Die Hauptcharakteristika des jährlichen Temperaturganges. *Gerlands Beitr. Geophys.*, 33: 406–428.

MACHADO, F. P., 1950. *Contribuição ao Estudo do Clima do Rio Grande do Sul*. Instituto Brasileiro de Geografia e Estatística, Rio de Janeiro, 91 pp.

MATTOS, J. N. B., 1925. *O Clima de São Paulo*. Secretaria da Agricultura, Comércio e Obras Públicas do Estado de São Paulo, 95 pp.

MINTZ, Y. and GORDON, D., 1952. The observed mean field of motion of the atmosphere. *Air Force Cambridge Res. Center, Geophys. Res. Pap.*, 17: 65 pp.

MONTEIRO, C. A. F., 1963. Clima. In: D. M. CATALDO (Editor), *Grande Região Sul*. Conselho Nacional de Geografia, Rio de Janeiro, 4(1): 117–169.

MORIZE, H., 1927. *Contribuição ao Estudo do Clima do Brasil*. Serviço de Informação do Ministério da Agricultura, Rio de Janeiro, 114 pp.

PAES DE CAMARGO, A., 1960. *Balanço Hídrico no Estado de São Paulo*. Instituto Agronómico de Campinas, São Paulo, 15 pp.

RIEHL, H., 1950. On the role of the tropics in the general circulation of the atmosphere. *Tellus*, 2(1): 1–17.

RIZZINI, C. T. and HERINGER, E. P., 1962. *Preliminares açerca das Formações Vegetais e do Reflorestamento no Brasil Central*. Serviço de Informação Agrícola do Ministério da Agricultura, Rio de Janeiro, 79 pp.

RIZZINI, C. T. and PINTO, M. M., 1964. Areas climático-vegetationais do Brasil segundo os métodos de Thornthwaite e Mohr. *Rev. Brasil. Geogr., Rio de Janeiro*, 26(4): 523–547.

SAMPAIO FERRAZ, J., 1931. *Causas Prováveis das Sêcas do Nordeste Brasileiro*. Diretoria de Meteorologia, Ministério da Agricultura, Rio de Janeiro, 30 pp.

SAMPAIO FERRAZ, J., 1950. Iminência duma "grande" seca nordestina. *Rev. Brasil. Geogr., Rio de Janeiro*, 12(1): 3–15.

SCHMIDT, J. C. J., 1942. O clima da Amazônia *Rev. Brasil. Geogr., Rio de Janeiro*, 3: 465–500.

SEREBRENICK, S., 1945. *Notas sôbre o Clima do Brasil*. Serviço de Documentação do Ministério da Agricultura, Rio de Janeiro, 38 pp.

SEREBRENICK, S., 1953. *Condições Climáticas do Vale do São Francisco*. Departamento de Imprensa Nacional, Rio de Janeiro, 134 pp.

SERRA, A., 1939. *La Circulation Générale de L'Amérique du Sud*. Serviço Nacional de Meteorologia, Ministério da Agricultura, Rio de Janeiro, 170 pp.

SERRA, A., 1945. *Climatologia Equatorial*. Serviço de Meteorologia, Ministério da Agricultura, Rio de Janeiro, 48 pp.

SERRA, A., 1946. *As Sêcas do Nordeste*. Serviço de Meteorologia, Ministério da Agricultura, Rio de Janeiro, 148 pp.

SERRA, A., 1955, 1956, 1960. *Atlas Climatológico do Brasil*. Conselho Nacional de Geografia e Serviço de Meteorologia do Ministério da Agricultura, Rio de Janeiro. Vol.I and II, 433 pp.; Vol.III, 350 pp.

SERRA, A. and RATISBONA, L., 1941a. *O Clima do Rio de Janeiro*. Serviço de Meteorologia, Ministério da Agricultura, Rio de Janeiro, 75 pp.

SERRA, A. and RATISBONA, L., 1941b. *As Ondas de Frio da Bacia Amazônica.* Serviço de Meteorologia, Ministério da Agricultura, Rio de Janeiro, 65 pp.

SERRA, A. and RATISBONA, L., 1942. *As massas de ar da América do Sul.* Serviço de Meteorologia, Ministério da Agricultura, Rio de Janeiro, 137 pp.

SETZER, J., 1946a. A new formula for precipitation effectiveness. *Geogr. Rev.,* 36(2): 177–221.

SETZER, J., 1946b. *Contribuição para o Estudo do Clima do Estado de São Paulo.* Escolas Profissionais Salesianas, Sao Paulo, 239 pp.

SETZER, J., 1954. Isolinhas de umidade do clima no estado do Rio de Janeiro e no distrito federal. *Rev. Brasil. Geogr., Rio de Janeiro,* 16(3): 315–325.

THORNTHWAITE, G. W., 1948. An approach toward a rational classification of climate. *Geogr. Rev.,* 38: 55–94.

VUORELA, L. A., 1950. *Synoptic Aspects of Tropical Regions of the Atlantic Ocean, West Africa and South America.* University of Helsinki, Institute of Meteorology, Helsinki, 130 pp.

Additional references[1]

ALDAZ, L., 1971. Meteorological Conditions. Accompanying the Alagoas (Brazil) Floods of March 10–20, 1969. SUDENE/DNMET/WMO, 70 pp.

ALDAZ, L., 1971. A Partial Characterization of the Rainfall Regime of Brazil. SUDENE/DNMET/WMO, 30 pp.

BARROS, L., 1967. Esboço climático da região leste Brasileira. *Rev. Brasil. Geogr.,* 29 (2,3,4); 30(1).

BECKER, B. K., 1968. Aplicação dos indices climáticos ao nordeste do Brasil. *Rev. Brasil. Geogr.,* 27: 3–21.

DEAN, G. A., 1971. The three-dimensional wind structure over South America and associated rainfall over Brazil. *Fa. State Univ., Dept. Meteorol. Rept.,* 71–4: 122 pp.

DEAN, G. A., NUNES, G. S. and NUNES, H. T., 1971. Frequency of cloud types over South America east of the Andes Mountains. *Dept. Meteorol., Florida State Univ., Rept.,* 71–1: 26 pp.

DEPARTAMENTO NACIONAL DE METEOROLOGIA, 1972. *Balanço Hídrico do Brasil.* Rio de Janeiro, 94 pp.

GALVÃO, M. V., 1967. Regiões biclimáticas do Brasil. *Rev. Brasil. Geogr.,* 29(1): 3–36.

MONTEIRO, C. A. F., 1971. *Análise Ritmica em Climatologia, Problemas da Atualidade Climática em São Paulo e Achegas para um Programa de Trabalho. Climatologia, 1.* Instituto de Geografia, Universidade de São Paulo, São Paulo, pp. 1–21.

MONTEIRO, C. A. F., 1971. *Comparação da Pluviosidade nos Estados de São Paulo e Rio Grande do Sul nos invernos de 1957 e 1963. Climatologia, 3.* Instituto de Geografia, Universidade de São Paulo, pp. 1–8.

NEWELL, R. E., KIDSON, J. W., VINCENT, D. G. and BOER, G. J., 1972. *The General Circulation of the Tropical Atmosphere.* M.I.T. Press, Cambridge, Mass., 258 pp.

NIMER, E., 1972. Climatologia da região norte do Brasil—Introdução à climatologia dinâmica. *Rev. Brasil. Geogr.,* 34(3).

NIMER, E., 1972. Climatologia da região centro-leste do Brasil—Introdução à climatologia dinâmica. *Rev. Brasil. Geogr.,* 34(4).

NIMER, E., 1972. Climatologia da região sueste do Brasil—Introdução à climatologia dinâmica. *Rev. Brasil. Geogr.,* 34(1): 3–48.

NIMER, E., 1972. Climatologia da região norte do Brasil—Introdução à climatologia dinâmica. *Rev. Brasil. Geogr.,* 34(2): 3–51.

STRANG, D. M. D., 1973. *Climatological Analysis of Rainfall Normals in Northeastern Brasil.* Centro Técnico Aeroespacial, 200 pp.

VULQUIN, A., 1968. Os tipos de climas de verão no sul do Brasil. *Bol. Geogr.,* 27: 202.

---

[1] These came to the attention of the author after completion of his manuscript.

TABLE III

Month	Mean sta. press. (mbar)	Temperature (°C)				Mean vap. press. (mbar)	Precipitation (mm)	
		daily mean	mean daily range	extreme max.	min.		mean	max. in 24 h
Jan.	1000.5	25.5	8.9	36.1	20.0	28.7	283.7	113.4
Feb.	1000.5	25.8	9.0	36.1	19.8	28.7	261.0	81.2
Mar.	1000.4	25.6	8.7	36.8	19.1	28.4	283.5	156.4
Apr.	1001.2	25.4	8.5	36.4	19.6	28.5	262.8	97.2
May	1002.0	25.1	7.9	36.9	17.7	28.5	328.9	105.0
June	1002.9	24.7	8.0	36.8	17.6	27.9	244.4	79.6
July	1003.5	24.3	8.1	35.8	16.0	27.1	234.1	157.0
Aug.	1002.5	25.0	9.2	35.8	18.5	27.6	186.2	60.0
Sept.	1001.6	25.5	9.7	38.0	18.8	28.0	160.3	65.2
Oct.	1000.6	25.8	9.5	39.0	18.7	28.5	163.8	113.3
Nov.	999.5	26.1	9.6	37.5	19.4	29.0	190.0	106.5
Dec.	999.9	25.7	8.8	36.9	19.3	28.9	269.9	99.1
Annual	1001.3	25.4	8.8	39.0	16.0	28.3	2868.6	157.0

Month	Mean evap. (mm)	Number of days with				Mean cloud-iness (tenths)	Mean sun-shine (h)	Wind	
		precip. >0.1 mm	thunder-storm	fog	haze			preval. direct.	mean speed (m/sec)
Jan.	44.3	21	10	8	1	7.7	160.5	C–N–NW	0.9
Feb.	47.4	17	10	6	1	7.1	154.4	C–SE	1.0
Mar.	54.4	18	12	7	2	7.7	156.6	C–SE	1.1
Apr.	44.3	19	9	7	1	7.8	142.8	C–SE	1.1
May	40.4	23	7	8	1	7.9	144.0	C–N	1.0
June	37.3	21	8	9	1	7.6	143.6	C–N	0.9
July	40.1	21	10	9	0	7.4	162.5	C–N	1.0
Aug.	48.7	18	15	9	1	6.8	195.9	C–N	1.1
Sept.	57.5	16	13	7	1	6.8	195.5	C–N	1.1
Oct.	58.5	15	12	6	1	7.4	188.1	C–N–W	1.2
Nov.	56.8	15	12	6	1	7.2	186.7	C–SE	1.1
Dec.	49.9	19	10	7	1	7.5	164.5	C–N–W	1.0
Annual	579.6	223	128	89	12	7.4	1995.1	C–N	1.0

TABLE IV

CLIMATIC TABLE FOR SENA MADUREIRA
Latitude 9°08′S, longitude 68°40′W, elevation 135 m

Month	Mean sta. press. (mbar)	Temperature (°C)				Mean vap. press. (mbar)	Precipitation (mm)	
		daily mean	mean daily range	extreme max.	min.		mean	max. in 24 h
Jan.	993.9	25.4	11.8	37.6	16.4	30.5	316.6	96.7
Feb.	994.4	25.3	12.2	37.0	15.2	30.4	285.4	90.1
Mar.	994.3	25.2	11.8	37.4	15.8	30.1	266.4	70.5
Apr.	994.8	24.9	12.1	37.0	14.2	29.7	231.3	112.2
May	995.9	24.1	12.6	37.0	8.8	28.0	125.1	87.1
June	996.9	23.3	13.4	37.3	7.3	26.9	66.0	61.2
July	997.3	22.9	14.7	37.4	8.0	26.3	36.0	52.7
Aug.	996.1	23.9	15.0	37.8	7.9	27.3	45.5	63.5
Sept.	995.1	24.9	14.1	38.4	8.0	31.0	126.3	91.2
Oct.	994.7	25.3	13.3	38.8	10.3	30.0	173.2	63.0
Nov.	993.8	25.5	13.1	37.7	14.0	28.8	193.2	69.0
Dec.	994.1	25.5	12.5	38.0	13.6	30.0	273.9	70.2
Annual	995.1	24.7	13.1	38.8	7.3	29.1	2138.9	112.2

Month	Number of days with				Mean cloud-iness (tenths)	Wind	
	precip. >0.1mm	thunder-storm	fog	haze		preval. direct.	mean speed (m/sec)
Jan.	18	7	3	0	6.5	C–N–NW	0.5
Feb.	16	6	2	0	6.2	C–N	0.4
Mar.	17	7	4	1	6.5	C–N–NW	0.4
Apr.	14	5	5	0	6.1	C–SE	0.5
May	9	4	7	0	5.0	C–S	0.5
June	8	3	8	0	4.4	C–S	0.5
July	5	2	7	0	3.8	C–S	0.6
Aug.	5	4	5	1	3.6	C–S	0.6
Sept.	9	7	3	0	4.6	C–N	0.5
Oct.	11	8	2	0	5.6	C–N	0.5
Nov.	11	7	2	0	5.8	C–N–NW	0.4
Dec.	15	6	2	0	6.4	C–N	0.4
Annual	138	66	50	2	5.4	C–N	0.5

TABLE V

CLIMATIC TABLE FOR MANAUS
Latitude 3°08′S, longitude 60°01′W, elevation 48 m

Month	Mean sta. press. (mbar)	Temperature (°C)				Mean vap. press. (mbar)	Precipitation (mm)	
		daily mean	mean daily range	extreme			mean	max. in 24 h
				max.	min.			
Jan.	1005.2	26.2	7.4	37.4	20.4	28.5	265.5	96.4
Feb.	1005.5	26.2	7.4	37.6	20.0	28.6	246.6	101.9
Mar.	1005.4	26.4	7.2	36.1	19.4	29.1	269.2	128.5
Apr.	1005.6	26.2	7.0	34.6	20.2	28.9	266.9	121.3
May	1006.3	26.3	7.2	35.0	20.0	28.8	193.9	103.1
June	1007.3	26.6	7.7	35.0	19.0	28.3	100.3	74.0
July	1007.7	26.8	8.5	35.2	17.6	27.6	63.7	68.8
Aug.	1006.6	27.5	9.4	36.7	19.2	27.6	37.6	50.6
Sept.	1006.1	27.9	9.5	37.2	20.0	28.4	59.6	51.8
Oct.	1005.4	27.8	9.1	37.8	20.2	28.8	124.4	92.0
Nov.	1004.2	27.6	8.7	37.2	20.2	29.1	151.6	99.7
Dec.	1004.8	26.8	8.0	38.6	19.6	29.0	216.3	85.2
Annual	1005.8	26.9	8.1	38.6	17.6	28.6	1995.6	128.5

Month	Mean evap. (mm)	Number of days with				Mean cloudiness (tenths)	Mean sunshine (h)	Wind	
		precip >0.1 mm	thunderstorm	fog	haze			preval. direct.	mean speed (m/sec)
Jan.	48.0	20	4	2	1	7.1	125.5	C–E	1.5
Feb.	43.9	19	3	1	0	7.3	108.4	C–E	1.7
Mar.	48.1	21	4	1	0	7.3	122.6	C–E	1.7
Apr.	42.3	20	5	1	0	7.2	124.6	C–E	1.4
May	48.7	18	3	1	1	6.8	160.6	C–E	1.6
June	58.3	12	4	2	2	6.2	202.5	C–E	1.6
July	76.7	8	3	2	4	5.7	238.2	C–E	1.6
Aug.	93.9	6	4	2	4	5.7	259.0	C–E	1.8
Sept.	91.9	8	5	2	4	6.1	225.0	C–E–S	1.8
Oct.	86.6	11	7	1	3	6.6	208.1	C–E–S	1.7
Nov.	73.5	12	7	2	3	6.7	187.7	C–E	1.7
Dec.	60.6	16	6	2	3	7.0	162.8	C–E	1.7
Annual	772.5	171	55	19	25	6.6	2125.0	C–E	1.6

TABLE VI

CLIMATIC TABLE FOR BELÉM

Latitude 1°28′S, longitude 48°27′W, elevation 24 m

Month	Mean sta. press. (mbar)	Temperature (°C)				Mean vap. press. (mbar)	Precipitation (mm)	
		daily mean	mean daily range	extreme			mean	max. in 24 h
				max.	min.			
Jan.	1008.7	25.2	8.6	34.6	20.3	28.4	339.4	97.5
Feb.	1009.1	25.0	7.8	33.9	20.2	28.6	407.5	120.9
Mar.	1009.1	25.1	7.9	34.5	19.8	28.8	435.9	101.5
Apr.	1009.4	25.5	8.2	34.1	21.1	29.2	343.5	87.4
May	1009.7	25.8	8.9	34.2	20.4	29.0	287.9	125.6
June	1010.7	25.8	9.6	34.2	19.9	28.1	174.8	70.8
July	1011.3	25.8	10.0	33.7	18.5	27.5	145.0	72.4
Aug.	1010.8	25.9	10.4	35.1	19.1	27.8	126.5	51.3
Sept.	1010.0	25.8	10.4	34.6	19.4	27.8	118.3	49.4
Oct.	1009.3	26.1	10.5	34.6	18.9	28.1	92.0	48.2
Nov.	1008.3	26.3	10.6	35.1	19.4	28.1	86.0	40.2
Dec.	1008.5	25.9	9.9	35.4	19.3	28.3	175.2	77.7
Annual	1009.6	25.7	9.4	35.4	18.5	28.3	2732.0	125.6

Month	Mean evap. (mm)	Number of days with				Mean cloud-iness (tenths)	Mean sun-shine (h)	Wind	
		precip. >0.1 mm	thunder-storm	fog	haze			preval. direct.	mean speed (m/sec)
Jan.	42.1	27	8	3	0	7.1	156.5	C–NE	0.9
Feb.	31.0	26	6	2	0	7.8	105.0	C–NE	0.8
Mar.	34.6	28	7	3	0	7.8	118.3	C–NE	0.9
Apr.	35.7	27	7	3	0	7.3	136.7	C–NE	0.9
May	47.9	26	7	2	0	6.7	201.0	C–NE	1.0
June	57.6	22	7	2	0	5.7	242.0	C–NE	1.1
July	67.5	17	7	1	0	5.0	275.4	C–NE	1.2
Aug.	67.7	15	7	1	0	4.5	278.8	C–NE	1.3
Sept.	61.6	18	7	1	0	4.5	255.0	C–NE	1.2
Oct.	66.6	15	5	1	0	4.6	265.8	C–NE	1.3
Nov.	68.1	13	5	2	1	4.9	247.8	C–NE	1.4
Dec.	58.9	19	7	3	0	5.7	230.7	C–NE	1.1
Annual	639.3	253	80	24	1	6.0	2513.0	C–NE	1.1

TABLE VII

CLIMATIC TABLE FOR TAPERINHA (SANTAREM)
Latitude 2°25′S, longitude 54°42′W, elevation 20 m

Month	Mean sta. press. (mbar)	Temperature (°C)				Mean vap. press. (mbar)	Precipitation (mm)	
		daily mean	mean daily range	extreme			mean	max. in 24 h
				max.	min.			
Jan.	1008.3	25.7	7.7	35.4	19.8	27.8	167.6	83.5
Feb.	1008.7	25.3	7.1	34.2	20.1	28.1	270.3	112.6
Mar.	1008.7	25.4	6.9	33.2	20.3	28.6	331.1	175.0
Apr.	1008.9	25.4	6.9	32.8	20.6	29.0	335.7	129.0
May	1009.4	25.3	7.2	34.3	20.4	29.0	285.8	87.6
June	1010.1	25.1	8.0	34.2	19.5	28.3	187.5	103.1
July	1010.7	25.1	9.1	34.8	18.5	27.3	100.7	71.1
Aug.	1010.0	26.0	9.8	35.7	19.0	27.8	43.3	30.2
Sept.	1009.0	26.6	10.0	37.1	20.2	27.8	35.7	39.8
Oct.	1008.1	26.9	10.0	36.6	20.4	27.7	46.7	97.1
Nov.	1007.2	26.9	9.6	36.0	20.6	27.8	60.2	88.2
Dec.	1007.6	26.4	8.8	35.6	19.9	27.7	108.7	98.4
Annual	1008.9	25.8	8.4	37.1	18.5	28.1	1973.3	175.0

Month	Mean evap. (mm)	Number of days with				Mean cloud-iness (tenths)	Mean sun-shine (h)	Wind	
		precip. >0.1 mm	thunder-storm	fog	haze			preval. direct.	mean speed (m/sec)
Jan.	47.3	21	7	0	2	6.3	132.0	C–NE	1.4
Feb.	32.9	23	8	0	0	6.9	96.4	C–NE	1.2
Mar.	34.9	25	11	0	0	7.0	96.6	C–NE	1.2
Apr.	30.8	26	11	0	0	6.8	110.1	C–NE	1.2
May	30.6	26	12	0	0	6.2	135.6	C–NE	1.1
June	34.0	22	11	0	0	5.0	163.0	C–NE	1.1
July	45.2	15	8	1	0	4.2	200.2	C–NE	1.4
Aug.	58.3	10	7	2	1	3.4	233.4	C–NE	1.6
Sept.	68.3	8	6	1	3	3.6	219.5	NE	1.7
Oct.	77.9	6	6	1	8	4.1	215.6	NE	1.7
Nov.	72.2	7	6	2	10	4.5	190.5	C–NE	1.5
Dec.	64.9	12	7	1	7	5.2	173.8	C–NE	1.5
Annual	597.3	201	100	8	31	5.3	1966.7	C–NE	1.4

TABLE VIII

CLIMATIC TABLE FOR BARRA DO CORDA
Latitude 5°30′S, longitude 45°16′W, elevation 81 m

Month	Mean sta. press. (mbar)	Temperature (°C)				Mean vap. press. (mbar)	Precipitation (mm)	
		daily mean	mean daily range	extreme			mean	max. in 24 h
				max.	min.			
Jan.	1000.4	25.5	9.5	37.3	19.0	27.6	189.9	168.0
Feb.	1000.7	25.4	9.2	37.0	17.6	27.8	208.2	63.0
Mar.	1000.7	25.4	9.2	35.2	18.0	28.0	213.9	151.4
Apr.	1000.8	25.6	9.4	38.6	18.8	28.3	144.4	88.0
May	1001.4	25.2	10.7	36.4	13.6	27.0	60.3	55.0
June	1002.5	24.6	13.1	36.8	14.2	24.6	16.4	32.0
July	1003.1	24.2	15.6	37.2	12.2	22.8	7.0	27.5
Aug.	1002.4	25.4	16.1	37.8	12.0	23.3	7.1	41.0
Sept.	1001.3	27.2	13.5	39.0	14.8	25.4	22.6	52.6
Oct.	1000.7	27.4	12.2	39.4	13.6	26.4	40.9	62.5
Nov.	999.8	27.0	11.5	38.0	12.5	27.0	70.0	67.7
Dec.	1000.0	26.2	10.6	38.6	18.6	27.4	116.6	74.6
Annual	1001.1	25.8	11.7	39.4	12.0	26.3	1097.3	168.0

Month	Mean evap. (mm)	Number of days with				Mean cloud-iness (tenths)	Mean sun-shine (h)	Wind	
		precip. >0.1 mm	thunder-storm	fog	haze			preval. direct.	mean speed (m/sec)
Jan.	56.7	17	16	3	0	7.2	132.2	C–N	0.7
Feb.	44.9	17	13	3	0	7.5	120.9	C–N	0.7
Mar.	47.1	19	15	5	0	7.5	138.4	C–N	0.7
Apr.	49.4	16	12	7	0	6.6	148.9	C–N	0.7
May	64.2	9	4	11	1	5.2	180.5	C–E	0.8
June	89.1	3	1	9	1	3.9	214.7	C–E	0.9
July	116.4	2	1	4	1	3.2	243.3	C–E	0.9
Aug.	132.7	1	1	2	0	3.7	227.1	C–E	0.9
Sept.	120.7	4	3	1	0	4.9	178.8	C–N	0.9
Oct.	113.9	6	7	0	0	5.5	169.0	C–N	0.9
Nov.	95.0	8	10	1	0	6.0	148.8	C–N	0.8
Dec.	75.7	12	13	2	1	6.4	150.4	C–N	0.7
Annual	1005.8	114	96	48	4	5.6	2053.0	C–N	0.8

TABLE IX

CLIMATIC TABLE FOR QUIXERAMOBIM
Latitude 5°12′S, longitude 39°18′W, elevation 198 m

Month	Mean sta. press. (mbar)	Temperature (°C)				Mean vap. press. (mbar)	Precipitation (mm)	
		daily mean	mean daily range	extreme			mean	max. in 24 h
				max.	min.			
Jan.	987.9	28.6	7.8	36.7	21.2	22.9	66.8	118.7
Feb.	988.2	27.8	7.0	36.3	19.8	24.4	107.9	113.4
Mar.	988.2	27.1	6.4	36.4	19.7	25.0	187.8	179.8
Apr.	988.8	26.8	6.1	35.0	20.4	25.9	168.8	102.5
May	989.4	26.5	6.2	34.8	18.4	24.6	110.8	137.8
June	990.7	26.2	6.6	34.4	18.4	22.7	54.0	52.6
July	991.5	26.4	7.6	33.5	19.0	21.1	25.6	63.8
Aug.	990.8	27.3	8.4	34.5	20.2	20.5	9.1	32.0
Sept.	989.9	28.0	8.8	35.6	21.4	20.6	3.3	22.0
Oct.	988.8	28.4	9.0	35.8	21.8	20.8	2.3	8.2
Nov.	987.7	28.5	8.9	36.3	20.7	21.2	5.8	26.9
Dec.	987.7	28.8	8.4	36.4	20.7	22.2	20.8	81.9
Annual	989.1	27.5	7.6	36.7	18.4	22.6	763.0	179.8

Month	Mean evap. (mm)	Number of days with				Mean cloud-iness (tenths)	Mean sun-shine (h)	Wind	
		precip. >0.1 mm	thunder-storm	fog	haze			preval. direct.	mean speed (m/sec)
Jan.	141.7	9	6	0	4	5.7	229.2	E	3.6
Feb.	99.9	12	10	1	2	6.4	189.3	E	2.7
Mar.	84.6	16	12	1	1	6.5	200.6	E	2.1
Apr.	74.2	16	10	1	0	6.5	200.3	E	2.1
May	78.9	13	4	1	0	5.8	225.0	E	2.1
June	90.3	10	1	0	0	5.1	231.1	C–E–SE	2.3
July	122.1	5	0	0	1	4.4	261.7	E	2.5
Aug.	150.8	3	0	0	4	3.6	286.3	E	3.5
Sept.	161.9	1	0	0	5	3.3	277.1	E	4.1
Oct.	173.8	2	0	0	4	3.7	280.0	E	4.7
Nov.	169.7	3	1	0	3	4.2	261.5	E	4.5
Dec.	165.5	5	2	0	4	4.9	251.7	E	4.3
Annual	1513.4	95	40	4	28	5.0	2893.8	E	3.2

TABLE X

CLIMATIC TABLE FOR OLINDA (RECIFE)
Latitude 8°01′S, longitude 34°51′W, elevation 57 m

Month	Mean sta. press. (mbar)	Temperature (°C)				Mean vap. press. (mbar)	Precipitation (mm)	
		daily mean	mean daily range	extreme			mean	max. in 24 h
				max.	min.			
Jan.	1007.8	27.1	5.4	33.2	21.3	27.8	46.8	109.7
Feb.	1007.6	27.1	5.5	33.4	20.2	28.4	108.8	150.1
Mar.	1007.8	27.0	5.7	33.8	20.8	28.2	156.5	105.0
Apr.	1007.7	26.6	5.8	33.9	20.1	28.3	225.6	133.9
May	1008.7	25.9	5.5	32.2	19.4	27.3	260.4	154.0
June	1010.3	25.0	5.4	31.8	18.8	26.1	256.6	228.3
July	1011.7	24.3	5.3	30.5	17.8	24.9	186.2	84.0
Aug.	1011.7	24.4	5.3	31.0	18.0	24.6	116.4	70.7
Sept.	1010.9	25.3	5.1	32.3	18.7	25.2	52.3	52.4
Oct.	1009.6	26.2	5.1	32.8	18.9	26.0	30.3	71.4
Nov.	1008.3	26.6	5.2	33.0	20.3	26.6	27.6	67.8
Dec.	1008.0	26.9	5.2	33.2	18.7	27.3	33.3	153.6
Annual	1009.2	26.0	5.4	33.9	17.8	26.7	1500.8	228.3

Month	Mean evap. (mm)	Number of days with				Mean cloud-iness (tenths)	Mean sun-shine (h)	Wind	
		precip. >0.1 mm	thunder-storm	fog	haze			preval. direct.	mean speed (m/sec)
Jan.	117.1	12	0	1	1	5.2	256.3	E–SE	3.0
Feb.	102.3	14	1	2	1	5.6	199.7	SE	2.8
Mar.	104.4	16	1	2	0	5.5	211.9	SE	2.6
Apr.	91.8	21	1	2	0	5.9	194.8	SE	2.9
May	85.6	23	0	2	0	6.2	184.9	SE	2.8
June	82.4	24	0	1	1	6.2	171.6	SE	3.6
July	93.1	25	0	1	0	6.3	159.2	SE	4.0
Aug.	99.4	22	0	1	1	5.9	204.4	SE	3.3
Sept.	104.7	15	0	1	3	5.0	223.7	SE	3.4
Oct.	112.2	11	0	0	2	4.6	238.6	E–SE	3.2
Nov.	113.3	9	0	0	2	4.6	234.7	NE	2.1
Dec.	116.5	10	0	0	2	4.7	263.2	E	2.0
Annual	1222.8	202	3	13	13	5.5	2543.0	SE	2.9

TABLE XI

CLIMATIC TABLE FOR ALTO TAPAJÓS
Latitude 7°20′S, longitude 57°30′W, elevation 140 m

Month	Mean sta. press. (mbar)	Temperature (°C)				Mean vap. press. (mbar)	Precipitation (mm)	
		daily mean	mean daily range	extreme max.	min.		mean	max. in 24 h
Jan.	997.5	24.7	8.6	36.2	19.0	28.0	397.8	96.4
Feb.	997.5	24.7	8.3	35.2	19.8	28.2	415.5	107.3
Mar.	997.7	24.9	8.7	35.7	19.0	28.5	378.9	84.9
Apr.	998.0	25.1	9.3	35.4	17.0	28.7	301.9	81.0
May	998.6	25.3	11.0	36.5	15.0	28.0	111.9	76.5
June	999.4	24.6	14.2	35.6	8.8	25.7	19.6	28.5
July	1000.1	24.0	15.9	36.8	9.4	23.9	21.6	32.5
Aug.	998.9	25.2	15.9	37.8	12.6	25.6	50.7	65.4
Sept.	998.3	25.3	12.3	37.7	16.0	27.4	143.8	71.8
Oct.	997.7	25.3	10.9	37.4	18.8	28.1	246.7	83.0
Nov.	996.4	25.1	9.8	36.6	19.8	28.4	314.9	88.0
Dec.	996.7	24.9	9.1	36.0	19.8	28.4	335.4	136.9
Annual	998.1	24.9	11.2	37.8	8.8	27.4	2738.7	136.9

Month	Mean evap. (mm)	Number of days with				Mean cloudiness (tenths)	Mean sunshine (h)	Wind	
		precip. >0.1 mm	thunderstorm	fog	haze			preval. direct.	mean speed (m/sec)
Jan.	32.2	26	7	7	0	7.7	107.8	C–E	0.4
Feb.	29.5	24	6	5	0	8.0	90.5	C–E	0.4
Mar.	33.4	25	7	7	0	7.7	110.2	C–E	0.4
Apr.	33.7	23	7	11	0	7.2	129.7	C–E	0.4
May	51.7	14	5	12	1	5.4	199.1	C–E	0.6
June	71.6	4	4	11	1	3.3	247.8	C–E	0.8
July	96.7	2	3	10	4	3.1	270.2	C–E	0.9
Aug.	91.8	6	7	10	8	3.6	261.4	C–E	0.8
Sept.	58.2	15	13	9	1	5.8	173.4	C–E	0.6
Oct.	47.6	19	14	8	0	6.6	160.7	C–E	0.6
Nov.	36.7	22	14	8	0	7.3	126.5	C–E	0.4
Dec.	35.4	24	10	7	0	7.7	116.9	C–E	0.3
Annual	618.5	204	97	105	15	6.1	1994.2	C–E	0.6

TABLE XII

CLIMATIC TABLE FOR PORTO NACIONAL
Latitude 10°31'S, longitude 48°43'W, elevation 237 m

Month	Mean sta. press. (mbar)	Temperature (°C)				Mean vap. press. (mbar)	Precipitation (mm)	
		daily mean	mean daily range	extreme			mean	max. in 24 h
				max.	min.			
Jan.	983.9	25.2	10.3	38.0	14.1	27.2	298.0	101.6
Feb.	984.3	24.9	9.6	37.7	18.1	27.2	290.0	70.1
Mar.	984.1	25.2	9.8	37.7	18.6	27.5	291.8	92.7
Apr.	984.5	25.7	10.8	39.7	16.5	27.6	152.1	68.4
May	985.4	25.6	13.3	38.2	10.9	25.4	43.6	57.6
June	986.6	24.6	16.8	39.9	11.2	22.1	0.4	2.6
July	987.4	24.3	18.0	38.0	9.5	20.3	2.8	24.0
Aug.	986.0	26.1	18.6	42.8	10.5	20.6	8.9	65.1
Sept.	984.5	27.6	15.6	40.1	13.1	23.2	42.1	65.0
Oct.	983.8	26.7	12.3	41.0	14.5	26.2	149.6	69.2
Nov.	983.2	25.8	10.4	38.4	15.4	27.2	241.8	80.6
Dec.	983.5	25.3	10.0	38.4	18.1	27.0	292.0	77.4
Annual	984.8	25.6	13.0	42.8	9.5	25.2	1813.1	101.6

Month	Mean evap. (mm)	Number of days with				Mean cloud-iness (tenths)	Mean sun-shine (h)	Wind	
		precip. >0.1 mm	thunder-storm	fog	haze			preval. direct.*	mean speed (m/sec)
Jan.	57.3	20	20	3	0	8.7	177.0	C–SW	0.6
Feb.	44.2	20	16	4	0	9.0	129.8	C–V	0.7
Mar.	51.4	20	16	4	0	8.6	156.8	C–V	0.6
Apr.	62.8	13	12	5	0	7.4	204.9	C–S	0.6
May	98.2	4	4	2	1	6.7	271.6	C–E	0.6
June	125.2	0	0	0	5	5.7	293.1	C–E	0.7
July	153.7	0	1	0	6	5.9	307.2	C–E	0.8
Aug.	188.9	1	2	0	10	6.5	314.4	C–E	0.8
Sept.	155.2	5	8	0	10	7.8	239.3	C–E	0.7
Oct.	100.5	12	18	1	2	8.4	194.9	C–V	0.7
Nov.	60.0	17	19	1	0	8.7	158.9	C–V	0.7
Dec.	56.9	19	18	2	0	8.8	164.1	C–V	0.7
Annual	1154.3	131	134	22	34	7.7	2612.0	C–S	0.7

* V = various.

TABLE XIII

CLIMATIC TABLE FOR FORMOSA
Latitude 15°32′S, longitude 47°18′W, elevation 912 m

Month	Mean sta. press. (mbar)	Temperature (°C)				Mean vap. press. (mbar)	Precipitation (mm)	
		daily mean	mean daily range	extreme			mean	max. in 24 h
				max.	min.			
Jan.	909.2	21.7	9.5	32.8	12.9	21.1	272.5	107.0
Feb.	909.5	21.8	9.6	33.0	13.4	21.3	227.4	103.0
Mar.	909.7	21.8	9.9	32.7	13.6	21.2	194.2	79.4
Apr.	910.5	21.5	10.8	32.8	9.9	19.8	99.4	63.6
May	911.9	20.1	12.2	31.4	7.8	16.9	18.8	31.6
June	913.0	19.0	13.7	30.2	5.1	14.7	4.2	41.7
July	913.9	18.7	14.2	30.8	5.5	13.1	4.1	25.2
Aug.	912.7	20.7	14.9	34.0	6.1	12.9	8.8	43.8
Sept.	911.0	22.7	14.1	35.5	8.8	14.8	44.9	63.6
Oct.	909.8	22.7	11.8	35.8	11.0	18.5	138.3	85.4
Nov.	908.8	21.8	9.6	35.5	10.7	21.0	232.6	107.5
Dec.	908.9	21.5	8.8	34.9	13.0	21.3	349.5	101.7
Annual	910.7	21.2	11.6	35.8	5.1	18.0	1594.7	107.5

Month	Mean evap. (mm)	Number of days with				Mean cloud- iness (tenths)	Mean sun- shine (h)	Wind	
		precip. >0.1 mm	thunder- storm	fog	haze			preval. direct.	mean speed (m/sec)
Jan.	69.6	21	19	2	0	7.8	188.5	C–N	2.1
Feb.	58.8	18	14	5	0	7.6	163.8	C–N	2.0
Mar.	70.7	18	15	5	0	7.2	204.7	C–N	2.1
Apr.	74.9	11	9	5	0	6.1	220.5	C–S	2.1
May	97.2	4	3	6	3	4.9	275.5	C–S	2.1
June	112.4	1	1	6	8	3.9	284.5	C–S	2.2
July	134.4	1	1	5	10	3.5	295.1	C–E	2.3
Aug.	181.7	1	2	4	10	3.3	308.6	C–E	2.4
Sept.	168.2	5	6	4	12	4.5	228.3	C–E	2.4
Oct.	133.8	12	13	3	10	6.7	213.2	C–N	2.3
Nov.	68.8	19	18	2	2	8.1	144.3	C–N	2.2
Dec.	63.9	22	16	3	0	8.4	145.3	C–N	2.1
Annual	1234.4	133	117	51	55	6.0	2672.3	C–N	2.2

TABLE XIV

CLIMATIC TABLE FOR REMANSO
Latitude 9°41′S, longitude 42°04′W, elevation 411 m

Month	Mean sta. press. (mbar)	Temperature (°C)				Mean vap. press. (mbar)	Precipitation (mm)	
		daily mean	mean daily range	extreme			mean	max. in 24 h
				max.	min.			
Jan.	967.9	27.2	14.8	38.7	10.8	18.6	78.0	67.4
Feb.	967.9	25.9	15.1	38.4	16.4	16.2	82.5	63.8
Mar.	967.9	27.1	15.2	38.7	15.8	17.9	87.9	98.4
Apr.	968.1	27.3	15.4	39.1	16.0	17.9	34.6	64.3
May	968.5	26.7	15.5	37.7	15.6	17.0	21.9	63.4
June	968.9	26.2	15.8	37.7	14.6	16.1	10.2	2.4
July	968.8	25.7	16.1	37.7	14.0	15.8	10.5	4.2
Aug.	969.0	26.2	16.2	38.1	14.8	15.8	0.0	0.0
Sept.	968.5	27.2	16.2	38.5	15.0	16.8	7.6	32.2
Oct.	968.2	28.0	15.9	39.5	16.4	17.3	13.8	56.4
Nov.	968.0	27.9	15.2	39.1	14.4	17.6	55.5	53.2
Dec.	968.0	27.2	14.7	39.3	9.8	17.7	91.8	64.3
Annual	968.3	26.8	15.5	39.5	9.8	17.0		

Month	Number of days with		Mean cloud-iness (tenths)	Wind	
	precip. >0.1mm	thunder-storm		preval. direct.	mean speed (m/sec)
Jan.	5	4	3.7	C–E	1.2
Feb.	6	3	3.6	C–E	1.4
Mar.	6	4	3.7	C–E	1.1
Apr.	2	2	3.2	C–E	1.2
May	1	1	2.8	C–E	1.4
June	0	0	2.7	C–E	1.3
July	0	0	2.6	C–E	1.6
Aug.	0	0	2.4	C–E	1.5
Sept.	2	1	2.7	C–E	1.5
Oct.	1	1	3.0	C–E	1.5
Nov.	4	3	3.4	C–E	1.3
Dec.	6	3	3.5	C–E	1.4
Annual	33	22	3.0	C–E	1.4

TABLE XV

CLIMATIC TABLE FOR CAETITE
Latitude 14°03'S, longitude 42°37'W, elevation 878 m

Month	Mean sta. press. (mbar)	Temperature (°C)				Mean vap. press. (mbar)	Precipitation (mm)	
		daily mean	mean daily range	extreme			mean	max. in 24 h
				max.	min.			
Jan.	915.1	22.6	10.4	36.9	11.4	20.9	119.7	87.4
Feb.	915.0	22.5	10.2	34.3	13.2	21.0	116.9	80.2
Mar.	915.1	22.6	10.3	34.8	10.2	20.8	97.2	70.0
Apr.	915.9	22.0	10.1	33.9	11.4	20.2	52.7	82.1
May	916.9	21.0	10.3	33.8	9.2	18.5	15.8	16.6
June	918.2	19.5	11.1	32.2	8.0	16.7	8.8	19.3
July	918.8	18.9	11.5	31.7	7.6	15.7	9.0	12.7
Aug.	918.2	20.0	12.5	34.6	7.6	15.7	10.4	42.5
Sept.	916.9	21.5	13.0	35.4	9.8	17.0	16.4	53.5
Oct.	915.7	22.7	12.1	36.7	10.8	18.3	61.5	52.2
Nov.	914.7	22.4	10.6	35.8	10.2	20.2	147.3	76.6
Dec.	914.8	22.3	10.2	35.6	11.8	20.7	151.6	82.6
Annual	916.3	21.4	11.0	36.9	7.6	18.7	807.3	87.4

Month	Mean evap. (mm)	Number of days with				Mean cloud-iness (tenths)	Mean sun-shine (h)	Wind	
		precip. >0.1 mm	thunder-storm	fog	haze			preval. direct.	mean speed (m/sec)
Jan.	99.9	9	8	2	2	6.1	197.5	SE	2.8
Feb.	99.1	9	8	2	2	6.3	184.7	SE	3.1
Mar.	107.7	8	6	3	2	5.9	211.6	SE	3.2
Apr.	106.0	8	3	2	2	6.7	194.9	SE	3.9
May	114.4	6	0	2	2	5.3	206.2	C–E	0.8
June	119.5	5	0	2	1	4.8	206.9	C–E	0.6
July	136.2	5	0	2	2	4.9	224.8	SE	1.3
Aug.	161.7	3	0	1	3	4.1	251.9	C–E	1.2
Sept.	180.4	2	1	1	6	4.4	234.4	C–E	1.1
Oct.	177.9	6	5	1	6	5.3	209.3	SE	4.4
Nov.	116.3	12	8	2	2	6.4	159.4	SE	3.4
Dec.	102.3	11	8	3	2	6.4	170.9	SE	2.7
Annual	1521.4	84	47	23	32	5.6	2452.5	C–SE	2.4

TABLE XVI

CLIMATIC TABLE FOR SALVADOR
Latitude 12°55'S, longitude 38°41'W, elevation 45 m

Month	Mean sta. press. (mbar)	Temperature (°C)				Mean vap. press. (mbar)	Precipitation (mm)	
		daily mean	mean daily range	extreme max.	min.		mean	max. in 24 h
Jan.	1008.2	26.1	6.8	34.4	20.0	27.0	73.5	80.2
Feb.	1009.8	26.3	6.6	34.6	19.0	27.4	115.7	90.2
Mar.	1008.1	26.2	6.4	35.2	19.9	27.6	165.3	158.0
Apr.	1009.1	25.8	5.8	34.2	20.6	27.5	278.4	123.5
May	1010.8	24.8	5.3	33.2	19.8	26.1	296.3	156.1
June	1012.8	23.8	5.3	30.7	18.0	23.4	224.5	110.3
July	1014.2	23.0	5.2	30.0	18.4	23.0	203.6	103.0
Aug.	1013.6	23.1	5.5	29.9	16.8	22.6	115.6	66.4
Sept.	1012.1	23.8	5.9	31.0	17.0	24.0	97.8	95.3
Oct.	1010.0	24.6	6.1	34.8	19.0	25.0	101.5	80.6
Nov.	1008.9	25.1	6.2	33.5	19.6	26.0	116.4	120.0
Dec.	1008.8	25.6	7.5	33.1	19.5	26.1	124.1	128.7
Annual	1010.5	24.8	6.0	35.2	16.8	25.5	1912.7	158.0

Month	Mean evap. (mm)	Number of days with				Mean cloud-iness (tenths)	Mean sun-shine (h)	Wind	
		precip. >0.1 mm	thunder-storm	fog	haze			preval. direct.	mean speed (m/sec)
Jan.	95.2	14	3	2	5	5.0	263.4	E	2.8
Feb.	86.8	16	4	3	4	5.4	224.9	E	2.7
Mar.	89.5	19	4	4	4	5.3	237.2	SE	2.6
Apr.	79.5	22	2	3	2	5.5	211.2	SE	2.8
May	81.1	25	1	3	2	6.2	189.9	SE	3.1
June	80.4	24	0	3	2	6.0	189.8	SE	2.6
July	84.6	24	0	3	1	6.5	189.8	SE	3.2
Aug.	91.3	20	0	4	2	5.8	222.7	SE	3.1
Sept.	87.2	16	0	4	4	5.3	217.9	SE	3.0
Oct.	91.6	14	1	3	6	5.1	238.7	E	3.3
Nov.	84.2	15	2	2	5	5.4	226.4	E	3.3
Dec.	88.1	14	3	3	4	5.1	248.6	E	3.0
Annual	1039.5	223	20	37	41	5.5	2660.5	SE	3.0

TABLE XVII

CLIMATIC TABLE FOR VITORIA
Latitude 20°19′S, longitude 40°20′W, elevation 31 m

Month	Mean sta. press. (mbar)	Temperature (°C)				Mean vap. press. (mbar)	Precipitation (mm)	
		daily mean	mean daily range	extreme max.	min.		mean	max. in 24 h
Jan.	1010.2	25.3	7.3	37.3	17.9	26.0	150.8	136.2
Feb.	1010.9	25.6	7.5	35.2	18.8	26.4	117.6	100.6
Mar.	1011.1	25.3	7.3	35.4	19.0	26.2	150.4	147.7
Apr.	1013.0	24.2	7.2	34.4	16.5	24.8	129.7	92.6
May	1014.9	22.6	7.4	34.0	14.7	22.1	89.9	86.0
June	1016.9	21.5	7.9	31.8	12.7	20.7	66.3	70.8
July	1018.5	20.5	7.5	33.1	10.3	19.3	71.9	60.8
Aug.	1017.4	21.0	7.6	32.3	11.8	19.6	47.8	73.9
Sept.	1015.7	21.9	7.0	33.7	14.5	21.0	84.6	71.9
Oct.	1013.2	22.7	6.5	33.9	9.3	22.2	135.1	110.8
Nov.	1011.4	23.6	6.4	34.4	16.4	23.7	169.0	77.8
Dec.	1009.9	24.6	6.9	36.5	16.8	25.0	191.3	125.7
Annual	1014.2	23.2	7.2	37.3	9.3	22.9	1409.7	147.7

Month	Mean evap. (mm)	Number of days with				Mean cloud-iness (tenths)	Mean sun-shine (h)	Wind	
		precip. >0.1 mm	thunder-storm	fog	haze			preval. direct.	mean speed (m/sec)
Jan.	88.2	14	6	1	1	5.9	224.8	N	3.4
Feb.	84.1	12	4	1	1	5.3	223.8	N	3.3
Mar.	84.7	15	4	1	0	5.5	217.9	C–N	3.1
Apr.	74.9	14	2	0	0	5.5	199.0	C–E	3.0
May	77.0	11	0	1	0	5.0	215.8	C–SW	2.9
June	71.6	8	0	1	1	4.5	208.1	C–N–SW	2.7
July	76.7	12	0	1	0	5.2	198.9	C–SW	3.0
Aug.	87.3	9	0	1	1	4.8	223.6	C–N	3.3
Sept.	85.4	10	1	2	2	6.1	175.9	N	3.7
Oct.	84.3	16	2	0	2	6.9	161.2	N	4.0
Nov.	78.1	17	3	0	1	7.2	160.0	N	3.9
Dec.	84.7	16	4	0	0	6.9	183.0	N	3.8
Annual	977.0	154	26	9	9	5.7	2392.0	C–N	3.3

TABLE XVIII

CLIMATIC TABLE FOR CUIABÁ
Latitude 15°35'S, longitude 56°06'W, elevation 171 m

Month	Mean sta. press. (mbar)	Temperature (°C)				Mean vap. press. (mbar)	Precipitation (mm)	
		daily mean	mean daily range	extreme			mean	max. in 24 h
				max.	min.			
Jan.	990.9	26.4	8.7	37.9	17.5	27.8	212.8	99.2
Feb.	990.3	26.2	8.6	38.4	15.0	28.0	200.2	118.0
Mar.	990.7	26.2	8.8	39.2	15.6	28.2	222.3	95.4
Apr.	991.9	25.9	9.6	38.2	13.3	27.2	105.8	80.0
May	993.7	24.3	11.2	38.8	6.4	23.5	46.3	74.6
June	994.8	23.0	12.6	36.0	1.2	20.7	13.6	41.3
July	995.6	22.5	14.3	37.8	5.0	18.1	8.9	53.9
Aug.	993.9	24.8	14.5	40.0	5.6	18.7	26.7	133.6
Sept.	992.0	26.6	12.6	40.6	7.4	21.5	48.2	106.2
Oct.	991.0	27.0	11.0	42.2	12.3	24.5	124.3	126.4
Nov.	987.0	26.8	9.7	38.7	12.8	26.7	161.5	106.7
Dec.	989.5	26.5	8.8	39.4	16.7	27.4	207.7	123.6
Annual	991.8	25.5	10.9	42.2	1.2	24.4	1378.3	133.6

Month	Mean evap. (mm)	Number of days with				Mean cloud-iness (tenths)	Mean sun-shine (h)	Wind	
		precip. >0.1 mm	thunder-storm	fog	haze			preval. direct.	mean speed (m/sec)
Jan.	61.1	19	13	0	0	7.4	137.2	N	1.7
Feb.	50.4	18	13	1	0	7.5	122.8	C–N	1.5
Mar.	52.7	19	15	2	0	7.1	143.3	C–N	1.4
Apr.	54.3	12	11	4	0	5.7	180.8	C–N	1.2
May	68.8	6	3	5	0	4.5	206.0	C–S	1.2
June	77.1	3	1	5	2	3.9	199.9	C–S	1.2
July	104.2	2	1	4	10	3.1	225.3	C–S	1.3
Aug.	135.7	2	2	4	16	3.1	210.4	C–S	1.5
Sept.	129.2	6	7	3	13	4.7	155.9	C–N	1.7
Oct.	107.0	11	11	1	4	6.1	168.6	C–N	1.7
Nov.	79.0	15	13	0	0	6.7	157.9	C–N	1.6
Dec.	66.2	19	13	0	0	7.5	129.3	N	1.7
Annual	985.7	132	103	29	45	5.6	2037.4	C–N	1.5

TABLE XIX

Latitude 19°00′S, longitude 57°39′W, elevation 138 m

Month	Mean sta. press. (mbar)	Temperature (°C)				Mean vap. press. (mbar)	Precipitation (mm)	
		daily mean	mean daily range	extreme			mean	max. in 24 h
				max.	min.			
Jan.	994.8	26.7	10.8	39.2	15.1	28.3	177.5	180.0
Feb.	995.2	26.5	10.9	40.0	14.0	28.0	146.9	97.0
Mar.	996.1	26.1	10.9	39.0	11.6	27.5	118.5	118.0
Apr.	997.8	24.3	10.6	38.0	10.8	24.5	83.2	90.0
May	999.8	22.6	10.8	36.3	5.4	21.6	66.5	104.0
June	1000.6	21.2	10.5	35.8	0.8	19.4	32.1	69.5
July	1001.7	21.0	11.7	36.3	3.8	17.3	17.8	73.6
Aug.	1000.2	23.1	12.1	38.6	5.8	17.7	24.1	98.7
Sept.	997.9	24.8	12.1	40.6	6.2	20.6	66.1	77.8
Oct.	996.6	25.7	12.1	41.8	10.0	22.7	103.3	114.3
Nov.	995.0	26.6	12.0	40.2	11.0	25.3	122.4	104.0
Dec.	994.5	26.8	11.3	41.0	12.0	27.5	162.9	131.0
Annual	997.5	24.6	11.3	41.8	0.8	23.4	1121.3	180.0

Month	Mean evap. (mm)	Number of days with				Mean cloud- iness (tenths)	Mean sun- shine (h)	Wind	
		precip. >0.1 mm	thunder- storm	fog	haze			preval. direct.	mean speed (m/sec)
Jan.	72.8	13	5	1	0	5.9	218.3	C–E	1.2
Feb.	66.0	12	5	0	0	5.6	198.9	C–SE	1.3
Mar.	70.3	11	4	0	1	5.2	216.9	C–SE	1.3
Apr.	67.7	7	2	0	0	4.5	217.9	C–S	1.4
May	74.7	5	1	1	0	4.2	221.5	C–E	1.5
June	71.6	3	1	1	0	4.4	209.7	C–E	1.6
July	97.3	2	1	0	2	3.4	244.1	E	1.8
Aug.	131.5	2	1	0	7	3.2	255.8	C–E	1.9
Sept.	126.7	5	3	0	9	4.2	219.1	C–E	1.8
Oct.	120.0	8	4	0	3	4.8	229.3	C–E–S	1.6
Nov.	102.5	9	4	0	1	4.9	236.6	C–E–S	1.5
Dec.	82.0	11	5	0	1	5.5	227.7	C–E	1.3
Annual	1083.1	88	36	3	24	4.6	2695.8	C–E	1.5

TABLE XX

Latitude 20°47′S, longitude 51°42′W, elevation 312 m

Month	Mean sta. press. (mbar)	Temperature (°C)				Mean vap. press. (mbar)	Precipitation (mm)	
		daily mean	mean daily range	extreme			mean	max. in 24 h
				max.	min.			
Jan.	973.0	25.4	10.8	40.0	12.5	26.2	201.7	86.0
Feb.	973.6	25.5	10.7	39.0	14.5	26.7	168.3	75.1
Mar.	974.5	25.1	11.3	39.0	9.8	25.8	128.6	72.0
Apr.	976.1	23.8	11.7	38.8	8.0	23.7	90.3	68.8
May	978.0	20.8	12.8	35.6	3.8	19.8	62.0	73.4
June	979.1	19.5	13.5	37.0	−0.8	18.0	49.6	55.0
July	980.0	19.2	14.9	35.5	1.5	16.7	20.7	42.5
Aug.	978.4	20.6	15.8	38.7	3.5	17.3	35.2	57.0
Sept.	976.4	22.6	14.0	41.0	4.1	19.2	73.3	57.4
Oct.	974.9	24.1	13.0	39.3	8.2	21.3	109.3	57.0
Nov.	972.9	25.0	12.2	39.5	9.8	23.4	146.7	73.6
Dec.	972.2	25.3	10.6	39.8	14.0	25.4	199.6	69.1
Annual	975.8	23.1	12.7	41.0	−0.8	22.0	1285.3	86.0

Month	Mean evap. (mm)	Number of days with				Mean cloud-iness (tenths)	Mean sun-shine (h)	Wind	
		precip. >0.1 mm	thunder-storm	fog	haze			preval. direct.	mean speed (m/sec)
Jan.	67.2	15	6	0	0	5.9	206.0	NE	2.1
Feb.	55.7	13	5	0	0	5.8	189.3	NE	2.0
Mar.	71.0	10	5	1	0	5.0	229.2	NE	1.9
Apr.	65.9	6	2	2	0	4.2	226.9	NE	1.9
May	59.9	4	1	4	0	3.5	242.0	SE	1.8
June	61.6	4	2	3	0	3.7	229.4	NE	1.8
July	83.0	3	0	3	0	3.2	245.1	C–NE	1.8
Aug.	100.8	3	2	1	6	3.5	254.3	C–NE	2.0
Sept.	93.8	6	3	1	7	4.8	195.3	C–SE	2.1
Oct.	104.6	8	5	1	3	5.1	213.8	SE	2.1
Nov.	89.3	10	4	0	0	5.2	222.5	SE	2.0
Dec.	70.0	14	6	0	0	6.2	195.9	NE	2.2
Annual	922.8	96	41	16	16	4.7	2649.7	NE	2.0

TABLE XXI

CLIMATIC TABLE FOR BELO HORIZONTE
Latitude 19°56'S, longitude 43°56'W, elevation 915 m

Month	Mean sta. press. (mbar)	Temperature (°C)				Mean vap. press. (mbar)	Precipitation (mm)	
		daily mean	mean daily range	extreme			mean	max. in 24 h
				max.	min.			
Jan.	913.7	22.5	9.3	35.3	12.6	21.5	319.4	154.9
Feb.	914.2	22.8	9.6	34.2	12.2	21.5	201.6	105.4
Mar.	915.0	22.4	9.7	32.5	11.8	21.1	157.2	110.2
Apr.	915.8	21.3	10.3	32.8	10.2	19.5	79.4	73.4
May	917.3	19.2	11.4	32.9	2.5	16.7	20.2	35.2
June	917.9	17.7	12.5	30.8	2.4	15.1	8.5	35.6
July	919.7	17.2	12.3	30.0	3.4	14.1	8.4	21.2
Aug.	918.3	18.9	13.0	34.0	5.8	14.8	18.2	32.8
Sept.	916.4	20.5	11.8	34.2	7.5	17.0	42.2	56.3
Oct.	914.8	21.4	10.4	35.2	9.8	18.5	135.5	75.1
Nov.	913.3	21.7	9.3	35.2	12.0	19.9	225.3	78.1
Dec.	913.1	21.9	8.4	35.5	12.6	21.5	345.6	87.9
Annual	915.8	20.6	10.7	35.5	2.4	18.4	1561.5	154.9

Month	Mean evap. (mm)	Number of days with				Mean cloudiness (tenths)	Mean sunshine (h)	Wind	
		precip. >0.1 mm	thunderstorm	fog	haze			preval. direct.	mean speed (m/sec)
Jan.	69.5	17	8	3	2	6.4	175.9	C–NE	1.1
Feb.	64.8	13	6	4	2	5.9	171.0	C–NE	1.1
Mar.	71.0	13	6	3	2	5.7	192.9	C–NE	1.1
Apr.	64.9	7	3	4	3	4.6	211.9	C–NE	1.0
May	68.4	3	1	7	6	3.7	245.3	C–NE	1.0
June	67.9	1	0	7	7	3.2	244.8	C–NE	1.0
July	70.8	2	0	7	6	3.2	259.7	C–NE	1.2
Aug.	86.4	2	1	5	8	2.8	259.1	C–NE	1.4
Sept.	89.9	5	2	4	8	4.1	204.2	C–NE	1.5
Oct.	85.6	11	5	1	4	5.6	182.8	C–NE	1.5
Nov.	70.8	15	7	2	1	6.6	163.3	C–NE	1.3
Dec.	62.9	20	7	1	2	7.4	126.9	C–NE	1.2
Annual	872.9	109	46	48	51	4.9	2437.8	C–NE	1.2

TABLE XXII

CLIMATIC TABLE FOR CAMPINAS
Latitude 22°53′S, longitude 47°05′W, elevation 663 m

Month	Mean sta. press. (mbar)	Temperature (°C)				Mean vap. press. (mbar)	Precipitation (mm)	
		daily mean	mean daily range	extreme			mean	max. in 24 h
				max.	min.			
Jan.	937.9	22.4	10.7	34.8	10.5	20.4	240.6	110.5
Feb.	938.8	22.3	9.9	35.8	10.4	20.4	198.7	88.8
Mar.	939.4	21.9	10.7	33.3	11.5	19.6	148.4	83.0
Apr.	940.8	20.0	11.4	32.8	4.3	17.1	60.8	85.0
May	942.4	17.5	12.2	31.5	1.5	15.0	56.2	74.0
June	943.6	16.1	12.4	35.9	−1.5	13.3	52.7	70.0
July	943.9	16.2	14.1	30.9	0.2	12.4	28.7	81.0
Aug.	943.1	17.0	13.6	33.0	0.2	12.8	35.6	39.5
Sept.	941.5	18.8	12.6	35.6	1.8	14.5	75.3	100.0
Oct.	939.9	20.1	11.8	36.7	5.2	15.7	121.2	79.6
Nov.	938.2	21.9	11.4	35.5	8.0	18.2	159.7	97.0
Dec.	938.0	22.1	10.9	36.7	9.5	18.7	215.2	97.5
Annual	940.6	19.7	11.8	36.7	−1.5	16.5	1393.1	110.5

Month	Mean evap. (mm)	Number of days with			Mean cloudiness (tenths)	Mean sunshine (h)	Wind	
		precip. >0.1 mm	thunderstorm	fog			preval. direct.	mean speed (m/sec)
Jan.	41.5	17	9	1	6.6	216.9	C–SE	1.0
Feb.	35.7	15	9	3	6.6	177.9	C–SE	1.3
Mar.	42.9	12	7	3	5.5	220.9	C–SE	1.4
Apr.	45.4	6	3	5	4.3	230.4	C–SE	1.4
May	41.1	6	2	6	4.1	250.8	C–SE	1.1
June	34.7	6	1	8	4.3	219.1	C–SE	1.1
July	38.8	3	1	8	3.6	249.2	C–SE	1.2
Aug.	52.3	5	2	8	3.9	251.1	C–SE	1.4
Sept.	52.8	8	4	5	5.7	199.1	C–SE	1.7
Oct.	52.8	10	5	2	6.0	219.7	C–SE	1.9
Nov.	48.5	12	6	1	5.8	223.2	C–SE	1.8
Dec.	44.0	16	8	1	6.2	205.0	C–SE	1.7
Annual	530.5	116	57	51	5.2	2663.3	C–SE	1.4

TABLE XXIII

CLIMATIC TABLE FOR CURITIBA
Latitude 25°26′S, longitude 49°16′W, elevation 949 m

Month	Mean sta. press. (mbar)	Temperature (°C)				Mean vap. press. (mbar)	Precipitation (mm)	
		daily mean	mean daily range	extreme max.	min.		mean	max. in 24 h
Jan.	910.8	20.1	10.5	34.3	7.5	19.2	182.7	79.0
Feb.	911.6	20.1	10.3	33.4	7.0	19.4	148.9	68.0
Mar.	912.6	19.2	10.1	33.2	5.5	18.5	105.5	67.9
Apr.	914.1	17.1	10.2	30.3	0.8	16.2	75.5	63.6
May	914.8	14.3	10.9	28.4	−3.6	13.4	88.1	97.6
June	915.5	12.9	11.0	26.4	−6.3	12.3	104.2	163.5
July	916.7	12.1	12.0	28.1	−6.2	11.4	69.3	86.0
Aug.	915.5	13.5	12.1	30.7	−4.5	12.1	84.6	57.7
Sept.	914.2	14.5	10.8	31.0	−1.9	13.4	123.8	93.0
Oct.	917.4	15.9	10.5	33.4	1.5	14.5	122.1	111.9
Nov.	911.0	17.7	11.0	34.4	5.3	16.1	119.9	70.7
Dec.	910.5	19.3	10.9	34.3	6.2	17.9	138.3	80.0
Annual	913.7	16.4	10.9	34.4	−6.3	15.4	1362.9	163.5

Month	Mean evap. (mm)	Number of days with				Mean cloud-iness (tenths)	Mean sun-shine (h)	Wind	
		precip. >0.1 mm	thunder-storm	fog	haze			preval. direct.	mean speed (m/sec)
Jan.	72.4	20	9	2	2	7.3	174.6	C–E	2.6
Feb.	63.0	17	6	4	1	7.1	166.7	C–E	2.4
Mar.	62.3	17	5	6	1	7.0	167.4	C–E	2.5
Apr.	55.3	14	3	7	2	6.7	160.8	C–E	2.2
May	54.5	12	2	10	3	6.1	167.7	C–NE–NW	2.1
June	48.9	11	2	8	3	6.0	153.6	C–NW	2.0
July	57.7	10	1	8	6	5.6	184.9	C–NE	2.1
Aug.	71.4	10	3	6	9	5.6	184.2	C–NE–E	2.4
Sept.	62.7	14	4	4	8	6.9	144.1	E	2.8
Oct.	71.3	14	4	2	6	7.2	163.9	E	2.9
Nov.	75.9	14	5	2	5	7.1	178.2	E	3.0
Dec.	79.5	16	5	2	3	7.3	186.7	E	2.9
Annual	774.9	169	49	61	49	6.7	2032.8	C–E	2.5

TABLE XXIV

CLIMATIC TABLE FOR RIO DE JANEIRO
Latitude 22°54'S, longitude 43°10'W, elevation 30.5 m

Month	Mean sta. press. (mbar)	Temperature (°C)				Mean vap. press. (mbar)	Precipitation (mm)	
		daily mean	mean daily range	extreme			mean	max. in 24 h
				max.	min.			
Jan.	1004.1	25.1	5.8	39.1	15.5	24.9	156.6	171.6
Feb.	1004.9	25.6	5.9	37.8	17.0	25.7	124.7	135.8
Mar.	1005.8	24.3	5.8	36.4	17.6	24.2	134.3	143.7
Apr.	1008.1	23.6	5.6	35.0	15.3	22.9	101.5	223.0
May	1009.9	22.1	6.0	35.2	13.8	20.9	63.2	216.6
June	1011.7	21.1	6.3	32.6	10.9	19.6	56.4	205.7
July	1013.0	20.2	6.5	34.1	11.3	18.0	50.8	58.5
Aug.	1011.7	20.8	6.7	36.4	11.5	18.8	39.7	50.9
Sept.	1009.6	21.0	6.0	37.6	10.2	19.4	63.4	71.6
Oct.	1007.8	21.8	5.4	39.0	13.4	21.0	80.2	51.5
Nov.	1005.5	22.9	5.5	37.5	15.0	22.1	92.2	98.5
Dec.	1004.3	22.4	6.1	39.0	13.4	21.1	129.9	151.3
Annual	1008.1	22.7	6.0	39.1	10.2	17.4	1092.9	223.0

Month	Mean evap. (mm)	Number of days with				Mean cloud-iness (tenths)	Mean sun-shine (h)	Wind	
		precip. >0.1 mm	thunder-storm	fog	haze			preval. direct.	mean speed (m/sec)
Jan.	108.1	12	7	10	4	6.3	221.7	C–SSE	3.1
Feb.	96.7	12	5	10	8	5.8	206.3	C–SSE	3.0
Mar.	99.3	13	5	12	8	6.0	215.2	C–SSE	3.2
Apr.	89.1	11	2	15	7	5.7	207.2	C–SSE	2.7
May	92.0	10	1	18	12	5.4	209.7	C–SSE	2.8
June	89.2	8	0	19	10	5.1	194.4	C–SSE	2.7
July	95.1	7	0	19	10	4.8	209.2	C–SSE	2.7
Aug.	102.5	8	1	19	14	5.3	204.8	SSE	3.1
Sept.	89.8	11	2	15	14	7.0	153.1	C–SSE	3.5
Oct.	91.6	12	3	11	10	7.0	151.4	SSE	3.9
Nov.	98.5	13	4	9	7	6.7	180.9	SSE	4.0
Dec.	106.1	14	5	7	5	6.6	197.4	SSE	3.8
Annual	1158.0	131	35	164	109	6.0	2351.3	C–SSE	3.2

TABLE XXV

CLIMATIC TABLE FOR ALEGRETE
Latitude 29°46′S, longitude 55°47′W, elevation 104 m

Month	Mean sta. press. (mbar)	Temperature (°C)				Mean vap. press. (mbar)	Precipitation (mm)	
		daily mean	mean daily range	extreme			mean	max. in 24 h
				max.	min.			
Jan.	997.8	24.7	12.8	39.4	8.2	20.0	156.3	100.4
Feb.	998.3	24.0	12.3	38.3	10.0	21.6	124.7	82.0
Mar.	999.4	22.3	12.2	38.6	4.3	20.1	147.7	105.4
Apr.	1001.5	18.0	11.2	36.2	0.8	16.2	191.3	145.0
May	1003.5	15.0	10.7	30.6	−0.6	13.4	171.4	145.0
June	1003.2	13.7	10.0	29.5	−5.0	12.7	109.3	124.1
July	1004.6	12.9	10.6	30.4	−5.0	11.8	109.5	74.3
Aug.	1004.0	14.2	11.3	31.4	−2.4	12.5	112.7	86.8
Sept.	1002.5	15.6	11.7	35.7	−0.6	13.4	140.2	94.1
Oct.	1001.4	17.6	11.7	34.3	0.7	15.0	165.6	109.8
Nov.	1000.4	20.9	12.7	38.8	3.0	17.2	122.5	96.2
Dec.	997.2	23.5	13.0	38.6	7.2	19.8	130.8	153.6
Annual	1001.1	18.6	11.7	39.4	−5.0	16.1	1682.0	153.6

Month	Mean evap. (mm)	Number of days with				Mean cloud-iness (tenths)	Mean sun-shine (h)	Wind	
		precip. >0.1 mm	thunder-storm	fog	haze			preval. direct.	mean speed (m/sec)
Jan.	160.0	9	6	0	10	4.6	276.7	C–E	4.5
Feb.	122.9	8	6	1	7	4.7	253.1	C–E	4.1
Mar.	115.8	9	6	1	10	4.8	244.3	C–E	4.0
Apr.	77.6	9	5	2	6	4.7	204.2	C–E	3.9
May	60.7	10	5	3	4	5.5	183.9	C–E	3.9
June	50.4	11	4	4	3	6.1	151.9	C–E	3.9
July	62.5	11	5	3	7	5.6	169.3	C–E	4.1
Aug.	84.4	10	5	2	10	5.3	189.6	C–E	4.5
Sept.	93.0	9	5	1	8	5.4	195.5	C–E	4.9
Oct.	106.4	10	6	1	7	5.1	226.9	C–E	4.7
Nov.	128.6	8	5	1	8	4.2	258.3	C–E–SE	4.5
Dec.	155.0	8	6	0	9	4.7	288.5	C–E	4.3
Annual	1217.3	112	64	19	89	5.1	2642.2	C–E	4.3

TABLE XXVI

CLIMATIC TABLE FOR PORTO ALEGRE
Latitude 30°02′S, longitude 51°13′W, elevation 10 m

Month	Mean sta. press. (mbar)	Temperature (°C)				Mean vap. press. (mbar)	Precipitation (mm)	
		daily mean	mean daily range	extreme			mean	max. in 24 h
				max.	min.			
Jan.	1008.6	24.7	10.7	39.2	10.4	22.1	92.7	145.0
Feb.	1009.5	24.3	10.3	40.4	11.3	22.4	90.4	67.5
Mar.	1010.9	23.0	10.2	38.9	9.0	21.2	91.1	63.8
Apr.	1012.9	20.2	9.8	35.9	5.8	18.7	108.5	115.2
May	1014.5	17.1	9.4	32.1	0.4	15.8	124.6	70.3
June	1015.1	14.8	9.3	31.4	−2.0	13.8	127.8	91.1
July	1016.8	14.3	9.4	32.9	−1.5	13.2	127.4	86.5
Aug.	1015.8	15.1	9.6	33.3	−0.9	13.7	122.7	92.2
Sept.	1014.3	16.5	9.1	36.0	2.4	14.5	133.6	103.6
Oct.	1012.8	18.6	9.5	35.3	4.7	16.1	107.1	100.5
Nov.	1010.1	21.0	10.6	38.0	6.4	17.6	85.4	92.2
Dec.	1008.5	23.6	11.0	39.6	7.8	20.3	86.2	86.4
Annual	1012.5	19.5	9.9	40.4	−2.0	17.4	1297.5	145.0

Month	Mean evap. (mm)	Number of days with				Mean cloud-iness (tenths)	Mean sun-shine (h)	Wind	
		precip. >0.1 mm	thunder-storm	fog	haze			preval. direct.	mean speed (m/sec)
Jan.	93.4	10	4	2	4	5.1	247.5	SE	1.3
Feb.	78.6	8	3	3	4	5.2	223.3	SE	1.1
Mar.	71.7	9	3	4	4	5.0	209.0	SE	1.0
Apr.	52.5	10	2	7	4	5.3	183.9	SE	0.8
May	42.8	11	2	8	4	5.5	172.0	SE	0.9
June	32.2	11	3	7	3	5.8	149.5	SE	0.6
July	36.5	10	3	8	3	5.6	160.9	SE	0.8
Aug.	42.3	11	4	7	4	5.9	165.1	SE	1.1
Sept.	49.7	11	4	4	4	6.0	146.1	SE	1.3
Oct.	65.9	10	3	3	4	5.8	194.8	SE	1.5
Nov.	78.1	9	3	2	5	5.1	234.5	SE	1.4
Dec.	89.4	9	4	2	4	5.0	258.3	SE	1.4
Annual	733.1	119	38	57	47	5.4	2344.9	SE	1.1

Chapter 6

The Climate of Northern South America

J. W. SNOW

Preliminary remarks

The area and climate in general

The area treated in this chapter is essentially all of South America north of the Equator which therefore serves as a very approximate but convenient southern boundary. The equatorial North Atlantic, North Pacific and the Caribbean Sea constitute nearly all of the remaining boundary. The area lies well within the Tropics, the extreme latitude being only 12°N. The climate is everywhere tropical in the sense that: (*1*) the mean annual temperature is relatively high, ≥ 25°C at sea level; (*2*) the diurnal range of temperature exceeds the annual range (defined as the difference between the mean temperatures of the warmest and coolest months); and (*3*) the annual range of temperature is minimal, specifically, < 5°C. The overall tropical climate of northern South America can be divided into five types which are easily recognized from the annual courses of rainfall and temperature: two marine, two continental and one mountain type.

Considering that coastline constitutes over half of the total boundary length and is found on three sides, it is not surprising that marine conditions are common. Two wet and two dry periods are distinguishable in this type of climate which is found up to 400 km inland. Furthermore, subtypes of wet marine and dry marine are present. The wet marine climate normally receives rainfall in excess of 2,000 mm/yr. (occasionally much in excess), and the annual course of mean monthly temperature manifests two maxima occurring during the dry periods. All of the Guianas, except for southwest Surinam and the southern one-third of Guyana, are of the wet marine type. Also the perennially rainy Pacific coast of Colombia shows these characteristics, though weakly. The dry marine areas rarely receive rainfall in excess of 1,500 mm/yr. and the course of temperature has a single maximum occurring during the high-sun (Northern Hemisphere summer in the present case) dry period. The Caribbean portions of Venezuela and Colombia, north and west of the mountain barriers, possess this type.

Low elevation ($H < 300$ m) characterizes the majority of the area of northern South America and much of this is not marine, in the sense outlined. The vast Orinoco basin in central Venezuela and adjacent eastern Colombia, and the smaller area of low savanna in southern Guyana both experience a single, very prominent wet season, occurring during the high-sun months, and a single dry season during low-sun. The course of mean monthly temperature has two maxima, one before and one after the wet season. Since this combination of rainfall and temperature patterns is found in a large part of the low interior it is appropriately called continental.

Additionally, the area under consideration contains two major orographic features, the Guyana Highlands in southeastern Venezuela, and the northern portion of the Andes Mountains in western Colombia and northwestern Venezuela. Within these parts the typical vertical stratification and variety of mountain climate is found, even to the extreme of some glaciered peaks. However, throughout the elevated regions the restricted annual range of temperature and its relationship to the diurnal range, referred to above, prevail and thus the mountain climate here is differentiated from that occurring at higher latitudes; it too is tropical. Also characteristic of the mountain areas in northern South America are rainfall and temperature patterns which both have double maxima. Interestingly, the maxima in mean monthly temperature occur nearly in conjunction with the rainy periods of the year, the minima during the dry periods. This mountain type is typically confined to elevations above 1,000 m.

The fifth climate type is found over low elevation interior areas at very low latitudes. It is characterized by abundant rainfall, usually in excess of 3,000 mm/yr. Two especially wet periods are distinguishable but the temperature profile has essentially one maximum which occurs during the high-sun months of the Southern Hemisphere. The southernmost interior of Colombia and Venezuela is the only area within the countries here treated which has this wet continental climate.

In summary, the climate is universally tropical but wet marine, dry marine, continental, mountain and wet continental regions can be differentiated based upon the month-to-month courses of rainfall total and mean temperature.

Rainfall is the basis for the specification of seasons and not temperature. Sociologically and from the point of view of atmospheric science this is entirely appropriate. The annual range of temperature is small and human activity is therefore influenced more by rainfall, which for most locations does have a large annual variation. The meridional translation of the equatorial trough essentially determines the variation in the area's large-scale atmospheric circulation during the course of the year. The presence of the trough within a particular area connotes the wet season(s).

"Equatorial trough" means, in general, the latitudinal (zonal) belt of relatively low surface pressure lying between the subtropical highs of the hemispheres. More particularly, it is the 1,000–2,000 km wide area of the tropical atmosphere wherein the trades (trade winds), which dominate the lowest 2,000–3,000 m of the atmosphere in the outer portions of the trough, are decreasing in speed as they approach the center of the trough. The mean annual position of the trough center is about 5°N, the so-called "meteorological equator" (HUSCHKE, 1959). The trough translates meridionally, following but lagging the zenithal position of the sun. The annual range of translation appears to be greater over land (∼ 20° of latitude) than over ocean (< 10°). Sufficient moisture is usually present in the levels between 3,000 and 7,000 m to sustain clouds. However, apart from local modifications, convection is concentrated within relatively small areas, often but not necessarily having the shape of zonal lines, which are themselves very transient. Only in a time-averaged view, such as given in Fig. 1, is the preferred location of these especially convergent features readily apparent; this is the center of the equatorial trough. The term "intertropical convergence" (I.T.C.) is variously applied either to the transient areas or to their time-averaged preferred location. In either sense, I.T.C. is not equivalent to equatorial trough. It is only a part, albeit a very active part, of the trough. Because of this ambiguity, use of the term I.T.C. is minimized in this chapter and phrases such as "center-

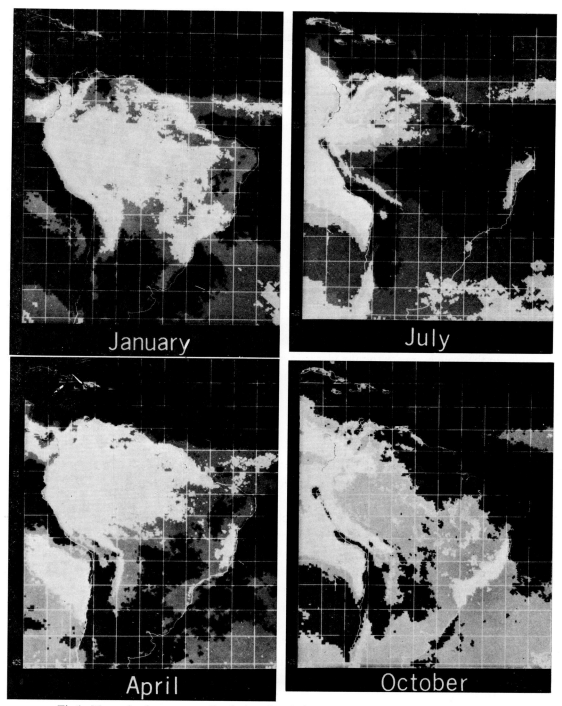

Fig.1. Mean cloud cover over South America during mid-afternoon. Five gradations of grey scale—octal correspondence: black—0 through 2; dark grey—3; grey—4; light grey—5; white—6 through 8. Period of mean: 1 January 1967 through 31 December 1970. Data source: United States meteorological satellites. (Modified from original charts prepared and presented by D. B. Miller, 1971.)

line of the equatorial trough" or "especially convergent area" are used to refer to the time-averaged or one-time view, respectively, of the convection within the trough.

The absence of the trough, resulting from its movement northward or southward, allows domination by the trade winds. "Trade winds" refers to that portion of the atmosphere lying between the subtropical highs and the equatorial trough. Moisture and therefore cloudiness is typically confined below 3 km, sometimes well below, where the very steady trades blow obliquely toward the Equator; above, subsidence and varying winds prevail. The distinction between trade winds and equatorial trough is based upon the downstream speed change of the trades. Presence of the "trade winds" implies downstream acceleration and therefore reinforced subsidence. Such a condition is naturally associated with the dry season(s).

Within the lowest latitudes of the Northern Hemisphere, which includes most of the area here being dealt with, it is expected that two transits of the center of the equatorial trough will occur annually. The resulting annual rainfall pattern of two wet and two dry (or at least drier) seasons is generally found near the coasts, in the mountains and within some of the interior low elevation areas which are very close to the Equator. However, it is completely absent within the majority of the interior lowlands.

Fig.1 reveals certain of these features. (It should be kept in mind that the mean cloud conditions portrayed are for the early to mid-afternoon.) The most ideal equatorial trough in the area of the figure, both in regard to meridional translations and attendant cloudiness, is found in the equatorial North Atlantic. Its centerline is readily apparent near 4°N in January. Over the continental lowlands major excursions and alternations in the trough appear to take place; in particular note the western Llanos (approximately 4°–10°N, 65°–73°W) in January versus July and the reverse situation simultaneously occurring over much of the Amazon basin.

Chapter outline and terminology

The climate of northern South America is treated by country. Although this procedure is artificial since climate does not conform to political borders, it is followed for two reasons.

(*1*) The amount, type and quality of climatic data available from each of the five countries varies widely. Thus, uniform coverage of the entire area would allow the least satisfactory data to determine the level of treatment. By dealing with each country separately the level can be adjusted to make fuller use of all data. The advantage is that some insight into the climate of data-sparse regions can be gained from the study of climatically similar areas having better data coverage in neighboring countries.

(*2*) In many instances the reader will be interested only in a particular country, and his needs are better met. It is worthy of note, however, that certain generally applicable topics are treated more thoroughly in the Surinam and Venezuela sections than elsewhere. The order of coverage proceeds from east to west: French Guiana, Surinam, Guyana, Venezuela, Colombia. The general outline followed for each country or section thereof is similar: brief geographical description, climatic controls and/or factors, rainfall, other climatic elements, topics of special interest. Rainfall is the most widely measured and discussed element. Its role as determinant of seasons justifies its singular treatment.

Climatic tables are provided in the Appendix at the end of the chapter for the stations shown by circles in Fig.2 and listed by decreasing Northern Hemisphere latitude in Table I. Wherever in the text other stations are cited for the first time, the coordinates and elevation are there given.

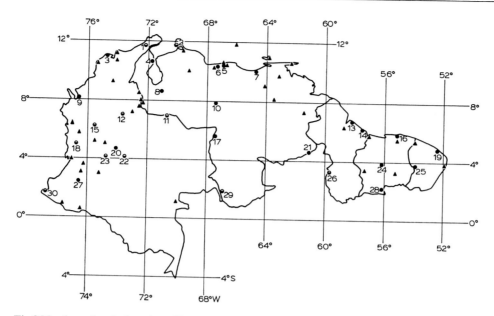

Fig.2.Northern South America. Climatic tables included for stations designated by circles; solid circle indicates full table, half-solid circle partial table. Numbering corresponds to listing in Table I. Triangles indicate additional stations for which climatic tables have been prepared and are on file at Dept. of Meteorology, University of Wisconsin, Madison, Wisc., U.S.A.

Certain terms, in addition to equatorial trough and trade winds discussed above, are here clarified. "Temperature", unless otherwise specified, refers to air temperature at normal instrument shelter height. The "true mean" of any intensive climatic element denotes the mean of 24-hourly values. Whenever relative humidity is mentioned it is an estimate of the true-mean value unless otherwise stated. "Precipitation", "rain" and "rainfall" are all taken to have the same meaning since frozen hydrometeors are confined to a few mostly uninhabited mountain areas. The vegetation cover in northern South America is described using three basic terms: "steppe", "savanna" and "rain forest". The first is taken to mean grass and/or low bush cover without trees. By "savanna" or "tropical savanna" is meant grassland with at most scattered trees and/or narrow forest bands immediately adjacent to river courses. "Rain forest" is used in a generic sense referring to any and all of the three traditional tropical forest types: rain forest, monsoon forest and savanna forest (RICHARDS, 1964). In most cases, rain forest so defined is equivalent to the more common term "jungle".

French Guiana

Geography

Of the three Guianas, French Guiana is the easternmost, smallest, lowest and wettest. Its area is nearly 100,000 km² and is located between 2° and 6°N, 52° and 54.5°W. On the east and south it is bounded by Brazil, on the west by Surinam, and its northeast boundary is 330 km of equatorial Atlantic coast. Three sections, differentiated basically by

TABLE I

INDEX OF STATIONS FOR WHICH CLIMATIC TABLES ARE INCLUDED (IN ORDER OF DECREASING LATITUDES)

Sta. no.	Station name	Country	Coordinates	Elev. (m)
1	Uribia	Colombia	11°44′N 72°14′W	23
2	Las Piedras	Venezuela	11°42′N 70°12′W	10
3	Barranquilla (Granja Exptl.)	Colombia	10°57′N 74°47′W	13
4	Maracaibo (Aeropuerto)	Venezuela	10°41′N 71°39′W	40
5	Caracas (Obs. Cagigal)	Venezuela	10°30′N 66°56′W	1,035
6	Colonia Tovar	Venezuela	10°25′N 67°17′W	1,790
7	Barcelona	Venezuela	10°07′N 64°41′W	7
8	Mérida (Aeropuerto)	Venezuela	08°35′N 71°10′W	1,495
9	Turbo	Colombia	08°06′N 76°44′W	2
10	San Fernando (Aeropuerto)	Venezuela	07°53′N 67°26′W	73
11	Arauca	Colombia	07°04′N 70°40′W	122
12	Barrancabermeja	Colombia	07°04′N 73°52′W	107
13	Georgetown (Botanic Gardens)	Guyana	06°48′N 58°08′W	2
14	New Amsterdam	Guyana	06°15′N 57°31′W	2
15	Medellín (Fac. de Agronomía)	Colombia	06°15′N 75°35′W	1,450
16	Paramaribo (Cultuurtuin)	Surinam	05°51′N 55°10′W	3
17	Puerto Ayacucho	Venezuela	05°41′N 67°38′W	99
18	Andagoya	Colombia	05°06′N 76°40′W	65
19	Rochambeau (Cayenne)	French Guiana	04°50′N 52°22′W	8
20	Bogotá (Obs. Met. Nac.)	Colombia	04°38′N 74°05′W	2,556
21	Santa Elena	Venezuela	04°36′N 61°07′W	907
22	Villaviciencio	Colombia	04°09′N 73°36′W	423
23	Espinal (C.C. de Tabaco)	Colombia	04°09′N 74°53′W	322
24	Tafelberg	Surinam	03°47′N 56°03′W	344
25	Maripasoula	French Guiana	03°38′N 54°02′W	104
26	St. Ignatius	Guyana	03°21′N 59°48′W	99
27	La Florida (Mpio. de Popayan)	Colombia	02°27′N 76°35′W	1,789
28	Sipaliwini	Surinam	02°02′N 56°07′W	253
29	San Carlos de Rio Negro	Venezuela	01°54′N 67°03′W	95
30	Tumaco	Colombia	01°49′N 78°47′W	4

distance from the coast, are outlined in Fig. 3. The coastal fringe is a coast-parallel band, 10–20 km wide made up of mangrove swamps, spots of open marshland and, farther from the coastline, rain-forest covered low (⩽10 m) coastal plains. Southeast of Cayenne a series of small ridges (< 300 m) are located. The inland and interior sections also have remarkably low elevations, with only small central areas and the western portion of the southern border rising above 200 m. The inland section is flat and densely covered with rain forest especially to the east. Toward the west some areas of savanna are found in the midst of rain forest (similar to that occurring in the region of the lower Berbice river in Guyana). The interior section consists of gently sloping hills, only the most exceptional of which extends above 300 m. Here too rain forest is the predominant vegetative cover. Numerous small south-to-north flowing rivers carry the excess of the copious rainfall to the ocean.

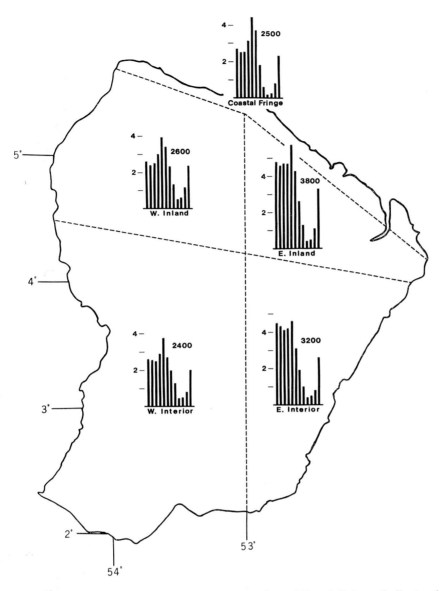

Fig.3. Annual distribution of rainfall for various sections of French Guiana. Ordinate values in mm/100, annual totals in mm. Based on period 1931–60 (certain data subjected to reduction).

Climate

General features and controls

In effect, two seasons comprise the climate of French Guiana; wet, from mid-November through mid-August, and dry during the remaining 13 weeks. The annual variation in climatological variables other than rainfall is minimal. Therefore, with the long wet season, the climate does resemble the popular (but not always appropriate) conception of the humid tropics, specifically invariant, high temperature and humidity, light winds, and abundant rainfall. The province is the landfall of the "meteorological equator", the annual mean latitude of the equatorial trough (HUSCHKE, 1959) for the eastern coast of the Americas. Thus, its overall wet and minimally varying climate can be better comprehended. The climate is tropical wet marine, in the sense outlined in the preliminary remarks of this chapter, despite the fact that the characteristic double maximum and minimum in the annual rainfall distribution is not immediately apparent.

The equatorial trough of the western Atlantic is the predominant large scale circulation control. It does translate meridionally and also undergoes an annual variation in the extent and intensity of associated convection. Referring to Fig.1, April is close to the time of maximum convective activity over the western margin of the Atlantic, as the extensive cloudiness near and over the coastal area between 3°S and 7°N testifies. The October conditions within the same area contrast markedly. The translations of the equatorial trough place its most convergent zone over French Guiana twice annually, in May and again in late December and January. During the months that the center is to the south, February through April, it is rarely far enough away to preclude convection sufficiently deep to produce rain, although thunderstorms are then not common.

The trade winds of both hemispheres are present at different times of the year; those with south Atlantic origin being present the longer period, approximately May through December. The switch from one source to the other, attending the passage of the trough centerline, results in a practically solid shift of 35° in the direction of mass flow within the lowest 2 km of the atmosphere (see Table II). In the layer above that level very little change indeed is noted. In the layers below the apparent separation which occurs between 1,800–3,000 m, an onshore component exists year round but with significantly varying strength. When the trough is located to the south, this component is 3 times stronger than during the months June through November. The absence of a secondary dry season, as is common within the other Guianas to the west, is largely attributable to this reinforced shoreward mass flow.

Rainfall

As a province-wide average, the rainfall approximates 2,800 mm/yr., which is received at a rate of roughly 290 mm/month during the wet season and 50 mm/month otherwise. Days-with-rain are numerous during the wet season, 22–27 days/month, and even during September on 7–10 days/month measurable rain occurs. Fig.5 provides further resolution

TABLE II

RESULTANT WINDS FOR CAYENNE, FRENCH GUIANA (10's of degrees and m/sec)

Level (m)	D.J.F.		M.A.M.		J.J.A.		S.O.N.	
3,000	09	5.4	09	6.2	09	8.3	10	6.8
1,800	08	6.4	08	6.9	11	6.8	12	6.5
900	07	7.8	07	8.0	11	6.5	11	5.9
600	06	7.3	06	7.7	10	5.6	10	5.2
Sfc (3)	06	5.9	05	5.5	09	3.2	09	4.3

Period of record: February 1937–August 1940.
Source: U.S. WEATHER BUREAU, 1945.

of the areal distribution. The rainfall total is seen to pass through a ridge-like maximum which clearly parallels the coastline, very close to the seaward boundary of the inland section. Within the coastal fringe, the gradient perpendicular to the coastline is particularly large. Low-level convergence is induced within the predominantly onshore flow as it passes from the ocean, of relatively small surface roughness, onto the mangrove swamps and the dense rain forest which is found along the inner coastal fringe section. The greatly increased surface roughness accounts for the ridge of maximum rainfall. Thermodynamic instability arising through heating of this air by the land probably contributes little to the feature because along the coast nocturnal and early daylight rains are very common. Furthermore, considering the true-mean of temperatures it can be shown that the ocean is actually a heat source for the coastal area nearly all year as is the case too for neighboring Surinam.

The absolute maximum rain total is received over the inland portion of the eastern rivers, particularly the Approuague. On certain hills of the small ridges within the area of absolute maximum, as much as 8,000 mm/yr. is received (FOUGEROUZE, 1965), which very probably gives the area the dubious distinction of the rainiest spot on the Atlantic watershed of the Americas. The fact that the eastern half of the coastline is more nearly perpendicular to the prevailing east to east-northeast wet season winds, contributes to the higher rainfalls there. The broad, low basins of the eastern rivers, especially the border river Oyapock, experience the farthest inland intrusion of large rainfall totals. The reason has some association with the confining of low-level flow between the Serra Lombard, nearby to the east, and the modest highlands in the center of French Guiana.

Counter to the expectation of more rainfall at higher elevation, the interior of the province receives progressively less rainfall as distance from the coast increases. Although the available data are inadequate to confirm the hypothesis, it is likely that an area exists in the southeast which receives not more than 2,000 mm/yr. (CAMPAN, 1959). Assuredly the increases in elevation toward the interior is not abrupt, but were other factors not active, a rainfall increase should nonetheless occur. The first counteracting factor is that the southeast area may find itself on the lee side of the Monts Tumuc-Humac, along the southern border, when the equatorial trough is north of latitude 3°N. Also, especially during the wet season, it appears that evaporation demands such a large portion of the energy available at the surface that sensible heating of the lower atmospheric layer over the interior is frequently insufficient to initiate the deep convection needed. Orographic lifting is likewise insufficient and the inducement of increased surface roughness has had its effect closer to the coast.

The annual march of rainfall for the various portions of the province is shown in Fig. 3. The mid-August to mid-November dry season, called summer (été), is found over the entire province but is best developed close to the coast. Also apparent throughout is the marked May maximum which attends the rapid northward advance of the equatorial trough centerline. Not so evident is a secondary maximum around January and the subsequent "petit été de mars", a 3 to 4 week respite in the long wet season which occurs within the period mid-February to mid-April of most years. The "petit été de mars" is difficult to detect from monthly mean data alone, especially away from the coast. However, there is a real differentiation of the long wet season which is evident when certain other statistics are examined. The monthly values of the coefficient of variation ($100 \, \sigma/\bar{x}$, σ = standard deviation, $\bar{x}$ = mean) is a convenient, although not entirely satisfactory,

TABLE III

COEFFICIENT OF VARIATION (%) OF MONTHLY RAINFALL FOR L'AEROPORT ROCHAMBEAU, 1946–72

Jan.	Feb.	Mar.	Apr.	May	June
35	49	63	50	25	29

July	Aug.	Sep.	Oct.	Nov.	Dec.
22	39	88	70	70	38

measure of the year-to-year consistency of a climatic element. In Table III the monthly values of this parameter are presented for 27 years of rainfall data recorded at Rochambeau (Île de Cayenne).

The greatest value of the coefficient is found during the dry season, as is anticipated, since an approximately constant standard deviation yields a larger coefficient for smaller means. The significant feature, however, is the clear second maximum from February through April. The explanation in this case is not a small value of the average monthly rainfall but rather the high probability of large deviations from the mean during those months when the equatorial trough is furthest removed to the south. From the table data it is not apparent whether the deviations are primarily toward wetter or drier conditions. Inspection of the individual monthly rainfall totals shows that out of the 27 years, 56% were below the mean; during 3 years the negative departure was more than 80% of the mean. The "petit été de mars" is real, but not a very consistent feature of the climate. In fact, the high coefficient also includes isolated, large positive departures, e.g., during March 1971 Rochambeau registered 883 mm, the second highest monthly value in the 27-year record. With such large year-to-year variation it is not surprising that the phenomenon of a secondary dry season appears only weakly in simple mean statistics. Some analysis of variation, or long actual residence in the area, is necessary to appreciate this refinement of French Guiana's climate, especially since rain can be expected on 20–25 days/month during the "petit été de mars". Detailed analyses for other stations reveal a similar pattern of the coefficient as shown in Table III.

Data on the diurnal distribution of rain for 1968 and 1969 indicate that at Rochambeau 55% of the rain total occurs at night (20h00–08h00 local time) during the months November through May, 30% during the other months. Stations farther inland specifically St. Georges (03°53′N 51°48′W, 2 m) and Maripasoula, also receive more nocturnal rain when the trough centerline is to the south, but the percentages of total are only about half the coastal values, 30% and 15%, respectively. Particularly from August through October, clear nights are common.

The annual distribution of days-with-thunderstorms also indicates a change in the manner of rainfall production, depending on which side of the equatorial trough the province lies. During the months January through April (years of record 1951–65), St. Laurent du Maroni (05°30′N 54°02′W, 4 m) experienced only one thunderstorm-day per month while receiving an average rain total of 430 mm/month which occurred on 23 rain-days per month. During July through October 12 thunderstorm-days per month were recorded while only 140 mm/month accumulated on 18 rain-days per month. The reduced rainfall received south of the trough emanates predominantly from afternoon thunderstorms, whereas north of the trough less intense, more frequent and more nocturnal convection produces the abundant rain. It is seen that there is a great difference between the two fundamental seasons in French Guiana.

Other elements

The mean annual temperature is approximately 1.0°C higher at the coast than away from it. The annual true-mean within 1 km of the coast is close to 26.0°C. At a distance of 15 km the value has dropped to 25.5°C, and the minimum for the near-sea-level portion of the province, 25.0°C, occurs within the inland section. In the interior small increases occur due primarily to less daytime cloudiness. However, the slight increase in elevation and the constant availability of water for evaporation counteract this effect in most places and the air temperature for the interior is also approximately 25.0°C.

The annual variation in mean temperature attains a primary maximum and minimum in October and January, but the total range is only 1.0° to 1.5°C. A weak secondary maximum and minimum occur in April and June, respectively. The diurnal range of temperature is greater, but still small, particularly near the ocean. Directly on the coastline the diurnal range is only 4.0°C during the wet season and 5.5°C during the dry. Within the inland and interior areas these values are doubled. Along the coastal fringe 6.5°C and 10.0°C are the comparable diurnal ranges.

Relative humidity is high, being greater than 85% near the coast and 80% away from it during the wet season. 80% and 70% are the respective dry season values. Minimum relative humidity occurs in early afternoon, 60 to 65% during the wet season, 55 to 60% during the dry season (CAMPAN, 1959).

Measured evaporation is low, as little as 700 mm/yr. As an estimate of the total water being brought into the atmosphere from the land surface, this value is deceptively small because the lush vegetation is adding moisture through transpiration at a rate which is comparable.

Surinam

Introduction and geography

The atmospheric circulation feature which primarily determines the annual variation in Surinam's climate is the equatorial trough. The meridional translations of the trough (more detail is given in the preliminary remarks of this chapter) result in the passage of its centerline over Surinam twice annually. Therefore most climatological elements manifest a strong semi-annual component of oscillation. The entire country spends roughly half of the year in the southern "meteorological hemisphere", and half in the northern. The influence of the trough is never totally absent and as a result the amplitudes of the variations, particularly those of temperature and atmospheric moisture content, are small. Rainfall shows the greatest variation during the year and is expectedly the basic determinant of the seasons.

Surinam is at the heart of the American humid tropics and offers singular possibilities for climatic studies of the region. The quality and quantity of climatic data already available and presently being recorded are extraordinary and reflect the country's interest in its own water resources and in the economic impact of its climate. This abundance of climatological information permits here the most detailed treatment of the wet marine climate type occurring throughout the more coastal portions of the Guianas and northward to the delta of the Orinoco river.

Surinam, the middle country of the Guianas, has an area of about 150,000 km² and is located within 2°–6°N and 54°–58°W. Its relief, in the northern half of the country, is mainly low and flat. This half is divided topographically into a flat coastal plain and farther inland low hills. The coastal plain increases in width from 25 km in the east to 80 km in the west (C. W. van Scherpenzeel, personal communication, 1973). Large grassy swamps cover the shoreward portion of the coastal plain while the higher ground, which tends to lie in bands paralleling the coastline, is typically covered by rain forest. From the climatological point of view it is meaningful to designate this shoreward 15–

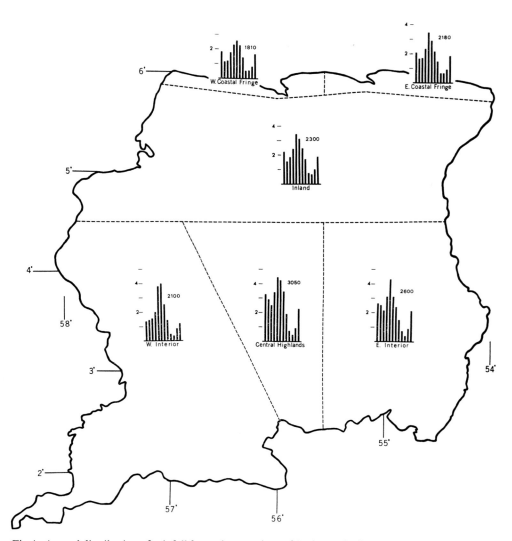

Fig.4. Annual distribution of rainfall for various sections of Surinam. Ordinate values in mm/100, annual totals in mm. Based on period 1931–60 (certain data subjected to reduction).

25 km as the coastal fringe (see Fig.4). A type of quite open tropical savanna is common over the more inland portion of the coastal plain. Rain forests become common over the hilly remainder of the northern half. The southern half of the country displays a gentle rise to 600 or 700 m along the southern border. The overall simple pattern is interrupted by the central highlands which occupy the south-central 18,000 km² and which rise at a few spots to slightly over 1,000 m. These highlands essentially divide the watersheds of the two rivers which constitute Surinam's borders with its neighboring Guianas. Rain forest is the predominant vegetation in the southern half of the country.

Climate

Precipitation

Area-averaged mean annual rainfall for Surinam is close to 2,200 mm, increasing from about 1,900 mm along the coastline to a maximum of near 3,200 mm over the highest, north-facing slopes of the central highlands. Farther southward a decrease takes place, due basically to the increasing distance from the principle moisture source, the tropical Atlantic, without sufficient compensation being provided by increased elevation. The

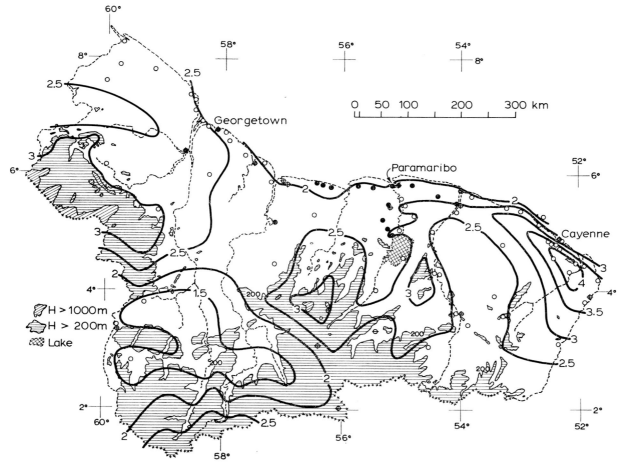

Fig.5. Annual total rainfall (m), the Guiana countries. Mean based on period 1931–60.

decrease is accentuated to the southwest of the central highlands by leeside subsidence and/or by the prevalence of low-level divergence of mass, a topic discussed more fully in the Guyana portion of this chapter. As shown in Fig.5, Surinam's minimum rainfall occurs in the southwest where rather open savanna is not uncommon.

Four seasons can in general be differentiated in Surinam: 2 wet, 2 dry. The "kleine droge-tijd" (short dry season) takes place in February and March and is easily distinguished in the long-term rainfall records of the coastal fringe and inland sections (MINISTERIE VAN OPENBARE WERKEN EN VERKEER, 1968; see Fig. 4). During these months the equatorial trough is at its southernmost position which, over the eastern edge of the continent, may result in a single broad centerline of the trough close to the geographic Equator or in a double I.T.C. (see Fig.1, the April insert especially). Thus, considerable rainfall can occur over the southern half of Surinam where the "kleine drogetijd" is ill defined.

Associated with the northward translation of the equatorial trough and particularly with the passage of its centerline over Surinam in May and very early June, the "grote regen-tijd" (long rainy season) occurs from April through late July. It begins first in the south-east, early in April, spreading thence northward and westward. All areas can expect at least 200 mm per month during this season. May is Surinam's wettest month with an area-averaged mean value of nearly 350 mm. The time of maximum rainfall shows a definite translation from east to west in addition to its basically northward movement. Those sectors receiving the highest annual totals also receive the highest monthly totals. Specifically, the central highlands and the Tapanahoni river basin can anticipate monthly amounts in excess of 400 mm during at least two months of the "grote regentijd".

Maximum recorded 24-h amounts, typically in the neighborhood of 200 mm, have attended this season's initial northward thrust for most recording stations. The trade winds from the North Atlantic are then still vigorous and provide persistent low level inflow and moisture for prolonged, deep convection. (In spite of this, the all-time record for 24-h rainfall is 261 mm which fell at Alliance, 35 km east of Paramaribo, between 08h00 on the 10th and 08h00 on the 11th of February 1954, when the remnants of a Northern Hemisphere cold front probably arrived at the coast.)

The principal dry season, "grote drogetijd", occurs from early August through November. The equatorial trough is then centered over the Atlantic north of Surinam thus placing the entire country in the southern "meteorological hemisphere". Additionally, convec-tive activity within the western Atlantic portion of the equatorial trough is remarkably suppressed during this time of year for reasons which are not yet resolved. Attention is directed to Fig.1, the October insert. September and October are the driest months with under 100 mm per month falling everywhere except over a few localities of the inland section. The typical mean value for the sections of the coastal fringe and inland is 65 mm per month. Over the central highlands and the interior even less, 50 mm per month, is normal. Dessication, even of the rain forests which exist in the central highlands and the east interior sections, can become quite extreme and forest fires have actually occurred there during the latter half of the "grote drogetijd".

December and January constitute the "kleine regentijd" (short rainy season) which follows the southward passage of the equatorial trough centerline. This transit is much more in the nature of a retreat ahead of the primary mass flow, the Northern Hemisphere trade winds, and therefore low level convergence is not as intense as that which accom-panied the trough's earlier northward advance. Monthly rainfall amounts near 200 mm are typical, but negative exceptions are not uncommon and are especially large over the southwest. There the annual course of rainfall becomes continental in the sense outlined earlier in the chapter and bears a strong resemblance to that found in southern Guyana and the Venezuelan Llanos.

Diurnal distribution of rainfall

Fig.6 depicts the percent-of-total rainfall which occurs during each hour of the day at various stations in Surinam. The predominant afternoon maximum is the most notable feature at all the lowland stations. This is caused essentially by the preferential heating of the lower atmosphere through absorption of the solar beam at the surface, and is a common characteristic of land areas in general. A modulation of the primary forcing is indicated by the coherent delay in the time of peaking as distance from the coast increases. Convergence in the lowest layers is reinforced by the daytime sea breeze, the effect of which penetrates further and further inland as the day's heating persists. It is not proposed that the same air which was at the coast in the morning is involved in the 16h00-peak at Afobaka. But, an impulse engendered by the reversal of the land–sea circu-

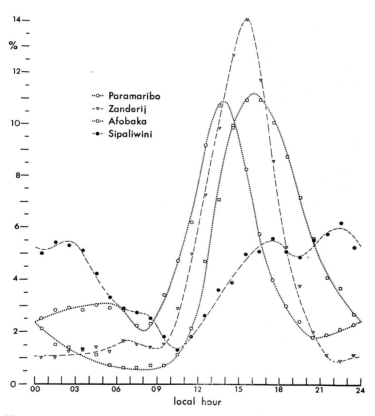

Fig.6. Hourly distribution of rainfall amount (percent of total, average of all months). Station (coordinates and elevation, period of record) kilometers from the coast (source of data): Paramaribo (Table I, 22–33) 12 (BRAAK, 1935). Zanderij (05°27′N 55°12′W, 16 m, 68 and 69) 50 (C. W. van Scherpenzeel, personal communication, 1973). Afobaka (05°00′N 55°00′W, 155 m, 64–66) 100 (HOWELL, 1967). Sipaliwini (Table I, 68 and 69) 400 (GROSH, 1971). Even-weight, 3-member moving averages plotted for stations having a period of record 3 years or less.

lation during the morning does appear to be transmitted inland. The identity of this wave is probably completely lost beyond about 100 km inland, being overwhelmed by the very widespread and often heavy rainshowers of the late afternoon, when solar heating has proceeded to its maximum. The similar transition of the time of minimum rainfall amount during the early daylight hours is also consistent with such a landward propagation. In fact, the entire daytime pattern manifests an evident translation inland with time.

Sunshine data support this notion as Table IV, particularly Paramaribo vs. Zanderij, indicates. The percent-of-possible sunshine during each daylight hour is given, averaged over all months. The approximate time when the percentage changes from above to below the overall daytime average is marked by an "×"; the hour of minimum daytime cloudiness is italicized. At Paramaribo the change to below average percent-of-possible sunshine (or to above average daytime cloudiness) occurs 2 h earlier than at Zanderij. A similar delay between the two stations is found in the time of maximum percent-of-total rainfall which is shown in Fig. 6. In an area such as Surinam's lowlands, where very small gradients in meteorological variables prevail, it is not surprising that a discontinuity as sharp as a coastline has a strong and enduring impact upon atmospheric dynamics even up to the regional scale.

The central highlands are too far removed from the coast for the land–sea circulation to have any effect on the diurnal distribution of rainfall there. Fig. 6 (Sipaliwini) reveals that a fundamentally different circulation is active since the preponderance of the rainfall there occurs during the night-time. The familiar up-slope by day and down-slope by night type of circulation (FLOHN, 1969) is active to some extent, but this alone is not an adequate explanation. For example, at Tafelberg, elevation 344 m, the time of minimum daytime cloudiness, as shown in Table IV, is during the mid-afternoon. Also in the vicinity of Sipaliwini the inclination angles of the terrain are very small which tends to minimize the impact of the up-slope and down-slope circulation. It is likely that a circulation such as proposed by LETTAU (1968), arising from the horizontal temperature gradients set up by a sloping surface undergoing radiative cooling, is present and can explain the strong tendency toward nocturnal rainfall at locations such as Sipaliwini.

The annual variation of the diurnal course of rainfall is cursorily examined in Table V, using the January and June data from Paramaribo. As long as the station is in the northern "meteorological hemisphere", i.e., northeast trade winds prevailing, the oceanic

TABLE IV

HOURLY DISTRIBUTION OF PERCENT-OF-POSSIBLE SUNSHINE* (mean of all months)

Station:	Local time: 0700		0900		1100		1300		1500		1700
Nieuw Nickerie	60	70	73	73	74	76	76	76	72	64	
Paramaribo	59	70	*72*	70	62 ×	58	57	55	53	47	
Zanderij	45	66	73	*75*	72	68	64 ×	56	46	34	
Tafelberg	26	38	46	53	53	52	61	*62*	60	50	

* Period of record, 1962–1970. Data prepared and provided by Dr. C. W. van Scherpenzeel (personal communications, 1973).

TABLE V

PERCENT OF TOTAL RAINFALL OCCURRING WITHIN EACH HOUR, JANUARY AND JUNE,* PARAMARIBO, SURINAM

Local time:	0000			0300			0600			0900		1200
January	3	5	3	4	5	7	6	4	5	5	6	6
June	2	2	3	2	2	3	1	1	1	1	2	6

Local time:	1200			1500			1800			2100		2400
January	7	4	4	6	2	2	1	4	2	2	3	4
June	12	17	16	10	6	4	3	2	1	1	1	1

* $10\frac{1}{2}$ years of data. Values from BRAAK, 1935, fig.1.

influence is reinforced. This is evidenced by the near-dawn peak being fully comparable with that of the early afternoon which, in the annual mean, predominates strongly.

The phenomenon of a near-dawn rainfall maximum is typical of tropical oceans and is due to the combination of several convection-related processes (MALKUS, 1963).

(*1*) Radiational cooling of the ambient atmosphere, in which the upper portions of the cumuli grow, has its maximum effect during the pre-dawn hours.

(*2*) Sensible heating of the lower atmosphere by the warm ocean attains one of its two diurnal maxima between 04h00 and 06h00, local time.

(*3*) Convergence due to tidal motions of the atmosphere (which is proportional to the pressure tendency, indicated by the slope of the pressure curve in Fig.7, p.316) passes through a maximum between 06h00 and 08h00, local time. The presence of the coastline offers an additional contribution to such low level convergence off-shore, but close to the coast.

(*4*) The seaward component of the land–sea circulation, which during the Northern Hemisphere low-sun months is more directly opposed by the North Atlantic trade winds, attains its maximum strength in the pre-dawn hours.

It is noteworthy that for January in Table V the peak occurs just before sunrise. This rainfall is rendered by showers which originate over the sea and are then carried overland by the onshore component of the deeper wind layer above the shallow seaward branch of the nocturnal land–sea circulation. These showers are in their dissipative stages since they are now cutoff from the sensible heat input (see (*2*) above), and are displayed from the site of maximum low level convergence, as per (*4*) above. Thus, in spite of the fact that spectacular lightning displays may accompany nocturnal rains over the nearby ocean, these near-dawn showers over land are seldom so attended. Such rainshower activity will not endure long after sunrise due to the reversal of processes (*1*) and (*4*) above. However, it may well reach an abrupt climax soon after sunrise along the coastline itself when the resistance to onshore movement offered by (*4*) collapses.

In spite of the afternoon conditions of processes (*1*) and (*3*) militating against an afternoon maximum, sensible heating by the warm land and the afternoon phase of (*4*) lead to a distinguishable peak in Paramaribo's diurnal distribution even at this time of year, when the rainfall character is strongly oceanic.

The profile during the time when Paramaribo finds itself south of the "meteorological equator", (Table V, June) strongly resembles the mean annual profile (Fig.6) but with a much sharper peak occurring during the 14th and 15th hours. Direct oceanic influence is minimized, but the daytime landward branch of the land–sea circulation now opposes the large scale off-shore flow of recurved, Southern Hemisphere trades. The strength of the large scale flow essentially determines when and where this confrontation will be most effective, since what forces the land–sea circulation is nearly invariant. The farther north of Surinam the equatorial trough centerline is, the stronger the large scale flow. Table VI lists the hour during which the afternoon peak occurs vs. month. As expected, the peak hour is progressively delayed from (local times) 12h00–13h00 in May, to around 16h00 in late August and early September when the trough has attained its maximum northward excursion. Thereafter the reverse trend accompanies its return.

The annual variation at Paramaribo is representative of the coastal fringe, at most. Farther inland from the coastline such a striking seasonal contrast in the diurnal distribution of rainfall is not anticipated.

TABLE VI

ANNUAL VARIATION OF THE HOUR (local time) OF MAXIMUM PERCENT-OF-TOTAL RAINFALL. PARAMARIBO, SURINAM*

Mar.	Apr.	May	June	July	Aug.	Sep.	Oct.	Nov.	Dec.	Jan.	Feb.
13th	13th	13th	14th	15th	16th	16–17th	15th	14th	14th	13 + 6th	5th

* $10\frac{1}{2}$ years data, from BRAAK, 1935, fig.1.

Pressure and wind

The passages of the equatorial trough centerline can be seen in both the pressure and wind data. However, in both cases the influence of modifying factors must be taken into account. The sea level pressure data for Paramaribo (see Appendix, Table XLIX) are dominated strongly by the annual variation in location and intensity of the Amazon low and the subtropical highs. However, when harmonic analysis is employed to remove this 12-month component of variation, then the transits in May and late November of the trough centerline become evident.

Pressure gradients are generally very slack at these lowest latitudes, as are the winds supported thereby. Before the movements of the equatorial trough can be seen in surface wind data, account must therefore be taken of strong local effects such as the land–sea circulation and the modifications engendered by sloping terrain. Small annual variation indeed is found in the direction of the surface winds at Paramaribo (Table XLIX) being east-northeast from December into June, and east the remainder of the year. The true extent of the diametric shift in air flow is masked by more local effects, especially the land–sea circulation and the fact that the observations are strongly biased toward daylight conditions. Table VII, containing the monthly wind directions recorded at a station 115 km inland, gives a more representative portrayal, for the country as a whole, of the change in wind regime accompanying the trough centerline passages.

TABLE VII

RESULTANT WIND DIRECTION FOR KABELSTATION, SURINAM*

Dec.	Jan.	Feb.	Mar.	Apr.	May	June	July	Aug.	Sept.	Oct.	Nov.
N	NNW	NW	NNW	N	SSW	S	S	S	SSE	SSE	SSE

* 5 years of 08h00 observations recorded before 1926 (from BRAAK, 1935).

Mean wind speeds, like most climatic variables here, show the strong semi-annual rather than annual oscillation, although the total range of mean monthly values is little more than 0.5 m/sec. Maxima occur during the dry seasons with the principal maximum, at all locations except the far interior, occurring during the "kleine drogetijd" (short dry season) and having a direction somewhat north of east. At the higher interior locations maximum speeds occur during the "grote drogetijd" (long dry season) and have a component south of east. Along the coastline itself daytime wind speeds of 3 to 4 m/sec are normal during these dry seasons. Away from the coast 2 m/sec or less is more typical of daytime winds. During the wet seasons, and especially at the time of the "grote regentijd" (long rainy season), a daytime mean wind speed of little more than 1 m/sec is representative of the country as a whole. Nocturnal calms, away from the immediate vicinity of the coast, are very common during all seasons (Dr. C. W. van Scherpenzeel, personal communication, 1973). This is especially true in the low, inland areas.

Storms

Thunderstorms are by no means an everyday occurrence here, contrary to what is believed even by many meteorologists trained in the mid-latitudes. Over the coastal fringe itself thunderstorms are uncommon during those months when the equatorial trough is centered to the south, and even during the "grote regentijd" and up into September, Paramaribo experiences thunderstorms only every 3rd or 4th day, on the average. The frequency increases inland to the extent that during the "grote regentijd" every other day, in the mean, is a thunderstorm-day. However, even here, when the trough is centered to the south, the norm is only 2 to 4 thunderstorm-days per month. Thunderstorms occur predominantly when Surinam is in the southern "meteorological hemisphere."

Also somewhat surprising is the fact that, in spite of being often accompanied by very heavy rainfalls, these thunderstorms normally occur without destructive winds. But this is not universally true. Maximum wind speeds are associated with an anomalous thunderstorm locally referred to as "sibiboesie". ("Sibiboesie", pronounced $\bar{c}$ - $\bar{b}$ - b$\bar{o}\bar{o}$ - $\bar{c}$, is a colloquialism derived from the combination of the words for broom and forest in the local idiom; thus connoting a wind which sweeps the forest clean.) Wind speeds of 22 m/sec have been recorded and gusts as high as 30 m/sec are possible during these severe thunderstorms (Dr. C. W. van Scherpenzeel, personal communication, 1973). Normally "sibiboesies" occur in July and August, but can be expected during any month that Surinam finds itself in the southern "meteorological hemisphere". The late afternoon seems to be the preferred time of day for these events (ALEWIJNSE et al., 1963, also MINISTERIE VAN OPENBARE WERKEN EN VERKEER, 1964), indicating that surface heating, if not the total forcing, is at least the triggering mechanism. "Sibiboesies" are very local phenomena and certainly are the exceptional thunderstorm. The duration of these violent episodes, which have caused structural damage and raised havoc among the citizenry, is usually less than 10 minutes; 5 minutes being more representative. During this very short time span the air temperature drops 6° to 8°C, the relative humidity increases by as much as 40%, and the wind speed proceeds from nearly calm to its maximum and back again to calm. Another interesting characteristic is that the "sibiboesie" typically produces only about 10 mm of rain.

These observations can at least partially be explained by the fact that air parcels, laden with liquid water, experience accelerated descent when they are cooled by evaporation of some of that water to temperatures below that of the ambient atmosphere at their own altitude. While this happens to some extent in all thunderstorms, it becomes strongly developed if: (*1*) the parcels themselves were originally rather dry, e.g., residing above the trade wind inversion, and subsequently received their liquid water load by being entrained into an active thunderstorm; and (*2*) the condensation level is higher than normal; under such conditions the substantial evaporational cooling necessary for well-developed downdrafts takes place. When such downdrafts impinge upon the earth's surface, strong gusty winds ensue. That the proper juxtaposition of requisite circumstance does not often exist relegates the "sibiboesie" to the rare-event classification, fortunately.

Waves-in-the-easterlies normally pass north of Surinam and their most notable effect is a day with abnormally calm and sunny weather followed by several cloudy days, on the first of which moderate to heavy rainshowers are wide spread and persistent (Dr. C. W. van Scherpenzeel, personal communication, 1973). Tropical storms invariably pass to the north, on their westward trek, and affect Surinam by strengthening its southerly wind and tending to minimize precipitation.

Temperature

A unique climatological record for the American tropics is the simultaneous series of air and sea water temperatures measured routinely aboard Surinam's lightship, stationed 20 km offshore. Table VIII contains these monthly means which were recorded at 7.4 m above the sea surface and within the first 0.5 m below it. The waters of the south equatorial current, augmented and heavily silted by the effluence of the Amazon river, continually flow past the coast of the Guianas. Reflecting very long residence within the

TABLE VIII

AIR AND SEA WATER TEMPERATURES (°C) RECORDED ABOARD THE SURINAM LIGHTSHIP (06°06'N 55°11'W)*

	Jan.	Feb.	Mar.	Apr.	May	June	July	Aug.	Sept.	Oct.	Nov.	Dec.	Ann.
Air	26.2	26.3	26.4	26.8	26.6	26.7	27.1	27.4	27.4	27.4	26.9	26.7	26.8
Water	26.3	26.3	26.7	27.1	27.2	27.4	27.9	28.3	28.0	28.1	27.8	27.3	27.4
W.—A.	0.1	0.0	0.3	0.3	0.6	0.7	0.8	0.9	0.6	0.7	0.9	0.6	0.6

* Means for 5-year period 1962, 1964–67. Both daily means computed by (0800 + 1400 + 1800) 1/3, local time.

lowest latitudes, the mean annual water temperature is 27.4°C, which is 0.6°C warmer than the corresponding mean air temperature. It is significant to note that the Surinam climatological means, rather than 24-h means, are shown here. This undoubtedly minimizes the true amount by which the water temperature exceeds that of the air.

Over Surinam itself, the diurnal range of air temperature is 4 to 5 times greater than the annual range (annual range defined as the difference between the highest and lowest mean monthly values). Such a range relationship is characteristic of the tropics. The annual range near sea level is 2°C or less, that at the higher stations ($150 < H < 400$ m) is slightly more, but still not more than 3°C. The diurnal range varies more with distance from the sea than with elevation, showing the moderation effected by the onshore component of the wind. Along the immediate coastline the diurnal range is as small as 5°C, along the coastal fringe it is generally 7° to 8°C, and 10° to 11°C inland. Interior valleys may have a diurnal range somewhat in excess of 11°C, but the highest interior ridge locations experience a diminution in diurnal range due to frequent afternoon cloudiness. The net effect of these variations is minimum temperature variability along the coastal fringe, maximum at interior valley stations.

The true-mean (mean of 24-h values) annual temperature of the coastal fringe is 26.3°C. It decreases, but by not more than 1.0°C, inland and throughout the interior, which suggests the role of the warm sea as a sensible heat source for the coastal fringe.

The annual course of the mean air temperature is determined by a number of competing factors, the two most influential of which are cloudiness and evaporation. Taking variables individually, percent-of-possible sunshine explains the largest part of the variance of monthly means about the annual mean. And in turn, variation in percent-of-possible sunshine depends almost exclusively on amount of daytime cloudiness. What can be anticipated is that evaporation from the surface varies little during the course of the year, i.e., surface moisture is rarely deficient. At most locations two air temperature maxima are distinguishable; they coincide well with the dry seasons which are also the periods of maximum percent-of-possible sunshine, specifically March–April and September–October. One minimum occurs in May–June when sunshine is limited by the frequent daytime rains of the "grote regentijd". However, the primary minimum occurs in January; it must be attributed not only to daytime cloudiness, which is considerable but nevertheless not as great as in May–June, but also to the minimum in duration of daylight and the occasional advection of cooler air from somewhat higher latitudes of the Northern Hemisphere.

The diurnal variation of temperature (and other variables of climatological interest) is available for Zanderij Airport due to the superior efforts of the Surinam Meteorological Service. Fig.7 shows such data which are hourly means for all months of the 12-year period 1952–1963. The dry-bulb temperature profile can be utilized to obtain the corrections needed to find the true-mean from that determined by the averaging of various hourly values. Of particular interest is the correction for the mean determined from the average of readings at the "climatological hours": 08h00, 14h00 and 18h00. Table IX (*a* and *b*), determined from the data for individual months, gives the true-mean monthly temperatures and the correction factors which must be added algebraically to the climatological mean to obtain the true mean. It is important to note that these correction factors can only be applied with precision to similarly located and similarly exposed stations. Zanderij Airport is located on the inland portion of the coastal plain, 50 km

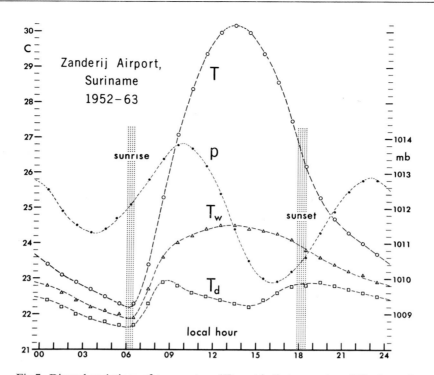

Fig.7. Diurnal variations of temperature (T), wet-bulb temperature (T_w), dew-point temperature (T_d), and sea-level pressure (p) for all hours from January 1952 through December 1963. Local time, GMT minus 3 h and 30 min. Annual variation in sunrise and sunset times shaded. Location: 5°27′N 55°12′W, 16 m.

from the coastline yet only 16 m above sea level. The landscape is flat, sandy savanna covered basically with moderately dense grass and bushes, interspersed with forested areas which grow to heights not more than 20 m and occur particularly along the water courses. Closely to the north the coastal fringe is found where swamps and bands of rain forest are common. The correction factors given in Table IX(*b*) are not appropriate here. For example, BRAAK (1935) suggests a mean annual correction to the climatological-mean temperature for Paramaribo of −0.9°C, which is 0.5°C less than at Zanderij. The influence of the ocean, especially the persistent daytime onshore wind component, re-

TABLE IX

MEAN MONTHLY TEMPERATURE STATISTICS* (°C), ZANDERIJ AIRPORT (1952–1963)

	Jan.	Feb.	Mar.	Apr.	May	June	July	Aug.	Sept.	Oct.	Nov.	Dec.	Ann.
(*a*)	25.0	24.8	25.2	25.5	25.5	25.4	25.6	26.2	26.7	26.8	26.2	25.4	25.7
(*b*)	−1.1	−1.2	−1.4	−1.3	−1.1	−1.1	−1.4	−1.7	−2.0	−1.9	−1.6	−1.1	−1.4
(*c*)	9.3	9.4	9.9	9.5	8.9	9.3	10.3	11.1	12.1	12.0	11.4	9.8	10.2
(*d*)	2.3	2.4	2.1	2.2	2.4	2.3	2.1	2.1	1.8	1.8	2.5	2.5	2.2

* (*a*) True-mean (24-h mean) temperature; (*b*) correction for the climatological mean to obtain (*a*); (*c*) daily range determined from max. and min. thermometers; (*d*) correction for range determined from thermograph to obtain (*c*).

shapes the diurnal temperature profile along the coastal fringe. Nevertheless, the entire inland section of Surinam and the low, eastern interior can be expected to have corrections similar to Zanderij.

The mean diurnal temperature range, using data for all months, is seen from Fig.7 to be 8.0°C, which contrasts with the corresponding value of 10.2°C in Table IX(c). The discrepancy arises from the fact that the mean hourly temperature values in Fig.7 were taken from thermograph records, whereas the values in Table IX(c) were derived from the daily readings of maximum-and-minimum thermometers. Thus, the fine-structure of the individual temperature records, which is either not recorded by the thermograph or is averaged out in the analysis of the record, amounts to somewhat more than 2°C in the annual mean diurnal temperature range.

The lesson inherent in these exegeses is that even in latitudes notorious for temperature invariance, serious errors are introduced if specifications of the manner in which means are calculated, or of the technique of measurement, are not made. The excellent data series from Zanderij illustrate well these points, specifically: (*1*) the annual range of mean monthly temperature values is exaggerated by 1/3 (2.7° vs. 2.0°C) if the climatological- rather than the true-mean is used in determining the monthly values; (*2*) the daily temperature range can vary by more than 2°C depending upon the measurement technique.

A very long series of reasonably homogeneous air temperature data is available for the Cultuurtuin (Botanical Gardens) in Paramaribo. The true-mean temperature for the pentad 1900–1904 was 25.86°C, for the pentad 1905–1909, 25.96°C, and for the last, 1965–1969, 26.06°C. The inter-pentad changes in true-mean temperature for the full series of 14 pentads is contained in Table X. It is noteworthy that the mean of the first 7 such differences, corresponding to the period 1900–1939, is +0.11°C per pentad, that of the final 6 values, −0.09°C. The large magnitude of the 5th inter-pentad difference along with the small, but persistent decrease evident from the 2nd to 4th values are explained by the gradual growth of trees around and above the instrument shelter. The state of affairs is revealed best by quoting directly the source of the earlier data (BRAAK, 1935, p.100): "In 1923 or thereabouts a change became necessary, the screen becoming overgrown by large trees". The relocation was only by a short distance and to an open lawn area. Increasing portions of the large 5th value thus rightfully belong to the sub-series 2 through 4. It is not justified, therefore, to consider individually any of the values 2 through 5 unless the impact of tree growth is ascribed.

Overall, the increase of 0.7°C during the first 40 years of this century is remarkably mono-

TABLE X

INTER-PENTAD CHANGES (°C) IN MEAN AIR TEMPERATURE, CULTUURTUIN, PARAMARIBO (1900 to 1969, inclusive)

1905–09 minus 1900–04:	+0.10	1940–44 minus 1935–39:	−0.08
1910–14 ,, 1905–09:	−0.04	1945–49 ,, 1940–44:	−0.26
1915–19 ,, 1910–14:	−0.02	1950–54 ,, 1945–49:	+0.14
1920–24 ,, 1915–19:	−0.06	1955–59 ,, 1950–54:	−0.22
1925–29 ,, 1920–24:	+0.52	1960–64 ,, 1955–59:	+0.06
1930–34 ,, 1925–29:	+0.16	1965–69 ,, 1960–64:	−0.18
1935–39 ,, 1930–34:	+0.08		

tonic. The decrease of 0.5°C during the latter 30 years is not as uniform. It is likely, none-theless, that this later overall downward trend is at least as real as the earlier upward trend, because instrument exposure has been more conscientiously considered during the latter period.

The soil temperature recorded at the Cultuurtuin in Paramaribo is of interest because of its high value and minimal variation. Table XI gives the mean monthly values at 3 depths, for a 3-year period during the early 1930's. The measurements were made be-neath an unshaded "spacious lawn near the thermometer screen. The grass is kept short, the soil consists of brown sand" (BRAAK, 1935, p.119). Simultaneous air temperature means, height 1.6 m above the surface, computed by the same method are also given in (*a*) of Table XI.

The correction to the annual climatological-mean air temperature for Paramaribo, needed to obtain the true-mean, has been mentioned, i.e., —0.9°C. The amplitude of the diurnal variation in soil temperature decreases with depth and by 160 cm no differences among the three daily, 3-year mean values are found (see Table XII). It is unlikely that the correction needed to obtain the true-mean soil temperature, at least for levels below 100 cm is more than —0.1°C. The annual, true-mean, unshaded soil temperature is thus seen to be at least 2.7°C warmer than the comparable shelter air temperature. Employing similar soil temperatures, but recorded in the shade, this excess is reduced by nearly 2°C to about 1.0°C (BRAAK, 1935).

TABLE XI

MEAN SOIL (AND AIR) TEMPERATURE (°C), CULTUURTUIN, PARAMARIBO (Sept. 1932–Apr. 1935)*

	Jan.	Feb.	Mar.	Apr.	May	June	July	Aug.	Sep.	Oct.	Nov.	Dec.	Ann.
(*a*) Air, 1.6 m, shelter	26.5	26.5	27.1	27.5	27.1	26.8	27.1	27.9	28.6	28.5	28.1	26.8	27.4
(*b*) Soil, 60 cm	28.0	28.1	28.9	29.7	29.4	29.0	29.3	29.7	30.6	30.7	30.4	28.9	29.4
(*c*) Soil, 110 cm	28.3	28.2	28.6	29.2	29.4	29.0	29.1	29.4	30.1	30.4	30.1	29.3	29.3
(*d*) Soil, 160 cm	28.6	28.4	28.7	29.1	29.4	29.1	29.2	29.4	29.9	30.3	30.1	29.5	29.3

* All means from daily values computed by (0800 + 1400 + 1800) 1/3. Soil temperatures from BRAAK, 1935, table 29. Air temperatures determined from data given in *Klimatologische Tabellen, Diverse Elemente, Paramaribo, Periode 1931–1960* (1965).

TABLE XII

MEAN ANNUAL SOIL TEMPERATURES (°C) AT EACH CLIMATOLOGICAL HOUR, PARAMARIBO*

Depth (cm)	Local time			Mean
	0800	1400	1800	
60	29.5	29.4	29.3	29.4
110	29.3	29.3	29.2	29.3
160	29.3	29.3	29.3	29.3

* Period of record and source of data same as Table XI.

Atmospheric moisture

Relative humidity undergoes a small but significant decrease as distance from the primary moisture source increases. Table XIII details this change by month for 3 stations having nearly the same annual rainfall total but increasing inland distance. For the coastal fringe the mean annual (climatological-mean) value is 80%, for the inland and interior 75%. A double maximum corresponding to the two rainy seasons is apparent in all sections except the southwest where the more continental climate has only one rainy season.

The mean vapor pressure values are high and remarkably constant, close to 28 mbar year round in the more populated lowlands. Paramaribo and the coastal fringe in general have mean annual vapor pressure values slightly in excess of 29 mbar.

In concord with the nearly invariant vapor pressure, the annual course of mean monthly dew point values shows an extremely small amplitude. Table XIV(*a*) gives these data for Zanderij Airport where the amplitude of variation is seen to be only 0.5°C about the annual mean of 22.4°C. The diurnal course of dew point at the same location, presented in Fig.7, shows no greater amplitude about the mean. Also shown in Fig.7 is the mean diurnal course of Zanderij's wet-bulb temperature which can be considered as a convenient, though very approximate, measure of human comfort.

Evaporation data given in the climatic tables for Tafelberg and Sipaliwini, Surinam (Appendix, Tables LVIII, LXIII) are values computed using a Penman-type evaporation formula (LENSELINK and VAN DER WEERT, 1973). Similarly computed values for Zanderij Airport and Paramaribo are given in Table XIV(*b* and *c*). The errors involved in

TABLE XIII

MEAN RELATIVE HUMIDITY (%); CLIMATOLOGICAL-MEAN (0800 + 1400 + 1800)1/3; TRUE-MEAN VALUES IN PARENTHESES

Station (record)	km from ocean	Jan.	Feb.	Mar.	Apr.	May	June	July	Aug.	Sep.	Oct.	Nov.	Dec.	Ann.
Paramaribo (1931–60)	12	82	80	78	81	85	85	82	78	75	76	78	82	80
Zanderij Airport	50													
climatic (1961–70)		82	78	76	76	82	83	78	75	72	70	75	82	77
true (1952–63)		(86)	(86)	(84)	(85)	(87)	(87)	(85)	(82)	(79)	(79)	(83)	(86)	(84)
Sipaliwini (1961–70)	400	77	76	78	78	80	78	76	73	68	62	68	71	74

TABLE XIV

MONTHLY MEANS RELATED TO ATMOSPHERIC MOISTURE

	Jan.	Feb.	Mar.	Apr.	May	June	July	Aug.	Sep.	Oct.	Nov.	Dec.	Ann.
(*a*)	22.2	21.9	21.9	22.4	23.0	22.8	22.5	22.4	22.3	22.4	22.5	22.6	22.4
(*b*)	143	145	162	159	147	151	175	189	190	190	164	145	1,960
(*c*)	138	140	165	155	146	143	162	182	191	189	159	138	1,908

(*a*) True-mean dew point temperature (°C), Zanderij Airport (1952–63).
Computed total evaporation using a Penman equation (in mm; from LENSELINK and VAN DER WEERT, 1973): (*b*) Zanderij Airport, 1961–70, and (*c*) Paramaribo, 1931–60.

the measurement of evaporation in the humid tropics have recently been studied by PORTIG (1972), and, indeed, valid doubt exists as to the representativeness of measured values. These computed values probably represent, better than actual measured values, the mean monthly total evaporation from a vegetation covered surface having an adequate moisture supply.

Certainly local evaporation is an important source of atmospheric moisture, and in the lowest atmospheric layers is probably comparable to that advected from the sea. Local evaporation insures the damped amplitude in the annual course of moisture variables. As can be seen from the entries in Table XIV, a dew point decrease is countered by increased evaporation which arrests the decline; refer particularly to March and to August–September.

Apparently the dry seasons are not normally long enough to critically deplete the local moisture reserves of the vegetation and the soil. Modification of the vegetative cover could conceivably alter this condition since the ability of the soil to store water is reduced by deforestation.

The high moisture content gives to the air an intangibility which is most agreeable and consoling to the long-time resident, but which is tedious and enervating to the newcomer from higher latitudes. Contributing greatly to this imperceptibility is the relatively small water vapor gradient between the human and the ambient air which, along with the prevalent low wind speed, impedes the evaporation process. The cooling arising from heat loss at the outermost skin layer during the evaporation of body moisture is reduced. Moreover, the moist lower atmosphere is very active in the long wave (thermal) radiation processes, and, as a source thereof, tends to minimize the chill even of a clear night-time sky. This blanketing effect is what the native finds comforting since the heat regulation mechanisms of his body are conditioned to operate over a small range and to deal most effectively with heat dissipation, as opposed to heat conservation. The neophyte's mechanisms, on the other hand, are geared to operate over a greater range and appear to thrive on variety of external conditions, a feature which the high vapor pressure always tends to minimize. Also, his regulation compensates more effectively for heat deficiencies than for excesses. For example, the solar radiation is absorbed at a greater depth in his flesh than in that of the well-tanned, long-time tropical resident (STONE, 1941). Although most efficient when heat is needed to maintain proper body temperature, this only serves to obstruct heat dissipation through radiation, when heat is surplus, since conduction to the outer skin surface must first be accomplished. The normal reaction to the inability to deal effectively with excess body heat is early exhaustion. And it is high atmospheric moisture content rather than high temperature which leads to the inability to cope with body heat excesses. For example, the absolute maximum temperature recorded at Paramaribo in over 70 years of careful records keeping is 37.3°C (little more than 99°F). After a short period of adjustment, however, most of those from more variable climates can work and live as efficiently and comfortably, in these areas where high atmospheric moisture content tends to minimize variability, as the long-time resident.

In conclusion it may be recalled that for all of Surinam except the southwest, the seasons are four in number, but not the mid-latitude set of four. Rainfall rather than temperature is the basis for the differentiation of seasons. The climate is described, in the sense specified early in this chapter, as tropical wet marine. The latitudinal displacements of the equatorial trough result in Surinam being in the northern "meteorological hemisphere" from December to May and in the opposite hemisphere otherwise. The wet seasons are associated with trough centerline proximity, the dry seasons with its maximum displacements. The area-mean rainfall is 2,200 mm per year, and a typical monthly total for the 6 wet months is 250 mm, for the dry months 100 mm. The annual variation of most meteorological elements, except rainfall, is small, and in most instances the diurnal variation exceeds the annual.

Guyana

Geography

Guyana is bounded by the equatorial North Atlantic to the northeast, Surinam to the east, Brazil to the south and southwest, and Venezuela to the northwest. It is the largest of the three Guyana countries, having an area somewhat more than 200,000 km². The north–south extent (800 km) is much greater than its average east–west span ($\sim$ 250 km). The terrain is low, two-thirds of the area being below 200 m elevation, and low elevations extend even to the areas most removed from the coast. The numerous rivers run generally south to north.

Four sections are easily distinguished which are approximately outlined in Fig.8. The coastal fringe is a flat plain of 20–40 km width which is mostly covered with rain forest. The exceptions are spots of wet, natural grasslands near the rivers and areas cleared for agriculture, principally sugar cane growing. Northwest of the Essequibo river the coastal fringe is outlined by mangroves and the forest cover itself is here quite dense over the inner portion of the section. Southeast of the river the wet grasslands are more common. The inland section is a 150–250 km wide band, paralleling the coast, composed of broad, low ($20 < H < 100$ m) hills. It is almost totally covered with rain forest except in the vicinity of the lower Berbice river. There open areas of grass or "muri bush" (RICHARDS, 1964) are frequently found amid the general evergreen rain forest. This anomalous savanna type probably owes its existence to man's impact by way of crop raising and cattle grazing. The highlands are actually only the east slope of the Guyana Highlands, the bulk of which lie in Venezuela. The portion within the Guyana border is made up of the Pakaraima Mountains, located along the central-western border parallel to the coast (mean ridgeline elevation not more than 1,000 m), and the smaller Merume Mountains, approximately 100 km shoreward from and parallel to the main mountains. These highlands are also mostly rain-forested except in the border area with Brazil, south of about 4°N. There more open upland savanna is found. Very little climatological information is available for the highlands.

The southern third of Guyana, which may be termed the "southern plains", possesses a climate considerably different from the remainder of the country. The elevation of the southern plains is low, in spite of the distance from the coast, and the terrain is generally level. North of 3°N, an elevation of less than 200 m is typical and even south of that latitude the broad river basins remain lower than 200 m. Only very limited areas rise above 300 m. Southward across the section tree growth thins, becomes confined to river courses and finally disappears. The majority of the section is true tropical savanna. The origin and reason for the continued existence of this savanna is not resolved (RICHARDS, 1964), but its distinctive annual rainfall pattern suggests that climate itself at least sustains the open grassland.

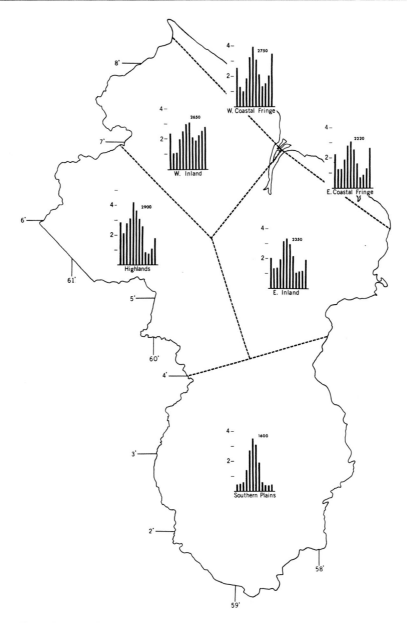

Fig.8. Annual distribution of rainfall for various sections of Guyana. Ordinate values in mm/100, annual totals in mm. Based on period 1931–60 (certain data subjected to reduction).

Climate

Rainfall

As is characteristic of the low elevation tropics, rainfall in Guyana is the element manifesting the most recognizable annual variation and therefore is the basis for the differentiation of seasons. In all but the southern plains section two wet and two dry seasons occur. Also, the annual rainfall total and the temperature characteristics conform to the definition of a tropical wet marine climate given in the preliminary remarks of this chapter. It is generally true that the primary wet season is centered on June, occurring from mid-April to mid-August; the secondary wet season consists of December and January. The primary dry season is from mid-August through November; the secondary, February through mid-April. As can be seen from Fig. 8, the primary dry season is longer and more intense in the eastern and highland areas. In the western region it is longer but the driest months occur during the secondary dry season.

The seasonal distinction is directly related to the annual shifts of the equatorial trough and its associated zone of most intense convection, the I.T.C. The change in strength and location of the North Atlantic subtropical high and the Amazon low pressure systems mask the movements of the equatorial trough as seen through sea level pressure data alone. However, the contrast between April and July in Fig. 1 indicates what the rainfall confirms; i.e., the northward passage of the zone of maximum convergence during June. Likewise, a comparison of October and January reveals a regeneration and southward transit of the maximum cloud area during that time. The center of the trough probably passes south during November but, curiously, the major convection occurs north of the center. The primary dry season must be associated with trade winds of the South Atlantic, the secondary one with those of the North Atlantic, although the variation in winds during the year is slight.

Surface winds over all but the southern plains are quite invariant seasonally, having a prevailing northeast to east-northeast direction year round. Surface speeds, as may be seen from the climatic table for Georgetown (see Appendix), are somewhat reduced when Guyana is on the southern side of the equatorial trough. The short record of resultant wind data presented in Table XV indicates that this reduction extends up to above 1,800 m but not as high as 3,000 m. Above, the speeds are greater to the south of the trough than to the north. In fact, within the layer 1,800–3,000 m a decoupling seems to occur which allows in the lowest 2 km of atmosphere an approximately 20° shift in the direction of mass flux from one side of the trough to the other, whereas in the higher layers, the direction is independent of season. Particularly within 100 km of the coastline it is expected that these small seasonal variations in wind will be obscured by a strong and perennial land–sea circulation. But, in spite of the small shift in direction (east-northeast to east in the lowest 2 km) and the small speed change (only about 1 m/sec) brought about by the transits of the equatorial trough, the annual pattern of rainfall over all of the

TABLE XV

RESULTANT WINDS FOR GEORGETOWN, GUYANA (10's of degrees and m/sec)

Level (m)	D.J.F.		M.A.M.		J.J.A.		S.O.N.	
3,000	08	3.7	09	6.0	09	8.9	09	6.6
1,800	07	6.2	08	7.0	10	6.5	11	5.5
900	07	7.7	07	8.0	09	6.4	09	6.0
600	06	6.7	07	7.7	09	5.9	09	6.4
Sfc (2)	06	3.3	06	3.9	09	2.7	09	3.8

Period of record: March 1937–November 1940.
Source: U.S. WEATHER BUREAU, 1945.

northern two-thirds of the country is essentially determined by these small adjustments in the large scale circulation.

More local climatic factors such as orography and the presence of a coastline influence rainfall but apparently more in regard to amount received rather than annual distribution. The maxima in June and December are higher close to the primary moisture source, the ocean, than inland. The annual total, however, is largest over the highlands section where additional lift is given to the atmosphere by orography. The modest increase in annual total with distance from the sea, found in Surinam (see Fig.5, p.307) continues in northeastern Guyana. However, along the northwestern part of the coast a band of maximum rain appears to be present less than 50 km inland. The origin of this and the similar but more pronounced shore-parallel band of maximum rainfall found in French Guiana is to be attributed to the large increase in surface roughness when passing from the ocean to the mangroves and rain forests which cover the coastline. Although direct observations are not available, there is little reason to doubt the existence of a well developed land–sea circulation which should yield a diurnal distribution of rainfall within 100 km of the coast not unlike that found in Surinam (see Fig. 6).

The mean rainfall amount over the northern sections of Guyana is 2,500 mm/yr. Somewhat more is expected in the highlands section and a mean of 1,600 mm/yr. is representative of the southern plains. Days-with-rain are frequent during the wet seasons in the northern section, as is also true during the single wet season in the south. Rain is then recorded on 20–25 days per month. During the primary dry season, along the coast rain occurs on only 5–10 days per month, but inland the number remains high, near 15. The season with the least days-with-rain (10–15 days per month) in the inland section is the secondary dry season. During the long dry season in the southern plains the value is 4–8 days per month but on these days the accumulation is not large, only 5–10 mm per rain-day.

The distinctive feature of the annual rainfall pattern of the southern plains, its single wet and single dry season, along with two maxima in mean monthly temperatures determine the climate of this section as continental. During the wet season, mid-April through mid-August, much of the low western part of the plains is frequently covered by water (IM THURN, 1883). During the long dry season airborne dust and haze may be present (SECOND WEATHER GROUP, 1962). The pattern and subsequent events are similarly found over the large, low Llanos of Venezuela.

The explanation of this monsoonal pattern over northern South America is not complete, but it certainly involves the Amazon basin and connected low basins to the north; reference is made to the more detailed remarks in the Venezuela portion of this chapter. The most frequently given reason for the relative dryness of the southern plains is subsidence on the lee side of the central highlands of Surinam. That subsidence is occurring during the September to mid-April dry season cannot be denied. However, to ascribe the origin of the sinking wholly and simply to these highlands is not a sufficient explanation. With the relatively low wind speeds found in the tropical atmosphere, why is the center of the dryness displaced 200 km or more downstream from the barrier? Also, given the prevailing northeast flow near the surface over northern Guyana, why is the rain deficiency best developed directly between the highlands of Guyana to the northwest and those of Surinam to the southeast, rather than more directly behind them?

A more satisfactory explanation takes into account the fact that only very low ($\lesssim 200$ m) divides separate the Amazon basin from the basin of the southern plains. Also, the annual course of pressure difference between Guyana and the Amazon basin displays a pattern not unlike the inverse of the rainfall pattern of the southern plains itself, i.e., maximum differences occur from November through April, minimum during the other months. During the months of maximum pressure difference subsidence in southern Guyana is reinforced by the divergent flow of mass out of the area. The mass induced by the pressure gradient to enter the Amazon basin via the low divides exceeds that entering through the gap between the highlands of Guyana and Surinam. No actual wind information is presently available from the area, but such a mechanism would require moderate east to northeast surface winds during the dry season. During the early portion of the dry season, September and October, there is evidence (DEAN, 1971) of a relative maximum in southeasterly winds near the 3-km level over the area. Such a speed maximum could also lead to low level divergence. From May to August, when the Amazon–Guyana pressure gradient is minimal (or reversed), low-level convergence of mass over the southern plains occurs. Surface winds then should be very light easterlies, but even light winds from the west to southwest may frequently be present.

Other climatic elements

In the annual mean, temperatures for the lowlands of Guyana are higher along the coast and in the southern plains than in the inland section. Estimates for the annual true-means are 26.5°C near the coast, 27.0°C for the plains and 25.5°C for the inland section. The adjacent ocean is an overall heat source for the coastal area, and the prevalence of subsidence and limited cloudiness over the plains provide for the higher temperatures there. The range between warmest and coolest month increases monotonically from 1.0°–1.5°C along the coast to 2.0°–3.0°C in the southern plains section. Diurnal range undergoes a similar increase with distance from the coast, being about 5.0°–6.0°C near the coastline and 9.0°–10.0°C for the plains. The month-to-month progression of mean temperature shows a double peaked curve which is the mirror image of the marine rainfall pattern. The primary maximum occurs in October, the secondary in April. Minima occur in June and January, with the primary value being recorded in January for northern Guyana, but near June in the south.

Relative humidity is high and varies little with season in all sections except the southern plains. Estimates of the annual mean values, based upon syntheses of the mean diurnal variation from hourly data given by CLEARE (1961), are 80–85% for the coastal and inland sections. For the more continental south, such data are not available but a range of 70–75% is likely. The month-to-month progression corresponds directly to the rainfall pattern throughout the country. The annual range of monthly means near the coast is only 4–6% due to the ocean proximity. Inland the variation is 6–9% but in the southern plains 15–20%. The mean diurnal range is more uniform throughout Guyana, 14–20%; the exception being the highlands section from which no data exist but where the daily range must be lowered by frequent daytime cloudiness or near-condensation conditions. The maximum average daily cloud cover occurs close to the coast and along the east faces of the mountains of the highland section. Average daily cloudiness is less in the region between these areas being confined mainly to the daytime and early evening hours. However, the inland cloudiness is typically more developed, and especially from July through October afternoon thunderstorms are expected on one-third of the days. The greatest seasonal variation in cloudiness occurs over the southern plains. Minimal variation occurs within about 100 km of the coast which may be verified from Fig.1 (p.297).

Venezuela

Introduction and geography

The map of Venezuela describes, roughly, a triangle; the shortest side being the border with Guyana and Brazil, the other two longer sides being the coastline and the border with Colombia. The total area is somewhat more than 900,000 km² and lies between the latitudes 01° and 12°N, longitudes 60° and 73°W. The topography is diverse as is the climate which is strongly influenced thereby. This is particularly true for temperature, which is a very consistent function of elevation, and for rainfall, the areal distribution of which is strongly influenced by orographic lifting, lee side subsidence and other interference to air flow caused by slopes and obstructions.

For the following discussion it is convenient to divide the country into three major portions of very roughly equal size (letters refer to sections outlined in Fig.9).

(*1*) The north and west portion (*A–I*), containing the Maracaibo basin, the essentially east–west orientated Cordillera de la Costa and the very prominent southwest–northeast Cordillera de Mérida. In this part resides the large majority of Venezuela's population, nearly 85%. Vegetation varies from semi-desert along the east coast of the Gulfo de Venezuela to true rain forests along some slopes. At intermediate elevations (1,000 < H < 2,500 m) persistent agriculture and population pressures have superseded much of the original rain forest cover with less substantial tree growth. At higher elevations of the Cordillera de Mérida, mountain bush-and-grassland (páramo) is found which, with increased elevation, gives way to the total absence of vegetation on the highest peaks where small areas of glacier actually exist.

(*2*) The central portion (*J–M*), more descriptively referred to as the "Llanos". This is the vast, low (H < 300 m) and flat basin of the Orinoco–Apure–Meta river complex. Vegetation is predominately open savanna with tree growth typically confined to the banks of rivers and creeks. Chaparral is found intermittently along the higher periphery of Llanos, while mangrove swamps and wet marshes are common over the delta of the Orinoco river.

(*3*) The southeast portion (*N–Q*), called the Guyana Highlands, which constitutes between 1/3 and 1/2 of Venezuela's area. These highlands lie generally between 400 and 1,400 m elevation with only very small portions extending above 2,000 m. The southernmost area sheds its abundant rainfall into the Negro river and thence to the Amazon. Rain forests are common at the lower elevations (H < 600 m), especially in the northeast and over the west- and south-facing slopes. However, highland steppe and savanna are more typical in the highlands proper (*Atlas Agrícola de Venezuela*, 1960).

The following essay treats in general the climatic factors of solar radiation and the seasonal changes in the atmospheric circulation. The climate types as specified in the preliminary remarks of the chapter are located. In some detail precipitation and temperature are then dealt with. An outline of the rainfall producing disturbances is made and finally a discussion of the intriguing dry zone along the Caribbean coast is included. Specific treatment of such elements as cloudiness, atmospheric moisture, wind, etc. is foregone since the coverage available in the accompanying climatic tables for Venezuela is relatively comprehensive.

General climatic factors and climate types

Solar radiation

The essential input on which climate depends is the short wave radiation from the sun. The atmosphere through its hydrodynamics and thermodynamics mollifies this input in such a way that conditions at the earth's surface are life sustaining. The annual variation of the radiation incident upon the top of the atmosphere at the Equator is almost totally described by a second harmonic (period equals 6 months) oscillation. The small first harmonic (period equals 12 months) component which is present at the Equator is due to the annual variation in the earth–sun distance. Away from the Equator, the semi-annual oscillation gives way to another first harmonic component which is due to the increased time of daylight attending high-sun, coupled with the decreased time at low-sun. Because of the variation in duration of sunlight, by 12° of latitude no second harmonic component remains in the extra-atmospheric radiation, even though the sun's declination amplitude is 23.5° of latitude. These facts are indicated by the data in Table XVI,A. In spite of the fundamental change in the month-to-month distribution of solar radiation incident upon the top of the atmosphere, the mean annual total (the last column of Table XVI multiplied by 12) is nearly constant over Venezuela retaining 99.7% of its equatorial value at 5°N, 98.6% at 10°N and even at 15°N, 96.9% (LIST, 1958, table 133).

At the earth's surface, as can be seen from Table XVI,B, little more than half of the short

TABLE XVI

MONTHLY MEANS OF SHORT WAVE RADIATION (ly/day)

	Station	Latitude	Jan.	Feb.	Mar.	Apr.	May	June	July	Aug.	Sep.	Oct.	Nov.	Dec.	Mean
A		Equator	857	888	898	870	824	791	801	841	878	884	863	844	853
		10°N	756	821	876	899	890	874	877	887	878	835	771	734	842
		20°N	635	730	827	900	930	935	930	907	850	760	661	608	806
B	Santa Elena	04°36′N	448	479	509	512	427	416	442	465	502	479	456	415	462
	San Fernando	07°53′N	424	464	470	410	344	334	364	376	399	407	414	413	402
	Mérida	08°35′N	509	537	562	565	501	487	521	544	542	492	492	492	520
	Barcelona	10°07′N	467	509	548	543	507	478	502	507	516	487	453	438	496
	Coro	11°25′N	478	543	590	580	546	543	568	581	576	491	467	453	535

A: Computed solar radiation incident upon horizontal surface at top of atmosphere, solar constant = 1,940 ly/min. Data according to BERNHARDT and PHILIPPS, 1958; also available from LIST, 1958, table 132.
B: Measured solar radiation direct plus diffuse, incident upon horizontal surface at ground level. Recorded with Robitzsch bimetallic actinograph. Data source and (period or record): all stations except San Fernando from *Promedios Climatológicos de Venezuela* (1955–60), San Fernando from LÖF et al. (1966).

wave radiation incident on the top of the atmosphere arrives, the absorptive and reflective properties of the atmosphere significantly depleting that input. Also, the slow but monotonic decline with distance from the Equator, present in the annual total extra-atmospheric input at these low latitudes (Table XVI,A last column) is no longer evident at the surface. In fact, Coro (*11°25'N 69°41'W, 20 m*) receives the greatest total, San Fernando (*07°53'N 67°26'W, 73 m*) the least of the 5 stations listed. Altitude is an important consideration, as shown by Mérida (08°35'N 71°10'W, *1,495 m*) vs. San Fernando, which are both at approximately the same latitude. Also the ratio of the amplitude of the second harmonic component to that of the first does not fall rapidly to zero as latitude increases, which was the case for the extra-atmospheric solar radiation. This can be seen in the data for Santa Elena (04°36'N) but is particularly noteworthy for Barcelona (10°07'N). At San Fernando a first harmonic component (period equal to one full year) has become present but with its phase actually reversed from that of the extra-atmospheric annual component, thus making June the month of minimum short wave radiation at the surface. The influence of the atmosphere on the incident solar input is debilitating, but not simply so. In most cases the variation associated with change in declination angle (at low latitudes, primarily a variation of period 6 months) is passed relatively unattenuated by the atmosphere, whereas that associated with change in duration of daylight (primarily of period 12 months) is strongly affected. Furthermore, the atmosphere may introduce an annual variation in attenuation of its own which, as in the case of San Fernando, is actually influential enough to invert the phase of the annual variation in the solar input.

Large scale circulation

Throughout the year, but to varying degrees, Venezuela is under the influence of the equatorial trough and the Northern Hemisphere trade winds (both terms are discussed in the preliminary remarks of the chapter). The trade winds are present along the Caribbean coast year round and the equatorial trough never completely relinquishes the southern Guyana Highlands, although recurved Southern Hemisphere winds prevail there when the trough is farthest north. Otherwise the generalization can be made that the trades occupy the country during the dry and warm part of the year (December–April) and the equatorial trough dominates during the wet and cool months (May–November). These two seasons are referred to by the Spanish terms "verano" (literally meaning summer) and "invierno" (winter). These terms originated in the mid-latitudes to apply specifically to the Mediterranean climate which does have a warm, dry summer and a cool, wet winter. When applied by Latin Americans of the tropical latitudes to their climate, the warm, dry "verano" is enjoyed when the trades prevail, even though this occurs mostly during the Northern Hemisphere winter months. The period December through April is here the "verano". Likewise, wet and cool conditions generally attend the presence of the equatorial trough. The months May through November are therefore called the "invierno". The confusion arising from this reversal of the time of the year to which these terms apply in tropical Latin America can be avoided if the literal translations are not insisted upon; rather "verano" means dry and warm, "invierno" means wet and cool.

The simple, two-season year although quite extensively found, is not an adequate classification of the climate in every part of Venezuela. Also, even at locations which strongly manifest the two-season (first harmonic) pattern in rainfall, other climatic variables will be seen to possess semi-annual (second harmonic) oscillations, i.e., two maxima and two minima are attained annually. The presence of a similar variation in solar radiation incident on the surface is immediately recalled and is certainly active. But in addition, the annual variation in the central pressure of the North Atlantic subtropical high contains a strong semi-annual component. Maxima occur in February and July, the primary being July, minima are reached in April and November. Similarly the low-level pressure gradient over the south-central Caribbean executes a semi-annual variation which follows very closely that of the central pressure itself. As a consequence the resultant gradient-level wind speed is maximum in February and early-July, minimum in late-April and early-November (ATKINSON and SADLER, 1970). The relationship between the second harmonic oscillations associated with the North Atlantic subtropical high and that found in solar radiation is not the subject of this climatology. Nevertheless, the presence of such oscillations, especially over northern Venezuela, should be noted because many climatic variables will be seen to possess this type of variation.

Climate types

In the sense defined in the preliminary part of the chapter, the climate throughout Venezuela is tropical. More specifically, all five types, based upon the relationships of the annual courses of rainfall and temperature, are found (letters refer to sections seen in Fig.9). The dry marine type (basically all of *A–G*) is well developed along the Caribbean coast and will be discussed in the section "South Caribbean dry zone" (p.348). Also present is a small area of the wet marine type (most of *K*, also *N*) which was treated more completely in the Surinam part of the chapter. The continental type (*I, J*, inland *K, L, M*, also *O*), which is the "verano"–"invierno" climate par excellence, is quite extensive and will be detailed in the Llanos portions on pp.333 and 334. Small areas of the wet continental (*Q*) and the mountain (*H*, also to some extent *P*) types are present; in addition to the coverage of these given here, the Columbia discussion offers pertinent information.

Precipitation

Basic annual distributions

Amid the variety of month-to-month distributions of rainfall found in Venezuela only two generic patterns emerge: (*1*) the Llanos pattern which has a single maximum occurring during the high-sun months and distinct low-sun dryness; and (*2*) the semi-annual pattern which during the course of the year completes two oscillations between wet and dry.

The Llanos pattern is characteristic of the entire central section of Venezuela, as implied by the name, but also of a large portion of the Guyana Highlands. This pattern is to be identified with the two-season year and with the notion of continentality. Its prototype is insert *L* of Fig.9.

The semi-annual pattern is most prevalent over the coastal areas and there it exhibits two varieties which correspond to the marine climate types; dry, e.g. insert *A* of Fig.9, and wet, insert *N*. The distinguishing characteristic of the *A*-variety is rainfall minimum

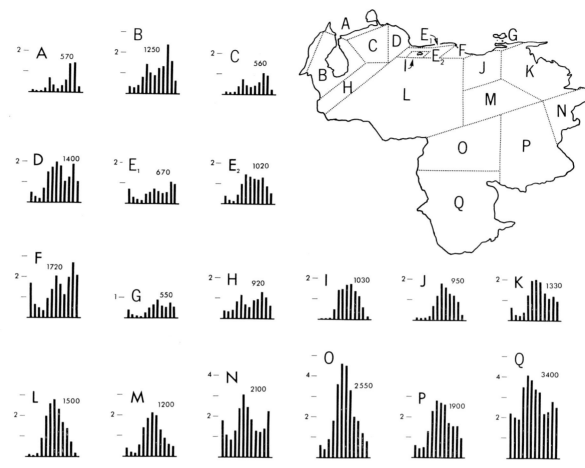

Fig.9. Annual distribution of rainfall for various sections of Venezuela. Ordinate values in mm/100, annual totals in mm. Based on period 1931–60 (certain data subjected to reduction).

attending the high-sun months of the Northern Hemisphere. The simple two-season year is obviously less applicable to this double pattern and the colloquial name "verano de San Juan" (the celebration day of St. John falling on June 24th) has been given to the secondary dry period centered on July in the *A*-variety (Röhl, 1946). The *N*-variety is directly associated with the changes in the equatorial trough as it is found over the western equatorial Atlantic.

The semi-annual pattern also is found at elevations above 1,000 m where the mountain climate type is expected, e.g. insert *H* of Fig.9, and too in the wet continental type, insert *Q*. Some evidence of a direct response to the solar input exists in the case of the *H*-variety, whereas advection from the Southern Hemisphere appears to be active in accounting for the primary rainfall maximum in the *Q*-variety.

Annual totals, areal apportionment

In general, Venezuela's annual rainfall total increases from north to south, ranging from a mere 300 mm/yr. on small islands of the southern Caribbean to somewhat more than 3,500 mm/yr. over the southernmost extension of the country. Mountainous terrain complicates this basically meridional gradient of rainfall, as can be seen from Fig.10.

Nevertheless, the simple relationship:

$$R \text{ (mm/yr.)} = 600 + 3\,D \text{ (km)}$$

approximates very closely the rainfall increase with distance (D) southward from the Caribbean coast along the longitude of Venezuela's greatest north–south dimension. This strictly empirical equation is invalid within 100 km of the coast because of the complex rainfall pattern engendered by the Cordillera de la Costa. For the remaining 900 km across the Llanos and thence along the upper Orinoco river, and on down to San Carlos de Rio Negro, it is applicable and is amazingly valid. The linear relationship serves also to point out a quizzical feature of Venezuela's rainfall. Usually maximum rainfall is associated with the principal source and ultimate sink of atmospheric water vapor: the sea. Here the antithesis obtains—the area most removed from the ocean is wettest, that surrounded by the sea is driest.

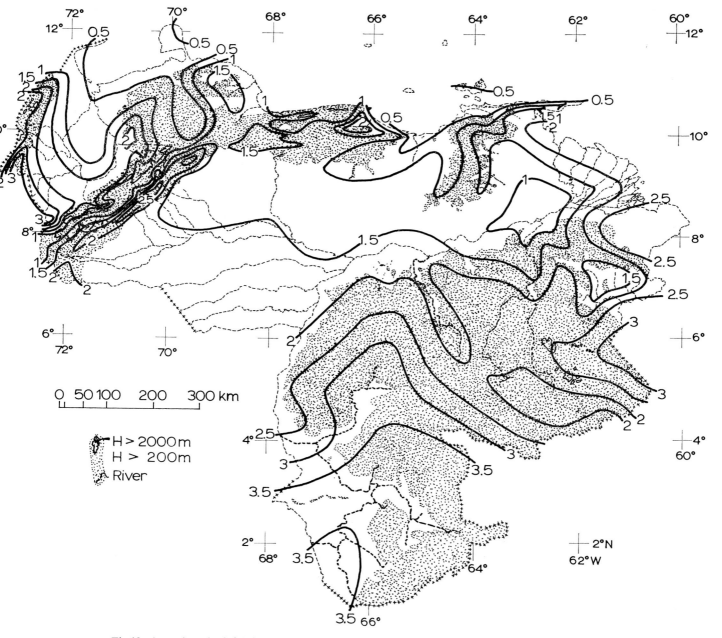

Fig.10. Annual total rainfall (m), Venezuela. Mean based on period 1931–60.

North and west section

An annual total of less than 1,000 mm/yr., frequently much less, is received along the east–west orientated stretches of the Caribbean coast line, the offshore islands and the lee (western or southwestern) sides of obstructions to the trade-wind current. The rainfall along these dry coastal areas is depicted particularly in Fig.9,A, E_1 and G. Aridity such as this is truly anomalous and is discussed in more detail on pp.348–358.

Windward slopes and basins of the Cordillera de la Costa which open eastward can expect from 1,500 to 2,000 mm/year at elevations below 500 m. These distributions are shown in Fig.9,D and F.

Within the Maracaibo basin rainfall increases outward from the minimum over the lake, in all directions except to the north. As shown in Fig.10, these increases reach and exceed 2,000 mm/yr. mainly along the more windward slopes of the Sirrea de Perijá to the west, and over the lowlands themselves southwest of the lake. A small area of the lowest slopes to the east also receives up to 2,000 mm/yr. This is seen to be the northeast end of a line of maximum precipitation which occurs, for the most part, just before the base of the Cordillera de Mérida is reached and is orientated parallel to it. Stowing of the wind ahead of the barrier is evident. Fig.9,B indicates that the basin manifests the semi-annual, or double, rainfall pattern of the dry marine climate type, in spite of rather high totals in the south and west. In the extreme south of the basin the second maximum is delayed until November or even into December. Also rainfall during the generally driest months of the "verano", January and February is not unusual here.

The modest highlands ($H \lesssim 1,000$ m) north and west of Barquisimeto (10°04′N 69°19′W, 590 m), the Segovia Highlands, are extremely dry, receiving in much of the area less than 500 mm/yr. The aridity is attributable to a combination of large scale dynamics and local causes. General subsidence over the southern Caribbean stabilizes the atmosphere and tends to confine atmospheric moisture within the lowest 3 km, i.e., below the trade-wind inversion. Thus, the rainshowers which do occur produce on the average less rain than at comparable elevations elsewhere. For example, Barquisimeto receives in the mean less than 6 mm of rain per rain-day, whereas Santa Elena receives nearly 8 mm. Concerning smaller scale influences, atmospheric pressure over the Maracaibo basin, downwind, is low and over these highlands compensating low-level divergence is present. Also upwind of the highlands are the slopes of the Tocuyo, Aroa and Yaracuy river basins. These are sufficient to trigger rainshowers within the conditionally unstable trade-wind layer and therefore a "rain shadow" is cast over the highlands proper. A more sophisticated form of subsidence may be active near the northern coast, as will be seen on pp.351–357. Within the Segovia Highlands, as in the Maracaibo basin and in all areas north of the Cordilleras, except eastward facing basins (D and F of Fig.9), the dry marine rainfall pattern prevails.

The discussion of rainfall in the mountain areas of Venezuela is complex because of diametric changes over very short distances. Inserts E_2 and H of Fig.9 give only average values derived from 19 and 27 stations, respectively, and much detail is therefore lost. It should also be noted that these inserts are strongly biased toward the more populated elevations, 800 to 2,000 m. Rather than treating in detail individual mountain regions (a procedure of limited practicality or possibility) certain generalizations will be made.

(*1*) Maximum rainfall totals, along slopes which extend above 1,000 m, occur at some intermediate elevation. The orographic lift (and during daylight the preferential thermal forcing) which reinforces convergence along inclines initially, is eventually thwarted by depletion of the total precipitable water present in the atmosphere. So, even though rainfall frequency and cloudiness may be as great or greater at higher elevation stations, the total rainfall received is less. Whereas Mérida and Santa Elena receive on the average approximately 8 mm per rain-day, San Fernando and Puerto Ayacucho receive 12 mm. Intensity of rainfall in mm/rain-day or mm/rain-hour both appear to have their maximal values well below the 1,000 m elevation. The elevation at which the rainfall total (mm/ year) reaches its maximum varies somewhat with location but in general lies between 500 and 1,500 m. Along the northwest-facing slopes of the Cordillera de Mérida, i.e., in the southeast Maracaibo basin, the maximum is found at lower elevations, possibly as low as 400 m, which indicates high incidence of nocturnal rains. On the slopes of the Tuy river basin maximum amounts also occur well below 1,000 m. Along the Llanos-facing slopes of the principal Cordilleras this elevation is in the neighborhood of 1,000 m. Notwithstanding local exceptions, above 2,000 m smaller precipitation totals are well established and universal. It is likely, therefore, that the minimum annual total in Venezuela is received not over the small offshore islands but rather on the highest portions of the Cordillera de Mérida.

(2) Orientation and slope are important local determinants of precipitation in the mountains. Totals along windward slopes of barriers are reinforced, along the leeward slopes suppressed, the "rain shadow" effect. However, isolated promontaries, which allow the atmosphere to flow *around* as opposed to *over*, may actually lead to rainfall augmentation over localities on their leeward sides (LÓPEZ and HOWELL, 1961). The presence of an atmospheric discontinuity such as the trade-wind inversion appears to reinforce this lee side enhancement.

(3) Valleys and canyons are prone to night-time precipitation, whereas ridges tend to receive their rainfall during the daylight hours (FLOHN, 1969).

Llanos

An annual total of 1,500 mm/year and a late-April through October wet season are characteristics of the Llanos which show remarkably little variation, considering the vastness of this tract. Some increase in total is nearly universal around the periphery but it extends to 2,000 mm/yr. only in 3 small areas: (1) close to the delimiting elevation of the Llanos ($H = 300$ m) against the steep Cordillera de Mérida; (2) near the upper Arauca river; and (3) over the coastal fringe of the Orinoco delta. Less than 1,000 mm/yr. occurs only: (1) on the lee side of the Cordillera de la Costa, south of the Tuy basin; (2) over the depression of the Unare river (Fig.9,*J*); and (3) in a small area northeast of Ciudad Bolivar (08°09′N 63°33′W, 50 m). Irrespective of the low total, all these areas exhibit the distinctive Llanos pattern. July is generally the month of maximum rainfall in the Llanos, the only widespread exception being found in the southwest and west where June is the wettest month. February–March is ubiquitously the driest period. The climate type is everywhere in the Llanos what is herein termed continental.

The contrast between the "verano" (dry, warm) season and the "invierno" (wet, cool) is striking throughout the Llanos, but is especially so in the far southwest, the region of the "médanos" (dunes), between the Arauca and Meta rivers (RAMIA, 1959). These are to a large degree inundated during the "invierno" and the area is traversable only by shallow boat. A large portion of the water has not fallen locally, but rather is en route to the Orinoco river from higher elevations to the west. The dry season reveals the "médanos" which support only sparce vegetation. But the numerous streambeds between these sand banks, though empty of flowing water, are then filled with tender grasses. Cattle grazing in the extreme southwest region of the Llanos is therefore superior during the "verano". This is not the norm for the Llanos, however. In the major portion of the extensive grasslands the reverse is true. The wet season does not result in widespread inundation and the higher ground between streams is then covered with grasses. During the "verano" ground water levels are generally too low to maintain a verdant grass cover even in the empty stream beds. Fortunately, the technology of man is now being applied to impound the superabundance of wet season runoff not only to control "invierno" flooding but also to permit irrigation of these rich grazing lands during the "verano". The first dam in a planned series created the 240 km² Embalse de Guárico which now supports a successful irrigation project (BLANCHARD, 1964).

At one time or another during the "verano" nearly all the grasslands of the Llanos are deliberately burnt (RAMIA, 1959). The benefit or damage attending this practice is a controversial question. However, there is little possibility of curtailment since such burning is deeply entrenched in the tradition of the indigenous inhabitants and the probability of accidental ignition of unburnt dry grass is high. The resulting haze and smoke is widespread and is a definite modification of the atmosphere wrought by man.

Further investigation of the wet season in the Llanos shows that the duration varies somewhat with location, beginning earlier and ending later in the south and over the foothills of the Cordillera de Mérida than elsewhere. The months April through November inclusive constitute the "invierno" for these locations. In the central Llanos it is late-April to the end of October, along the northern portion May through early-November. In the lower Orinoco the rains do not begin until mid-May but persist well into November. Further eastward still, as the equatorial North Atlantic is approached, the double rainfall pattern, wet marine variety, begins to be evident (Fig.9,*K*) with a weak secondary maximum appearing in November.

Llanos pattern

The Llanos rainfall pattern, as opposed to the double pattern, is more difficult to explain considering the strong semi-annual variations in solar input and of the North Atlantic subtropical high. The single peak pattern, though ubiquitous over the large, uniform area of the Llanos, is the exception rather than the norm throughout the extensive region treated in this chapter. It is indicative of persistent and prolonged residence of the most active portion of the equatorial trough counterpoised by its complete absence. This phenomenon of a single, high-sun rainfall maximum is too remindful of the large majority of the monsoon areas in Africa, India and southeast Asia (RAMAGE, 1971, fig. 2.6) to forebear from further comment.

Fundamentally, the monsoon climate is based on differential heating, usually of continent vs. ocean. Quoting from the *Glossary of Meteorology*, "The primary cause is the much greater annual variation of temperature over large land areas compared with neighboring ocean surfaces, causing an excess of pressure over the continents in winter and a deficit in summer . . ." (HUSCHKE, 1959). "The replacement of one current of air by another coming from a different direction" (WATTS, 1955) is the normal consequence and underlies the etymology of the word. RAMAGE (1971) outlines more stringent wind criteria for an authentic monsoon, requiring in part that: (*1*) at least a 120° shift in direction occurs between January and July; and (*2*) "the mean resultant winds in at least one of the months exceed 3 m/sec" (RAMAGE, 1971, p.6).

In the case of the Llanos the motivating differential heating cannot be between land and sea because the advance is *toward* the Caribbean during high sun and *away* from it at low. Additionally, the "replacement of one current by another" is not universally found in the Llanos as may be seen from the resultant winds of San Fernando and Ciudad Bolivar in Table XVII. The annual march of monthly departures from the annual mean of sea level pressure for San Fernando is shown in Table XVIII. (Valid criticism has been made of the magnitude of sea level pressures published for Venezuelan stations. Observed winds support the conclusion that these data are in most cases excessive. For this reason, in the

TABLE XVII

MEAN RESULTANT WINDS (1961, 1962, 1963)

	Months*	Direction (10's of deg.)	Speed (m/sec)	Steadiness	Calm (%)
San Fernando	1	07	2.9	0.83	11
	2	07	0.9	0.39	16
Ciudad Bolivar	1	05	2.0	0.91	14
	2	05	0.7	0.66	44
Puerto Ayacucho	1	01	0.9	0.62	38
	2	18	0.4	0.33	42
Santa Elena	1	06	0.9	0.62	40
	2	16	0.6	0.61	61
San Antonio	3	33	0.3	0.28	69
	4	15	4.1	0.85	16

* Period over which mean values computed: 1 = Dec., Jan., Feb., Mar.; 2 = June, July, Aug., Sep.; 3 = Jan. only; 4 = July only.

TABLE XVIII

MONTHLY DEPARTURES OF SEA LEVEL PRESSURE FROM THE ANNUAL MEAN (mbar)*

	Jan.	Feb.	Mar.	Apr.	May	June	July	Aug.	Sep.	Oct.	Nov.	Dec.
San Fernando	+0.3	−0.4	−1.1	−1.2	0.0	+1.4	+1.9	+1.3	+0.2	−0.7	−1.2	−0.7
Plesman Airport, Curaçao	+1.3	+1.1	+0.6	−0.1	−0.4	+0.4	+0.9	−0.1	−1.2	−1.4	−1.2	0.0
$\Delta_1 = $ (SF−PA,C)	−1.0	−1.5	−1.5	−1.1	+0.4	+1.0	+1.0	+1.4	+1.4	+0.7	0.0	−0.7
Uaupés, Brazil	−0.9	−0.9	−1.0	−0.2	+0.7	+1.8	+2.3	+1.6	+0.3	−0.6	−1.6	−1.5
$\Delta_2 = $ (SF−U,B)	+1.2	+0.5	−0.1	−1.0	−0.7	−0.4	−0.4	−0.3	−0.1	−0.1	+0.4	+0.8

* Period of record 1951–60. Annual mean values: Plesman Airport 1011.6 (sea level), 1010.7 (station). Uaupés 1001.0 (station). $\Delta > 0$, excess at San Fernando.

accompanying climatic tables, such data are given for Maracaibo only and in the present discussion pressure departures alone are dealt with.) It is seen from the table that the maximum departure coincides precisely with the maximum rainfall. Also entered are the mean monthly departures for Dr. A. Plesman Airport, Curaçao, N.W.I. (12°12′N 68°58′W, 8 m). Assuming approximately equal annual mean values at both stations (if published magnitudes are used the following statement is further reinforced), the "excess of pressure" is found over the Llanos during the high-sun months. Clearly, no classical monsoon is occurring here.

Nevertheless, remarkably monsoon-like phenomena are generated by the translations of the equatorial trough's principal area of convection between the Amazon basin and the Llanos. Among these is the single-peaked Llanos rainfall pattern itself and its antithesis found near the Amazon during the Northern Hemisphere low-sun months. For example, Fig.11A shows San Fernando versus Manaus, Brazil (03°08′S 60°01′W, 44 m). Likewise, Fig.1 indicates very convincingly the high-sun cloudiness maximum and the low-sun minimum over the Llanos and the Amazon basin, especially the area south of the Amazon

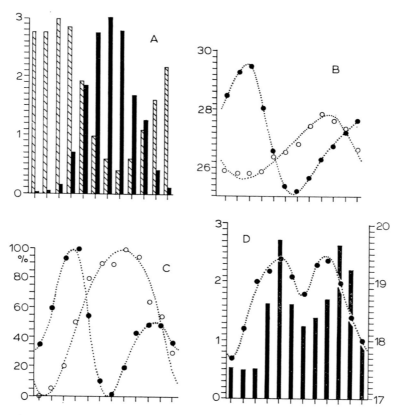

Fig.11. A. Monthly rainfall (ordinate values mm/100) at San Fernando (solid, annual total 1,491 mm) and at Manaus, Brazil (diagonal, 2,095 mm); period of mean 1931–60. B. Monthly temperature (°C) at Puerto Ayacucho (black circles, mean computed from 24 hourly values, period of mean 1958–64) and at Manaus, Brazil (open, mean computed by {ave. noon temp. + 2 (ave. midnight temp.) + ave. max. + ave. min.} 1/5, CLINO). C. Monthly relative temperature (%) at San Fernando (black circles, max. month 29.0°C, range 3.5°C) and at Maracaibo (open, max. month 28.8°C, range 2.0°C). D. Mérida monthly data: rainfall (black columns, left ordinate in mm/100, annual total 1,770 mm, period of mean 1931–60) and temperature (°C, annual mean 18.8°C, period of mean 1951–64).

river itself (refer insets January versus July, particularly). A temperature differential between the Llanos and Amazon can be found which operates in the required manner (Fig.11B). And furthermore, the resultant winds over southern Venezuela (see Puerto Ayacucho and Santa Elena in Table XVII) display a definite monsoon character. Although actual data are not available, streamline analysis of surface winds in adjacent areas suggests that in the southwest Llanos and over portions at the low mesas of eastern Colombia winds are truly monsoonal. The resultant winds at San Antonio (07°51'N 72°27'W, 404 m), a station 60 km from the Llanos proper (and probably subject to some orographic interference), are also given in Table XVII. These seem to satisfy even Ramage's strict criteria. Regarding the "excess of pressure", Table XVIII contains a pressure departure analysis for Uaupes, Brazil (00°08'S 67°05'W, 85 m), a station in the northern Amazon basin, directly south of San Fernando. Under the assumption of approximately equal annual means, the pressure excess is now in the Llanos during the Northern Hemisphere winter. In this Llanos–Amazon monsoon the roles of "continent" and "ocean" change. "Ocean" is played by the low-sun basin.

It may well be maintained that the use of the term "monsoon", if not totally inappropriate, is at least unnecessary. The translation of the equatorial trough may provide an adequate explanation. The contention is defensible as long as it allows for certain abnormalities in the translation, specifically: (*1*) the most convective area, the I.T.C., executes a rapid transit from one basin to the other after which it shows little movement; and (*2*) passage of the I.T.C. is followed by large changes in wind direction over the more interior regions.

Guyana Highlands

Moving southward and southeastward from the Llanos, up into the Guyana Highlands, rainfall totals increase, but not rapidly. The eastern half of the highlands generally receives between 2,000 and 2,500 mm/yr., but less in localities surrounded by higher terrain. La Gran Sabana, north and west of Santa Elena (refer to Fig.9,*P*) is such an instance. There open steppe and highland savanna are found. The Alto Cuyuni river valley is another example of suppressed rainfall occasioned by the "rain shadow" of adjacent ridges and subsidence over valleys during daylight hours. Tumeremo (07°10'N 61°27'W, 180 m), a station located in this basin, receives on the average only 1,250 mm/yr. The pattern of rainfall for the eastern highlands is essentially of the Llanos type, but transitions to the semi-annual pattern, the wet marine variety, occur as the coast is approached (refer insert *N* of Fig.9).

The western and southern extent in general receives 3,000 to 3,500 mm/yr. The rainiest area of Venezuela is the region of the headwaters of the Orinoco and Negro rivers. There, somewhat in excess of 3,500 mm/yr. probably falls. Long-term data are not available from the area, but 8 years of data from San Carlos de Rio Negro indicate no real dry season, only slight reductions in the abundant year-round rainfall occurring in March and September, just ahead of the northward advance and southward retreat of the Llanos–Amazon monsoon. The climate of the southern portion (roughly insert *Q* of Fig.9) is what was called wet continental in the preliminary remarks. The origin of the majority of this moisture is probably farther southward. And the occurrence of such large total rainfall this far removed from the ocean requires that a considerable portion derives from re-evaporation and evapotranspiration of prior rainfall which occurred farther down in the Amazon basin.

Temperature

Dependences

Solar radiation is the most basic determinant of temperature and as such, the measure of its annual variation at a particular location might be expected to correlate well with the annual march of temperature there. As seen from Table XVI,A, Venezuela lies within the latitudes of maximum semi-annual (second harmonic) variation of the solar radiation incident upon the atmosphere. And air temperatures at the higher elevations do respond directly to this forcing, but those near sea level generally do not.

The mean annual air temperatures throughout Venezuela span an enormous range, from below 0°C on the desolate, permanently snow-covered peaks above Mérida to more than 28°C in the vicinity of Maracaibo. Elevation is a vital and ever-present factor governing mean temperatures.

Another factor, which is especially influential for time scales on the order of one month and below, is the rain process. (By "rain process" is meant not only the receipt of rain at the surface but also the requisite cloudiness and the increased evaporation, from falling raindrops and from the earth's surface, made possible by the readily available liquid water.) Where and when the rainfall amount is excessive, i.e., runoff is produced, the rain process appears to have its greatest effect on temperature.

Along the Caribbean coastal fringe, where rainfall is generally sparse, and therefore has little impact on temperature, advection, especially the oceanic trades, becomes a factor modulating the air temperature.

In the mid-latitudes, a negative correlation is normally found between geographic latitude and mean annual air temperature. In the tropics this is not generally true. The data of Table XVI,B suggest, if any at all, a direct relationship between temperature and latitude. The correlation is very weak, at least for the area covered in this chapter, and does not merit formulation.

The relationship between temperature and elevation is, on the other hand, so predominant and consistent (being founded on the hydrostatic nature of the atmosphere itself) that a differentiation of climate in the northern Andes, based solely upon elevation was formulated (JAHR, 1934) and has since been widely used (MILLER, 1943; RöHL, 1952). Table XIX details this climatic stratification.

In the annual mean, the decrease in air temperature with elevation, recorded at surface locations within Venezuela below 2,000 m, is very close to −6.0°C/km. The rate undergoes a small annual variation from −5.8°C/km in the Northern Hemisphere low-sun months to around −6.2°C/km during the months of August and September. The mean annual rate is seen to be consistent with the values given in Table XIX.

Statements regarding the dependence of air temperature on ocean proximity are at the present time mainly qualitative. It remains a problem for the new discipline of "climatonomy" (analytic climatology) to quantify such impacts (LETTAU and LETTAU, 1974 and 1975, especially chapters X and XI). The most evident and not surprising effect is the suppression of the diurnal range. Table XX, generated from the climate tables for Vene-

TABLE XIX

CLIMATIC STRATA OF THE VENEZUELAN ANDES[1]

Designation	Limiting elevations (m)	Temperature (°C)	
		mean for limits	extremes for stratum
Tierra helada (land of frost)[2]			−8 17
...............................3,000 .... 9.0...			
Tierra fría (cool land)			0 24
...............................2,000 ....15.0...			
Tierra templada (temperate land)			8 32
.... 800 ....22.5...			
Tierra caliente (hot or warm land)			15 36
...............................sea level ....27.5...			

[1] Limits of elevation and temperature after Röhl, 1952.
[2] Translations to English by Huschke, 1959.

zuela, demonstrates the effect. The suppression is not excessive. A notable feature of the diurnal range at the low lying coastal stations, all within 15 km of the coast, is its large magnitude when compared with that of similarly located mid-latitude stations (Landsberg, 1969, p.289). This is due to the generally reduced cloudiness along the Caribbean coast where most of the stations used in the coastal area are located. (The reversal of the increasing trend in diurnal range evident in the stratum above 1,000 m is attributable to the high incidence of clouds or near-saturation conditions there.)

On the other hand, the annual range is much smaller than is found at mid-latitude counterparts. As is evident from Table XX, the annual range serves poorly as an index of continentality for Venezuela.

Often the phase of an annual oscillation of air temperature which is determined most basically by the solar forcing, is altered by ocean proximity. The ocean's great thermal

TABLE XX

TEMPERATURE RANGES (°C)

Elevation (m):	< 100 coastal	< 100 Llanos	100–1,000 Country-wide	1,000–2,000 Mountain
location:				
Range:				
Diurnal[1]	7.3	9.8	10.9	9.8
Annual[2]	2.5	2.9	2.4	2.1
Number of stations averaged	7	5	7	3

[1] Average value of mean monthly ranges.
[2] Range of mean monthly temperatures.

inertia (a deep layer of the ocean is involved in the annual heat cycle and the specific heat of water is large) is primarily responsible. In the tropics the effect is not easily seen because the annual ranges of mean monthly air and sea temperatures are small, which makes precise determinations of times of maximum and minimum difficult. Nevertheless, the delay in the minimum air temperature values on the island of La Orchila (11°48′N 66°11′W, 3 m) and at Porlamar, Isle de Margarita (10°57′N 63°51′W, 9 m) to February is attributable to this inertia.

The ocean also has some effect upon the magnitude of the annual mean air temperature of land adjacent to it; due essentially to the ocean's ability to circulate. But even a qualitative, unequivocal statement concerning the impact is encumbered by the following: (*1*) the roles of ocean and adjacent land as source or sink of heat likely change in the course of the year; (*2*) what the temperature would be without the ocean nearby is difficult to determine; and (*3*) in the tropics land–sea temperature gradients are small. The Venezuelan littoral of the Caribbean, dealt with in the section "South Caribbean dry zone" (p.348), probably acts as a source, i.e., gives heat to the Caribbean water mass, during the greater part of the year. As was indicated in the Surinam section of this chapter, the adjacent ocean is there a heat source during nearly the entire year. That state of affairs likely persists only up to the Gulf of Paria.

As with ocean proximity the effect of the rain process on air temperature is not straight forward, being dependent upon the time scale under consideration and the diurnal distribution of the rain itself. The short-term ($\leqslant$ one month) impact in regions where the bulk of the rain occurs during the daylight hours, is reductive due to the increased albedo of the clouds and the ready expenditure of surface heat for the evaporation of available water. Where night-time rains are prevalent the reverse is probably true. The long-term impact, the modification of annual mean values, is more subtle. If rainfall occurs as in the Llanos, as a purely first harmonic monsoon, then the reductions realized during the wet season must be weighed against the augmentations brought about by the paucity of clouds during the dry season. It may be qualitatively stated, however, that if rainfall occurs primarily during daylight hours in sufficient quantity to produce runoff, and is modestly well distributed throughout the year, then the mean annual temperature will be reduced.

Areal means

In concord with the aforementioned climatic stratification, as per Table XIX, and other possible modifying factors, Venezuela's highest mean air temperatures are found in the low lying areas where the influences of excess rainfall and/or the role of the ocean as a heat sink are minimized. Specifically, the more leeward lands adjacent to the Golfo de Venezuela and the northern half of the Maracaibo basin record the highest mean annual values, between 28.0° and 28.5°C. Along the remainder of the Caribbean coastal fringe mean annual temperatures are held within the range 26.5° to 27.5°C by the maritime influence. Given the relative dearth of clouds along this coast, temperature would be higher were it not for the adjacent sea.

The second warmest area is found in the Llanos, at elevations below 150 m. There the long-term mean annual values fall between 27.5° and 28.0°C. Individual years, especially in the northern Llanos, may experience means somewhat in excess of 28.0°C.

The impact of the Llanos–Amazon monsoon (discussed earlier) on the individual mean monthly temperatures is profound and evident. Its effect on the annual mean, however, is not certain. In little more than $2\frac{1}{2}$ months, as Fig.11C shows, the temperature at San Fernando falls through its entire annual range, and the mean minimum for the year actually occurs near the time of the Northern Hemisphere solstice. The first harmonic (period, one year) component of temperature variation associated with the Llanos–Amazon exchange reinforces the second harmonic component present in the solar input during the months March through July. This leads to the principal maximum and minimum occurring in April and July, respectively. Thereafter the influences of the two oscillations begin to counteract each other and a depression of temperature continues into October. The second temperature maximum of the year, characteristic of the continental climate type, nonetheless, is attained in October or November throughout the Venezuelan Llanos. During the Northern Hemisphere winter a second minimum is reached but the low temperatures which would be anticipated from the solar input alone do not materialize because at that time the monsoon temperature oscillation attains its maximum. In spite of great influence on the monthly means, the net effect of the Llanos–Amazon exchange on the annual mean temperature is probably quite small.

In the Llanos south of the Aruca–Orinoco line, the monsoon component is even stronger, so much so that the second harmonic component is scarcely visible in the temperature profile of Puerto Ayacucho (refer Fig.11B). In fact, the profile resembles that of the wet continental climate found proximately to the south. It is remarkable indeed that the atmosphere, through its hydro- and thermodynamics, can so transmute the fundamental input. Again, the annual-mean impact of the monsoon on the air temperature is not evident. The measured annual mean in this southern area is between 27.0° and 27.5°C. But somewhat higher elevation accounts for at least part of the decrease from the Llanos proper.

Southward, into the Guyana Highlands, the temperature naturally decreases due to increased elevation. Climatological data from the area are scarce, but an estimate of mean temperatures is possible by applying the surface temperature change-rate of 6°C/km to the data of the station nearest to the locality of interest. In the area of La Gran Sabana temperatures are pleasant, e.g., Santa Elena (907 m) has an annual true-mean temperature of 22.0°C (based on hourly data, 1961–64). Toward the west and south rainfall is considerably heavier and the short records from both San Fernando de Atabapo (04°03′ N 67°41′W, 124 m; RöHL, 1952) and San Carlos de Rio Negro indicate lower temperatures, specifically 26.0°–26.5°C. In fact, when all values are reduced to a common level, Venezuela's rainiest area also stands out as its coolest. Here the rain process definitely has a reductive impact which is evident even in the annual mean.

At elevations above the "tierra caliente" air temperature manifests the direct influence of solar forcing. The temperature trace for Mérida in Fig.11D demonstrates well the tendency. The impacts of ocean proximity and of the rain process are progressively diminished with elevation, leaving the large variety of temperature values recorded in Venezuela's mountains determined primarily by elevation, and secondarily by exposure.

Basic patterns

The classification of the distributions of mean monthly temperature into 2 basic patterns similar to those employed for rainfall is possible, i.e., single-peaked or double-peaked. Examples of the two basic patterns are shown, in relative terms together in Fig.11C.

A curious juxtaposition of patterns occurs within the "tierra caliente". Three types of tropical climate which together encompass nearly all of Venezuela, specifically continental, dry marine and wet continental, have temperature and rainfall patterns which differ. If rainfall attains one maximum annually, temperature attains two, and vice versa. Such a situation is quite understandable in the continental case, i.e., in the central Llanos, since on a short-term (monthly) basis rainfall and air temperature have been seen to be inversely correlated. However, the conspicuous weakness or total absence of the double-peaked temperature profile at the western stations of the Caribbean coast, where the dry marine type is best developed, cannot be explained so readily. The rainfall pattern is double-peaked but the association with temperature is not consistent. Along this dry coast, even during the wettest months, evaporation does not exhaust the supply of excess heat available at the surface (the "net radiation"). And too, Table XVI,B indicates that the solar radiation reaching the surface ("global radiation") has a semi-annual component which actually increases in strength from the Llanos to the dry coast. As will be seen in more detail in sub-part 6, dealing specifically with the dry coast anomaly, surface wind speeds respond directly to the ground level solar forcing. The months of maximum global radiation are also those of maximum windspeeds, thereby efficiently replacing the air heated by contact with the surface with cooler, maritime air. When the zenithal sun is north of the coast, daytime cloudiness increases and, expectedly, wind speeds slacken. However, this cloudiness is not productive of rain and therefore the maximum monthly temperatures of the year are then attained.

It is interesting to note that, in a sense, a full circuit has been made. The fundamental input, extra-atmospheric solar radiation, is almost totally described by a first harmonic variation at the latitudes of the northern coast. The "global radiation", however, has a strong semi-annual component. Finally, air temperature, which might well be envisioned as the final output variable, responds with a basically first harmonic pattern which parallels closely the fundamental input.

Once above the "tierra caliente" the association between rainfall and temperature becomes more straight forward, considering the essentially convective origin of the rainfall. This is shown in Fig.11D, which is a fine example of the tropical mountain climate. Here at Mérida in the upper "tierra templada" the temperature profile follows with only very small lag the solar input. Likewise precipitation varies directly but with somewhat greater lag.

Hourly distribution

In Table XXI are given the mean hourly temperatures of 4 stations, for approximately a 3-year period. The maximum longitude difference among the stations is less than 1°, while the total latitudinal span is 5°. The astronomical variation in times of sunrise and sunset are therefore less than 0.1 h. A great many discerning analyses can be performed using such data, especially if some adjustment for differing elevations is applied. The most obvious deviation from the average profile occurs at Maiquetia (10°36′N 66°59′W, 43 m) during the time of afternoon maximum. A curtailment on the order of 3.0°C is effected by the reinforced onshore winds. Also evident, when such data are analyzed in

TABLE XXI

HOURLY TEMPERATURES, MEAN ANNUAL VALUES (°C)*

Station	1st	2nd	3rd	4th	5th	6th	7th	8th	9th	10th	11th	12th	Elev.(m)
Maiquetia	24.5	24.3	24.2	24.0	23.9	23.8	24.0	25.5	27.3	28.0	28.3	28.6	43
Caracas (La Carlota)	18.9	18.5	18.2	17.9	17.7	17.5	17.6	19.1	21.7	23.6	24.9	25.9	842
Maracay	21.2	20.8	20.4	20.0	19.7	19.5	19.5	21.6	24.5	26.4	27.8	29.0	442
Puerto Ayacucho	25.0	24.6	24.3	24.0	23.8	23.6	23.5	24.7	26.3	27.7	29.0	30.1	99

Station	13th	14th	15th	16th	17th	18th	19th	20th	21st	22nd	23rd	24th	Mean
Maiquetia	28.6	28.6	28.4	28.0	27.4	26.6	25.8	25.5	25.3	25.2	24.9	24.7	26.1
Caracas (La Carlota)	26.5	26.7	26.6	26.2	25.4	24.1	22.7	21.7	20.9	20.3	19.8	19.3	21.7
Maracay	29.6	30.0	30.0	29.7	29.0	27.8	26.2	24.8	23.8	23.1	22.4	21.8	24.5
Puerto Ayacucho	31.0	31.5	31.7	31.3	30.6	29.5	28.2	27.3	26.7	26.2	25.8	25.4	27.2

* All stations except Caracas: Jan. 1961–Oct. 1964.
Caracas: Jan. 1961–Sep. 1962, Nov. 1963–Oct. 1964.

detail, is a distinct slackening of the nocturnal cooling rate beginning between midnight and 02h00 local time.

Another important item available from these data is the correction factor to be added to the "climatological-mean" temperature to obtain the "true-mean" (24-h) temperature. Table XXII gives this correction for each of the four stations of Table XXI. Two climatological means are generally used in Venezuela. The corrections to both are given. It is immediately noticed that the commonly used combination (01h30 + 07h30 + 13h30 +

TABLE XXII

TRUE-MEAN TEMPERATURES AND CORRECTIONS TO THE "CLIMATOLOGICAL" MEANS (°C)*

Station		Jan.	Feb.	Mar.	Apr.	May	June	July	Aug.	Sep.	Oct.	Nov.	Dec.	Ann.
Maiquetia	T	24.7	24.5	24.6	25.5	26.3	26.2	26.4	27.2	27.5	27.6	26.7	25.5	26.1
	A	+0.3	+0.2	+0.1	0.0	0.0	0.0	+0.1	0.0	0.0	0.0	0.0	+0.1	+0.1
	B	−0.3	−0.4	−0.4	−0.5	−0.5	−0.5	−0.4	−0.6	−0.6	−0.6	−0.6	−0.4	−0.5
Caracas	T	20.1	20.3	21.6	22.4	23.3	22.3	21.9	22.2	22.2	22.1	21.6	20.9	21.7
(La Carlota)	A	+0.4	+0.5	+0.3	+0.1	+0.1	+0.1	+0.1	+0.2	+0.2	+0.2	+0.2	+0.3	+0.2
	B	−0.8	−0.9	−0.9	−1.1	−1.0	−0.7	−0.7	−0.7	−0.7	−0.7	−0.7	−0.7	−0.8
Maracay	T	23.3	23.9	25.1	26.0	26.0	24.8	23.9	24.3	24.5	24.4	24.3	23.8	24.5
	A	+0.4	+0.4	+0.3	+0.2	+0.1	+0.1	+0.1	+0.2	+0.2	+0.2	+0.2	+0.3	+0.2
	B	−1.1	−1.0	−1.0	−1.0	−0.8	−0.8	−0.8	−0.8	−0.9	−0.9	−1.0	−1.1	−0.9
Puerto	T	28.7	29.4	29.3	28.0	26.3	25.3	25.1	25.6	26.3	26.5	27.1	27.9	27.2
Ayacucho	A	+0.3	+0.3	+0.3	+0.1	0.0	+0.1	+0.1	+0.1	+0.1	+0.1	−0.1	+0.1	+0.1
	B	−0.7	−0.6	−0.6	−0.7	−0.6	−0.6	−0.5	−0.6	−0.7	−0.7	−1.0	−0.8	−0.7
H < 400 m	C	−0.7	−0.6	−0.8	−0.8	−0.7	−0.8	−0.7	−0.6	−0.6	−0.6	−0.7	−0.9	−0.7

* Period of record same as Table XXI. T = mean of 24-h values. Correction to be added to mean computed by: A = (01h30 + 07h30 + 13h30 + 19h30)1/4, local time; B = (07h30 + 13h30 + 19h30)1/3, local time; C = (max. + min.)1/2, taken from *Promedios Climatológicos de Venezuela*, 1965.

343

19h30) 1/4 (local time) gives a very satisfactory approximation to the true-mean. The combination which excludes the first observation of the day is decidedly less appropriate. In general, the lower the station elevation the smaller the correction. Also included in Table XXII is a suggested correction to the mean determined from the maximum and minimum daily values. This correction is most applicable to stations of low elevation.

Rainfall producers

Percent of rainfall per rain-day

The preponderance of the total rainfall received in Venezuela occurs during a remarkably few days of the year. The common notion that all days are alike in the tropics is countered by the facts. Table XXIII provides a rough quantization of this statement. A more direct statement of the table's content is, for example, that only 10% of the days produced 50% of the total rainfall at Puerto Ayacucho, 90 % at Coro. These percentages of total rain are diluted somewhat when it is considered that not every day is a rain-day. The expected numbers of rain-days, for each month and station, are available from the appropriate climatic tables. During the "invierno" (May–November) only 6 days/month at Coro, but 21 days/month at Puerto Ayacucho are rain-days. The ratio, 3-days over the expected number of rain-days can be used in conjunction with the data of Table XXIII. The portions of the total rainfall, given in the table, were produced on the percent of the total rain-days determined by the ratio. Thus, at Coro, 90% of the total rain was received on 50% of the rain-days. In the southern extreme of the Llanos, at Puerto Ayacucho, where during the "invierno" rainfall is as evenly distributed among the rain-days as anywhere in Venezuela, the reader can verify that 50% of the rain total was received from only 14% of the rain-days. The underlying facts remain clear, irrespective of how expressed; the majority of rain incidents contribute little to the overall rainfall total.

TABLE XXIII

PERCENT OF TOTAL RAINFALL OCCURRING ON THE 3 WETTEST DAYS OF EACH MONTH, MAY–NOV., 1968

Coro	90
Maracaibo	80
San Fernando	60
Santa Elena	55
Puerto Ayacucho	50

Recent studies of the rain characteristics on the Mesa de Guanipa, in the northeast Llanos between Barcelona and Ciudad Bolivar, have revealed "that half of the precipitation from June to September (1969) came in only 3 h, falling in short bursts that seldom lasted more than 30 min, scattered over 14 days of the entire season." (RIEHL, 1973). The conclusion to be drawn, then, is that not the very-large-scale features which seasonally translate or change intensely, but rather smaller, transient disturbances in the atmosphere are most directly responsible for the bulk of the rainfall. And even more specifically, a few heavy rainshowers or thunderstorms within these transient disturbances are the true rainfall producers.

It is not surprising that Venezuela, located approximately astride the meteorological equator, is subject to such disturbances originating in both hemispheres. Additionally, atmospheric features with origin in, and which travel along and out of the equatorial trough affect the area.

Northern Hemisphere disturbances

The most easily specified disturbances are those incoming from the Northern Hemisphere mid-latitudes. In great part this is due to the relative abundance of synoptic data available therefrom. Two such disturbances can be distinguished: the cold front and the cold pool. These have importance only during the "verano" (December–April), only over northern Venezuela and only infrequently. However, their impact can be indeed costly.

In spite of the very low latitude and also of the intervening Caribbean, occasionally a cold front does arrive at the Venezuelan littoral. The months of February and March, the heart of the Venezuelan "verano", are preferred for such an incident. The break in the dry season which ensues is impressive and easily remembered. The singularity has occurred with sufficient frequency during the second week of March to be given the colloquial name "invierno de las chicharras", since at the same time the sound of the cicada is first heard (GOLDBRUNNER, 1963). The impact of the frontal intrusion is felt almost exclusively along the north-facing slopes of the Cordillera de la Costa where low-level convergence is assured when the reinforced onshore wind component impinges against the barrier. Even during the dry "verano" the atmospheric moisture content below the trade-wind inversion is high. The "fuel" for moist convection is always present; what is normally inadequate is low-level convergence. Probably the reinforcement of this convergence in the vicinity of the Caribbean coast of South America is the most consequential function of these fronts, as is suggested by the fact that they can pass quite inconspicuously over the open Caribbean or small, low-relief islands.

The rain induced along the Cordillera de la Costa is normally continuous in nature and the maximum amount is received at intermediate elevations along the northern slopes, specifically between 500 and 1,500 m (GOLDBRUNNER, 1963). These facts plus the condition that the soil and vegetation cover existing along these slopes are not conducive to water infiltration and retention combine to produce rapid and destructive river rises at the lower elevations. Numerous lives have been lost and considerable property destroyed by these sudden floods (RÖHL, 1950). Any effort to forest these slopes would certainly act to diminish this hazard.

The true "invierno de las chicharras" is not an every-year event, but once such a colloquialism is coined its use tends to be exaggerated. The records show that only on the order of once in 7 years do these coastal slopes record more than 50 mm/month in either February or March; once in 20 years more than 100 mm/month. Fortunately, the truly catastrophic flood is rare. During the first 70 years of the present century only one, that in February 1951, has occurred. Maiquetia recorded 470 mm during that month.

The other Northern Hemisphere disturbance, the cold pool, also traces its origin to the mid-latitudes. In this case, as opposed to the cold front, the initial effect on the tropical atmosphere takes place aloft rather than near the surface. During the Northern Hemisphere winter, cold air within a well developed and deep trough in the westerlies can become separated from its source region and continue in existence within the tropical atmosphere. The cold pool augments instability and serves as a sink for released latent heat once moist convection has begun. Such pools may develop downward into "subtropical cyclones" (SIMPSON, 1952), but probably do not proceed to this stage at latitudes as low as northern Venezuela. Nevertheless, the undeveloped cold air pools themselves do affect the rainfall here, especially over the Cordilleras and the northern Llanos.

The months January through April are most subject to the cold pool disturbance, and the frequency of occurrence is about once every 2 years (GOLDBRUNNER, 1963). The associated precipitation is intermittent in time, spotty in areal coverage, and occurs in the form of thunderstorms. Very localized areas have received as much as 150 mm/day, but more typical values are 10–20 mm/day.

Southern Hemisphere disturbances

Two disturbances, somewhat similar to those which reach Venezuela from the Northern Hemisphere, arrive from the south, especially during the height of the wet season. Ordinary synoptic data for southern Venezuela are particularly rare, mostly non-existent, and therefore little can be stated regarding the impact of such disturbances. There is no doubt, nevertheless, that they modulate the Llanos rainfall.

During the months June through October, cold fronts from the Southern Hemisphere mid-latitudes do penetrate into the central Amazon basin and occasionally even cross the Amazon river. Very convincing time series of satellite photographs exist which demonstrate that such "fronts" also cross the geographical Equator, reach the Guyana Highlands and actually enter the western Llanos via the mesas of eastern Colombia (BRINKMAN et al., 1971). RIEHL and LÜCKEFEDT (1973), from an analysis of satellite data covering 6 months during recent wet seasons, suggest a frequency of one to two instances per month for such cross-equatorial penetration of Southern Hemisphere disturbances.

Cold air pools in the mid- and upper troposphere also cross Venezuela from the Guyana Highlands moving toward the northwest. These pools, which may appear during any month of the year, are associated with abnormally high pressure over the Guyana countries and, through increased instability of the tropical atmosphere, are accompanied by characteristically unorganized but very deep thunderstorms (GOLDBRUNNER, 1963). The origin of these cold pools is not certain but it is improbable that they are the detached pendants of even well developed mid-latitude upper troughs. It is difficult to conceive of continued existence for pools with such an origin, nearly 10 degrees of latitude into the opposite hemisphere. Also if this were the case, they would show a definite preference for the Southern Hemisphere winter months. More likely, these pools originate in association with areas of organized convective activity over the northern Amazon basin and southern Guyana Highlands, possibly by the evaporation of precipitation products into unsaturated air aloft.

Equatorial trough disturbances

In spite of suspiciously high surface pressures reported by many inland Venezuelan stations, the asymptote of surface streamlines indicates that during much of the "invierno"

(May–November) the equatorial trough centerline, i.e., the I.T.C., is located not far from the line formed by the Orinoco–Apure–Masparro rivers. During the "verano" (December–April) the overland portion of the trough centerline apparently is located somewhat south of the Amazon. Thus, especially from May to November, it is to be expected that disturbances having a strong zonal component of propagation will influence Venezuela's rainfall.

The specification of all such disturbances which have as their origin, or as their region of influence the equatorial trough is, at the present stage of tropical meteorological knowledge, incomplete. Two have been formalized and these, to a limited extent, have an effect on Venezuelan weather. They are tropical storms (or hurricanes) and easterly waves. Hurricanes, in the true sense of the word (closed warm-core circulation with surface winds in excess of 33 m/sec) can be expected to strike Venezuela not more frequently than once per century (GOLDBRUNNER, 1963), and then only on the northeastern peninsulae and islands. More often, on the average of twice in 10 years, hurricanes traveling westward over the southern Caribbean exert an indirect influence on Venezuela's rainfall. Low-level mass convergence into such circulation results in abnormally strong offshore winds which result in augmented rainfalls along the Llanos-facing slopes of the cordilleras. Such induced rainfall, 10–50 mm/day, is to be expected particularly when the hurricane center has progressed to a position northwest of the Golfo de Venezuela.

The more common disturbance of low latitude origin is the westward propagating easterly wave (RIEHL, 1954). These have an effect on the rainfall over the northeastern and north-central portions of the country. The influence of such waves, which tend to manifest a repetition period of from 3 to 5 days, is probably strongest during the early and late "invierno", May–June and again October–November. During the central months of the "invierno" the primary impact of these waves is felt more over the Caribbean itself. They are also present, no doubt, during the dry "verano" months, but the strongly subsiding atmosphere then present over northern Venezuela tends to erode the convergence associated with them.

The impact on rainfall of these waves is surprisingly small, generally less than 10 mm per wave for an area mean. However, considering the frequency of these disturbances and the relative dryness of the northern coastal area, this is substantial. Where topography augments convergence, specifically over the Tuy and Tocuyo–Yaracuy basins, the passage of these waves may net 25 mm and, locally, more.

Thus far, the discussion of interdiurnal variations in rainfall, or what might more succinctly be termed storminess, has treated mostly peripheral matters, peripheral in the sense that some account has been made for only the infrequent or the very infrequent event, and also in the sense of area treated and amount of rain accounted for. Specifically, only in the case of the relatively dry northern portion of Venezuela, where disturbances similar in nature to the easterly wave may well be responsible for much of the variability, has the accounting been anything but cursory. Concerning storminess over the majority of the country, namely the Guyana Highlands and the Llanos where nearly all Venezuela's runoff is generated, practically nothing definitive has been said. Storms are hardly less a feature of the rainfall there either, e.g., it is known that at Santa Elena over 50% of the rain total occurs in less than 15% of the rain-days. Storminess is no less important in these rich and developing sectors than it is in the more populated north and west sectors.

The unfortunate fact is that little can now be said conclusively of storminess within the pith of the equatorial trough which overlies these more southern sectors during the "invierno". Dynamic theory is offering numerous and varied mechanisms, most of which are in the nature of atmospheric waves (YOUNG et al., 1972). But a precise specification of phenomena which are presently covered by omnibus terms like "equatorial wave" or "cloud cluster" has not yet distilled. And even more uncertain is the association of these with rainfall over particular areas of the tropical continents.

The future holds great promise, however, because: (*1*) the inherent nature of wave phenomena allows the possibility of long-term forecasts; and (*2*) precise monitoring of events within the equatorial trough is available through satellite surveillance, especially that provided by geostationary satellites, and by sensing in the infrared and even microwave ranges of the spectrum. However, the final test of all hypotheses and the absolutely essential ingredient in the development of usable relationships between equatorial disturbances as seen through remote sensing and their impact at a particular location on the earth's surface is accurate, representative and continuous climatological observation. Few long-term investments which a country can make in itself return higher dividends than those accrued by the climatological data effort.

South Caribbean dry zone

Extent and characteristics

A remarkable feature of the climate of northern South America is the aridity along the Caribbean coast and over the adjacent southern Caribbean itself. This dry zone is characterized as abnormal primarily by its limited annual rainfall amount, but also by the tendency for a suppression of rainfall during the Northern Hemisphere high-sun months. Table XXIV demonstrates the unusually low rainfall total received here, by comparison with similarly located areas of the Northern Hemisphere and with the entire 10°–12°N latitudinal belt. Given the low latitude of the area, the proximity of the ocean, the orographic obstruction of the predominant trade-wind flow, and the location at the western end of a subtropical high; then lush rain forests might well be anticipated. The flora is, in fact, generally scant. On the lower, coastal promontories and lee side exposures, only sparse grass or even mesquite and/or cactus are found (*Atlas Agrícola de Venezuela*, 1960, map 12).

TABLE XXIV

AVERAGE ANNUAL RAINFALL (mm) 10°–12°N

Location	Amount	Source
Central America	3,000	PORTIG, this volume, Ch.7, fig.7
Central Philippines (Visayas)	2,400	FLORES and BALAGOT, 1969, table VI
Indonesia	2,300	WATTS, 1955, fig.28
All longitudes	1,500	RIEHL, 1954, fig.3.1
Northern Venezuela and Colombia	900	Fig.10 and 13, this chapter.

Defining the dry zone as that essentially contiguous area receiving, in the annual mean, 1,000 mm or less of rainfall (LAHEY, 1958), its longitudinal extent is from the eastern tip of the Península de Paria, Venezuela to Cartagena, Colombia (62°–75°30′W). The southern limit is approximately 10°30′N, but exhibits considerable local departure from this latitude, determined primarily by topography (attention is directed to Fig.10 and Fig.13). Where mountain barriers are adjacent to the coast and aligned basically with the wind, the dry zone is confined to the narrow seaward slopes, for example insert E_1 of Fig.9. The coastal lowlands are usually more arid and are subject to greater inland intrusion of the dry zone, inserts *A* of Figs.9 and 12. This is not the case, however, for eastward facing gulfs or river basins which favor confluence and therefore increased rainfall. Inserts *D* and *F* of Fig.9 are such cases. Lee sides of barriers to the persistent east to northeast wind also display inland extension of the dry zone, manifesting the well-known "rain shadow" effect. Intermountain valleys account for the southernmost extensions of the 1,000 mm/ yr. isohyet. There, in addition to lee side subsidence, elevation itself and local mountain-valley circulations are active which lead to rainfall suppression. Such dry pockets are common throughout the higher mountains (as is more evident in Colombia), and are thus not properly a part of this arid anomaly.

The seaward extension is largely a matter of speculation since direct measurements of precipitation over the Caribbean are lacking. Nevertheless, from the percent of ship reports which indicate rainfall, LAHEY (1958, map 4) has estimated that the area receiving less than 1,000 mm/yr. of rainfall extends to the vicinity of 15°N on the western extreme of the overland dry zone, and diminishes in width east-southeastward to barely include Isle de Margarita and the eastern tip of the Península de Paria. A similarly shaped, but somewhat larger area is considered by PORTIG (this volume, Chapter 7, p.419; also fig.6, p.422, the 5% iso-line) to be the best estimate presently available for the seaward extent of the dry zone. The world charts of atmospheric heat budget components of BUDYKO (1963, chart 68) show a substantial portion of the Caribbean, north of the over-water dry area described above, wherein the atmosphere receives less than 20 kly/year (equivalent to about 340 mm of rain) from latent heat release. Recent measurements of rainfall on the small, low-relief island of La Orchila, 150 km northeast of Maiquetia, yield an annual mean of only 325 mm, which supports the mappings mentioned. On Curaçao, roughly 75 km north of the mainland, a long-term mean annual total of 532 mm is received (this volume, Chapter 7, p.461), which is comparable to the amount falling on the adjacent odd-shaped Peninsula de Paraguaná. The dry zone phenomenon, then, does exist over the sea to include at least the offshore islands. Where the impact of land is absent, it may be even more severe and extensive than present rainfall data can verify. General subsidence on a moderately large scale is suggested by the overall size of the dry zone.

Another distinctive characteristic is the inclination toward a rainfall minimum attending the Northern Hemisphere high-sun months. Inserts *A* of Fig.9 and Fig.12, representing the northernmost promontories of the Caribbean coast, exhibit best the high-sun rainfall suppression. Coastal stretches farther south and east also undergo a similar suppression, but the impact is not so immediately evident since some influence of the Llanos–Amazon monsoon (cf. the Precipitation-section, p.333) persists there. By comparison of adjacent rainfall distributions (e.g., insert *G* with *J*, or E_1 with *I* of Fig.9) the suppression during high-sun is seen to be active along these coastal areas also.

Tropical latitudes in general (WATTS, 1955, fig. 24–27) and the western oceanic areas thereof in particular (RIEHL, 1954, p.86), tend toward maximum precipitation in conjunction with the high-sun months; the time when the influence of the sun-following equatorial trough is strongest. This is the opposite tendency to that found in the maritime climates of the mid-latitudes, especially on the windward exposures of continents, where low-sun precipitation maxima are typical. The difference has led to the well-known reversal of the seasons to which the Spanish terms "verano" and "invierno" generally apply when used in the tropical Americas. However, here in the dry zone a reversion to the original, astronomical meaning of the seasons is evident. For example, both Porlamar, on the east side of Isla de Margarita, and Maiquetia, low on the face of the Cordillera de la Costa, have their maximum monthly rainfall in December and their minimum monthly temperature in February. By its anomalous nature the dry zone is in this one sense normal.

Two additional features, not so immediately evident but nevertheless unusual, warrant mention: (*1*) the marked changes in rainfall amount and/or distribution which can occur over very short distances; and (*2*) the hourly rainfall distribution.

Wherever the local orography is such that the low-level flow becomes nearly confined, steep rainfall gradients are present. An example is found in the Tuy basin (insert *F* of Fig. 9; also refer to Fig.10). Local mitigation of a larger-scale rainfall suppression mechanism is present here.

The distribution of rainfall may also change radically over short distances. The month-to-month distribution of rainfall for the vast Llanos is characterized by its pronounced maximum attending high-sun. The Llanos distribution extends all the way to the Caribbean coastline, within the Unare basin. Barcelona, at the northeast extreme of the basin and little more than 5 km from the coastline, exhibits well the Llanos pattern, with a July maximum. However, at Porlamar, only 130 km further northeast, the pattern has undergone radical change and the primary maximum occurs in December. The full comparison is given in Table XXV. While the size of the dry zone alone indicates that the anomaly is on a scale larger than local, such extraordinary differences over very short distances point out the strong impact of smaller scale processes.

TABLE XXV

COMPARATIVE MEAN MONTHLY RAINFALL (mm)

Station (years)	J.	F.	M.	A.	M.	J.	J.	A.	S	O.	N.	D.	Ann.
Barcelona (31–60)	10	4	6	7	43	99	137	111	73	66	49	24	629
Porlamar (14 < 60)*	51	38	22	15	14	30	41	51	23	31	68	100	484

* Discontinuous record of 14 years prior to 1960.

Another feature is the diurnal distribution of rainfall. Available hourly data are not extensive and therefore only a compilation of frequency of rainfall by hour is given. The data in Table XXVI are from the island of Curaçao. The diurnal land–sea circulation is especially prevalent within the tropics (FLOHN, 1969, p.140) and, therefore, it is expected

TABLE XXVI

RELATIVE FREQUENCY (%) OF RAINFALL OCCURRENCE FOR EACH HOUR (LOCAL TIME), HATO FIELD, CURAÇAO*

Hour	Mdnt.	1	2	3	4	5	6	7	8	9	10	11
Percent	3.8	4.8	5.6	5.8	5.8	5.9	5.9	5.4	4.9	4.8	4.9	4.8
Hour	Noon	13	14	15	16	17	18	19	20	21	22	23
Percent	4.6	4.1	3.5	3.4	3.5	3.1	2.4	1.9	2.0	2.6	3.1	3.4

* Hato Field same location as Dr. Plesman Airport (12°12′N 68°58′W, 8 m). Period of record, Sept.1942 to Jan.1946. Data from LAHEY, 1958, graph 3, compiled by N. Delver. Given values re-normalized to accumulated total of 100% and subjected to 3-member filtering using weights: 1/4, 1/2, 1/4.

that maximum rainfall frequency (and amount) be found over coastal land areas by day and over adjacent ocean by night. (Such a diurnal variation was seen to occur over northern Surinam.) The profile shown in Table XXVI is surely not the anticipated land portion of this local circulation. Since the data are from a relatively small island not too far distant from a large continent, it might be conjectured that the station responds as the ocean portion. But even the smallest tropical islands typically have afternoon cumulus masses associated with them and therefore something more than the weak arrest, around 16h00, of the steady daytime decline in rainfall frequency should occur. Also, if this were the ocean portion, the absolute minimum frequency would not be found at 19h00. In the tropics, and especially over oceans, low-level convergence, associated with the semi-diurnal pressure oscillation, attains one of its maxima very near the local time, 19h00 (MALKUS, 1964, fig.2). Table XXVI is the profile of neither the land nor the ocean portion of a classical land–sea circulation.

The downward trend from the maximum near sunrise proceeds, with only slight arrests, throughout the daylight hours; at night the trend is monotonically upward, advancing at nearly a constant rate until about 02h00 and thereafter continuing at a reduced rate up to the near-sunrise maximum. This nocturnal profile is very similar to the nocturnal trend of hourly temperature values, e.g., Maiquetia in Table XXI; only reflected. The suggestion of an overriding diurnal atmospheric oscillation associated directly with the solar input by day and terrestrial cooling by night is strong indeed.

Cause(s)

Dry climates are caused either by subsidence, which suppresses convection, or by lack of atmospheric moisture, which renders any convection non-productive of rain (LETTAU, this volume, pp.188–192). The second possibility can be summarily dismissed in this maritime tropical situation.

Subsidence can occur on various scales and is induced by various mechanisms. That occurring within the dry zone is not of a single scale nor from a single source. The largest scale, but not necessarily the most intense subsidence is related to the upper-level mass convergence into the North Atlantic subtropical high. Although present in the region of the dry zone, this atmospheric sinking is not a sufficient explanation because: (*1*) areas similarly located with respect to subtropical highs do not experience such aridity, e.g., Table XXIV; and (*2*) diametric changes within short distances, such as shown in Table

XXV, cannot be accounted for by such ubiquitous subsidence. The causative role of the subtropical high should not be overemphasized simply because of its large scale. It is a factor, one which is especially active during the Northern Hemisphere winter, but it cannot stand alone, not even as a cursory explanation.

Topography is perhaps the most frequently invoked explanation and rightly so. It is quite impossible to conceive of an arid anomaly having the location, shape and intensity of the one in question without the irregular, but nonetheless imposing Cordillera de la Costa and the highlands to the north and northwest of Barquisimeto (the Segovia Highlands). Such coastal inclines induce subsidence in subtle but very effective ways.

Three processes, not necessarily independent or all-inclusive, will here be mentioned. For convenience of reference these are termed: (*1*) compensation; (*2*) interruption; and (*3*) modified response.

The atmosphere seems to compensate for upward mass movement as quickly and as proximately as possible. In the case of the widespread convection over the Llanos area during its wet season, the cordilleras to the west and north serve as nodal lines for regional-scale circulations which compensate for a part of the upward motion. During the "invierno" (May to November), therefore, the dry zone (as well as the Maracaibo basin) is under the descending arm of this vertical circulation. The observed suppression of high-sun rainfall north and west of the cordilleras is in concord with such a process. During the remaining months the compensatory circulation reverses and the dry zone would then experience its wet season were it not for reinforced large-scale subsidence attending the seasonal changes in the subtropical high. As was previously pointed out, some evidence exists within the dry zone of a reversion to the original concept of "invierno" as the winter season. Indeed, there are some stations in the southern extreme of the Maracaibo basin for which December, and in certain years even January, is the month of maximum rainfall.

LAHEY (1958, ch.3) deduced from the mean wind field the vertical motion within the lower half of the troposphere over the Venezuelan portion of the dry coast and the adjacent waters, for each of the four mid-latitude seasons. He found mean subsidence during all seasons except the Northern Hemisphere winter. His deductions are likely a measure of one arm of this compensatory circulation.

In the low levels of the atmosphere such a compensatory circulation may take the form of a "foehn" if lower pressure exists downwind of the descending branch. Certain areas along the northwest-facing slopes of the Cordillera de Mérida undoubtedly experience such an effect, an example being San Antonio. However, along the Caribbean slopes of the Cordillera de la Costa it is not likely that the "foehn" condition is more than rare, due to the adverse pressure distribution which prevails.

Another process attributable to topography is the straightforward hydrodynamic interruption of the trade-wind current by the coastal cordillera. The effect upon the trades in the vicinity of the coast is increased speeds accompanied by some veering of direction. Although the process is direct, most of the consequent atmospheric sinking comes about more circuitously. Upwelling of colder waters from the Caribbean, adjacent to the off-shore islands, is greatly enhanced by the altered trade-wind flow. Subsidence then occurs over water due to the presence of a heat sink below.

An estimation of the extent and strength of the upwelling can be made from the charts of monthly sea surface temperature given by FUGLISTER (1947). In the annual mean, the surface temperature of the east-central and southwest Caribbean is approximately 27.5°C. The 27.0°C isotherm has a shape very similar to the 5% iso-line of PORTIG (this volume, Chapter 7, fig.6) for the southern Caribbean. The location of the isotherm departs from the 5% iso-line only by being displaced eastward approximately 100 km and by passing somewhat closer to the north coast of Isla de Margarita. A 26.5°C isotherm, again very similar in shape, lies shoreward of, parallel to, and at a distance of about 200 km from the 27.0°C isotherm. The continental termini of the 26.5°C isotherm lie near 66°30'W and 73°30'W. Not all waters south of these isotherms are cool, specifically the waters within the relatively shallow Golfo de Venezuela and the Lago de Maracaibo are warm. However, year round, a shield of somewhat cooler water separates the dry zone coastline from the warmer waters of the central and western Caribbean. The temperature differential between coastal and mid-Caribbean waters is not large; nevertheless the difference is significant.

The upwelling is in the nature of a positive feedback response and should not therefore be invoked as primary cause of the dry zone anomaly. Certainly, over water, the upwelling substantially reinforces atmospheric sinking and extends the area of the dryness. But over land, especially at the higher elevations, it can have no direct causal relationship to aridity. Cool upwelling and landward dryness are well correlated; not as cause and effect but rather because both are, to a great degree, effects of the same cause: topography. Whereas the upwelling arises from the orographic interruption of the trade-wind current, the landward dryness owes its existence partly to compensatory subsidence and partly to a more insidious consequence of the fact that much of the coastal area is inclined; which consequence is herein referred to as the "modified response".

The normal response of the atmosphere to an extended heat source and an extended heat sink lying side-by-side in a horizontal plane is the classical land–sea breeze. The different heating rates of the land vs. adjacent ocean results in a gradient of temperature (or, inversely, density between land and ocean), the land being the warmer area by day and the cooler by night. Near the surface the air motion is toward the warmer area. The land-sea breeze is a quasi-balanced circulation so that at some higher level, normally between 1,000 and 3,000 m in tropical latitudes, an oppositely directed return branch is expected (WEXLER, 1946). The upper return branch, directed away from the heat source, does not operate as effectively as the lower branch; being further removed from the forcing of the circulation which is wholly from below. Thus, a net transfer of mass landward by day and seaward by night occurs. The afternoon maximum in rainfall normally experienced over low elevation coastal lands is thereby explained, as is also the preferential low-level convergence offshore at night.

If the land surface is inclined, then a new dimension is added to the forcing and the response of the atmosphere is modified. The land–sea circulation is now forced not simply from below but also from the side. It is easily visualized that the upper return branch is modified more than the lower since the additional forcing is present above the horizontal plane of the classical land–sea breeze situation. The return branch is reinforced whereas the lower is relatively unaffected. While the above explanation is essentially heuristic, it is understandable that the result of the modified response can be a net mass transfer opposite to that attending the normal land–sea circulation; most importantly, low-level convergence over land during the daytime is at least impeded, if not wholly reversed.

This modified onshore–offshore circulation is one component of the total response of the lower atmosphere to the inclined heat source (or sink). The other component operates parallel to the coastline. The theory of such an atmospheric adjustment has recently been developed by H. Lettau (LETTAU and LETTAU, 1975, ch.10) and an application to the dry coast of Peru is included in this volume as Appendix I to Chapter 4. Essentially, a low-level jet is generated by the strong and persistent daytime differential in the rate at which the air immediately over the land is being heated, as opposed to the rate over the adjacent sea; in a word, by differential heating. In lieu of the phrase "differential heating" the more common term "temperature gradient" could be used, but this might be misleading. Indeed, a temperature gradient exists but to determine its strength would not reveal the true intensity of the forcing for this intriguing secondary circulation. The whole function of the circulation is to minimize temperature (or density) gradients which gave it birth and sustain it. Obviously, what should be determined is the rate at which heat is being added to (or, during the night, subtracted from) the air in contact with the land as opposed to the air in contact with the sea.

The jet parallels the coastline and, in the case of the northern coast of South America, is directed approximately from east to west. Around the jet axis a circulation across the coastline takes place which in the Northern Hemisphere operates in a clockwise sense when viewed in the direction of the jet. The function of the cross-coastline component is to transfer excess sensible heat from land to sea. The two components do not operate independently but form together a single circulation, the intensity of which is determined by the strength of the differential heating; the larger the differential the greater the wind speed and vice versa. Both the shore-parallel and the cross-coastline components accelerate and relax together. Although inseparable in reality, the cross-coastline component is here segregated and referred to as the modified response in order to distinguish it from the normal response, the land–sea breeze circulation, which is also basically a cross-coastline circulation.

The discussion which follows primarily presents empirical evidence that such a modified circulation is occurring along the dry zone coastline. The reader interested in the more theoretical aspects of such baroclinically supported secondary circulations is referred to the original publications of Lettau.

Coro, located at the base of the Segovia Highlands (and incidentally, Venezuela's oldest, continuously occupied European settlement), and Maiquetia, low on the Caribbean face of the Cordillera de la Costa, are especially interesting because of their open exposure, their dryness, and their relatively detailed climatic records. Mean rainfalls are: Coro 478 mm/yr. (1931–60); Maiquetia 551 mm/yr. (1951–66). The prevailing wind direction varies little from east-northeast year round at either location. Steadiness of the wind is remarkably high at Coro, more than 0.9, and only somewhat less, about 0.8, at Maiquetia. In such cases the terms "prevailing wind direction" together with "mean wind speed" are to a great extent interchangeable with "resultant wind". Table XXVII gives mean annual values of prevailing wind directions for four hours of the day and the ratios of mean wind speeds at each of the hours to the 13h30 value. Also included is the 13h30 mean value for all months used to generate the table. What is immediately noticed is the considerable increase in wind speeds at Coro and Maiquetia during the afternoon. The east component changes at each by nearly 4.0 m/sec between 01h30 and 13h30.

A portion of the diurnal change is due to the transfer downward of higher-level westward momentum when the lapse rate in the lower layers is steepest. Data concerning the low-

TABLE XXVII

DIURNAL WINDS, PREVAILING DIRECTION AND RATIO OF MEAN SPEED TO 13h30 VALUE (m/sec)*

Station	Local hour:			
	01h30	07h30	13h30 (m/sec)	19h30
Coro	E	E–ENE	ENE	E–ENE
	0.52	0.54	1.00 (9.6)	0.64
Maiquetia	ESE	E	ENE	E
	0.10	0.16	1.00 (4.4)	0.55
La Orchila	E	E	ESE	ENE
	0.82	0.94	1.00 (6.2)	0.95
Willemstad,	–	087°	088°	085°
Curaçao	–	0.93	1.00 (4.0)	0.98

* Data for Venezuela stations from *Boletín Bimestral*, Jan.1961 through Dec.1963. Willemstad data as given by BRAAK (1935); period of record 1910–1921, hours of observation 08h00, 14h00 and 18h00 or 19h00.

level wind structure within the dry zone are not abundant but the mean annual resultant wind speeds within the lowest 2,000 m, recorded once daily at the port of La Guaira (10°36′N 66°56′W, 5 m), less than 10 km from Maiquetia, have been published for the years 1938–41 (U.S. WEATHER BUREAU, 1945). The shear in this wind is only +1.0 m/sec within the first 1,000 m and +1.6 m/sec within the next kilometer. Also, the 4-month mean trade-wind profile for the open Pacific, given by MALKUS (1956), suggests a surface-to-1,000 m shear in the neighborhood of +2.0 m/sec and a nearly equal negative shear from 1,000 to 2,000 m. Additional insight into the impact of downward momentum transfer on the wind speed near the coastline can be gained by a comparison of the diurnal data from the island of La Orchila with the coastal stations. From Table XXVII it is seen that the 01h30 to 13h30 change on La Orchila is only about 1.0 m/sec. (In this regard, the appropriate climatic tables in the Appendix show that in the annual mean, the diurnal temperature ranges for La Orchila and Maiquetia are the same to within 0.1°C, and even for the individual months the difference never exceeds 0.4°C. Low-level instability due to buoyancy is similar at either station.) It is concluded from the limited but pertinent evidence presented that the low-level wind speeds in the vicinity of the coastline undergo a diurnal variation considerably larger than can be explained by changes in the effectiveness of vertical mixing. The excess variation (at least 2 m/sec) is due essentially to the intense and persistent daytime heating differential between the inclined land and the adjacent sea.

Further evidence exists that such a low-level jet as theorized by Lettau, is actually present along the dry coast. The resultant wind data determined from 40 months of daily wind observations at La Guaira can be decomposed into the shore-parallel and cross-coastline components. (Unfortunately, the time of day for the observations is not given in the data source (U.S. WEATHER BUREAU, 1945). However, there is strong reason to believe the time to be after sunrise, but prior to local noon.) The shore-parallel components (units of m/sec, easterly flow < 0; elevations in parentheses) are: −1.7 (surface), −3.0 (500 m), −2.8 (1,000 m). The corresponding cross-coastline components (onshore > 0) are: 0.8, 0.0, —0.9. The jet is seen to be centered near 500 m and the cross-coastline circulation operates in the sense predicted, specifically clockwise when viewed in the direction of the jet.

A significant investigation (KELLEY et al., 1968) of the atmospheric dynamics over the conspicuous Península de Paraguaná, north-northwest of Coro, discovered a telling feature in the diurnal variation of surface winds. The peninsula is perennially swept by strong, essentially easterly winds. Sunshine there is abundant, as can be seen from Table XVI, B. The diurnal variations in surface winds that would be anticipated from the straightforward heating of the lower atmosphere over the peninsula are: during the daytime, increased speeds along the east coast and decreased speeds along the west coast; during the nighttime, vice versa. Precisely the opposite variations were found. The stronger winds on the west coast occurred during the daytime, the stronger winds on the east coast occurred at night (period of the study was July and August, 1967). A persistent discontinuity was found over the peninsula within the layer 400–1,000 m, the atmosphere above being drier and decidedly more stable. This inversion-base executed a diurnal variation in height, reaching its maximum late in the forenoon and attaining its minimum height just prior to midnight. The dynamics of the lower atmosphere over the peninsula are not as expected. Though further investigation is certainly recommended, it appears that the descending arm of a modified circulation is active here during the afternoon and evening.

Other relevant features of the wind flow seen from Table XXVII are: (*1*) the suggestion of downstream acceleration over land but not over sea; comparison of differences, Coro minus Maiquetia vs. Willemstad (12°06'N 68°56'W, 10 m) minus La Orchila, suggests this; (*2*) the reluctance of the winds on the offshore island to blow from a direction more north of east during the afternoon, as a well developed land–sea circulation for the South American north coastal area would require. On La Orchila the winds actually veer to the south of east during the afternoon. The modified response can account for such a direction change since it operates most vigorously during the afternoon and, while transferring sensible heat seaward, southeasterly momentum is also imported to the offshore areas.

Two positive feedback mechanisms can be distinguished which intensify subsidence in the vicinity of the dry coast: (*1*) daytime cloud suppression; and (*2*) cool upwelled water.

The modified response operates in either direction, but its intensity varies directly with the strength of the heating differential. It is, therefore, most active during the afternoon when insolation maximizes the land–ocean heating contrast. The daytime phase of the modified response initiates a self-accelerating sequence of events. The circulation results in afternoon subsidence over land, minimizing cloudiness there. This allows more insolation to reach the land surface, thus strengthening the overland heating, which in turn reinforces the circulation, etc. etc. In contrast, the normal response during the daytime initiates a negative feedback sequence. The sea breeze results in net low-level convergence over land by day, maximizing cloudiness, which in turn restricts insolation and thus retards the circulation.

The upwelled waters intensify the cooling, from below, of the mass moving seaward during the daytime. Subsidence is thereby extended further seaward than would otherwise be the case. Attention is redirected to the rainfall frequency profile from Curaçao (Table XXVI) which can now be better understood.

Not only on the time scale of 24 h do the low-level wind speeds respond directly to the solar input (recall Table XXVII), but the same is true when means of days are considered, specifically on the time scale of 12 months. The coefficient of correlation between the mean monthly values of solar radiation incident on the surface (the so called "global radiation") at Coro (Table XVI,B) and the mean monthly wind speeds there, is no less than $+0.96$. Maximum mean speeds occur in March and August, minima in May and December, in close correspondence with the "global radiation". Insolation does not directly accelerate the wind. But by establishing and maintaining a heating differential it forces the atmosphere to adjust its pressure field. The adjusted pressure field sustains the wind. Each day the sequence takes place and occurs in proportion to the intensity of the "global radiation". (Apparently, there is little annual variation in the portion of the global radiation which is absorbed at the surface or in the mechanisms through which it is expended.) The effect carries through consistently into the monthly means.

That subsidence over the dry zone is reinforced by the lower atmosphere's response to a differential in heating which it cannot deal with through the normal sea breeze, may be appreciated from the fact that the correlation coefficient between the 12 mean monthly wind speeds at Coro, and the simultaneous mean monthly rainfalls (for the 10-year period, 1951–60) is -0.72. (The coefficient is significant at the 1% level, employing the analysis of variance testing technique. Were individual monthly values correlated, the magnitude of the coefficient and its significance would undoubtedly increase.)

Dry zone résumé

One area along the northern coast of South America and over the adjacent Caribbean is remarkably rainless in comparison to similarly located regions of the world. It extends from 62°W westward for at least 13° of longitude. Meridionally, it begins at roughly 10°30′N and continues northward over the southern Caribbean for a mostly unknown distance, but for at least 100 km. Local-scale influences can alter the amount and month-to-month distribution of rainfall markedly. Both the annual and diurnal distributions are similar and unusual for the tropics because of suppressed high-sun (June through August, or afternoon) occurrence of rainfall. The cause of the dry zone is not fully resolved. It does result from subsidence, i.e., atmospheric sinking motion, active on various scales and due to various processes. The very large scale subsidence associated with the North Atlantic subtropical high is present, but is of insufficient intensity to stifle the normal convective processes of the tropics except during the heart of the winter. The role of topography is paramount. It is very doubtful that such a dry zone would exist if a large portion of the south Caribbean coastline were not an inclined barrier. The coastal cordilleras provide a convenient nodal line about which a regional-scale compensatory circulation between the Llanos and the arid coast occurs. The peculiar suppression of coastal rainfall during the high-sun months is partly accounted for thereby. The re-direction and acceleration of the trade-wind current, caused by the interruption of the coastal barrier, strengthens upwelling of cooler waters which in turn induces and extends subsidence over the south-central Caribbean.

The crucial mechanism in the landward aridity derives from the atmosphere's response to an inclined heat source. A "desert wind" (LETTAU, Appendix I to Chapter 4 of this volume) appears to be present at least along the most arid portions of the coast. The resulting motion contains both shore-parallel and transverse components which respond directly to solar forcing. The function of this secondary circulation is to redistribute excesses of sensible heat between the land and the sea. The direction of transport is preponderantly from land to sea, as is the requisite flux of mass. The consequent subsidence is most intense during the afternoon when the rain processes over tropical land areas are usually most active. Net rainfall suppression is thus assured. Such a thermally forced circulation can have the feature of a much larger shore-parallel scale than transverse dimension. The small-scale, but distinct changes such as occur between the rainfall patterns of Barcelona and Porlamar (see Table XXV) may be based on the fact that a jet core passes between them much of the time. In spite of such local features, the large zonal extent of the dry anomaly is also allowed for.

The explanation of the entire phenomenon is presently incomplete. But theory is at the stage where guidance from well planned and carefully executed measurements is needed. Simultaneous, hourly monitoring of air pressure at the base and at an intermediate level of the Cordillera de la Costa is especially recommended. Also a precise determination of the wind structure, at all hours of the day, within the lowest 2 km near Coro would be particularly revealing. Any attempt to ameliorate the aridity, even on a local scale, must be cognizant of intermediate-scale thermally forced circulations. For example, if a low-level jet as described above is active, modification schemes with the sole objective of decreased surface albedo are predisposed to failure. On the other hand, projects which provide more water (even unpalatable water) on land, or which result in increased surface roughness may have a positive impact on rainfall. Any attempt to augment the vegetative cover appears most promising.

Colombia

Introduction

Colombia is the country of climatic extremes for the area treated in this chapter. Here are found the rainiest and the driest areas, the warmest lowlands and the coolest mountain peaks, the windiest coast and probably the calmest interior.

The total area of Colombia, a bit over 1,100,000 km², is to a large degree surrounded by aquatic boundaries. Its span of latitudes is from the Caribbean to the Amazon (12°31'N to 04°14'S); longitudes range from the Orinoco–Negro rivers to the Pacific Ocean (66°51' W to 79°03'W). The Andes reach their northern terminus in Colombia, forming a massive and complex-structured series of parallel ranges which cut across the country from its Ecuadorian border toward the north-northeast. This imposing barrier serves to naturally sectionalize Colombia. Pertinent details of the division are given in Table XXVIII, and, making use of the sub-sections listed therein, the delineation can be seen in Fig.12.

The partitioning is appropriate for the description of Colombia's climate and in the following essay each section will be dealt with separately. The climate types given in Table XXVIII are described in the preliminary remarks of the chapter. Within every section the climate is determined to some extent by orography. A great deal concerning the climates is also revealed by the vegetative cover. Therefore, as each section is discussed, a brief treatment of its geography will be presented.

TABLE XXVIII

SECTIONS OF COLOMBIA*

Section	Percent of total		Climate type	Fig.12 subsections
	area	population		
Caribbean lowlands (Llanura del Caribe)	10	15	dry marine	*A–D*
Andes (Región Andina)	20	80	mountain	*G, I–N,R₁,R₂*
Llanos (Orinoquia)	30	2	continental	*P*
Amazon (Amazonia)	30	1	wet continental	*O,S*
Pacific coast (Costa del Pacifico)	10	2	very wet marine	*F,H,Q*

* Spanish titles and certain data from *Atlas de Economía Colombiana*, 1959.

The factors which everywhere influence the climate of Colombia are, of course, the sun and the general atmospheric circulation of the tropics; but more regionally, the coastlines, the vegetation, and the mountains have very large impacts. An evaluation of the amount and annual distribution of sunlight incident upon the top of the atmosphere can be gained from Table XVI,A. Measurements of solar radiation received at the surface are not abundant but the data in Table XXIX can be used to arrive at a cursory understanding of this element as it occurs in the more populated highlands.

The equatorial trough, discussed in the preliminary remarks, is at all times present over some section(s) of Colombia, as it executes an annual translation between about the Equator and 6°N. The I.T.C. (defined as the zone of the equatorial trough where convection is most prevalent) undergoes a greater meridional excursion, particularly over the Llanos and Amazon sections. The impact over land of the equatorial trough and the presence of its most active inhabitant, the I.T.C., is often different from its effect over ocean. Much of Colombia's eastern lowlands experience the continentally modified form of I.T.C. presence which was described in the Venezuela portion of this chapter as the Llanos–Amazon monsoon and which may be seen in Fig.1, January vs. July.

The trade winds of the southern Caribbean are present over the northern lowlands and the Atrato valley during January through April. These reach abnormally far inland within the Magdalena valley. Naturally, presence of the trades is associated with a dearth of rainfall.

TABLE XXIX

MEASURED SOLAR RADIATION (ly/day)* AT GROUND LEVEL, BOGOTÁ, COLOMBIA, 1956–60

J.	F.	M.	A.	M.	J.	J.	A.	S.	O.	N.	D.	Mean
453	445	418	375	377	393	407	413	415	355	386	401	403

* Data from *Anales del Observatorio Meteorológico Nacional*, 1960.

The effect of coastlines in the tropics, where temperature and pressure gradients are typically weak, is frequently marked. Such is especially the case along the northern portion of the Caribbean and the southern portion of the Pacific coastline. Evaporation (including any process which transmits water vapor back to the atmosphere) is highly significant in the rainfall which occurs over the Amazon section and the east-facing slopes of the Cordillera Oriental. Of the more regional factors determining the climate of Colombia, by far, the most important is the Andes themselves. These not only offer direct interference with or channeling of the otherwise existing atmospheric flows, but by their effective absorption and emission of short- and long-wave radiation they serve as elevated heat sources and sinks and thereby generate important atmospheric circulations of their own. In the equatorial atmosphere where the wind speeds due to large scale circulations are small, such smaller scale circulations need always be considered. Unfortunately, wind is the common climatic element about which the least information is readily available.

Caribbean lowlands section

The pyramid-shaped Sierra Nevada de Santa Marta and the narrow northern extreme of the Cordillera Oriental, along the border with Venezuela, are the only two localities within the Caribbean lowlands section which extend above 500 m. The Sierra Nevada is, in fact, a topographic oddity, rising abruptly to nearly 5,800 m from generally level and low surroundings. The vegetation of the southern third of the section is that of swampland, the water being supplied by the converging Cauca and Magdalena rivers. Otherwise, southwest of the Sierra Nevada the character of a tropical savanna generally prevails. The vegetation becoming increasingly xerophytic as the coastline is neared and the area northeast of the Sierra Nevada massif is steppe. Desert flora, in the presence of an abundance of atmospheric moisture, is present on the outer Guajira Península. On the north and west slopes of the Sierra Nevada, limited areas of rain forest are present.

The mean annual rainfall for the section is only 1,500 mm which occurs principally from May to October, with separate maxima in May–June and September–October (see Fig. 12). The areal distribution is uniform, at least in comparison to adjacent sections of Colombia (see Fig.13). The highest annual total, possibly as high as 4,000 mm/yr., occurs in a narrow strip below 2,000 m along the west face of the massif. The minimum rain (excluding, of course, the extreme peaks of the massif), less than 300 mm/yr., occurs over the most remote and ocean-surrounded portions of the Guajira Península. Days-with-rain are few, 30 per year in the driest area, 70–80 per year in the southern part and certainly over 100 in the rain-forested bands of the Sierra Nevada. Data concerning the diurnal distribution of rain are not available.

An interesting feature of the somewhat increased rainfall on the slopes of the Sierra Nevada is that the maximum occurs on the lee slope. The explanation is based upon the fact that to a large degree the atmosphere flows *around* rather than *over* the massif (LÓPEZ and HOWELL, 1961). In combination with the presence of an inversion, such flow leads to upward motion preferentially along the lee slope. Comparison of the rainfalls for the more exposed station, Santa Marta (11°08′N 74°13′W, 1 m), and the adjacent but leeward location, Aracataca (10°35′N 74°11′W, 30 m), reveals more than a three-fold increase in annual total; from 460 to 1,600 mm/yr.

Air temperatures are high, 28°–30°C for the annual mean near sea level. Diurnal range is 9°C versus a range of 2°C between the warmest and coolest months. Mean relative humidity is in the range 75–85% except over the driest area where 65% is more representative. With minimal day-time cloudiness, insolation is strong, especially along the coast (the data for Coro, Venezuela, Table XVI,B, is indicative.) This combination of conditions, including a high rate of ventilation, permits the retrieval of salt by the evaporation of sea water on a commercial basis along the coastline.

Winds are strong for tropical latitudes over the coastal portions of the section. An annual mean resultant speed of 9 m/sec occurs at ~ 1,500 m over the peninsula with a direction varying little from east (ATKINSON and SADLER, 1970). The mean surface winds for Barranquilla (Appendix, Table XXXVI), although substantial year-round, are about 3 m/sec stronger during mid-December through mid-April than in the remainder of the year. Such an annual profile, but with lower mean speeds, is likely representative of the majority of the section. It does not apply over the peninsula or along the coastline east of the city of Santa Marta, however. There a double peak in the annual profile of wind speed will be found, maxima occurring March–April and again in August–September (parallelling closely the annual course of "global radiation"). An interesting and telling observation has been made by LÓPEZ and HOWELL (1961) regarding the winds in the driest area. At certain times of the day wind speeds up to 18 m/sec were recorded in the layer ~ 1,000 to 2,000 m. Further upward sharp decreases were encountered so that by 3,000 m, speeds were down to about 5 m/sec.

The dryness of the section during the normal dry season of the southern Caribbean, mid-December through April, may be attributed to the presence of the trade winds and their associated broad scale subsidence. During the remainder of the year the general increase in pressure, particularly from June through August, over the mountains closely to the south and east, contributes to subsidence, at least over the inland portion. The coastal aridity is the real enigma. Along the coast and over the peninsula a rather large offshore wind component can contribute to sinking motion over land, but this is not a sufficient reason; the east end of the peninsula would be very rainy if it were. Some other feature of the lower atmosphere daily prevents development of the anticipated and normal land–sea circulation which would augment rainfall. The feature is not the trade winds themselves since the annual rainfall distribution is not in concord with the variation in the strength of the trades. It is likely that a circulation on a much smaller scale than the trades and similar to the low-level coastal jet stream mentioned in the final pages of the Venezuela part is active here.

Andes section

Geography

Topographically the Andes section consists of three nearly parallel and nearly north–south orientated mountain ranges which merge into one very large range near the Ecuadorian border. Between the ranges are two river valleys, the high and narrow Cauca valley to the west, and the low and broad Magdalena valley to the east. The least imposing range is the westernmost, the Cordillera Occidental which has average ridgeline elevation of ~ 2,000 m and an average width at the 1,000 m contour of 40 km. The

Cordillera Central is the highest (ridgeline elevation 3,000–3,500 m) and most prominent of the ridges. Certain of its peaks extend above 5,000 m and are glacier capped. Its typical width is 80–100 km; shear slopes are found along both sides south of latitude 6°N. North of that latitude the Cordillera Central decreases gradually in elevation and broadens before terminating in the lowlands near the confluence of the Cauca and Magdalena rivers. The easternmost range, the Cordillera Oriental is wide, 100–200 km, and generally lower than the middle ridge, specifically about 2,500 m for an average ridgeline elevation. Much of its area, particularly between 4° and 8°N consists of a wide, high-elevation (1,500–2,500 m) plateau, the Altiplano.

Forests cover little of the Andes section due to a combination of factors: (*1*) at high elevations, the natural precariousness of forests near the treeline; (*2*) at lower elevations large areas of steep slopes and the deforestation carried out by the population; and (*3*) at all elevations apart from the peripheral slopes of the section, the limited, quickly disposed of and poorly distributed rainfall. There is little doubt that within the last 3 centuries, particularly the present, the forest-covered area of this section has greatly decreased under the impact of population expansion. The removal of forests, which had required multitudes of years to become established along steeply inclined terrain, has introduced a severe soil erosion problem which now plagues much of the section.

The vegetation cover consists primarily of grasses and low bush growth. Trees are typically confined close to rivers or on consistently windward slopes. The only extensive rain forest area is the lower Magdalena valley and its adjacent slopes. Considerable areas of the lower and middle slopes are under cultivation, principally coffee growing.

Rainfall

The Andes section is much like a "dry island" in a "sea of rain". Excluding the exterior slopes which are taken to include insert *G* of Fig.12, the mean rainfall for the entire section is near to, but less than 1,500 mm/yr. Along the exterior slopes, however, the value exceeds 3,000 mm/yr. to the east and to the north (insert *G*), and is nearly 6,000 mm/yr. to the west (see Fig.13). The reduction over the "dry island" can be attributed to the depletion of precipitable water due to increased height in the atmosphere and to the priority of the exterior slopes in the extraction of water from moist air advecting into the section.

The complex topography makes all generalization concerning the areal and diurnal distribution of rainfall subject to exceptions. Nevertheless, similarities do occur throughout the section with sufficient frequency to make a few generalizations possible and useful in understanding the complex rainfall pattern.

(*1*) Along slopes the maximum rainfall is received at some intermediate elevation, not normally above 1,500 m. If the rainfall is primarily nocturnal the elevation of maximum is much lower; if primarily afternoon rain then the maximum occurs near to or above 1,000 m.

(*2*) Comparing opposite slopes of major valleys, if windward and leeward can be definitely identified, the windward slope receives approximately double the rain amount of the leeward slope.

(*3*) Slopes located across a valley from a major saddle in the peripheral ridgelines receive considerably more rainfall than would otherwise be expected; whereas the saddle itself is unexpectedly dry.

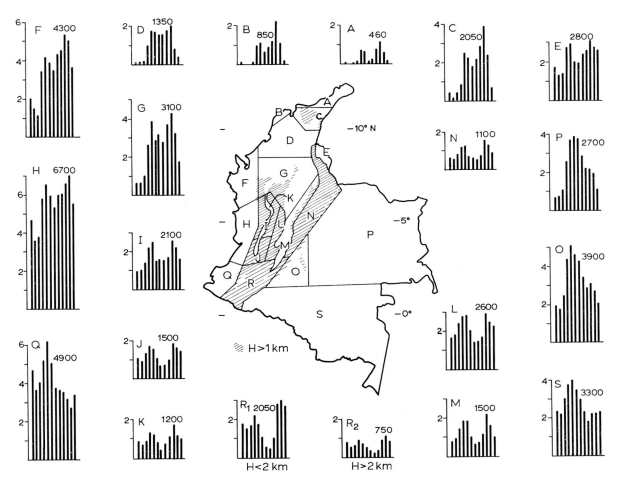

Fig.12. Annual distribution of rainfall for various sections of Colombia. Ordinate values in mm/100, annual totals in mm. Based on period 1931–60 (certain data subjected to reduction).

(4) Valley rain occurs primarily at night, slope or Altiplano rain during the daytime, especially the afternoon (see Fig.14). (Also, an illustration for the Cauca valley is given by FLOHN, 1969, fig.13).

(5) The higher the elevation of the floor of a major valley, the lower the rain total; specifically if the floor elevation is $\geq$ 1,000 m, the annual rainfall is $\lesssim$ 1,000 mm.

(6) Above 3,000 m the annual rain total is less than 1,000 mm.

The "dry island" would be considerably drier were there not avenues by which moist air penetrates deep into the section. The principal avenue is the Magdalena valley and secondarily the Cauca valley. As a rule the precipitation deriving from such moist air intrusions occurs along the slopes of the valleys rather than over the valley floors. Examples are found in both major valleys between 6° and 7°N where more than 3,000 mm/yr is received, particularly along the west (windward) side of the valleys. Saddles in the exterior ridgeline provide an alternate source of relatively unmodified moist air. The results are contained in generalization (3) above; the phenomenon has been explained by LÓPEZ and HOWELL (1967). Examples of the maximum are found near 3°N in both major valleys, against the Cordillera Central's slopes. In the Magdalena valley case, both the

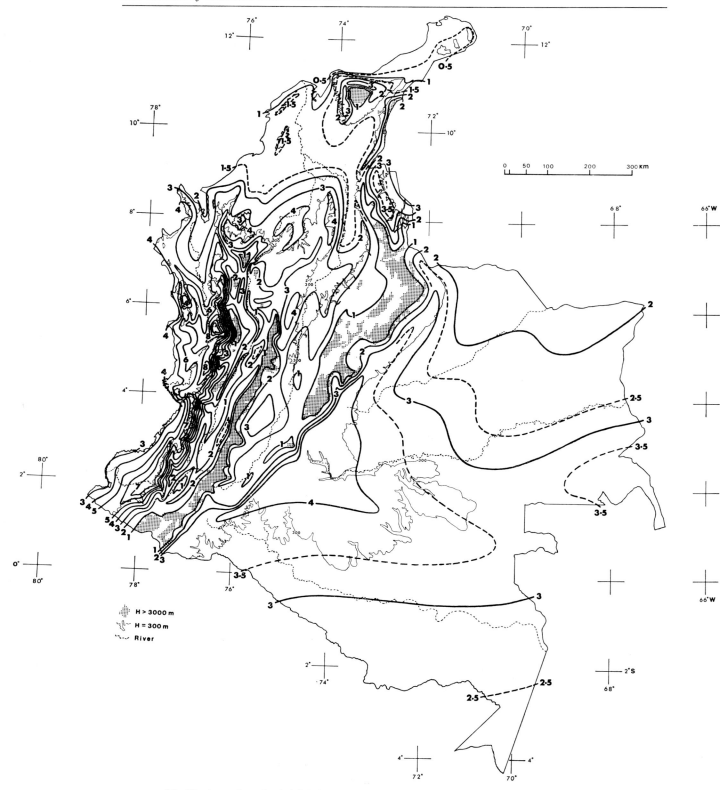

Fig.13. Annual total rainfall (m), Colombia. Mean based on period 1931–60.

saddle and up-valley sources combine to provide this far-inland area with more than 3,000 mm/yr. (see Fig.13).

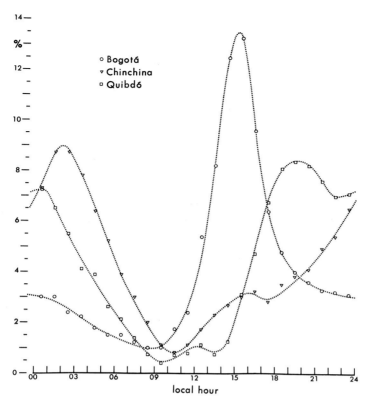

Fig.14. Hourly distribution of rainfall amount (percent of total), average of all months. Station (coordinates, elevation, period of record): Bogotá (Table I, 31–60); Chinchina (04°58′N 75°37′W, 1,360 m, 63–65); Quibdó (05°42′N 76°39′W, 54 m; TROJER, 1959, fig.14). Even-weight, 3-member moving averages plotted for Chinchina.

The Cauca valley being the higher and narrower is, not surprisingly, the drier, generally receiving 1,000 mm/yr. or less over its floor and 2,000 mm/yr. in narrow bands along both slopes. The floor of the Magdalena valley receives about 1,700 mm/yr., its windward (west side) slope 3,000 mm/yr., but its leeward slope little more than 1,500 mm/yr. The deep extension of the Caribbean lowland's dryness up the valley floor nearly to 8°N is due partially to increased low-level wind speeds in this throat of the valley. The second dry area of the valley floor, between 4° and 5°N, is due primarily to subsidence on the lee side of the imposing barrier to the east. The large cross-valley maximum between 6° and 7°N is due largely to nocturnal convection and owes its existence to the extensive areas of high plateau and slopes to the east. From these drain nightly large masses of cool air reinforcing low-level convergence in the valley.

The area of insert *G* of Fig.12 is actually a transition zone between the dry Caribbean lowlands and the Andes. Throughout, similar amounts and annual distributions of rainfall are found which qualify it as a separate entity. And by virtue of its considerable high terrain it is considered here as part of the Andes section. However, its high rainfall total excludes it from the "dry island" concept and shows it to be in fact part of the surrounding "sea of rain", i.e., an exterior slope. The origin of the considerable rainfall here is essentially the moist low-level northerly flow from the Caribbean. More particularly, two mechanisms initiate the heavy rain. By day orographic lifting over somewhat warmed slopes accounts for the areas in excess of 4,000 mm/yr. along the slopes of the northern extreme of the Cordillera Occidental and the west side of the entrance to the Magdalena valley. By night the third area receiving more than 4,000 mm/yr., the lowlands in the immediate vicinity of the junction of the Cauca and Nechí rivers, is strongly subject to cool air drainage from the large area of slopes north of Medellín, thus inducing low-level convergence.

The largest block of highland-area is that north and northeast of Bogotá, the Colombian Altiplano. Daytime cloudiness is abundant here but not especially productive of rain, at least by tropical standards. Not more than 1,000 mm/yr., and in much of the area less, is received. The mid-afternoon is the greatly preferred time of occurrence (see Fig.14, Bogotá).

The annual distribution of rainfall is similar throughout the Andes section. Two wet and two dry seasons are easily distinguished: mid-December through mid-March, dry; mid-March through May, wet; June through September, dry; October through mid-December, wet. The two wet seasons are of comparable length and intensity except in the upper Magdalena valley and in the mountains adjacent and south, where the first wet season is more protracted. The two dry seasons are differentiated by the greater length of the one accompanying the months of Northern Hemisphere high-sun. This is the true dry season for the Colombian Andes because the depletion of soil moisture is then maximum. An exception to the high-sun principal dry season is found in the area of insert *G*, Fig.12, where the Northern Hemisphere trade winds are so strongly influential that the low-sun dry season predominates.

A somewhat surprising feature of the annual distribution of rain in the Colombian Andes is the high-sun dry season, which contrasts so conspicuously with the distribution in the Llanos, closely to the east. Basic meteorological principles suggest the formation of thermally induced low pressure, and the continuity of the large scale circulation dictates that the center of the equatorial trough be in the vicinity during the high-sun months. These considerations are overridden essentially by the fact that the section is of high elevation ($\sim$ 2,000 m). Heating of the high terrain decreases the density of the air near the surface, which initiates a mass flow toward the mountain area within the lower atmospheric layers. Overall upward motion in the first kilometer or so of atmosphere above the mountain block is assured. However, an implication of the terrain's high elevation is that adjacent land is relatively low and slopes (the exterior slopes) must exist. Inspection of Fig.12 reveals that surrounding the Andes section, the high-sun rainfall is substantial, specifically in the areas of inserts *G*, *H*, *Q*, *P* and *O*. Deep convection is then occurring *around* the "dry island" and this leads to a compensating sinking motion in the upper troposphere over the mountains. Such an explanation is supported by the prevalence of non-rain producing clouds over the Altiplano and by the relatively high surface pressure values recorded at stations in the mountain area from May through August (see the climatic table for Bogotá). A speculative implication is that years of abnormally low high-sun rainfall over the exterior slopes should coincide with above average dry season rain over the Andes themselves.

Other climatic elements

Statements concerning temperatures without concurrent specification of elevation, can be deceptive due to the strong interrelation of the two. Table XXX contains the mean annual rate of decrease of temperature with terrain elevation for the Colombian Andes.

The occurrence of very high terrain requires a slight modification and extension of the climatic stratification detailed in Table XIX, for improved applicability to the Colombian Andes. The forestless stratum from 3,000 m to the lowest permanent snowline, approximately 4,700 m, is called "páramos". Here the mean annual temperature ranges from 12° to 0°C. The stratum of permanent snow and ice, mean temperature less than 0°C, is referred to as "nevado".

TABLE XXX

RATE OF TEMPERATURE DECREASE

Span of elevations (km)	Rate of decrease (°C/km)
4–5	7.0
3–4	6.0
2–3	5.0
1–2	6.5
< 1	4.5
	5.8 mean

Naturally, the highest mean temperatures are found at the lowest elevations of the section, specifically in the Magdalena valley. Annual means of 29°C are found from Barrancabermeja northward along the river and within a small area of the valley floor (elevation ~ 350 m) between 4° and 5°N, where the annual rainfall total is less than 1,500 m. In general, the floor of the Magdalena valley has an annual mean of 27°–28°C, while that of the Cauca valley is 24°–26°C. Slopes can anticipate 18°–24°C with typically larger diurnal ranges (~ 12°C) than are found in the valley bottoms (9°–10°C). The somewhat unexpected arrangement of diurnal ranges (GEIGER, 1969, table 7) may be linked to the high moisture content of the air, particularly in the valley bottoms. The Altiplano is cool for the tropics, 12°–17°C, and experiences the least annual variation in temperature of the section; no more than 1°C separates the means of the warmest and coolest months. The influence of year-round cloudiness is evident. Irrespective of the small range, the annual course of mean monthly temperatures exhibits two distinguishable maxima occurring during the wet seasons, which is characteristic of the tropical mountain climate.

The mean relative humidity is moderate in the most inhabited areas of the section, 70–75%. The annual range of mean monthly values is small, only about 10%, in spite of large variations in monthly rainfall. The cooler temperatures of the dry seasons along with only slightly reduced absolute moisture contents of the air during these months combine to prevent any substantial decrease in relative humidity. Additionally, local wind circulations within the lower atmosphere insure an effective mixing of water vapor originating from the surface. Vapor pressures show annual ranges as small as, or in some cases even smaller than, tropical oceanic stations. In addition to the surface as a moisture source, the year round rainfalls over the Pacific coast section, the high-sun rain maximum to the east and the invasions of maritime trade winds from the north all tend to maintain the nearly constant vapor pressure in the air immediately over the Andes.

Winds within the section can generally be described as light and predominantly diurnal in nature. In the atmospheric layers immediately above the mountains there is evidence that the resultant flow is from the east or southeast during most months, and is very light, only 1–2 m/sec. Combining this with the year-round large solar radiation input and the nearly universal presence of slopes, it is anticipated that topography influences air motions actively (defined by GEIGER, 1969, p.119) in the Colombian Andes, i.e., winds are primarily of local origin. The only portion of the section where passive topographic influence predominates is probably that outlined roughly by insert *G* of Fig.12. Wind data are conspicuously unavailable, but those from the coffee plantations of Blonay (07°35′N 72°37′W, 1,235 m), Chinchiná (04°58′N 75°37′W, 1,360 m) and La Florida (climatic table LXII) show clearly the strong diurnal character and the almost complete absence of annual variation in the air flow.

Not only the more well-known diurnal up-slope and down-slope, up-valley and down-valley winds (FLOHN, 1969) are present, but associated currents which basically parallel the contours of elevation, such as outlined by LETTAU (1968), are surely active. To understand the diurnal distribution of rainfall in the Colombian Andes in anything but a superficial manner, more extensive and more detailed wind information from near the surface and within the lowest atmospheric levels above the terrain is indispensible.

Llanos section

The large area of eastern Colombia eventually draining into the Orinoco river is the Llanos section. It consists primarily of low (elevation 100–250 m) mesas with intervening rivers and streams which are especially numerous in the western third of the section. Geographically and climatologically it is the extension of the Venezuelan Llanos, and therefore much of the earlier discussion relevant thereto is here applicable. Such an expedient is in fact necessary because observational data from the area are scarce. Of the accompanying climatic tables only Arauca is properly in the Colombian Llanos. However, the tables for Villavicencio and Puerto Ayacucho, Venezuela, contain pertinent information.

The section is by-and-large open grassland. Arboreal growth is typically found only near river and stream courses. Farther southward a narrow zone of tree-and-grass wet savanna soon gives way to the type of rain forest characteristic of the majority of the Amazon basin.

Prevalence of the grass type vegetation suggests what the annual rainfall pattern confirms, namely a single wet season, April through mid-November, and otherwise a distinct dry season (see Fig.12, *P* and Fig.9, *L* and *O*). The area displays the annual change from strong presence to total absence of a modified, yet active equatorial trough. Here the climate is continental, a type in many ways suggestive of a monsoon, but one occurring between the basins of the Orinoco and Amazon rather than between land and ocean. Fig.11 contains further details of this monsoonal character. In the south of the section the climate becomes more like the wet continental type so that no pronounced dry season occurs. Thus rain forest is sustained.

The average total rainfall for the section is close to 2,800 mm/yr., but the majority of the area receives considerably less. The rainfall gradient tightens appreciably toward the south and southwest of the section as is evident from Fig.13. Over the extensive grasslands themselves 2,300 mm/yr. is a more realistic average. Away from the slopes, the diurnal distribution shows a direct response to the solar heating cycle, i.e., rainfall is largely an afternoon event. Near the slopes the response is not so straightforward and especially south of about 4°N nocturnal rains predominate.

The mean annual temperature is 26°–27°C, with the principal minimum occurring in July; minima precede and follow the rainy season. The annual range is about 3°C, the diurnal range 9°–10°C. Humidity reflects the binary character of the rainfall, being approximately 65% during the dry season and 80–85% during the wet months.

Air motions over the mesas are not known directly since the few data which are available do not allow local diurnal variations to be eliminated. However, the streamlines of resultant winds over southern Venezuela suggest, for the Colombian Llanos, north-northeast flow during the dry season and south to southwest flow during the wet part of the year.

Amazon section

This large and largely uninhabited section, draining into the Amazon river system, is rain-forest covered and topographically featureless (elevation 100–300 m) except for the low remnants of a very ancient mountain range which now constitutes the divide between the Orinoco and Amazon watersheds.

Rainfall is substantial year-round with even the driest months receiving nearly 200 mm. Such an abundance of rain suggests the continuous presence of an active equatorial trough. But, the annual distribution also displays the double rainfall maxima (see Fig.12, inset *S*), which is typical of a narrow, basically east–west orientated band (200–800 km in width and located mainly between 2°S and 3°N) of the interior Amazon basin (west of 60°W) where the wet continental climate type, described in the preliminary remarks of the chapter, is found. Harmonic decomposition of the annual distribution reveals that both the 12-month and the 6-month cycles have significant amplitudes which are superimposed upon a large average monthly value. The annual oscillation is the monsoon component, while the semi-annual reflects the translation of the I.T.C., or more fundamentally, the dual maximum in solar radiation incident at these latitudes.

The area-average of rain is high, nearly 3,500 mm/yr., which is second only to the Pacific coast section. The maximum amounts, in excess of 4,000 mm/yr., occur along the slope of the Cordillera Oriental and in the vicinity of the Cordillera Macarena (inset *O*, Fig.12) which juts out toward the east between 2° and 3°N. Definite decreases of the average annual rainfall occur south of the Equator such that along Colombia's short Amazon river boundary a long-term mean of less than 2,500 mm/yr. is likely. Also, the annual rainfall pattern over this southernmost portion of Colombia exhibits a single broad maximum which occurs from November through May. Here the first indications of the southern complement to the Llanos rainfall pattern are seen.

Temperatures in the Amazon section are moderated by the year-round abundance of water, 24°–25°C being typical annual means. Also the single annual maximum in the month-to-month progression occurs during the Southern Hemisphere high-sun months. Relative humidity is perennially high, not usually less than 80% during any months.

Wind data are practically non-existent for the section. However, an interesting short record (probably 2 years) of prevailing direction is presented in Table XXXI for the river outpost of Tres Esquinas (00°43′N 75°14′W, 220 m). Evidently during the period of the record the monsoonal influence was well developed.

TABLE XXXI

MONTHLY PREVAILING WIND DIRECTION FOR TRES ESQUINAS, COLOMBIA*

Jan.	Feb.	Mar.	Apr.	May	June
NE	NE	NE	NE/SW	SW	SW
July	Aug.	Sep.	Oct.	Nov.	Dec.
SW	SW	SW	SW	NE	NE

* From U.S. WEATHER BUREAU, 1945.

Pacific coast section

Excessive rainfall

A glance at the maps of annual precipitation for the areas treated in this chapter reveals that the crescent-shaped Pacific coastal area of Colombia is by far the rainiest. In fact, at the center of the area of remarkable amounts of precipitation, 20–40 km further up the Atrato river from Quibdó (05°42′N 76°39′W, 54 m), the rainiest locality in all the Americas is found. Table XXXII details the amounts received during the early 1950's at a station located within this area. Reduction of the annual mean in the table to the period 1932–1960, based upon the very reliable data for Andagoya, 45 km south-southwest, yields a value of nearly 13,300 mm/yr.

Topping the list of impressive statistics is the annual total received at Quibdó during 1936, an astonishing 19,839 mm. This value is quoted by SCHMIDT (1952) and is not otherwise corroborated in readily available sources. However, the work of Schmidt pertaining to Colombian rainfall is outstanding and extensively quoted. Certain precipitation records now held by Mt. Waialeale, Kauai, Hawaii and Cherrapunji, India are challenged by the excessive amounts received along the Pacific coastal area of Colombia.

TABLE XXXII

RAINFALL (mm) FOR LLORÓ, COLOMBIA* (05°31′N 76°33′W, ∼ 90 m)

Jan.	Feb.	Mar.	Apr.	May	June
1,130	1,006	956	1,036	691	1,418
July	Aug.	Sep.	Oct.	Nov.	Dec.
1,354	1,096	984	1,351	1,295	1,156

Annual values:

1952	1953	1954	mean
13,476	18,449**	8,494	13,473

* Period of record 1952–54; data from GUTIERREZ, 1959.
** Verification of this value was sought by Dr. López, who has travelled extensively in the area. He received from the chief meteorologist for the national airline of Colombia assurance that the value is very probably authentic (M. E. López, personal communication, 1973).

Geography

The Pacific coast section comprises all of Colombia west of the Cordillera Occidental. The area is essentially the 60–150 km wide crescent linking Panamá and Ecuador which consists of a broad, low coastal plain and the west-facing slope of the Cordillera. Within the plain is found a low (300 m to generally less than 700 m), rain-forest-covered coastal range, the Serranía de Baudó, which extends from within Panamá southward nearly half the length of Colombia's Pacific coastline.

Between the Serranía de Baudó and the Cordillera lies the 30–100 km wide Atrato–San Juan axis. The valleys of the Atrato river, which empties into the Caribbean, and the San Juan river, emptying into the Pacific around the low, south end of the Serranía, make up the axis. A low saddle between Quibdó and Andagoya separates the two watersheds. One of the earliest concepts for an inter-ocean canal utilized this natural connection between the Caribbean and Pacific (TRAUTWINE, 1854). The lower Atrato is almost exclusively a palm-and-grass swamp, stippled by numerous shallow lakes. Otherwise the valleys consist of low hills covered by rain forest. The center of absolute maximum rainfall lies in this axis, somewhat north-northeast of the saddle at the base of the steep slope of the Cordillera Occidental.

Adjacent to the greater portion of the San Juan river, and continuing southward along the remainder of the crescent and on into Ecuador, lie the coastal lowlands consisting also of low (30–100 m), rain-forest-covered hills. The coastal lowlands make up the major portion of the section. Inland the terrain rises steeply along the Cordillera's slope, and even this is covered with dense rain forest except for spotty cleared areas, at elevations below 1,500 m, where subsistence farming is carried on. Within the Pacific coast section the predominant feature of topography is hills; of phytography, rain forest.

Climate

The area-average rainfall for the Pacific coast section is 5,500 mm/yr., based upon the period 1931–60. From Fig.13 it can be seen that the range is from about 3,000 mm/yr. at the northern and southern extremes of the crescent to more than 10,000 mm/yr. in the vicinity of the upper Atrato river. In general the pattern is bands which parallel the Cordillera Occidental. A band of maximum amounts occurs over the inland portion of the hills, at the base of the slope. The elevation of the belt of greatest rainfall is low, from 50 m to probably not more than 250 m. This is indicative of the priority of nocturnal rain.

All gradients from the band-maximum (vertical, cross-coast, and coast-parallel) are large. Along the Cordillera's slope, rates of decrease in the vertical are typically 5 mm/m, within the range of elevations 500–2,000 m. Although the band of maximum is at low elevation, it is easily seen that the slope in general receives a very large amount of rainfall. Rates of decrease with elevation as high as 15 mm/m are encountered in certain localities. Directed inland, rates of decrease of approximately 200 mm/km are found in the vicinity of certain dry pockets (WILHELMY, 1953) along the western face of the Cordillera. Seaward from the band-maximum, the decreases are more modest but nevertheless impressive: 50–100 mm/km. The coast-parallel gradient is not as large within Colombia as in either adjoining Panamá or Ecuador. Fig.15 presents selected data along the coastline it-

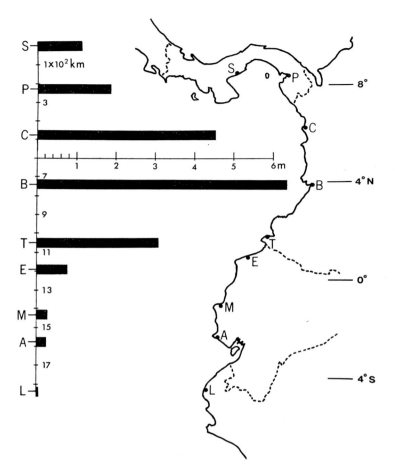

Fig.15.Annual rainfall totals (m) along the equatorial Pacific coastline of the Americas. Letter—station name (coordinates, elevation, period of record) annual total in mm: *S*—Santa Clara, Panamá (08°23′N 80°07′W, ~15 m, 24 < 67) 1,123. *P*—La Palma, Panamá (08°24′N 78°10′W, ~5 m, 14 < 69) 1,860. *C*—Ciudad Mutis, Colombia (06°13′N 77°24′W, 5 m, 6 < 46) 4,530. *B*—Buenaventura, Colombia (03°53′N 77°04′W, 12 m, 10–16 and 35–40) 6,375. *T*—Tumaco, Colombia (01°49′N 78°47′W, 4 m, 31–32 and 35–46) 3,075. *E*—Esmeraldas, Ecuador (00°59′N 79°39′W, 6 m, 44–54) 769. *M*—Manta, Ecuador (00°57′S 80°42′W, 6 m, 49–54) 283. *A*—Ancón, Ecuador (02°20′S 80°51′W, 6 m, 36–56) 256. *L*—Lobitos, Peru (04°27′S 81°09′W, 20 m, 11 < 61) 51. Inter-station distances determined from straight lines between major coastline features only.

self from northern Peru to the Gulf of Panamá. The maximum rate of change occurs along the northern coast of Ecuador; over 15 mm/km between Esmeraldas, Ecuador and Tumaco. The second greatest coastal rate of change is found along the southern coast of Panamá (cf. PORTIG, this volume, p.423, fig.7).

The area of excessive rainfall is not of great size; that receiving more than 4,000 mm/yr. is less than 1,000 km in length and typically not more than 100 km in width. The general pattern of the isohyets and available evidence (ALPERT, 1948) do not suggest that rainfall in excess of 4,000 mm/yr. extends far offshore. The cause of the abnormally high rainfall may also be of relatively small scale.

Regarding the annual distribution it can firstly be pointed out that year-round the monthly values are high, typically at least 350 mm/month. Only the northernmost extreme, e.g.

Turbo, can anticipate as little as 100 mm/month and that only during January through March. Particularly over the southern and central parts of the section, month-to-month variation is small in comparison to the average monthly value. Nevertheless, a consistent double maximum can be detected in Fig.12, insets *F, H* and *Q* which determines the basic climate type as wet marine. The essential characteristic, however, is that a very substantial portion of the rain is seasonally invariant.

Second only to the large amounts, the diurnal distribution of the rains has caused the most comment, beginning even before the first rainfall measurements were made (TRAUTWINE, 1854). DAY (1926), from an analysis of the 1914–25 record for Andagoya, found that nearly 80% of the rainfall amount was received between 19h00 and 07h00, local time. Also the mean amount falling per rain-night was 20 mm, that per rain-day (i.e., falling between 07h00 and 19h00) was 10 mm. SCHMIDT (1952) reports further that at Andagoya, on 92% of the days-with-rain, some rain is received during the night. Also 48% are exclusively rain-nights, while only 8% are exclusively rain-days. Fig.14 contains the hourly distribution of rainfall amounts for Quibdó, as extracted from TROJER (1959). Employing the same hours as used above, the proportions are 70% and 30%, preponderantly nocturnal.

Insight into the character of the rain itself is provided by LÓPEZ (1966, p.813). Of the lowlands rain he says, "Late evening or nocturnal rain predominates, usually taking the form of rather gentle showers of small drop size, which often begin and end gradually without marked wind increase or gustiness. A rainfall rate of 2 or 3 mm per hour is typical of the middle of the showers." The implication, applying the statistics of Andagoya, is that a rain-night is a 6–10 h precipitation episode. Regarding the slope of the Cordillera, López observes, "Much of the time (it is) blanketed with clouds that reach their maximum development in the late afternoon . . . the character of the rainfall is considerably more violent, and sudden showers with large drop sizes, strong cold outrushes, and high rainfall rates are common."

Daytime, at least from mid-morning to mid-afternoon, near the coastline and the outer lowlands stands in marked contrast to the rainy nights. "At midday clear or partly-clouded skies generally prevail, and bright sunlight gives a brilliant aspect to the green forested landscape", relates WEST (1957, p.37).

The number of days-with-rain, regardless of the hour at which it occurs, is high. It exceeds 200 per year throughout the crescent, with the exception of the lower Atrato valley. Close to the Pacific coastline the number is even higher than in the band of maximum rain farther inland. For these coastal areas, on at least 300 days per year rain can be expected, occurring mostly during the night. The absolute maximum number of days-with-rain was recorded at Bajo Calima (03°59′N 76°52′W, 40 m), not far from the port city of Buenaventura. There during the 3-year period 1952–54, less than 90 days, in total, were free of rain. The annual variation of days-with-rain is minimal over the central part of the crescent, e.g., at Andagoya the mean range is only from 21 to 25 days per month. Many other climatological variables exhibit very small annual variation. In most instances the prevalent tropical trait of the diurnal range much larger than the annual one is also strongly manifested. Air temperature is such an element. Mean annual values are not excessive, 26°–27°C over the greater portion of the lowlands. Slightly higher means probably occur within the middle and lower Atrato valley. Annual ranges are small, typically only 1.0°C separates the warmest from the coolest month. Diurnal ranges are in comparison large, apparently 7.5°–9.5°C. Only along the outer mangrove fringe and especially at insular locations such as Tumaco and Buenaventura will the diurnal range be less than 5.0°C.

Relative humidity is high and similarly shows minimal annual variation. An overall annual mean of 85% is representative and an annual range of mean monthly values of only 5% is likely.

Mean sea-level pressure can be estimated from a synthesis of close-by ship observations. Such an analysis by PROHASKA (1973) yields an annual mean of 1011.0 mbar for the southern end of the crescent with a 0.5 mbar decrease taking place northward across its extent. The annual range appears to be very small, only about 1.0 mbar with a broad maximum occurring during the months July–November.

The temperature of the adjacent sea is in no month less than that of the overlying air (U.S. NAVY, 1959). The difference is as much as 1.5°C during the period June–October and shrinks to near zero in February and March. In the annual mean the near-shore sea temperature is at least 27°C adjacent to the center of the crescent. This value decreases southward to about 26°C near the Ecuador border, then continues to decline at a much accelerated rate further southward. Within the Gulf of Panama the annual mean is also close to 27°C. The annual range is very small, not more than 1.0°C (PROHASKA, 1973) near the center of the crescent but shows a large increase in the Gulf of Panama and some smaller increase toward the south.

The crucial element in the understanding of the extreme wetness is the wind. Unfortunately, wind data are especially lacking and therefore recourse must be made largely to ship observations and the accounts of visiting researchers in order to construct an elementary picture of the air motions. Scant data from measurements are contained in Table XXXIII and the climatic table for Turbo.

Table XXXIII contains 3-months mean wind data derived from ship observations made in the Pacific east of 80°W, north (A) and south (B) of 5°N (the approximate latitude of Andagoya). The values north of 5°N are strongly biased toward the Gulf of Panama itself because of the terminus there of the Panama Canal. Nevertheless, it can be inferred that southwesterly winds prevail over the ocean adjacent to the southern half of the Colombian Pacific crescent, nearly year round. Over the waters to the north, the prevailing wind is from the north, but it is questionable whether this northerly flow persists near the coast

TABLE XXXIII

SURFACE WINDS, MOST FREQUENT DIRECTION AND MEAN SPEED (m/sec) IN THE VICINITY OF THE COLOMBIAN PACIFIC COAST*

	D.J.F.	M.A.M.	J.J.A.	S.O.N.	Annual
A. Gulf of Panama to 5°N; 80°W to the coast	NNW 4.6	NNW 3.1	NNW 2.7	SSW (NNW) 3.2	NNW 3.4
B. 5°N to the Equator; 80°W to the coast	SSW (NNW) 4.0	SW 3.4	SW 4.6	SW 4.3	SW 4.1

* Data from U.S. WEATHER BUREAU (1945). Period of record 1884–1933. Second-most frequent direction, if comparable, in parentheses.

as far south as 5°N during any month from May through December. A persistent south-west wind and associated oceanic current has been reported close to the coastline by MURPHY (1939) and WEST (1957).

Over the land itself, the surface wind data for Turbo (Table XLII), although of short record, clearly indicate that within the Atrato valley an annual reversal from northerly to southerly flow, from May to November, takes place. The duration of prevailing northerly flow over the land outside the Atrato valley is curtailed and probably such flow is present south of Buenaventura only temporarily during February and March. The reports of travelers and the wind data for Turbo suggest low-level convergence over the Pacific coast section during much of the year. The data of Table XXXIII indicate that this is the prevailing condition also over the adjacent sea. Even during the months when directional convergence is not present there, speed convergence is evident.

Factors effecting the raininess

Satellite cloud climatologies (such as Fig.1), atlases of ship observations, and the re-markably small annual ranges of numerous climatological variables all suggest that the equatorial trough never really vacates the eastern Pacific between the Equator and the Gulf of Panama. In addition, the annual translation of an especially convergent feature which resides in or near the equatorial trough, the I.T.C., may be detected in the sequence of insets *F, H, Q* of Fig.12. A slow southward migration can be pictured from October (*F*) to January (*Q*). Likewise a rapid northward advance occurs in May. During the months February through April, also during June through September it could be main-tained that the I.T.C. itself is located principally south, and north, respectively, of the Colombian Pacific coast.

However, it is evident from Fig.12, *H* and *Q* particularly, that a substantial portion of the rainfall received along the crescent is not attributable to seasonally translating large-scale features. Also the local circulation of great prominence in the tropics, the normal land–sea breeze, cannot be called upon to explain much of the seasonally invariant por-tion of the rain because at least 75% of it occurs during the night-time. The very large number of days-with-rain and its small annual variation suggest that the cause of the excessive rain is associated with a daily oscillation of the atmosphere. The semi-diurnal convergence–divergence cycle of the tropics, fundamentally linked to the semi-diurnal pressure oscillation, is expected to be overridden by other effects on land. However, the hourly distribution of rainfall here, especially near the coastline, does suggest that this oscillation has considerable impact. Dawn and sunset rains are favored, convection during the late forenoon and early afternoon is suppressed. Even some evidence of a respite in the nocturnal rains near midnight is seen at Quibdó in Fig.14.

It is not possible, nevertheless, to attribute the extraordinary rainfall wholly and simply to the semi-diurnal pressure oscillation. Certain important questions remain unanswered: Why should an oscillation which is generally quite ineffective in rain production over tropical lands have such an extraordinary impact here? Also, what prevents the normal land–sea circulation from developing, particularly when the forenoons are characterized as sunny? The answer to these, the final explanation of the wettest climate of the Ameri-cas, probably involves coast-parallel air currents and hinges upon the fact that a moun-tain barrier closely parallels a tropic coastline; the very same topographic arrangement which on the opposite side of the nearby Equator yields the driest climate of the Americas.

One may conjecture that H. Lettau's theory outlined in Appendix I of Chapter 4 (pp.188–192), also contains the key to the understanding of the phenomenon here discussed. However, a detailed treatment of this intriguing possibility cannot be included in the present text.

Recapitulation

All of the five climate types outlined in the initial pages of this climatology are present in Colombia, and extreme examples of each are found at surprising proximity. A straight-line distance of less than 1,000 km separates the extreme dry marine climate of the La Guajira Peninsula from the ultimate in wet marine conditions found over the upper Atrato–San Juan axis. Within these two areas, stations of comparable elevation can be found with mean annual rainfall totals of 250 mm/yr. vs. over 10,000 mm/yr. The latter area is the rainiest location in all the Americas. Glaciered peaks are found at these equatorial latitudes, practically within sight of the Magdalena river north of 7°N, where mean annual temperatures exceed 30°C. Surface winds which do exceed 15 m/sec (the mean is probably in the range 6 to 9 m/sec) over the Guajira Peninsula must be counterpoised with mean speeds as low as $\leqslant 1$ m/sec and frequent calm conditions in the Amazon section of the country. In the Andes section the most extensive area occurs of tropical mountain climate in northern South America. Also, the peculiar wet continental climate, which depends critically upon the re-evaporation and transpiration of prior precipitation for its own abundant rainfall and which has a cross-equatorial air flow as an essential ingredient, is found over the vast rain-forested Amazon section.

Appendix—Climatic tables

The climatic tables XXXIV–LXIII are shown on the pages 380–403 and are arranged from south to north.

References

ALEWIJNSE, C. P. J., D'HAMECOURT, J. L. A. and FUNG LOY, L. W., 1963. *Sibiboesi te Zanderij op 11 July, 1963*. Meteorol. Dienst, Paramaribo, 8 pp.

ALPERT, L., 1948. Notes on the areal distribution of annual mean rainfall over the tropical Eastern Pacific Ocean. *Bull. Am. Meteorol. Soc.*, 29:38–41.

ATKINSON, G. D. and SADLER, J. C., 1970. *Mean Cloudiness and Gradient-Level Wind Charts Over the Tropics*, Vol.II. Air Weather Serv., U.S. Air Force, 48 charts.

BANCO DE LA REPUBLICA, 1959 and 1962. *Atlas de Economia Colombiana*, Vol.1 and 3. Departamento de Investigaciones Economicas, Bogotá (pages unnumbered).

BERNHARDT, F. and PHILIPPS, H., 1958. Die räumliche und zeitliche Verteilung der Einstrahlung. *Abhandl. Meteorol. Hydrol., Berlin*, 45:227 pp.

BLANCHARD, W. (Editor), 1964. *Area Handbook for Venezuela*. Special Operations Research Office, The American University, Washington, D.C., 576 pp.

BRAAK, C., 1935. The Climate of the Netherlands West Indies. *Mededeel. Verhandl.*, 36:1–120.

BRINKMANN, W. L. F., WEINMAN, J. A. and RIBEIRO, M. N. G., 1971. Air temperature in central Amazonia, the daily record of air temperature in a secondary forest near Manaus under cold front conditions. *Acta Amazonica*, 1(2):51–56.

BUDYKO, M. I. (Editor), 1963. *Atlas of the Heat Balance of the Earth's Surface*. Geophysical Institute of the U.S.S.R. Academy of Sciences, Moscow, 69 charts.

CAMPAN, G., 1959. Note sur la climatologie des Antilles et de la Gyuane Française. *Monograph. Météorol. Natl., Paris*, 15:26 pp.

CLEARE, L. D., 1961. *The Climate of British Guiana*. The Government Printing and Stationary Office, Georgetown, Guyana, 65 pp.

DAY, P. C., 1926. Climatological data for Andagoya, Republic of Colombia, South America. *Mon. Weather Rev.*, 52:376–378.

DEAN, G. A., 1971. *The Three-Dimensional Wind Structure over South America and Associated Rainfall over Brazil*. Report No.71-4, Dept. of Meteorol., Florida State Univ., Tallahassee, Fla., 153 pp.

DIRECCIÓN DE PLANIFICACIÓN AGROPECUARIA, 1960. *Atlas Agrícola de Venezuela*. Litografia Miangolarra Hnos., S.A., Caracas, 107 mapas.

FLOHN, H., 1969. Local wind systems. In: H. E. LANDSBERG (Editor), H. FLOHN (Volume-editor), *General Climatology. World Survey of Climatology, Vol.2*. Elsevier, Amsterdam, pp.139–171.

FLORES, J. W. and BALAGOT, V. F., 1969. Climate of the Philippines. In: H. E. LANDSBERG (Editor), H. ARAKAWA (Volume-editor), *Climates of Northern and Eastern Asia. World Survey of Climatology, Vol. 8*. Elsevier, Amsterdam, pp.159–204.

FOUGEROUZE, J., 1965. Le climat de la Guyane Française. *Monograph. Météorol. Natl., Paris*, 38:36 pp.

FUGLISTER, F. C., 1947. *Average Monthly Sea Surface Temperatures of the Western North Atlantic Ocean—Papers in Physical Oceanography and Meteorology*. MIT and Woods Hole Oceanographic Institute, Cambridge, Mass., Vol.X, No.2, 25 pp.

GEIGER, R., 1969. Topoclimates, In: H. E. LANDSBERG (Editor), H. FLOHN (Volume-editor), *General Climatology. World Survey of Climatology, Vol.2*. Elsevier, Amsterdam, pp.105–138.

GOLDBRUNNER, A. W., 1963. *Las Causas Meteorológicas de las Lluvias de Extraordinaria Magnitud en Venezuela*, 2nd ed. Ministerio de la Defensa, Venezuela, 230 pp.

GROSH, R. C., 1971. *Cloud Photographs from Satellites as a Hydrological Tool in Remote Tropical Regions*. Master Thesis, University of Wisconsin, Madison, Wisc. 90 pp.

GUTIERREZ, M. M., 1959. *Información Climática*. Instituto de Fomento Algodonero, Bogotá, 251 pp.

HOWELL, W. E., 1967. *Rain Stimulation for the Brokopondo Watershed, Surinam*. W. E. Howell Associates, Inc., Lexington, Mass., 67 pp.

HUSCHKE, R. E. (Editor), 1959. *Glossary of Meteorology*. Am. Meteorol. Soc., Boston, Mass., 638 pp.

IM THURN, E. F., 1883. *Among the Indians of Guiana*. Kegan Paul, Trench and Co., London, 445 pp.

JAHR, A., 1934. Las temperaturas medias y extremas de las zonas altitudinales de Venezuela. *Bol. Soc. Venez. Cienc. Nat.*, 14 (as referenced by E. RÖHL, 1952).

KELLEY, J. I., HOSLER, C. L., DAVIS, L. G. and KELLEY, N. D., 1968. *The Paraguana Peninsula as a Potential Site for Weather Modification*. Dept. of Meteorology, The Pennsylvania State University, University Park, Pa., 114 pp.

LAHEY, J. F., 1958. On the origin of the dry climate in northern South America and the southern Caribbean. *Univ. Wisc., Dept. Meteorol., Sci. Rep.*, 10:290 pp.

LANDSBERG, H., 1969. *Physical Climatology*. Gray, DuBois, Pa. 2nd ed., 446 pp.

LENSELINK, H. J. and VAN DER WEERT, R., 1973. Estimating free water evaporation in Surinam. *Surinaamse Landbouw*, 21(2):22 pp.

LETTAU, H. H., 1968. *Small to Large-Scale Features of the Boundary Layer Structure over Mountain Slopes*. Colo. State Univ., Dept. Atmos. Sci., Pap.122:1–74.

LETTAU, H. H. and LETTAU, K., 1975. Regional climatonomy of tundra and boreal forests in Canada. *Proc. Alaskan Sci. Conf., 24th, Fairbanks, The Climate of the Arctic*, pp.209–221.

LIST, R. J. (Editor), 1958. *Smithsonian Meteorological Tables* (6th rev. ed.). Smithsonian Institution, Washington, D.C., 527 pp.

LÖF, G. O. G., DUFFIE, J. A. and SMITH, C. O., 1966. *World Distribution of Solar Radiation*. College of Engineering, Univ. of Wisconsin, Madison, Wisc., Rept. 21, 71 pp.

LÓPEZ, M. E., 1966. Cloud seeding trials in the rainy belt of western Colombia. *Water Resour. Res.*, 2:811–823.

LÓPEZ, M. E. and HOWELL, W. E., 1961. The campaign against windstorms in the banana plantations near Santa Marta, Colombia, 1956–57. *Bull. Am. Meteorol. Soc.*, 42:265–276.

LÓPEZ, M. E. and HOWELL, W. E., 1967. Katabatic winds in the equatorial Andes. *J. Atmos. Sci.*, 24:29–35.

MALKUS, J. S., 1956. On the maintenance of the trade winds. *Tellus*, 8(3):335–350.

MALKUS, J. S., 1964. Tropical convection: progress and outlook. In: J. W. HUTCHINGS (Editor), *Proc.*

Symp. Tropical Meteorol., *Rotorua*. New Zealand Meteorol. Serv., Wellington, N.Z., pp.247–277.

MILLER, D. B., 1971. *Global Atlas of Relative Cloud Cover 1967–70*. U.S. Dept. of Commerce (NOAA) and U.S. Air Force (AWS), Washington, D.C., 237 pp.

MINISTERIE VAN OPENBARE WERKEN EN VERKEER, 1964. *Sibiboesie te Moengo op 6 Oktober 1964*. Meteorologische Dienst, Paramaribo, 9 pp.

MINISTERIE VAN OPENBARE WERKEN EN VERKEER, 1965. *Klimatologische Tabellen, Diverse Elementen, Paramaribo, Periode 1931–1960, Maandgemiddelden*. Meteorologische Dienst, Paramaribo, 23 pp.

MINISTERIE VAN OPENBARE WERKEN EN VERKEER, 1966. *Klimatologische Tabellen, Zanderij, Periode 1952–1963*. Meteorologische Dienst, Paramaribo, 31 pp.

MINISTERIE VAN OPENBARE WERKEN EN VERKEER, 1968. *Klassificatie der Seizoenen*. Meteorologische Dienst, Paramaribo, 23 pp.

MURPHY, R. C., 1939. The Littoral of Pacific Colombia and Ecuador. *Geograph. Rev.*, 29:1–33.

OBSERVATORIO METEOROLÓGICO NACIONAL, 1961. *Anales del Observatorio Meteorológico Nacional, 1960*. Instituto Geográfico "Agustin Codazzi", Bogotá, 216 pp. (Contains 1931–60 means.)

PORTIG, W. H., 1972. *Evaporimetry in the Canal Zone, Part II, Comparison of Various Types of Evaporimeters on an Hourly Basis*. U.S. Army Tropic Test Center, Fort Clayton, Canal Zone, 23 pp.

PROHASKA, F. J., 1973. New evidence on the climatic controls along the Peruvian coast. In: D. H. K. AMIRAN and A. W. WILSON (Editors), *Coastal Deserts, Their Natural and Human Environments*. The University of Arizona Press, Tucson, Ariz., pp.91–107.

RAMAGE, C. S., 1971. *Monsoon Meteorology*. Academic Press, New York, N.Y., 296 pp.

RAMIA, M., 1959. *Las Sabanas de Apure*. Ministerio de Agricultura y Cria, Caracas, 132 pp.

RICHARDS, P. W., 1964. *The Tropical Rain Forest*. Cambridge University Press, London, 450 pp.

RIEHL, H., 1954. *Tropical Meteorology*. McGraw-Hill, New York, N.Y., 392 pp.

RIEHL, H., 1973. Controls of the Venezuela rainy season. *Bull. Am. Meteorol. Soc.*, 54(1):9–12.

RIEHL, H. and LÜCKEFEDT, W., 1973. *Tropical Convective Cloud Masses and Atmospheric Circulation*. Beilage zur Berliner Wetterkarte 96/73, July 1973, 15 pp.

RÖHL, E., 1946. *Climatología de Venezuela*. Tipografia Americana, Caracas, 95 pp.

RÖHL, E., 1950. *Los Diluvios en las Montañas de la Cordillera de la Costa*. Tipografia Americana, Caracas, 28 pp.

RÖHL, E., 1952. *Climatografía de Venezuela, Temperatura y Humedad*. Litografia del Comercio, Caracas, 223 pp.

SCHMIDT, R. D., 1952. Die Niederschlagsverteilung im Andinen Kolumbien. *Bonner Geogr. Abhandl.*, 9: 99–119.

SECOND WEATHER GROUP, 1962. *Climatological Studies, N-2, Weather and Climate of South America*. Air Weather Service (USAF), Langley Air Force Base, Va., 239 pp.

SERVICIO DE METEOROLOGÍA Y COMMUNICACIONES, SECCIÓN DE CLIMATOLOGÍA, 1961–1964. *Boletín Bimestral, No.67–90*. Ministerio de la Defensa, Venezuela.

SERVICIO DE METEOROLOGÍA Y COMMUNICACIONES, 1965. *Promedios Climatológicos de Venezuela, Período 1951–60*. Ministerio de la Defensa, Venezuela, 100 pp.

SIMPSON, R. H., 1952. Evolution of a "kona" storm, a subtropical cyclone. *J. Meteorol.*, 9:24–35.

STONE, R. G., 1941. Health in tropical climates. In: G. HAMBIDGE (Editor), *Climate and Man, 1941 Yearbook of Agriculture*. U.S. Government Printing Office, Washington, D.C., pp.246–264.

TRAUTWINE, JOHN C., 1854. Rough notes of an exploration for an interoceanic canal route by way of the rivers Atrato and San Juan, in New Granada, South America. *J. Franklin Inst.*, 57:145–54, 217–31, 289–99, 361–73; also 58:1–11, 73–84, 134–55, 217–26, 289–99.

TROJER, H., 1959. Fundamentos para una zonificación meteorológica y climatológica del trópico y especialmente de Colombia. *Cenicafé*, 10:289–373.

U.S. NAVY, CHIEF OF NAVAL OPERATIONS, 1959. *Marine Climatic Atlas of the World, Volume V, South Pacific Ocean*. NAVAER 50-1C-532, U.S. Government Printing Office, Washington, D.C., 267 charts.

U.S. WEATHER BUREAU, 1945. *Weather Summary, South America, Northern and Northwestern Part*. Hydrographic Office, U.S. Navy Department, Washington, D.C., H.O. Publ. 528, 166 pp.

WATTS, I. E. M., 1955. *Equatorial Weather*. Pitman, New York, N.Y., 224 pp.

WEST, ROBERT C., 1957. *The Pacific Lowlands of Colombia*. Louisiana State University Press, Baton Rouge, La., 278 pp.

WEXLER, R., 1946. Theory and observations of land and sea breezes. *Bull. Am. Meteorol. Soc.*, 6:272–287.

WILHELMY, H., 1953. Die Pazifische Küstenebene Kolumbiens. *Deut. Geographentag, Essen*, 25:96–100.

YOUNG, J. A., KRISHNAMURTI, T. N. and LINDZEN, R. S., 1972. *Dynamics of the Tropical Atmosphere*. Colloquium Notes: Summer 1972. Advanced Study Program and GARP Task Group, Nat. Center for Atm. Res., Boulder, Colo., 587 pp.

Other references and additional data sources

CLAYTON, H. H. and CLAYTON, F. L., 1947. *World Weather Records 1931–40*. Smithsonian Institution, Washington, D.C., 646 pp.

DEPARTAMENTO DE IRRIGACION, SECCIÓN DE METEOROLOGÍA Y AFOROS, 1944–48. *Anuario Meteorológico, 1937–47*. Ministerio de la Economía Nacional, Bogotá.

DEPARTAMENTO DE IRRIGACIÓN, SECCIÓN DE CLIMA E INGENIERÍA AGRÍCOLA, 1949–56. *Anuario Meteorológico, 1948–55*. Ministerio de Agricultura, Bogotá.

FEDERACIÓN NACIONAL DE CAFETEROS DE COLOMBIA, 1951–69. *Anuario Meteorológico, 1950–68*. Centro Nacional de Investigaciones de Café, Servicio Meteorológico, Chinchiná, Colombia.

GOLDBRUNNER, A. W., 1962. El clima de la Isla de Margarita. *Mem. Soc. Cienc. Nat. La Salle*, 21:146–155.

GONZALEZ, E., 1948. *Datos Detallados de Climatología de Venezuela*. Ministerio de Sanidad y Asistencia Social, Caracas, 639 pp.

INSTITUTO GEOGRÁFICO "AGUSTIN CODAZZI", 1968. *Boletín Climatológico de Colombia. Precipitación, Década 1951–1960, Vols. 1-4*. Servicio Meteorológico e Hidrológico Colombiano, Bogotá.

LÓPEZ, M. F. and HOWELL, W. E., 1965. Cloud seeding at Medellín, Colombia, in 1962–64 dry seasons. *J. Appl. Meteorol.*, 4:54–60.

METEOROLOGICAL OFFICE, AIR MINISTRY, 1958. *Tables of Temperature, Relative Humidity and Precipitation for the World, Part II. Central and South America, the West Indies and Bermuda*. Her Majesty's Stationary Office, London, 53 pp.

MINISTERIE VAN OPENBARE WERKEN EN VERKEER, (no date). *Regenkaarten voor Suriname, Periode 1931–1960*. Meteorologische Dienst, Paramaribo, 15 charts.

MINISTERIE VAN OPENBARE WERKEN EN VERKEER, 1962–1972. *Het Weer in 1961–1970, No.1–10*. Meteorologische Dienst, Paramaribo.

MINISTERIE VAN OPENBARE WERKEN EN VERKEER, 1963. *Klimatologische Tabellen, Regenval in Suriname, Periode 1931–1960, Maandsommen*. Meteorologische Dienst, Paramaribo, 40 pp.

MINISTERIE VAN OPENBARE WERKEN EN VERKEER, 1964. *Klimatologische Tabellen, Diverse Elementen, Nieuw Nickerie, Periode 1931–1960, Maandgemiddelden*. Meteorologische Dienst, Paramaribo, 23 pp.

MINISTERIE VAN OPENBARE WERKEN EN VERKEER, 1972. *Dagtotaal der Neerslag, Paramaribo: 1852–1971*. Meteorologische Dienst, Paramaribo, 92 pp.

MINISTRY OF WORKS AND COMMUNICATIONS, METEOROLOGICAL DIVISION, 1972 (Monthly mean values transmitted by personal communication.) Timehri Airport, Demerara, Guyana.

NELEMANS, B., 1970. *Suriname*. Ministerie van Onderwijs, Paramaribo, 86 pp.

REED, W. W., 1928. Climatological data for northern and western tropical South America. *Mon. Weather Rev., Suppl.*, 31, 21 pp.

SCHRÖDER, R., 1952. Die Verteilung der mittleren Lufttemperatur in Kolumbien. *Bonner Geograph. Abhandl.*, 9:120–123.

SERVICE MÉTÉOROLOGIQUE DU GROUPE ANTILLES—GUYANE FRANÇAISE. *Normales et Statistiques*, F.I&II. Cayenne, 80 pp.

TREWARTHA, G. T., 1961. *The Earth's Problem Climates*. The University of Wisconsin Press, Madison, Wisc., 334 pp.

U.S. DEPARTMENT OF COMMERCE, ESSA, 1966–70. *Climatological Data, West Indies and Caribbean, Vol. 40–46*. Natl. Climatic Center, Ashville, N.C.

U.S. DEPARTMENT OF COMMERCE, ESSA, 1966. *World Weather Records, 1951–60, Vol.3*. U.S. Government Printing Office, Washington, D.C., 357 pp.

U.S. NAVAL WEATHER SERVICE, 1968. *World-Wide Airfield Summaries, South America*. National Technical Information Service, Springfield, Va., 6(2):429–797.

U.S. WEATHER BUREAU, 1959. *World Weather Records 1941–50*. U.S. Government Printing Office, Washington, D.C., 1361 pp.

U.S. WEATHER BUREAU (later ESSA, later NOAA) and WORLD METEOROLOGICAL ORGANIZATION, 1956–1972. *Monthly Climatic Data for the World, Vol.9–25*. National Climatic Center, Ashville, N.C.

WERNSTEDT, F. L., 1961. *World Climatic Data, Latin America and the Caribbean*. Edwards, Ann Arbor, Mich., 87 pp.

TABLE XXXIV

CLIMATIC TABLE FOR URIBIA, COLOMBIA
Latitude 11°44′N, longitude 74°14′W, elevation 23 m

Month	Temperature (°C)				Mean relat. humid. (%)	Rainfall (mm)	
	daily mean	daily range	extreme			mean	rain days
			max.	min.			
Jan.	28.8	9.3	39.0	14.0	59.0	2.0	0.5
Feb.	29.4	9.5	39.5	14.0	58.0	1.0	0.1
Mar.	29.9	9.8	40.0	15.0	56.0	11.0	0.6
Apr.	30.4	9.9	40.0	16.0	58.0	15.0	0.8
May	31.0	9.6	41.0	16.0	61.0	51.0	2.9
June	31.4	9.0	41.0	15.5	60.0	26.0	1.5
July	31.6	9.8	40.0	17.0	54.0	5.0	0.3
Aug.	31.3	9.4	41.0	16.0	56.0	21.0	1.9
Sep.	30.7	9.8	39.5	16.5	61.0	51.0	4.3
Oct.	29.9	9.1	40.0	15.5	70.0	105.0	8.3
Nov.	29.5	8.9	40.0	14.0	71.0	59.0	5.1
Dec.	29.2	9.4	40.0	14.0	64.0	14.0	1.6
Annual	30.3	9.5	41.0	14.0	61.0	361.0	27.9
Record	7 < 55	7 < 55	11 < 55	11 < 55	11 < 55	16 < 60	16 < 60

Explanation. Days with rainfall equal to or exceeding 0.1 mm. Period of record: 16 < 60 indicates 16 years, not consecutive, before and including 1960. Missing data 1948–54; for 7 < 55, 1938 and 41–43 not used.
Mean temperature = (Mo. mean max. + mo. mean min.)1/2.

TABLE XXXV

CLIMATIC TABLE FOR LAS PIEDRAS, VENEZUELA
Latitude 11°42′N, longitude 70°12′W, elevation 15 m

Month	Temperature (°C)				Mean relat. humid. (%)	Rainfall (mm)	
	daily mean	daily range	extreme			mean	rain days
			max.	min.			
Jan.	26.9	9.0	35.0	18.9	13.0	41.0	4.4
Feb.	27.2	9.0	34.4	18.9	10.0	29.0	3.2
Mar.	27.7	9.7	35.6	20.0	5.0	9.0	1.7
Apr.	28.6	9.6	35.6	21.1	3.0	10.0	0.9
May	29.2	8.7	37.8	21.1	24.0	66.0	2.6
June	29.5	8.8	38.9	22.2	10.0	41.0	1.9
July	29.5	9.5	37.8	23.3	5.0	23.0	1.5
Aug.	30.2	9.4	38.9	21.1	19.0	102.0	1.7
Sep.	30.3	9.7	37.8	21.1	31.0	55.0	2.6
Oct.	29.5	9.3	37.8	21.7	74.0	84.0	6.4
Nov.	28.6	8.8	37.2	20.0	80.0	46.0	8.2
Dec.	27.2	8.5	35.6	20.0	39.0	35.0	6.9
Annual	28.7	9.2	38.9	18.9	313.0	102.0	42.0
Record	51–60	51–60	22 < 60*	22 < 60*	31–60	22 < 60	26 < 60

Explanation. Days with rainfall equal to or exceeding 0.1 mm. Period of record: 31–60 indicates 1931 through 1960 inclusive; 22 < 60 indicates 22 years, not consecutive, before and including 1960. Daily mean of temperature: (max. + min.)1/2.
* Data not available for 1947–1950.

380

TABLE XXXVI

Latitude 10°57′N, longitude 74°47′W, elevation 13 m

Month	Temperature (°C)				Mean relat. humid. (%)	Mean monthly rainfall (mm)	Days with			Wind	
	daily mean	daily range	extreme				rain	thunder	fog	mean speed (m/sec)	preval. direct.
			max.	min.							
Jan.	26.8	8.9	36.1	18.9	80.0	1.0	0.1	0.1	0.1	6.2	ENE
Feb.	26.9	9.1	36.7	17.8	77.0	0.0	0.0	0.0	0.0	6.8	ENE
Mar.	27.4	9.2	36.1	19.4	80.0	1.0	0.0	0.0	0.0	6.6	ENE
Apr.	28.4	8.9	37.8	20.0	79.0	11.0	1.1	0.4	0.0	4.9	ENE
May	28.9	8.9	37.8	18.3	80.0	87.0	7.0	5.4	0.1	3.4	NE
June	28.7	8.7	37.2	18.3	82.0	103.0	9.0	5.8	0.4	2.9	NE
July	28.7	8 7	38.9	21.7	80.0	54.0	6.1	3.0	0.1	3.8	NE
Aug.	28.9	9.0	37.8	20.6	81.0	102.0	6.9	5.5	0.0	3.4	NE
Sep.	28.6	9.0	37.8	21.1	83.0	138.0	11.7	8.6	0.6	2.9	E
Oct.	28.2	8.6	37.2	20.0	85.0	202.0	13.5	7.3	0.6	2.7	ENE
Nov.	28.1	8.9	35.6	20.6	83.0	82.0	7.0	5.4	0.8	3.9	NE
Dec.	28.0	9.1	36.1	19.4	81.0	65.0	1.1	1.6	0.6	5.3	NE
Annual	28.1	8.9	38.9	17.8	81.0	846.0	63.5	43.1	3.3	4.4	NE
Record	15 < 54	15 < 54	7 < 60	7 < 60	15 < 54	23 < 60	20 < 60	7 < 42	7 < 42	34–37	34–37

Explanation. Days with rainfall: equal to or exceeding approx. 0.5 mm. Visibility threshold 1,600 m. Period of record: 34–37 indicates 1934 through 1937, inclusive; 23 < 60 indicates 23 years, not consecutive, before and including 1960. Wind, thunderstorm- and fog-days recorded at Soledad Aeropuerto (10°54′N 74°47′W, 14 m). Mean temperature = (mo. mean max. + mo. mean min.)1/2; missing: 1943, 50, 51, 56 and 57.

TABLE XXXVII

CLIMATIC TABLE FOR MARACAIBO (AEROPUERTO), VENEZUELA
Latitude 10°41'N, longitude 71°39'W, elevation 40 m

Month	Mean sea level press. (mbar)	Temperature (°C) daily mean	Temperature (°C) daily range	Temperature (°C) extreme max.	Temperature (°C) extreme min.	Mean relat. humid. (%)	Rainfall (mm) mean	Rainfall (mm) max. in 24 h	Days with rain	Days with thunder	Days with fog	Mean cloudiness (tenths)	Mean sunshine	Mean monthly evap. (mm)	Wind mean speed (m/sec)	Wind preval. direct.
Jan.	1011.7	26.8	8.4	35.0	18.9	75.0	2.0	64.0	1.2	0.2	0.1	4.3	76.0	148.0	3.1	NE
Feb.	1011.4	26.9	8.1	35.4	20.9	75.0	1.0	5.0	0.7	0.1	0.0	4.8	74.0	147.0	3.7	NE
Mar.	1010.9	27.2	7.9	35.6	20.0	74.0	9.0	66.0	1.1	0.9	0.0	6.0	66.0	173.0	3.7	NE
Apr.	1010.5	27.8	7.6	36.3	20.2	76.0	24.0	95.0	3.5	2.7	0.0	6.7	54.0	161.0	3.8	NE
May	1010.3	28.4	7.7	37.6	20.5	77.0	78.0	93.0	8.8	7.3	0.3	7.5	48.0	132.0	2.4	NE
June	1010.6	28.6	8.1	36.4	20.6	75.0	47.0	69.0	8.1	9.7	0.1	7.7	54.0	124.0	2.4	NE
July	1010.9	28.6	8.4	37.0	21.0	74.0	44.0	94.0	7.1	9.1	0.1	7.5	65.0	141.0	2.5	NE
Aug.	1010.4	28.8	8.5	38.0	19.5	73.0	56.0	62.0	8.0	11.4	0.1	7.6	65.0	145.0	2.4	NE
Sep.	1010.1	28.7	8.6	38.7	19.0	74.0	49.0	53.0	8.9	13.0	0.1	7.8	58.0	129.0	2.1	NE
Oct.	1009.9	28.1	8.3	36.5	18.9	77.0	133.0	109.0	10.9	13.7	0.0	7.9	52.0	111.0	2.0	NE
Nov.	1010.0	27.9	7.8	35.9	20.4	77.0	77.0	166.0	7.8	6.7	0.0	6.9	61.0	108.0	2.3	NE
Dec.	1010.8	27.4	8.0	35.4	20.7	76.0	13.0	39.0	1.6	0.9	0.0	5.5	69.0	134.0	2.6	NE
Annual	1010.6	27.9	8.1	38.7	18.9	75.0	533.0	166.0	67.7	75.7	0.8	6.7	62.0	1,653.0	2.8	NE
Record	51–64	51–64	51–64	51–64	51–64	51–67	31–60	27 < 64	51–64	51–64	51–64	51–64	51–64	14 < 68	51–64	51–64

Explanation. Days with rainfall equal to or exceeding 0.1 mm. Visibility threshold unspecified. Sunshine expressed in % of possible. Evaporation from measurements only. Period of record: 31–60 indicates 1931 through 1960, inclusive; 22 < 60 indicates 22 yrs., not consecutive, before and incl. 1960. Max. rain in 24 h includes data from oil companies at Bella Vista (10°40'N 71°36'W, 4 m), 1934–46, 1951–64. Daily means of temperature: (sum of 24 obs.)1/24; of all other elements: (01h30 + 07h30 + 13h30 + 19h30)1/4, local time.

TABLE XXXVIII

CLIMATIC TABLE FOR CARACAS (OBS. CAGIGAL), VENEZUELA
Latitude 10°30'N, longitude 66°56'W, elevation 1,035 m

Month	Mean sta. press. (mbar)	Temperature (°C) daily mean	daily range	extreme max.	extreme min.	Mean relat. humid. (%)	Rainfall (mm) mean	max. in 24 h	Days with rain	thunder	fog	Mean cloudiness (tenths)	Mean sunshine	Mean monthly evap. (mm)	Wind mean speed (m/sec)	preval. direct.
Jan.	900.2	19.2	10.6	30.8	8.3	80.0	22.0	40.6	6.1	0.0	0.3	5.8	64.0	81.0	2.7	SE
Feb.	900.1	19.7	11.4	31.6	7.8	78.0	15.0	73.0	4.0	0.0	0.8	5.5	69.0	91.0	3.1	SE
Mar.	899.7	20.7	11.9	32.8	7.1	76.0	10.0	35.6	3.3	0.3	2.2	5.4	64.0	114.0	3.5	SE
Apr.	899.2	21.7	11.1	33.1	10.6	77.0	32.0	73.7	5.9	0.5	3.8	6.1	53.0	100.0	3.1	SE
May	899.8	22.0	9.5	32.4	11.1	80.0	95.0	78.8	12.4	2.2	0.8	7.0	51.0	100.0	3.2	SE
June	900.5	21.5	8.8	30.8	11.7	83.0	106.0	71.1	15.6	3.2	0.3	7.4	51.0	88.0	3.4	SE
July	900.7	21.1	9.0	29.8	11.1	83.0	97.0	71.1	17.0	6.5	0.0	7.1	56.0	87.0	3.1	SE
Aug.	900.2	21.6	9.8	31.0	11.7	82.0	112.0	91.9	15.7	5.2	0.3	6.8	58.0	82.0	2.8	SE
Sep.	899.2	21.8	10.2	31.2	11.7	81.0	94.0	83.8	14.1	9.5	0.3	6.6	59.0	82.0	2.6	SE
Oct.	898.9	21.5	9.9	31.5	12.2	82.0	122.0	81.3	14.8	8.8	0.3	6.6	58.0	78.0	2.6	SE
Nov.	898.7	20.8	9.9	30.5	10.6	84.0	86.0	69.7	14.6	6.0	0.7	6.5	58.0	60.0	2.6	NW
Dec.	899.4	19.9	9.7	30.0	8.3	82.0	44.0	48.3	10.6	0.0	1.0	6.4	59.0	71.0	2.7	SE
Annual	899.7	21.0	10.2	33.1	7.1	81.0	835.0	91.9	134.1	42.2	10.8	6.4	58.0	1,034.0	2.9	SE
Record	59–64	51–60	51–64	45 < 63	45 < 63	39–64	31–60	56 < 68	31–60	61–64	61–64	51–60	30 < 64	18 < 64	30 < 64	30 < 64

Explanation. Days with rainfall equal to or exceeding 0.1 mm. Visibility threshold unspecified. Sunshine expressed in % of possible. Evaporation from measurements only. Period of record: 31–60 indicates 1931 through 1960, inclusive; 22 < 60 indicates 22 yrs., not consecutive, before and incl. 1960. Entry 45 < 63 indicates continuous record except for 1924–34; 30 < 64, 1945–60 also missing; 18 < 64, no data 1947–50 and 1955–60. Daily means except wind: (07h30 + 13h30 + 19h30)1/3, local time. Hours for wind mean unknown. Barometer elevation 1,042 m. Rainfall (1891–1964): mean 828.0; max. 1,255.0 (33); min. 459.0 (14); max. month 327.0 (Oct. 54); min. month 0.0 (various).

TABLE XXXIX

CLIMATIC TABLE FOR COLONIA TOVAR, VENEZUELA
Latitude 10°25′N, longitude 67°17′W, elevation 1,790 m

Month	Mean sta. press. (mbar)	Temperature (°C) daily mean	daily range	extreme max.	extreme min.	Mean relat. humid. (%)	Rainfall (mm) mean	max. in 24 h	Days with rain	thunder	fog	Mean cloud-iness (tenths)	Mean sun-shine	Mean monthly evap. (mm)	Wind mean speed (m/sec)	preval. direct. day	night
Jan.	824.9	15.7	10.8	25.3	3.7	80.0	48.0	53.0	8.9	0.4	3.9	4.7	54.0	55.0	1.0	ENE	NNW
Feb.	824.7	16.1	11.4	26.8	2.9	79.0	21.0	83.0	5.1	0.3	2.4	4.9	61.0	60.0	0.9	ENE	NNW
Mar.	824.6	17.0	11.3	28.2	4.1	77.0	25.0	58.0	4.9	0.6	2.0	5.4	52.0	74.0	0.9	ENE	NW
Apr.	824.7	17.5	9.6	28.0	6.7	80.0	61.0	38.0	9.6	2.4	7.6	6.9	39.0	55.0	0.8	ENE	NNW
May	825.1	17.5	8.3	26.1	8.5	85.0	142.0	80.0	16.0	4.4	12.6	7.8	34.0	40.0	0.7	ESE	NW
June	825.7	16.7	7.7	24.0	8.3	88.0	135.0	66.0	20.2	6.3	12.8	7.8	34.0	31.0	0.8	ESE	NW
July	825.8	16.5	8.1	23.0	7.1	88.0	165.0	56.0	22.5	7.4	12.7	7.6	35.0	33.0	0.8	ESE	NW
Aug.	825.4	16.9	8.4	23.3	7.1	88.0	166.0	56.0	21.8	8.2	14.2	7.6	41.0	32.0	0.7	ESE	NNW
Sep.	824.8	17.1	8.9	24.1	8.1	88.0	143.0	53.0	21.7	11.8	11.9	7.4	43.0	35.0	0.7	ESE	NNW
Oct.	824.4	17.1	9.0	24.1	8.1	87.0	185.0	88.0	20.8	10.4	10.0	7.0	42.0	33.0	0.7	ESE	NNW
Nov.	823.9	16.9	9.3	24.5	7.4	86.0	140.0	51.0	19.8	4.2	9.6	6.5	40.0	34.0	0.8	ESE	NNW
Dec.	824.3	16.2	9.4	26.6	2.8	83.0	66.0	63.0	13.4	1.3	8.8	5.7	49.0	42.0	0.8	ESE	NNW
Annual	824.8	16.8	9.3	28.2	2.8	84.0	1,297.0	88.0	184.7	57.7	108.5	6.6	44.0	524.0	0.8	ESE	NNW
Record	54-64	51-64	51-64	26 < 64	26 < 64	51-64	31-60	26 < 64	30 < 64	51-64	51-64	14 < 68	17 < 64	10 < 68	51-64	51-64	51-64

Explanation. Days with rainfall equal to or exceeding 0.1 mm. Visibility threshold unspecified. Sunshine expressed in % of possible. Evaporation from measurements only. Period of record: 31–60 indicates 1931 through 1960, inclusive; 22 < 60 indicates 22 yrs., not consecutive, before and incl. 1960. Elements with period of record XX < 64 have continuous series except for years 1947–50, XX < 68 continuous except for 1965–67. Daily means of all elements: (07h30 + 13h30 + 19h30)1/3, local time.

TABLE XL

CLIMATIC TABLE FOR BARCELONA, VENEZUELA

Latitude 10°07'N, longitude 64°41'W, elevation 7 m

Month	Temperature (°C)				Mean relat. humid. (%)	Rainfall (mm)		Days with			Mean cloud-iness (tenths)	Mean sun-shine	Mean monthly evap. (mm)	Wind	
	daily mean	daily range	extreme			mean	max. in 24 h	rain	thunder	fog				mean speed (m/sec)	preval. direct.
			max.	min.											
Jan.	25.2	11.3	34.8	15.7	74.0	10.0	17.0	1.7	0.2	0.0	4.4	84.0	156.0	2.1	N
Feb.	25.5	11.6	35.0	16.6	72.0	4.0	17.0	1.1	0.1	0.2	4.3	86.0	162.0	2.3	N
Mar.	26.4	11.5	35.4	16.9	70.0	6.0	10.0	1.1	0.3	0.0	4.7	82.0	180.0	2.2	N
Apr.	27.2	10.8	36.0	19.4	71.0	7.0	23.0	1.8	0.7	0.1	6.1	73.0	175.0	2.3	N
May	27.8	10.1	37.0	19.0	73.0	43.0	44.0	6.0	3.5	0.1	6.5	67.0	147.0	2.2	N
June	26.7	9.6	37.0	19.0	80.0	99.0	71.0	15.2	10.4	0.1	7.2	56.0	105.0	1.7	N
July	26.0	9.6	35.2	20.0	82.0	137.0	86.0	17.4	13.5	0.4	6.9	61.0	90.0	1.6	N
Aug.	26.2	10.4	37.0	19.8	82.0	111.0	73.0	17.3	12.3	0.0	6.6	62.0	82.0	1.6	N
Sep.	26.7	10.7	36.2	19.7	80.0	73.0	56.0	13.0	12.8	0.0	6.2	69.0	93.0	1.7	ESE
Oct.	27.0	10.9	36.8	19.0	78.0	66.0	73.0	9.5	9.6	0.0	5.9	73.0	109.0	1.7	ESE
Nov.	26.5	11.0	35.4	19.2	78.0	49.0	42.0	8.8	6.1	0.0	5.5	75.0	106.0	1.7	N
Dec.	25.9	10.8	34.9	17.4	76.0	24.0	49.0	4.8	1.2	0.0	4.9	78.0	123.0	1.8	N
Annual	26.4	10.7	37.0	15.7	76.0	629.0	86.0	97.7	70.7	0.9	5.8	72.0	1,528.0	1.9	N
Record	55-64	10 < 64	13 < 64	13 < 64	55-67	31-60	20 < 68*	11 < 68*	11 < 68*	51-64	55-64	55-64	55-64	55-61	55-61

Explanation. Days with rainfall equal to or exceeding 0.1 mm. Visibility threshold unspecified. Sunshine expressed in % of possible. Evaporation from measurements only. Period of record: 31–60 indicates 1931 through 1960, inclusive; 22 < 60 indicates 22 yrs., not consecutive, before and incl. 1960. Daily means of all elements: (01h30 + 07h30 + 13h30 + 19h30)1/4, local time.

* Data not available for 1965–67.

TABLE XLI

CLIMATIC TABLE FOR MÉRIDA (AEROPUERTO), VENEZUELA
Latitude 08°35′N, longitude 71°10′W, elevation 1,495 m

Month	Mean sta. press. (mbar)	Temperature (°C) daily mean	daily range	extreme max.	extreme min.	Mean relat. humid. (%)	Rainfall (mm) mean	max. in 24 h	Days with rain	thunder	fog	Mean cloud-iness (tenths)	Mean sun-shine	Mean monthly evap. (mm)	Wind mean speed (m/sec)	preval. direct. day	night
Jan.	851.3	17.7	10.4	28.2	9.2	79.0	53.0	56.0	8.9	0.1	6.4	7.2	69.0	84.0	1.8	WSW	ENE
Feb.	851.1	18.2	10.4	29.7	10.0	77.0	50.0	68.0	8.5	0.5	3.2	7.2	67.0	84.0	1.9	WSW	ENE
Mar.	850.8	19.0	10.2	31.6	10.0	78.0	52.0	61.0	10.3	1.6	4.0	7.7	60.0	91.0	2.4	WSW	ENE
Apr.	850.7	19.2	9.5	31.2	10.5	82.0	164.0	82.0	17.3	4.7	5.1	8.6	51.0	70.0	2.2	WSW	ENE
May	850.8	19.4	9.1	31.6	11.5	82.0	272.0	117.0	22.9	4.9	5.9	8.7	46.0	65.0	1.7	WSW	ENE
June	851.1	19.1	9.2	30.8	11.1	82.0	163.0	68.0	21.8	3.4	5.1	8.8	44.0	58.0	1.6	WSW	ENE
July	851.3	18.8	9.8	30.1	10.7	81.0	125.0	59.0	20.6	2.6	3.9	8.5	52.0	70.0	1.6	WSW	ENE
Aug.	851.0	19.3	10.2	30.5	10.3	78.0	138.0	50.0	20.6	2.6	2.6	8.2	54.0	81.0	1.9	WSW	ENE
Sep.	850.8	19.4	10.4	31.4	12.2	78.0	172.0	87.0	20.2	5.8	2.2	8.3	52.0	80.0	1.9	WSW	ENE
Oct.	850.6	19.0	9.3	29.9	11.4	82.0	265.0	117.0	23.4	5.9	4.8	8.7	50.0	67.0	1.7	WSW	ENE
Nov.	850.4	18.4	9.1	29.3	10.8	84.0	221.0	104.0	20.2	1.2	8.6	8.5	55.0	57.0	1.5	WSW	ENE
Dec.	850.8	18.0	9.8	27.9	9.7	82.0	95.0	60.0	14.0	0.5	8.2	7.8	66.0	69.0	1.8	WSW	ENE
Annual	850.9	18.8	9.8	31.6	9.2	80.0	1,770.0	117.0	208.7	33.8	60.0	8.2	56.0	876.0	1.8	WSW	ENE
Record	51–64	51–64	51–64	26 < 64	26 < 64	51–67	31–60	27 < 68	51–64	51–64	51–64	51–64	24 < 64	14 < 68	51–64	51–64	

Explanation. Days with rainfall equal to or exceeding 0.1 mm. Visibility threshold unspecified. Sunshine expressed in % of possible. Evaporation from measurements only. Period of record: 31–60 indicates 1931 through 1960, inclusive; 22 < 60 indicates 22 yrs., not consecutive, before and incl. 1960. Included in extreme temperatures and 24-h rainfall maxima are data recorded in city (08°36′N 71°09′W, 1,623 m), 1935–46. Daily means of all elements (01h30 + 07h30 + 13h30 + 19h30)1/4, local time. Data not available for years 1947–50, except for rainfall.

TABLE XLII

CLIMATIC TABLE FOR TURBO, COLOMBIA
Latitude 08°06′N, longitude 76°44′W, elevation 2 m

Month	Temperature (°C)				Mean relat. humid. (%)	Mean rainfall (mm)	Days with			Wind	
	daily mean	daily range	extreme				rain	thunder	fog	mean speed (m/sec)	preval. direct.
			max.	min.							
Jan.	26.4	7.8	32.0	20.0	85.0	91.0	7.1	0.8	0.4	3.8	NNW
Feb.	26.4	7.2	31.0	20.0	85.0	70.0	6.7	0.0	0.2	4.7	NNW
Mar.	26.4	7.7	32.4	20.2	85.0	48.0	6.3	1.0	0.0	4.0	N
Apr.	26.3	8.0	32.0	20.0	84.0	200.0	11.8	0.5	0.0	3.6	NNW
May	25.4	7.3	32.6	20.0	85.0	276.0	17.4	3.3	0.8	2.9	SW
June	26.1	7.9	33.0	20.0	86.0	236.0	15.7	3.8	2.0	3.8	SSW
July	26.3	7.9	33.2	20.2	85.0	278.0	17.1	5.5	0.6	2.9	SSW
Aug.	26.4	8.4	33.0	20.2	87.0	257.0	17.2	2.5	1.0	3.1	SSW
Sep.	26.6	7 3	33.5	21.0	87.0	244.0	15.9	5.0	0.2	3.1	S
Oct.	26.5	6.9	33.0	21.0	88.0	245.0	13.4	2.9	0.5	3.8	SW
Nov.	26.6	6.5	39.4	21.0	87.0	298.0	15.3	2.1	1.3	2.7	S
Dec.	26.2	7.3	31.0	20.0	87.0	242.0	13.5	1.5	1.4	4.7	NNW
Annual	26.3	7.5	39.4	20.0	86.0	2,485.0	157.4	28.9	8.4	3.6	NNW
Record	6 < 54	6 < 54	37–39	37–39	6 < 54	15 < 60	15 < 60	6 < 44	5 < 44	40–41	40–41

Explanation. Days with rainfall equal to or exceeding approx. 0.5 mm. Visibility threshold 1,600 m. Period of record: 37–39 indicates 1937 through 1939, inclusive; 15 < 60 indicates 15 years, not consecutive, before and incl. 1960; 6 < 54: 1936, 38, 52–54; 15 < 60: 1938–43, 51–54, 56–60. Fog at 07h00, local time. Mean temperature = (mo. mean max. + mo. mean min.)1/2.

TABLE XLIII

CLIMATIC TABLE FOR SAN FERNANDO (AEROPUERTO), VENEZUELA
Latitude 07°53'N, longitude 67°26'W, elevation 73 m

Month	Temperature (°C)				Mean relat. humid. (%)	Rainfall (mm)		Days with			Mean cloudiness (tenths)	Mean sunshine	Mean monthly evap. (mm)	Wind	
	daily mean	daily range	extreme max.	extreme min.		mean	max. in 24 h	rain	thunder	fog				mean speed (m/sec)	preval. direct.
Jan.	26.7	11.5	37.6	17.5	71.0	1.0	11.0	0.6	0.1	0.0	4.2	78.0	182.0	2.5	ENE
Feb.	27.6	11.8	40.2	17.2	65.0	4.0	69.0	0.7	0.1	0.0	3.9	81.0	213.0	2.8	NE
Mar.	28.8	11.8	39.9	17.9	62.0	14.0	40.0	1.1	0.4	0.0	4.9	77.0	270.0	2.6	ENE
Apr.	29.0	10.7	39.4	20.0	63.0	71.0	87.0	5.2	2.9	0.0	6.8	60.0	216.0	2.4	ENE
May	27.4	9.3	38.4	19.9	76.0	186.0	165.0	13.9	7.7	0.1	7.2	49.0	121.0	1.7	ENE
June	25.9	7.7	36.5	19.1	84.0	277.0	112.0	20.5	7.9	0.4	7.7	41.0	66.0	1.5	E
July	25.5	7.3	35.2	19.5	86.0	303.0	122.0	25.1	10.5	0.1	7.7	46.0	55.0	1.3	E
Aug.	26.2	7.0	34.3	19.8	86.0	282.0	140.0	21.9	11.6	0.4	7.4	48.0	54.0	1.3	E
Sep.	27.0	7.5	36.5	20.0	83.0	171.0	76.0	16.7	11.5	0.0	7.0	57.0	61.0	1.5	E
Oct.	27.2	8.6	36.3	19.4	81.0	129.0	99.0	11.1	7.9	0.1	6.9	66.0	78.0	1.7	E
Nov.	27.2	9.5	36.2	20.0	80.0	42.0	122.0	5.8	3.0	0.1	6.0	74.0	95.0	2.0	E
Dec.	26.8	10.3	35.9	19.2	76.0	11.0	99.0	2.8	1.1	0.1	4.7	76.0	138.0	2.5	ENE
Annual	27.1	9.4	40.2	17.2	76.0	1,491.0	165.0	125.4	64.7	1.3	6.2	63.0	1,549.0	2.0	E
Record	51–64	51–64	23 < 64*	23 < 64*	51–67	31–60	24 < 68*	51–64	51–64	51–64	51–64	20 < 64*	15 < 68*	51–62	51–62

Explanation. Days with rainfall equal to or exceeding 0.1 mm. Visibility threshold unspecified. Sunshine expressed in % of possible. Evaporation from measurements only. Period of record: 31–60 indicates 1931 through 1960, inclusive; 23 < 64 indicates 23 yrs., not consecutive, before and incl. 1964. Daily means of all elements: (01h30 + 07h30 + 13h30 + 19h30)1/4, local time.
* Data not available for 1947–50 and 1965–67.

388

TABLE XLIV

<small>CLIMATIC TABLE FOR ARAUCA, COLOMBIA</small>
Latitude 07°04'N, longitude 70°40'W, elevation 122 m

Month	Temperature (°C)				Mean relat. humid. (%)	Mean rainfall (mm)	Rain days
	daily mean	daily range	extreme				
			max.	min.			
Jan.	27.1	10.1	38.0	18.0	65.0	12.0	0.8
Feb.	27.4	10.7	38.5	17.1	61.0	8.0	1.3
Mar.	28.0	9.7	39.8	18.8	60.0	24.0	2.5
Apr.	27.6	8.2	40.5	18.0	72.0	191.0	8.6
May	26.5	7.2	38.5	19.8	81.0	249.0	14.1
June	26.1	6.9	34.2	19.0	82.0	289.0	14.7
July	25.7	6.6	36.0	18.0	82.0	273.0	14.9
Aug.	26.5	8.1	36.8	19.0	80.0	219.0	11.0
Sep.	26.7	8.3	37.2	18.5	79.0	214.0	9.9
Oct.	27.1	8.5	35.0	18.0	78.0	200.0	8.9
Nov.	27.2	9.1	35.5	18.5	80.0	87.0	6.3
Dec.	27.4	9.4	37.0	17.2	73.0	31.0	2.7
Annual	26.9	8.6	40.5	17.1	74.0	1,797.0	95.7
Record	10 < 54*	10 < 54*	39–44	39–44	10 < 54*	17 < 57*	13 < 57*

Explanation. Days with rainfall equal to or exceeding 1.0 mm. Period of record: 39–44 indicates 1939 through 1944 inclusive; 22 < 60 indicates 22 years, not consecutive, before and incl. 1960. Mean temperature = (mo. mean max. + mo. mean min.)1/2.
* Missing years: 1937, 38, 41, 44–47, 49, 50, 55.

TABLE XLV

<small>CLIMATIC TABLE FOR BARRANCABERMEJA, COLOMBIA</small>
Latitude 07°04'N, longitude 73°52'W, elevation 107 m

Month	Temperature (°C)				Mean relat. humid. (%)	Mean rainfall (mm)	Rain days
	daily mean	daily range	extreme				
			max.	min.			
Jan.	29.3	7.6	37.2	21.7	79.0	71.0	3.4
Feb.	29.6	8.1	37.8	20.6	78.0	81.0	5.3
Mar.	29.7	7.8	38.3	20.0	79.0	135.0	8.3
Apr.	29.3	7.1	38.0	21.7	77.0	243.0	13.1
May	28.9	7.2	37.8	21.1	77.0	312.0	14.0
June	28.9	7.5	37.8	21.7	75.0	269.0	13.2
July	29.1	8.3	37.0	21.1	75.0	186.0	11.9
Aug.	28.8	8.5	39.0	21.1	74.0	280.0	12.9
Sep.	28.6	7.7	37.2	20.6	74.0	347.0	13.7
Oct.	28.2	7.4	36.0	20.0	80.0	446.0	16.0
Nov.	28.6	7.0	36.0	20.6	77.0	295.0	12.9
Dec.	29.1	7.6	36.7	21.1	75.0	121.0	6.2
Annual	29.0	7.7	38.3	20.0	77.0	2,786.0	130.9
Record	32–54	32–54	43–55	43–55	32–54	27 < 59*	27 < 59*

Explanation. Days with rainfall equal to or exceeding 1.0 mm. Period of record: 31–60 indicates 1931 through 1960 inclusive; 22 < 60 indicates 22 yrs., not consecutive, before and incl. 1960. Mean temperature = (mo. mean max. + mo. mean min.)1/2.
* Record continuous from 1932 except for 1948.

TABLE XLVI

CLIMATIC TABLE FOR GEORGETOWN (BOTANIC GARDENS), GUYANA

Latitude 06°48'N, longitude 58°08'W, elevation 2 m

Month	Mean sea level press. (mbar)	Temperature (°C)				Mean vap. press. (mbar)	Rainfall (mm)		Days with			Mean cloud-iness (tenths)	Mean sun-shine	Mean monthly evap. (mm)	Wind	
		daily mean	daily range	extreme			mean	max. in 24 h	rain	thunder	fog				mean speed (m/sec)	preval. direct.
				max.	min.											
Jan.	1013.7	26.3	4.8	31.1	20.0	27.4	250.7	182.9	20.0	1.0	0.1	6.2	51.0	105.0	3.1	NE
Feb.	1013.7	26.4	4.7	31.7	20.6	26.8	122.2	138.4	16.0	0.0	0.2	5.8	56.0	118.0	3.3	NE
Mar.	1014.0	26.8	4.9	31.7	20.6	26.9	113.3	152.4	16.0	1.0	0.3	6.0	56.0	144.0	3.5	NE
Apr.	1013.9	27.1	4.8	32.2	20.6	28.1	178.3	211.3	16.0	0.0	0.6	6.2	53.0	136.0	3.4	NE
May	1013.7	27.0	4.9	32.2	21.1	29.0	296.4	160.0	23.0	3.0	1.5	6.3	47.0	116.0	3.0	NE
June	1014.4	26.7	5.2	31.7	20.6	29.3	345.9	129.5	24.0	5.0	1.8	6.4	44.0	94.0	2.5	NE
July	1014.5	26.7	5.9	32.2	21.1	29.0	280.9	109.2	23.0	6.0	2.2	6.0	56.0	108.0	2.2	NE
Aug.	1013.9	27.2	6.3	32.2	21.1	29.4	185.2	116.8	17.0	8.0	2.2	5.7	64.0	128.0	2.2	NE
Sep.	1013.3	27.7	6.3	33.9	20.6	29.2	87.9	128.8	9.0	6.0	1.5	5.3	66.0	144.0	2.5	NE
Oct.	1012.5	27.7	6.2	33.9	21.1	29.2	98.3	114.3	8.0	6.0	0.1	5.6	61.0	142.0	2.6	NE
Nov.	1011.8	27.4	5.8	32.8	20.6	29.2	147.1	132.6	13.0	4.0	0.1	5.8	60.0	119.0	2.6	NE
Dec.	1012.6	26.7	5.2	32.2	21.1	28.3	313.4	195.6	21.0	2.0	0.1	6.3	50.0	97.0	2.8	NE
Annual	1013.5	27.0	5.4	33.9	20.0	28.5	2,419.6	211.3	206.0	42.0	10.7	6.0	55.0	1,451.0	2.8	NE
Record	41–60	31–60	38–67	57 < 67	57 < 67	21 < 70	31–60	73 < 53	82–55	5 < 60	5 < 45	10–22	31–60	16–55	16–55	16–55

Explanation. Days with rainfall equal to or exceeding 0.3 mm. Visibility threshold 1,600 m. Sunshine expressed in % of possible. Evaporation from measurements only. Period of record: 31–60 indicates 1931 through 1960, inclusive; 22 < 60 indicates 22 yrs., not consecutive, before and incl. 1960. Daily mean of temperature (max. + min.)1/2; of sea level and vapor pressure (08h15 + 14h15)1/2, local time. Rainfall (1880–1966): mean 2310.6, max. 3747.8 (54), min. 1338.6 (99), max. mo. 1022.1 (Dec. 42), min. mo. 0.0 (Oct., Nov. 84).

390

TABLE XLVII

CLIMATIC TABLE FOR NEW AMSTERDAM, GUYANA
Latitude 06°15′N, longitude 57°31′W, elevation 2 m

Month	Temperature (°C)				Mean vap. press. (mbar)	Mean rainfall (mm)
	daily mean	daily range	extreme			
			max.	min.		
Jan.	26.8	7.1	32.8	19.4	26.0	219.0
Feb.	27.0	7.2	33.3	20.6	25.5	134.6
Mar.	27.1	7.1	33.9	20.6	24.9	138.7
Apr.	26.9	7.0	33.3	18.9	26.0	210.3
May	27.2	7.1	33.9	21.1	27.7	277.1
June	27.1	7.1	33.3	21.1	27.1	295.7
July	27.2	7.4	33.3	21.1	27.5	265.9
Aug.	27.5	7.8	33.9	21.1	27.3	160.0
Sep.	27.9	7.9	35.0	21.1	28.3	65.0
Oct.	27.8	8.0	36.1	21.7	28.1	77.5
Nov.	27.7	7.8	36.7	21.7	27.6	112.0
Dec.	27.2	7.4	35.0	20.6	27.3	237.5
Annual	27.3	7.4	36.7	18.9	26.9	2,193.3*
Record	30 < 67	30 < 67	18 < 66	18 < 66	16–22	31–60

Explanation. Period of record: 31–60 indicates 1931 through 1960 inclusive; 22 < 60 indicates 22 years, not consecutive, before and incl. 1960. Daily mean of temperature: (max. + min.)1/2.
* Annually, days with at least 0.3 mm rainfall, 165.

TABLE XLVIII

CLIMATIC TABLE FOR MEDELLÍN (FAC. DE AGRONOMIA), COLOMBIA
Latitude 06°15′N, longitude 75°35′W, elevation 1,450 m

Month	Temperature (°C)				Mean relat. humid. (%)	Mean rainfall (mm)	Rain days
	daily mean	daily range	extreme				
			max.	min.			
Jan.	21.2	13.7	32.1	9.0	65.0	41.0	8.2
Feb.	21.7	13.4	32.5	7.6	64.0	49.0	8.6
Mar.	21.9	13.5	33.5	5.5	66.0	75.0	11.3
Apr.	21.8	12.4	32.0	10.5	71.0	147.0	19.3
May	21.7	11.8	31.7	10.5	73.0	178.0	20.7
June	21.7	13.0	32.0	11.0	68.0	112.0	15.6
July	21.5	14.0	31.6	10.0	65.0	105.0	15.2
Aug.	21.4	13.5	32.0	8.5	64.0	133.0	17.5
Sep.	21.3	13.1	31.6	10.0	67.0	129.0	17.7
Oct.	21.1	12.2	31.5	11.0	73.0	160.0	21.7
Nov.	21.1	12.2	30.5	11.2	73.0	114.0	18.6
Dec.	20.9	12.6	31.8	7.5	70.0	82.0	13.5
Annual	21.4	12.9	33.5	5.5	68.0	1,325.0	187.9
Record	37–52	37–52	37–54	37–54	37–55	37–60	37–60

Explanation. Days with rainfall equal to or exceeding 0.1 mm. Period of record: 31–60 indicates 1931 through 1960. Mean temperature = (mo. mean max. + mo. mean min.)1/2.

TABLE XLIX

CLIMATIC TABLE FOR PARAMARIBO (CULTUURTUIN), SURINAM
Latitude 05°51′N, longitude 55°10′W, elevation 3 m

Month	Mean sea level press. (mbar)	Temperature (°C) daily mean	daily range	extreme max.	extreme min.	Mean vap. press. (mbar)	Rainfall (mm) mean	max. in 24 h	Days with rain	thunder	fog	Mean cloud- iness (tenths)	Mean sun- shine	Mean monthly evap. (mm)	Wind mean speed (m/sec)	preval. direct.
Jan.	1011.2	26.4	7.2	35.9	15.6	28.4	192.9	178.0	22.1	1.0	0.0	6.6	47.0	116.0	1.4	ENE
Feb.	1011.5	26.6	7.4	35.2	16.6	27.9	150.0	202.2	17.5	0.0	0.0	6.3	50.0	132.0	1.6	ENE
Mar.	1011.5	27.0	7.8	35.1	17.1	28.0	162.2	107.5	18.6	3.0	0.0	6.4	52.0	156.0	1.8	ENE
Apr.	1011.4	27.2	7.4	35.3	17.4	29.1	231.9	148.2	20.3	3.0	0.0	6.8	48.0	150.0	1.6	ENE
May	1011.4	26.8	6.8	34.8	17.9	29.9	321.0	144.0	25.9	8.0	0.0	7.2	44.0	133.0	1.3	ENE
June	1012.3	26.8	7.1	34.7	17.8	29.9	302.7	126.5	26.8	9.0	0.3	6.9	51.0	107.0	1.1	ENE
July	1012.5	27.1	7.9	34.6	18.2	29.3	225.6	95.5	24.7	10.0	0.3	6.1	62.0	129.0	1.2	E
Aug.	1011.8	27.9	8.6	35.2	18.3	29.5	166.8	111.8	18.5	8.0	0.0	5.4	71.0	151.0	1.4	E
Sep.	1011.3	28.5	9.4	35.9	18.5	29.3	85.9	76.0	11.5	4.0	0.5	4.8	77.0	162.0	1.6	E
Oct.	1010.4	28.5	9.5	36.7	19.4	29.5	86.7	82.0	11.8	5.0	0.8	5.1	76.0	165.0	1.6	E
Nov.	1009.5	28.0	9.0	36.0	18.0	29.6	108.6	89.4	15.4	5.0	0.8	5.7	64.0	137.0	1.5	E
Dec.	1010.2	26.9	7.8	37.3	17.5	29.2	173.9	81.0	22.2	4.0	0.8	6.4	49.0	112.0	1.4	ENE
Annual	1011.2	27.3	8.0	37.3	15.6	29.1	2,208.2	202.2	235.3	60.0	3.5	6.1	58.0	1,650.0	1.5	ENE
Record	31–60	31–60		04–70	04–70	31–60	31–60	01 < 70	53 < 33	99–11	4 < 44	31–60	31–60	61–68	31–60	31–60
Covar.	3.5	1.1	10.4	2.3	6.6	****	14.9	31.8	****	****	****	7.0	5.6	****	12.9	****
Range	5.8	3.7	7.0	7.7	7.5	****	608.8	212.9	****	****	****	5.1	58.0	****	2.2	****

Explanation. Days with rainfall equal to or exceeding 0.1 mm. Visibility threshold "dense"; **** no value available. Evaporation from measurements only. Period of record: 31–60 indicates 1931 through 1960, inclusive; 22 < 60 indicates 22 yrs., not consecutive, before and incl. 1960. Coefficient of variation (%), defined as 100 (std. dev./mean) of annual means, totals or extreme values. Range, highest–lowest of all monthly values within period of record. Rainfall (1847–1970): mean 2253.8; max. 3,227.2 (71); min. 1,244.4 (99); max. mon. 611.1 (May 46); min. mon. 1.8 (Oct. 84). Daily means of all elements: (08h00 + 14h00 + 18h00) 1/3, local time. Sunshine data only from 07h00 to 17h00, solar time.

TABLE L

CLIMATIC TABLE FOR PUERTO AYACUCHO, VENEZUELA
Latitude 05°41'N, longitude 67°38'W, elevation 99 m

Month	Temperature (°C)				Mean relat. humid. (%)	Rainfall (mm)		Days with			Mean cloud-iness (tenths)	Mean sun-shine	Mean monthly evap. (mm)	Wind		
	daily mean	daily range	extreme max.	extreme min.		mean	max. in 24 h	rain	thunder	fog				mean speed (m/sec)	preval. direct. day	preval. direct. night
Jan.	28.5	11.8	38.0	20.0	69.0	14.0	28.0	1.6	1.4	0.0	4.1	85.0	166.0	2.2	N	C
Feb.	29.3	11.9	39.6	21.0	66.0	17.0	33.0	4.2	2.9	0.0	4.7	78.0	165.0	1.9	N	C
Mar.	29.5	11.3	39.4	20.2	64.0	66.0	190.0	8.2	5.4	0.0	5.9	69.0	163.0	1.4	N	C
Apr.	28.1	9.6	38.6	21.8	75.0	156.0	149.0	13.1	11.1	0.0	7.6	53.0	94.0	1.2	N	S
May	26.6	8.3	37.2	20.6	83.0	337.0	137.0	22.4	12.6	0.1	8.3	40.0	58.0	0.9	S	S
June	25.4	7.5	33.9	19.9	86.0	437.0	115.0	26.6	9.4	0.0	8.6	35.0	43.0	0.8	S	S
July	25.2	7.7	33.6	19.3	86.0	436.0	133.0	26.9	8.2	0.0	8.3	37.0	43.0	0.8	S	S
Aug.	25.7	8.2	35.4	20.0	85.0	292.0	98.0	23.0	10.7	0.1	8.2	41.0	47.0	1.1	SSE	SSE
Sep.	26.3	9.3	36.1	20.3	83.0	175.0	66.0	21.4	18.2	0.0	7.5	51.0	51.0	0.9	C	SSE
Oct.	26.8	10.0	36.3	20.1	81.0	169.0	104.0	16.6	19.7	0.0	7.5	56.0	60.0	1.1	C	C
Nov.	27.3	10.3	36.5	21.0	80.0	110.0	116.0	12.1	11.6	0.1	6.9	66.0	75.0	1.3	N	C
Dec.	27.7	11.0	36.8	19.5	76.0	40.0	56.0	6.4	4.4	0.0	5.2	74.0	98.0	1.4	N	C
Annual	27.2	9.8	39.6	19.3	78.0	2,249.0	190.0	182.5	115.6	0.3	6.9	57.0	1,063.0	1.2	N	C,S
Record	58–64	58–64	9 < 64*	9 < 64*	58–67	16 < 68***	10 < 68*	8 < 68	8 < 68	8 < 68	58–64	58–64	58–64	58–62	58–63	

Explanation. Days with rainfall equal to or exceeding 0.1 mm. Visibility threshold unspecified. Sunshine expressed in % of possible. Evaporation from measurements only. Period of record: 31–60 indicates 1931 through 1960, inclusive; 22 < 60 indicates 22 yrs., not consecutive, before and incl. 1960. Daily means of temperature: (sum of 24 obs.)1/24; of all other elements (01h30 + 07h30 + 13h30 + 19h30) 1/4, local time.
* Data for period Sep. 1944–Dec. 1946, and years 1958–64, also 1968 if 10 < 68.
** Years used: 1942, 43, 45, 46, and 1957–68.

TABLE LI

CLIMATIC TABLE FOR ANDAGOYA, COLOMBIA
Latitude 05°06′N, longitude 76°40′W, elevation 65 m

Month	Temperature (°C)				Mean relat. humid. (%)	Mean rainfall (mm)	Rain days
	daily mean	daily range	extreme				
			max.	min.			
Jan.	26.8	9.4	36.7	20.0	74.0	554.0	24.0
Feb.	27.2	9.7	37.2	18.5	77.0	519.0	20.8
Mar.	27.6	9.9	37.8	20.0	75.0	557.0	21.2
Apr.	27.5	9.4	37.8	20.6	74.0	620.0	24.0
May	27.4	9.4	37.2	20.0	82.0	655.0	24.9
June	27.1	9.7	37.8	20.6	82.0	655.0	25.0
July	27.2	9.8	37.8	20.0	85.0	572.0	24.4
Aug.	27.3	10.0	37.2	20.0	85.0	574.0	25.2
Sep.	27.2	10.0	37.8	18.9	85.0	561.0	25.0
Oct.	27.0	9.6	38.5	20.0	86.0	563.0	24.6
Nov.	26.7	9.3	37.7	16.7	82.0	563.0	24.8
Dec.	27.0	9.4	37.7	18.9	82.0	512.0	24.1
Annual	27.2	9.6	38.5	16.7	81.0	6,905.0	288.0
Record	32–54	32–54	28 < 55	28 < 55	32–55	32–60	33–60

Explanation. Days with rainfall equal to or exceeding 0.01 in. Period of record: 31–60 indicates 1931 through 1960, inclusive; 22 < 60 indicates 22 years, not consecutive, before and incl. 1960; 28 < 55: 1917–25, 1937–55. Max. and min. thermometers used. Mean temp.: (max. + min.)1/2.

TABLE LII

CLIMATIC TABLE FOR TUMACO, COLOMBIA
Latitude 01°49′N, longitude 78°47′W, elevation 4 m

Month	Temperature (°C)				Mean relat. humid. (%)	Mean rainfall (mm)	Rain days
	daily mean	daily range	extreme				
			max.	min.			
Jan.	25.6	3.8	30.0	21.4	86.0	430.0	23.5
Feb.	25.9	3.8	30.5	22.6	86.0	299.0	19.3
Mar.	26.2	3.9	30.8	22.6	87.0	242.0	15.0
Apr.	26.4	3.9	32.0	22.4	85.0	371.0	21.1
May	26.2	3.9	31.0	22.0	86.0	442.0	26.3
June	26.1	4.3	30.0	22.0	86.0	303.0	25.2
July	26.2	3.4	30.0	22.6	85.0	203.0	21.1
Aug.	26.2	3.3	29.5	20.0	86.0	197.0	20.6
Sep.	25.6	3.7	30.0	19.6	86.0	169.0	19.0
Oct.	25.6	3.7	29.6	20.0	86.0	164.0	16.6
Nov.	25.3	3.7	29.5	21.6	85.0	136.0	13.2
Dec.	25.6	3.5	30.0	22.0	86.0	172.0	17.6
Annual	25.9	3.7	32.0	19.6	86.0	3,128.0	238.5
Record	17 < 54	17 < 54	37–47	37–47	17 < 54	38–47	38–47

Explanation. Days with rainfall equal to or exceeding 0.1 mm. Period of record: 31–60 indicates 1931 through 1960, inclusive; 22 < 60 indicates 22 years, not consecutive, before and incl. 1960; missing data: 1947–50. Mean annual rain 2,590.0 (26 < 60). Mean temperature = (mo. mean max. + mo. mean min.) 1/2.

TABLE LIII

CLIMATIC TABLE FOR ROCHAMBEAU (CAYENNE), FRENCH GUIANA

Latitude 04°50'N, longitude 52°22'W, elevation 8 m

Month	Mean sea level press. (mbar)	Temperature (°C)				Mean vap. press. (mbar)	Rainfall (mm)		Days with			Mean cloud-iness (tenths)	Mean sun-shine	Mean monthly evap. (mm)	Wind	
		daily mean	daily range	extreme max.	min.		mean	max. in 24 h	rain	thunder	fog				mean speed (m/sec)	preval. direct.
Jan.	1011.7	25.2	6.5	32.8	17.4	27.6	431.0	417.6	27.0	2.0	2.0	6.5	38.0	75.0	3.4	ENE
Feb.	1012.1	25.3	6.1	33.9	19.2	27.3	423.0	374.9	24.0	1.0	1.0	6.2	36.0	78.0	3.8	ENE
Mar.	1012.1	25.6	6.3	33.3	18.9	27.3	432.0	259.1	24.0	1.0	1.0	6.7	40.0	99.0	4.0	ENE
Apr.	1012.1	25.7	6.5	33.3	18.3	28.0	480.0	596.1	23.0	2.0	2.0	6.2	43.0	88.0	3.8	E
May	1012.3	25.5	6.4	33.3	19.9	28.6	590.0	223.0	29.0	5.0	6.0	6.9	37.0	66.0	3.2	E
June	1013.2	25.1	7.4	33.9	18.9	28.2	457.0	148.1	28.0	7.0	12.0	5.9	43.0	60.0	2.8	E
July	1013.7	25.2	8.5	33.9	19.2	27.5	274.0	168.2	24.0	10.0	13.0	5.1	55.0	76.0	2.8	E
Aug.	1013.1	25.6	9.2	35.6	19.3	27.5	144.0	67.6	18.0	9.0	9.0	4.3	62.0	92.0	3.2	E
Sep.	1012.6	26.2	10.0	36.1	18.8	27.3	32.0	61.0	8.0	6.0	6.0	3.9	71.0	115.0	3.7	E
Oct.	1011.7	26.2	10.4	36.1	19.1	27.2	42.0	95.0	9.0	5.0	6.0	4.0	72.0	126.0	3.9	E
Nov.	1010.7	25.9	9.4	35.0	17.2	27.6	122.0	137.9	15.0	5.0	7.0	4.8	65.0	105.0	3.6	E
Dec.	1010.9	25.4	7.4	33.9	19.5	28.1	317.0	152.4	24.0	3.0	5.0	5.5	50.0	80.0	3.4	E
Annual	1012.2	25.6	7.9	36.1	17.2	27.7	3,744.0	596.1	253.0	56.0	70.0	5.5	51.0	1,060.0	3.5	E
Record	41–60	51–60	51–65	37 < 65	37 < 65	51–65	31–60	37 < 65	51–65	51–65	51–65	8 < 28	51–65	51–65	51–65	51–65

Explanation. Days with rainfall equal to or exceeding 0.1 mm. Visibility threshold 1,000 m. Sunshine expressed in % of possible. Evaporation from measurements only. Period of record: 31–60 indicates 1931 through 1960, inclusive; 22 < 60 indicates 22 yrs., not consecutive, before and incl. 1960. Ville 31–60 mean, 2,954.0. All 51–65, 31–60 and 41–60: Aeroport Rochambeau only; 37 < 65: 1893–1914 (Ville), 1951–65 (Aeroport). Daily means of temp. and vapor press. (02h00 + 05h00 + ... + 23h00) 1/8; of wind (08h00 + 09h00 + ... + 19h00) 1/12; of clouds (09h00 + 15h00 + 21h00) 1/3; Ville rainfall (1894–1965): mean 3,178.0, max. 6,884.0 (07), min. 1,397.0 (24), max. mon. 2,152.0 (Apr. 06), min. mon. 0.0 (various).

TABLE LIV

CLIMATIC TABLE FOR BOGOTÁ (OBSERV. METEOROL. NACL.), COLOMBIA
Latitude 04°38′N, longitude 74°05′W, elevation 2,556 m

Month	Mean sta. press. (mbar)	Temperature (°C) daily mean	daily range	extreme max.	extreme min.	Mean vap. press. (mbar)	Rainfall (mm) mean	max. in 24 h	Days with rain	thunder	fog	Mean cloudiness (tenths)	Mean sunshine (h)	Mean monthly evap. (mm)	Preval. wind direct.
Jan.	751.9	12.8	11.0	25.0	−1.5	10.8	51.3	32.8	9.0	0.2	20.0	6.4	178.0	40.0	WNW
Feb.	753.1	13.2	10.8	25.0	−5.2	10.5	50.4	40.0	10.0	1.5	16.0	5.8	151.0	42.0	W
Mar.	753.1	13.7	9.9	25.0	2.4	11.3	69.1	61.3	13.0	4.8	23.0	7.8	138.0	42.0	WSW
Apr.	753.0	13.7	9.0	24.2	1.0	11.8	99.5	46.1	19.0	4.5	13.0	8.6	87.0	36.0	SSE
May	753.2	13.7	8.5	24.6	3.8	11.8	105.2	69.1	21.0	3.8	10.0	8.4	99.0	37.0	SSE
June	752.9	13.2	8.2	23.4	4.2	11.2	57.3	27.8	18.0	0.2	6.0	8.8	107.0	41.0	SSE
July	753.6	12.9	8.1	23.2	0.4	10.5	46.8	37.3	18.0	0.2	6.0	8.7	130.0	46.0	SSE
Aug.	753.5	12.9	8.7	23.2	1.4	10.7	41.2	30.1	17.0	0.5	14.0	7.9	135.0	45.0	SSE
Sep.	752.8	12.8	9.6	24.0	1.8	10.5	52.3	39.8	15.0	1.5	7.0	8.2	132.0	47.0	SSE
Oct.	751.7	12.9	9.0	24.2	2.0	11.7	144.4	58.8	20.0	8.2	17.0	8.4	105.0	34.0	SW
Nov.	751.6	13.1	9.1	23.4	0.5	11.8	138.3	72.4	19.0	5.8	18.0	8.3	123.0	33.0	WNW
Dec.	751.2	13.1	9.8	23.8	1.0	11.7	85.0	48.7	13.0	5.2	17.0	8.4	147.0	34.0	NW
Annual	752.6	13.2	9.3	25.0	−5.2	11.2	940.9	72.4	192.0	36.4	167.0	8.0	1,532.0	477.0	SSE
Record	56–60	12 < 69		31–60		56–60	31–60	31–60	31–60	4 < 28	5 < 60	12 < 45	41–60	56–60	31–60

Explanation. Days with rainfall equal to or exceeding 0.1 mm. Visibility threshold "light". Evaporation from measurements only. Period of record: 31–60 indicates 1931 through 1960, inclusive; 22 < 60 indicates 22 yrs., not consecutive, before and incl. 1960. Evaporation measured in the shade. Fog recorded at 07h00, local time. Daily means of station pressure, temperature and vapor pressure (sum of 24 obs.) 1/24. Rainfall (1866–1960): mean 1011.8, max. 1633.1 (79), min. 552.7 (41), max. mon. 394.5 (Mar. 79), min. mon. 0.0. (June 97, Jan. 24). Sunshine in hours per month.

TABLE LV

CLIMATIC TABLE FOR SANTA ELENA, VENEZUELA
Latitude 04°36'N, longitude 61°07'W, elevation 907 m

Month	Mean sta. press. (mbar)	Temperature (°C) daily mean	daily range	extreme max.	extreme min.	Mean relat. humid. (%)	Rainfall (mm) mean	max. in 24 h	Days with rain	thunder	fog	Mean cloud-iness (tenths)	Mean sun-shine	Mean monthly evap. (mm)	Wind mean speed (m/sec)	preval. direct. day	night
Jan.	915.7	21.6	12.4	33.3	9.0	79.0	68.0	50.0	13.1	2.4	0.1	6.3	54.0	72.0	1.3	ESE	NE
Feb.	915.8	22.0	12.5	33.0	10.4	77.0	69.0	132.0	11.0	3.1	0.1	6.2	58.0	72.0	1.4	ESE	NE
Mar.	915.8	22.4	13.0	35.0	11.4	75.0	78.0	120.0	12.0	3.4	0.1	6.4	54.0	92.0	1.5	ESE	ENE
Apr.	915.9	22.3	12.0	35.1	10.9	80.0	145.0	84.0	14.7	5.1	0.5	6.8	50.0	71.0	1.4	ESE	ENE
May	916.6	22.0	10.5	32.5	11.6	85.0	221.0	63.0	24.0	5.9	1.1	7.3	42.0	54.0	1.1	ESE	ENE
June	917.4	21.4	9.6	31.0	12.2	89.0	248.0	82.0	26.7	4.6	1.7	7.8	38.0	39.0	0.9	S	C
July	917.6	20.9	10.0	30.2	12.0	89.0	229.0	68.0	26.4	5.4	2.8	7.1	45.0	40.0	0.9	S	C
Aug.	917.1	21.2	10.7	31.2	12.3	88.0	182.0	113.0	24.3	8.3	4.3	6.6	50.0	48.0	0.9	S	C
Sep.	916.5	21.4	13.2	33.5	9.5	84.0	109.0	83.0	16.1	8.9	2.4	5.7	66.0	64.0	1.0	S	C
Oct.	915.6	21.8	13.3	34.2	11.0	81.0	106.0	86.0	13.8	9.4	1.7	6.3	65.0	72.0	1.1	SSE	C
Nov.	914.9	21.7	12.8	34.2	11.4	83.0	130.0	99.0	15.5	6.5	1.8	6.6	58.0	57.0	1.0	ESE	C
Dec.	915.2	21.5	12.6	32.0	10.9	82.0	115.0	60.0	15.7	4.3	0.9	6.2	53.0	61.0	1.1	ESE	ENE
Annual	916.2	21.7	11.9	35.1	9.0	83.0	1,700.0	132.0	213.3	67.3	17.5	6.6	53.0	742.0	1.1	ESE	C
Record	51–64	51–64	51–64	21 < 64*	21 < 64*	51–67	21 < 64*	21 < 64*	21 < 64*	51–64	51–64	51–64	19 < 64*	14 < 68**	51–64	51–64	

Explanation. Days with rainfall equal to or exceeding 0.1 mm. Visibility threshold unspecified. Sunshine expressed in % of possible. Evaporation from measurements only. Period of record: 51–64 indicates 1951 through 1964, inclusive; 21 < 64 indicates 21 yrs., not consecutive, before and incl. 1964. Daily means of temperature (sum of 24 obs.)1/24, of all other elements (01h30 + 07h30 + 13h30 + 19h30) 1/4, local time.

* Data not available for 1947–50.
** Data not available for 1965–67.

TABLE LVI

CLIMATIC TABLE FOR VILLAVICIENCIO, COLOMBIA
Latitude 04°09′N, longitude 73°36′W, elevation 423 m

Month	Temperature (°C)				Mean relat. humid. (%)	Mean rainfall (mm)	Rain days
	daily mean	daily range	extreme max.	extreme min.			
Jan.	26.6	10.6	37.0	16.0	75.0	61.0	4.7
Feb.	26.9	9.6	38.5	15.0	71.0	113.0	6.7
Mar.	27.1	9.8	38.0	15.0	72.0	159.0	9.7
Apr.	25.8	9.6	37.0	14.0	81.0	461.0	19.8
May	25.6	9.5	37.0	15.0	82.0	605.0	23.2
June	25.4	9.5	39.0	14.0	83.0	498.0	20.9
July	25.2	9.3	35.5	15.0	80.0	524.0	21.6
Aug.	25.5	10.4	38.0	14.0	79.0	360.0	18.5
Sep.	26.0	10.8	37.0	13.5	80.0	335.0	16.3
Oct.	26.3	11.2	39.5	15.0	81.0	443.0	19.3
Nov.	26.2	11.4	36.5	15.5	80.0	381.0	17.0
Dec.	26.0	11.7	39.5	15.0	78.0	156.0	9.4
Annual	26.0	10.3	39.5	13.5	79.0	4,096.0	187.1
Record	12 < 53	12 < 53	10 < 51	10 < 51	12 < 53	24 < 60	21 < 60

Explanation. Days with rainfall equal to or exceeding approx. 1.0 mm. Period of record: 12 < 53 indicates 12 yrs., not consecutive, before and incl. 1953; missing years: 1936–38, 42, 44, 47. Mean temperature = (mon. mean max. + mon. mean min.)1/2.

TABLE LVII

CLIMATIC TABLE FOR ESPINAL (C.C. DE TABACO), COLOMBIA
Latitude 04°09′N, longitude 74°53′W, elevation 322 mm

Month	Temperature (°C)				Mean relat. humid. (%)	Mean rainfall (mm)	Rain days
	daily mean	daily range	extreme max.	extreme min.			
Jan.	28.0	10.7	41.0	16.0	70.0	52.0	4.0
Feb.	28.2	11.1	38.8	17.4	70.0	94.0	5.0
Mar.	27.7	10.7	38.4	18.8	72.0	124.0	7.1
Apr.	27.3	10.3	37.0	17.0	76.0	192.0	10.0
May	27.3	10.2	39.8	17.0	78.0	190.0	9.2
June	27.6	11.2	38.8	18.0	75.0	97.0	5.0
July	28.1	12.8	38.9	17.2	66.0	45.0	2.0
Aug.	28.5	12.7	40.0	16.0	64.0	55.0	4.1
Sep.	28.5	11.9	39.6	17.4	69.0	120.0	6.0
Oct.	27.3	10.4	39.0	17.0	73.0	213.0	11.0
Nov.	27.1	10.4	38.0	17.2	74.0	124.0	7.4
Dec.	27.6	10.6	38.6	18.0	72.0	86.0	5.1
Annual	27.8	11.1	41.0	16.0	71.0	1,392.0	75.9
Record	33–57	33–57	37–55	37–55	33–57	33–57	33–57

Explanation. Days with rainfall equal to or exceeding 1.0 mm. Period of record: 33–57 indicates 1933 through 1957 inclusive. Mean temperature = (mon. mean max. + mon. mean min.)1/2.

398

TABLE LVIII

CLIMATIC TABLE FOR TAFELBERG, SURINAM
Latitude 03°47'N, longitude 56°03'W, elevation 344 m

Month	Temperature (°C)				Mean relat. humid. (%)	Rainfall (mm)		Days with rain	Mean cloud-iness (tenths)	Mean sun-shine	Mean monthly evap. (mm)	Mean wind speed (m/sec)
	daily mean	daily range	extreme max.	extreme min.		mean	max. in 24 h					
Jan.	24.9	7.9	32.4	14.9	83.1	306.0	79.8	17.3	7.4	35.5	125.0	1.3
Feb.	24.9	7.7	31.6	17.3	80.6	249.1	106.2	12.7	7.4	39.4	122.0	1.3
Mar.	25.3	8.0	32.9	17.6	80.1	253.0	101.2	12.9	7.2	41.7	148.0	1.1
Apr.	25.5	8.1	32.5	18.9	79.6	284.3	192.8	13.4	7.5	42.1	144.0	0.9
May	25.4	7.8	32.2	18.0	82.2	402.1	157.1	19.8	7.6	43.6	145.0	1.0
June	25.3	8.0	31.0	18.8	82.3	400.6	65.4	18.5	6.7	47.1	142.0	1.1
July	25.7	8.8	31.7	18.3	79.0	321.2	86.5	16.2	5.9	53.6	155.0	1.1
Aug.	26.4	9.5	32.2	18.3	76.2	188.4	76.3	10.9	5.7	65.5	178.0	1.2
Sep.	27.0	10.5	33.2	17.7	71.6	64.6	97.4	3.9	5.8	67.7	175.0	1.3
Oct.	27.3	10.8	33.4	17.4	69.5	47.7	53.8	3.1	6.1	67.9	177.0	1.3
Nov.	26.8	9.6	33.7	18.2	73.9	109.3	63.0	6.6	6.7	55.2	147.0	0.9
Dec.	25.7	8.4	33.0	17.5	78.3	187.8	70.0	12.6	6.8	46.2	135.0	1.1
Annual	25.9	8.8	33.7	14.9	78.0	2,814.1	192.8	147.9	6.7	50.4	1,793.0	1.1
Record	61–70	61–70	9 < 70	9 < 70	61–70	9 < 70	61–70	8 < 70	61–70	61–70	61–70*	61–70
Covar.	1.1	3.6	1.9	5.3	2.1	15.7	36.7	9.3	3.4	10.2	3.6	16.8
Range	4.1	4.7	3.8	6.3	29.0	565.1	187.6	24.0	4.2	62.0	89.9	2.2

Explanation. Days with rainfall equal to or exceeding 3.0 mm. Visibility threshold "no data". Sunshine expressed in % of possible. Period of record: 61–70 indicates 1961 through 1970, inclusive; 22 < 60 indicates 22 yrs., not consecutive, before and incl. 1960. Coefficient of variation (%), defined as 100 (std. dev./mean) of annual means, totals or extreme values. Range, highest–lowest of all monthly values within period of record. Daily means of all elements: (08h00 + 14h00 + 18h00) 1/3, local time; Sunshine data only from 07h00 to 17h00, solar time.

* Computed from Penman equation.

TABLE LIX

CLIMATIC TABLE FOR MARIPASOULA, FRENCH GUIANA
Latitude 03°38'N, longitude 54°02'W, elevation 104 m

Month	Temperature (°C)				Mean vap. press. (mbar)	Rainfall (mm)		Days with			Mean sunshine (h)	Mean monthly evap. (mm)	Wind	
	daily mean	daily range	extreme max.	min.		mean	max. in 24 h	rain	thunder	fog			mean speed (m/sec)	preval. direct.
Jan.	24.5	7.8	32.0	16.9	27.1	218.0	50.0	24.0	3.0	1.0	151.0	53.0	1.7	NE
Feb.	24.5	7.4	31.6	18.0	26.9	214.0	75.0	22.0	5.0	2.0	142.0	47.0	1.8	NE
Mar.	24.8	8.4	32.4	18.4	27.0	212.0	70.0	20.0	3.0	2.0	161.0	61.0	2.2	NE
Apr.	25.1	8.3	32.9	18.9	27.6	249.0	59.0	21.0	7.0	3.0	190.0	55.0	1.7	NE
May	25.0	7.7	33.0	19.8	28.2	399.0	115.0	27.0	11.0	7.0	151.0	46.0	1.5	NE
June	24.7	8.0	33.5	19.5	27.6	278.0	60.0	26.0	10.0	14.0	170.0	46.0	1.3	E
July	24.8	9.0	33.3	19.4	27.2	196.0	45.0	23.0	9.0	17.0	215.0	52.0	1.3	E
Aug.	25.3	9.8	33.5	19.4	27.1	136.0	69.0	17.0	9.0	12.0	248.0	65.0	1.3	E
Sep.	25.6	11.1	31.0	18.9	26.7	77.0	75.0	11.0	6.0	10.0	257.0	82.0	1.3	E
Oct.	26.0	11.5	35.4	17.9	26.4	63.0	70.0	9.0	6.0	6.0	266.0	99.0	1.6	E
Nov.	26.0	10.5	34.9	17.9	26.8	92.0	35.0	13.0	6.0	6.0	230.0	86.0	1.6	E
Dec.	25.2	8.6	33.1	19.0	27.5	234.0	137.0	22.0	7.0	3.0	181.0	60.0	1.5	E
Annual	25.1	9.0	35.4	16.9	27.2	2,368.0	137.0	235.0	82.0	83.0	2,362.0	752.0	1.6	E
Record	56–65	56–65	56–65	56–65	56–65	56–65	56–65	56–65	56–65	56–65	61–65	56–65	56–65	56–65

Explanation. Days with rainfall equal to or exceeding 0.1 mm. Visibility threshold 1,000 m. Evaporation from measurements only. Period of record: 56–65 indicates 1956 through 1965, inclusive. Daily means of temperature and vapor pressure (02h00 + 05h00 + ... + 23h00) 1/8; of wind speed (daylight hourly values) 1/12.

TABLE LX

CLIMATIC TABLE FOR ST. IGNATIUS, GUYANA
Latitude 03°21′N, longitude 59°48′W, elevation 99 m

Month	Temperature (°C)				Mean relat. humid. (%)	Mean rainfall (mm)
	daily mean	daily range	extreme			
			max.	min.		
Jan.	27.7	9.1	35.0	18.3	64.0	31.2
Feb.	27.9	9.2	35.5	17.2	63.0	27.9
Mar.	28.2	9.3	36.1	20.5	60.0	31.7
Apr.	28.1	8.6	37.8	20.0	64.0	109.5
May	27.4	7.8	38.9	17.2	72.0	299.5
June	26.7	7.4	34.4	18.9	80.0	375.7
July	26.5	7.4	33.9	18.3	79.0	343.2
Aug.	27.2	7.9	35.0	18.9	74.0	223.5
Sep.	28.5	8.9	36.7	18.3	64.0	80.3
Oct.	29.1	9.5	37.2	17.8	60.0	52.3
Nov.	28.8	9.1	37.2	17.8	64.0	34.8
Dec.	28.4	9.1	36.1	16.1	64.0	26.2
Annual	27.8	8.6	38.9	16.1	67.0	1,635.8
Record	18 < 67	18 < 67	10 < 66	10 < 66	59–67	22 < 60

Explanation. Period of record: 59–67 indicates 1959 through 1967 inclusive; 18 < 67 indicates 18 yrs., not consecutive, before and incl. 1967. Temp. extremes for Lethem (03°24′N 59°38′W, 82 m). Daily mean of temperature: (max. + min.)1/2.

TABLE LXI

CLIMATIC TABLE FOR SAN CARLOS DE RIO NEGRO, VENEZUELA
Latitude 01°54′N, longitude 67°03′W, elevation 95 m

Month	Temperature (°C)			Rainfall		
	daily mean	extreme		mon. mean (mm)	max. in 24 h (mm)	rain days
		max.	min.			
Jan.	26.3	33.3	19.1	222.0	70.0	20.5
Feb.	26.3	33.6	19.0	229.0	62.0	18.8
Mar.	26.5	33.7	19.3	206.0	83.0	20.9
Apr.	25.9	32.9	19.0	395.0	115.0	22.9
May	25.6	32.3	18.8	381.0	76.0	28.8
June	25.7	31.8	19.6	390.0	70.0	26.5
July	25.4	31.7	19.1	330.0	75.0	28.3
Aug.	25.9	32.7	19.1	328.0	107.0	27.5
Sep.	26.6	33.8	19.1	249.0	88.0	21.5
Oct.	26.7	33.8	19.6	257.0	97.0	20.8
Nov.	26.7	33.8	19.5	314.0	87.0	18.5
Dec.	26.2	33.2	19.3	220.0	69.0	20.1
Annual	26.2	33.8	18.8	3,521.0	115.0	275.1
Record	51–58	51–58	51–58	51–58	51–58	51–58

Explanation. Days with rainfall equal to or exceeding 0.1 mm. Period of record: 51–58 indicates 1951 through 1958 inclusive. Daily mean of temperature: (07h30 + 13h30 + 19h30)1/3, local time.

TABLE LXII

CLIMATIC TABLE FOR LA FLORIDA (MPIO. DE POPAYAN), COLOMBIA
Latitude 02°27'N, longitude 76°35'W, elevation 1,789 m

Month	Mean sta. press. (mbar)	Temperature (°C)				Mean vap. press. (mbar)	Rainfall (mm)		Days with		Mean cloudiness (tenths)	Mean sunshine (h)	Mean monthly evap. (mm)	Mean wind direction	
		daily mean	daily range	extreme max.	min.		mean	max. in 24 h	rain	thunder				day	night
Jan.	805.5	17.3	11.3	28.8	9.0	15.3	152.0	59.9	18.4	5.0	7.1	180.0	27.0	SW	SE
Feb.	805.2	17.5	11.3	30.0	9.0	15.2	143.0	64.7	16.2	4.0	7.0	174.0	27.0	SW	SE
Mar.	805.3	17.5	10.9	29.3	9.5	15.5	168.0	74.0	18.2	8.0	7.4	149.0	24.0	SW	SE
Apr.	805.4	17.4	10.6	29.9	10.0	15.7	187.0	75.0	21.1	11.0	7.7	125.0	21.0	SW	SE
May	805.6	17.5	10.8	29.5	10.0	15.7	140.0	62.9	19.9	9.0	7.3	131.0	24.0	SW	SE
June	805.8	17.3	11.5	29.0	8.5	14.9	98.0	49.6	17.1	3.0	6.9	155.0	34.0	SW	SE
July	805.5	17.5	12.6	29.5	8.5	13.7	40.0	46.4	12.7	2.0	6.6	174.0	52.0	SW	SE
Aug.	805.5	17.9	13.4	29.9	8.5	13.3	32.0	26.5	9.5	4.0	6.4	177.0	58.0	SE	SE
Sep.	805.4	17.9	13.4	31.0	9.5	13.7	89.0	50.9	14.1	4.0	6.9	167.0	52.0	SW	SE
Oct.	805.8	17.2	11.2	29.6	9.2	15.1	267.0	92.6	23.5	11.0	7.8	134.0	28.0	SW	SE
Nov.	805.5	16.9	10.4	27.7	9.5	15.5	311.0	80.0	25.4	8.0	8.1	131.0	27.0	SW	SE
Dec.	805.1	17.2	10.6	28.2	9.5	15.7	284.0	86.0	23.5	6.0	7.6	155.0	21.0	SW	SE
Annual	805.5	17.4	11.5	31.0	8.5	14.9	1,911.0	92.6	219.5	75.0	7.2	1,852.0	395.0	SE	
Record	16 < 68	51–68	17 < 68	51–68	51–68	51–68	51–68	51–68	53–68	7 < 66	51–68	51–68	56–68	51–67	
Covar.	9.6	2.3	4.3	2.7	8.0	2.8	16.1	19.5	7.7	****	14.8	10.8	49.3		
Range	5.8	3.1	6.8	6.5	4.6	6.4	464.8	91.9	27.0	****	6.0	192.0	146.0		

Explanation. Days with rainfall equal to or exceeding 0.1 mm. Visibility threshold "no data". Evaporation from measurements only. Period of record: 51–68 indicates 1951 through 1968, inclusive; 16 < 68 indicates 16 yrs., not consecutive, before and incl. 1968. Coefficient of variation (%), defined as 100 (std. dev./mean) of annual means, totals or extreme values. Range, highest–lowest of all monthly values within period of record.
Data from 1965 not used in computation of temp. range. Thunderstorm-days recorded at Popayan Aeropuerto (02°27'N 76°34'W, 1,725 m). Daily means of pressures, cloud cover and wind (07h00 + 14h00 + 20h00) 1/3; of temperature (07h00 + 14h00 + 20h00) 1/4. Evap. in shade.
**** No value available.

TABLE LXIII

CLIMATIC TABLE FOR SIPALIWINI, SURINAM
Latitude 02°02'N, longitude 56°07'W, elevation 253 m

Month	Temperature (°C)				Mean relat. humid. (%)	Rainfall (mm)			Mean cloudiness (tenths)	Mean sunshine (h)	Mean monthly evap. (mm)	Mean wind speed (m/sec)
	daily mean	daily range	extreme max.	extreme min.		mean	max. in 24 h	rain days				
Jan.	26.1	10.1	34.7	14.8	76.6	150.8	50.0	10.9	6.8	41.9	134.0	1.1
Feb.	26.1	9.7	34.6	14.9	76.3	140.0	71.8	10.2	6.9	38.4	125.0	1.1
Mar.	26.0	9.6	33.5	16.4	78.3	203.1	73.6	11.2	7.1	39.5	145.0	1.0
Apr.	26.4	9.6	34.3	17.8	78.1	247.2	91.8	12.0	7.3	37.3	137.0	1.1
May	26.3	9.1	34.0	17.9	80.4	429.9	199.9	17.3	7.4	35.7	140.0	0.8
June	26.3	9.8	33.4	18.0	78.0	334.1	144.6	16.1	6.6	32.9	149.0	0.7
July	26.5	10.9	34.9	17.2	75.6	206.5	67.1	11.7	6.1	61.8	168.0	0.9
Aug.	27.1	11.9	35.0	17.0	72.6	105.0	59.7	7.6	5.6	61.0	171.0	0.7
Sep.	27.9	13.6	36.5	15.8	67.7	45.0	23.8	4.3	5.0	77.3	192.0	1.0
Oct.	28.6	14.7	36.6	15.6	62.2	37.0	37.0	2.6	5.2	79.2	194.0	1.2
Nov.	27.7	13.4	36.6	16.1	68.5	88.9	57.7	5.5	6.3	55.9	151.0	1.1
Dec.	27.1	11.9	35.8	16.6	70.8	83.4	49.1	5.4	6.6	42.4	138.0	1.1
Annual	26.8	11.2	36.6	14.8	73.8	2,070.9	199.9	114.8	6.4	51.5	1,844.0	1.0
Record	61–70	61–70	61–70	61–70	61–70	61–70	61–70	61–70	61–70	8 < 70	61–70*	61–70
Covar.	1.2	4.2	2.6	4.2	3.6	20.1	39.3	13.6	4.4	8.7	3.9	27.2
Range	4.1	8.0	5.2	6.1	30.0	717.7	198.9	22.0	4.5	71.0	101.3	2.5

Explanation. Days with rainfall equal to or exceeding 3.0 mm. Visibility threshold "no data". Sunshine expressed in % of possible. Period of record: 61–70 indicates 1961 through 1970, inclusive: 8 < 70 indicates 8 yrs., not consecutive, before and incl. 1970. Coefficient of variation (%), defined as 100 (std. dev./mean) of annual means, totals or extreme values. Range, highest–lowest of all monthly values within period of record. Daily means of all elements: (08h00 + 14h00 + 18h00) 1/3, local time. Sunshine data only from 07h00 to 17h00, solar time.

* Computed from Penman equation.

Chapter 7

The Climate of Central America

W. H. PORTIG

The climatological elements

Wind

The wind is among the climatological elements that are most affected by local influences. In higher latitudes one can obtain a fair estimate of representative winds through isobaric charts, applying the gradient wind equation. The closer to the Equator the less applicable is this method, not only because of the decreasing Coriolis parameter but also because of the increasing need for accuracy of the barometric measurements.

While not enough wind observations are available to give an adequate description of the wind conditions of Central America, the mean wind flow over the waters is fairly well assessed through observations made aboard ships. Through courtesy of the National Weather Records Center in Asheville, N.C., mean monthly streamlines could be devised for the Caribbean Sea from unpublished data (PORTIG, 1966). Fig.2 shows mean streamlines for the months with the greatest mutual deviations, September and December. In no month is there any southerly component of the wind over the open Caribbean Sea. Fig.2 also contains a division of the region into several areas according to their wind regimes. Typical stations of these areas are listed in Table I which presents monthly and annual mean wind vectors. Letters indicate the seasonal variation of the most important element, rain.

The maximum mean wind speed is encountered throughout the entire winter near Barranquilla, Colombia, and amounts to 8–9 m/sec. The magnitude of the maximum mean wind vector diminishes to 5–6 m/sec in summer and early fall, and moves to the central Caribbean, up to 15°N 70°W in September. The location of the maximum surface wind moves in the same direction as do the cloudiness minimum and the rainfall minimum of which it stays somewhat to the north. Also, the speed has a high negative correlation with moisture.

Stations in the pure trade winds (see below, and Fig.2) show in all months a speed–direction distribution similar to that of San Juan, P.R.:

N	NE	E	SE	S	SW	W	NW
4.5	5.6	4.8	3.2	2.8	2.7	3.4	4.5 m/sec

The fastest quadrant is 5°–95° (approximately northeast), the slowest 146°–237° (approximately south). The climatic tables at the end of this chapter show that the wind

TABLE I

SEASONAL VARIATION OF WIND DIRECTIONS (360 = N, 90 = E)

	Jan.	Feb.	Mar.	Apr.	May	June	July	Aug.	Sept.	Oct.	Nov.	Dec.	Year
Area A, trade winds throughout the year:													
Barbados	78	85	86D	96	97	91	88	90R	99d	98r	94	78	90
Gustavia	70	80	82D	98	86	82r	67d	64R	95	93	87	78	81
San Juan	78	80D	90	81	90r	90d	82	86R	90	97d	83	81	84
Grand Turks	89	92	92D	96	108r	102d	101	100	107	97	75R	82	96
Port au Prince	93D	92	91	87	92R	96	94d	93	95r	103	105	95D	95
Curaçao	88	88	84	88D	91	92	91	93r	93d	97	85R	86	89
Area B, transition between trade winds and westerlies:													
Bermuda	227	214	217	214D	169	182	183	182r	106d	88R	225	257	180
Area A$_N$, trade winds throughout the year but with occasional northers in winter:													
Ocean field I	59	50	75D	69	69	96r	98	96d	85R	61R	46	55	70
Ocean field II	51	51	54D	60	72r	77	72d	74	66R	72R	44	47	61
Cruta, Hond.	70	89	93D	78	83	88r	83d	83	81	81R	81R	84	77
Belize 1	79	81	101D	103	100	102r	98d	95	113	335R	332	355	78
Belize 2	34	81	96D	94	96	96r	96d	95	92	40R	21	27	77
Area B$_N$, as B but with occasional northers in winter:													
Nassau	68D	74	79	81	98	110r	116	113d	89	69R	53	61	85
Area C, trade winds interrupted in summer:													
Colón	6	3D	1D	357	341	211	325r	262r	173d	170	258R	1	354
Area C$_N$, as C but with occasional northers in winter:													
Amapala	34	44D	131	140	197	178r	47	97d	214R	220	352	15	43

Legend: D = driest month; R = wettest month; d = month with a secondary rainfall minimum; r = month with a secondary rainfall maximum.
Ocean field I is positioned from Cuba to 25°N 82°–83°W; Ocean field II is positioned 18–20°N 74–76°W.
Belize 1: 07h00, 29 years: Belize 2; all observation hours, 17 years.

speed is at its minimum in fall throughout the region. The maximum is measured in the western part of the region in February–March, in the eastern part in mid-summer.

Fig.2 and Table I show that only the part of Central America that is influenced by the Pacific Ocean has substantial seasonal changes of wind direction. The areas A, B and B$_N$ (Table I) have the annual rainfall maximum at, or very near, the time when the winds blow from the most northerly point.

The average wind speed is generally small with relatively small deviations from the mean. Although hurricanes are typical for the region, they affect individual locations so seldom that they hardly show up in the statistics of many years. Raizet on Guadeloupe and Lamentin on Martinique, for example, had only 3 or 4 observations, respectively, with more than 11 m/sec out of a total of 23,300 observations (1954–1961). However, winds up to 51 m/sec were measured in Desaix, Martinique, on June 21, 1960 (SERVICE MÉTÉO-ROLOGIQUE DU GROUPE ANTILLES–GUYANE, 1962).

Inland stations have a surprisingly high number of calms. Catacamas, Honduras, for example, reports between 59 % in April and 86 % in October based on 24-hourly observations daily. Table II shows the change of annual variation of frequency of calms across

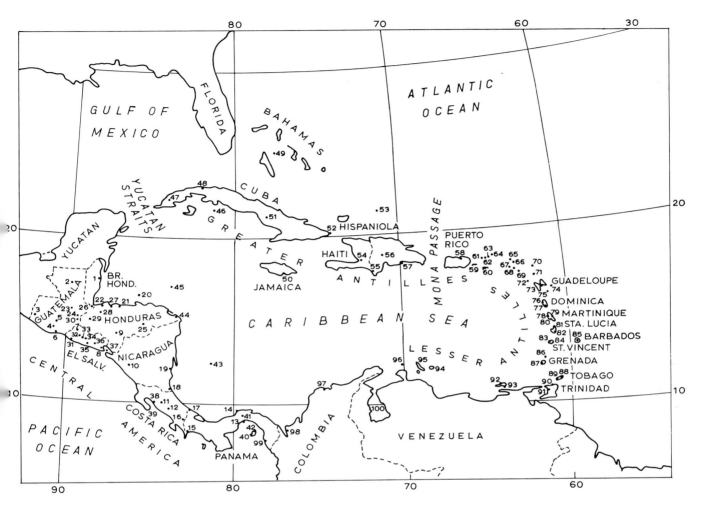

Fig.1. Location of stations and regions mentioned in the text. Acajutla *31*; Amapala 37; Antigua *71*; Aruba *96*; Balboa *13*; Barahona Peninsula *55*; Baranquilla *97*; Barbuda *70*; Batalie, Dominica *77*; Bayabou, St. Vincent *84*; Belize *1*; Blanchisseuse *90*; Bluefields *19*; Bonaire *94*; Bridgetown *85*; Camagüey *51*; Caracol *16*; Carriacou *86*; Catacamas *25*; Cayo Guano del Este *46*; Chiquimula *30*; Colón *14*; Cristóbal *14*; Cruta *44*; Curaçao *95*; Cutuco *8*; Desaix *78*; Désirade *74*; Flores *2*; Fonseca Bay *37*; Fort-de-France *80*; Government Farm, Tobago *89*; Grand Turks *53*; Grenada *87*; Greytown *18*; Guabito *17*; Guanaja *20*; Guatemala City *5*; Gustavia *6*; Havana *48*; Isla del Rey *42*; Isla San José *40*; Jaque *99*; Kingston *50*; Kingstown, St. Vincent *83*; La Asunción, Margarita *92*; La Ceiba *21*; La Habana *48*; Lake Maracaibo *100*; Lamentin *79*; Madden Dam, C.Z. *41*; Managua *10*; Margarita *92, 93*; Marie Galante *75*; Montserrat *72*; Moule à Chique, St. Lucia *82*; Nassau *49*; Nicoya peninsula *39*; Orotina *38*; Padre las Casas *56*; Panama City *13*; Paso Real *47*; Piarco *91*; Pinar del Rio *47*; Pointe-à-Pitre *73*; Port au Prince *54*; Port of Spain *91*; Puerto Armuelles *15*; Puerto Barrios *26*; Puerto Cortés *22*; Punta Maisí *52*; Raizet *73*; Road Town, Tortola *63*; Roxborough, Tobago *88*; Saba *67*; St. Barthélémy *66*; St. Croix *60*; St. Eustatius *68*; St. John (island) *62*; St. John's, Antigua *71*; St. Kitts *69*; St. Maarten (Martin) *65*; St. Thomas *61*; Salamanca *93*; San Andrés (island) *43*; San Andrés, El Salvador *32*; San Andrés Osuna *4*; San José, Costa Rica *11*; San José, Guatemala *6*; San Juan, Puerto Rico *58*; San Juan del Norte *18*; San Miguel *36*; San Pedro Sula *28*; San Salvador, C. A. *34*; Sta. Cruz Porrillo *35*; Sta. Rosa de Copán *29*; Sta. Tecla *33*; Sta. Teresa *3*; Sto. Domingo *57*; Sepacuité *23*; Shawford, Dominica *77*; Spanish Town, Virgin Gorda *64*; Swan Island *45*; Tegucigalpa *9*; Tela *27*; Texis Junction *7*; Tortola (island) *63*; Turbo *98*; Uplyme *81*; Vieques Island *59*; Villa Mills (elev. 3,000 m) *12*; Virgin Gorda (island) *64*; Virgin Islands *60–64*; Zacapa *24*.

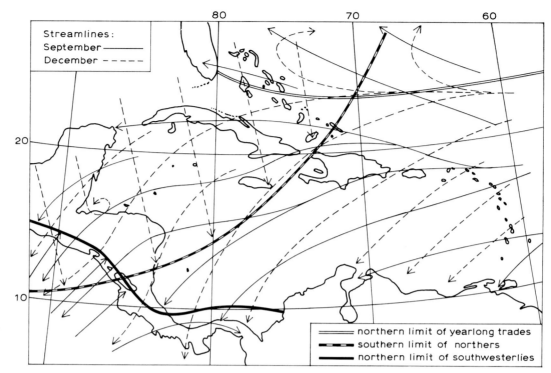

Fig.2. Mean streamlines. Over the Pacific coast of Central America two wind regimes alternate in September.

Central America. For readers without personal experience in continental tropics it is emphasized that the wind actually drops completely dead for hours, a fact that to the knowledge of this author has never been duly considered in theories concerning heat or moisture exchange. It may be added that balloon records occasionally reveal layers of several 1,000 m thickness without any horizontal motion.

TABLE II

PERCENTAGE OF CALMS

Region	Jan.	Feb.	Mar.	Apr.	May	June	July	Aug.	Sept.	Oct.	Nov.	Dec.
Honduras 1[*1]	35	31	29	26n	29	35	32	35	38	43	45x	43
Honduras 2[*2]	9n	15	17	17	21	27	24	25	32x	30	23	17
El Salvador 1[*3]	15n	19	20	28	26	31x	25	21	26	24	19	15n
El Salvador 2[*4]	6	9	8	10x	8	4	3	2n	3	3	5	3

Legend: n = minimum, x = maximum.
[*1] Guanaja, La Ceiba, Tela, San Pedro Sula, Cruta, and Santa Rosa de Copán.
[*2] Tegucigalpa and Amapala.
[*3] San Salvador, Santa Tecla, San Miguel, and San Andrés (inland stations).
[*4] Acajutla (coast) and Santa Cruz Porrillo (coastal plain).
Computation based on: SERVICIO METEOROLÓGICO NACIONAL DE HONDURAS, 1960; SERVICIO METEOROLÓGICO NACIONAL DE EL SALVADOR, 1954–1957.

TABLE III

MEAN DIURNAL WIND VARIATION AT DR. A. PLESMAN AIRPORT, CURAÇAO, IN SEPTEMBER 1960 AND 1961*

Hour (local time)	Wind vector		Deviation from mean wind vector	
	dir.(°)**	speed (m/sec)	dir.(°)	speed (m/sec)
0h30	99	5.6	273	1.1
1h30	103	5.4	256	1.3
2h30	107	5.2	251	1.7
3h30	108	5.1	250	1.9
4h30	107	5.0	253	1.8
5h30	108	4.9	253	2.0
6h30	108	4.9	253	2.0
7h30	106	5.9	232	1.2
8h30	107	7.1	171	1.1
9h30	107	7.4	157	1.4
10h30	103	7.6	133	1.1
11h30	105	7.8	141	1.4
12h30	102	7.8	121	1.3
13h30	98	7.9	95	1.3
14h30	95	8.0	78	1.4
15h30	91	8.4	64	2.0
16h30	90	8.3	61	2.0
17h30	88	7.9	47	1.8
18h30	86	7.6	33	1.8
19h30	88	7.4	32	1.4
20h30	90	7.3	36	1.2
21h30	93	6.9	24	0.6
22h30	93	6.5	353	0.6
23h30	95	5.9	298	0.8

* The sea is nearby to the north-northeast of the station. There is a low mountain ridge opposite to the sea and parallel to the coast line. The computation was based on data published by STATISTIEK- EN PLAN-BUREAU, CURAÇAO (formerly Bureau voor de Statistiek), 1960–61.
** East = 90°.

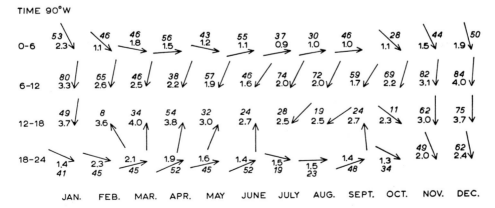

Fig.3. Diurnal wind variation at San Salvador. Roman numbers = mean scalar wind in m/sec; italics = steadiness in per cent. The wind vectors were computed from wind roses obtained from Servicio Meteoro lógico Nacional de El Salvador.

The annual variation of calms over the oceans (U.S. NAVY HYDROGRAPHIC OFFICE, 1955) near Central America follows for the Atlantic approximately "Honduras 2", for the Pacific, "El Salvador 2" (see Table II). The other ocean areas of the region have insignificantly small percentages of calms, whereas these numbers for the West Indian land stations greatly depend on their exposure to the trade winds and to the sea breeze. The majority of those stations report less than 2% calms, but since wind measuring stations are usually exposed more freely than the average location, this statistical result should not be generalized.

Tropical stations have distinct diurnal variations. Table III shows the diurnal variation of the wind at a typical trade-wind station. Each month presents the same pattern, so one is typical of all. The wind displays a systematic, gradual deviation from the daily mean, not the abrupt change of a "sea breeze front". Only when a station has a large land mass behind it, such as along the Pacific coast of Central America, do "sea breeze fronts" develop.

Fig.3 shows the diurnal wind variation at a station, San Salvador, which is alternatingly under the trade winds, the northers, and the southwesterlies. Each of them is superimposed by a strong sea breeze although San Salvador is separated from the coast by a 1,000 m high mountain range. As may be expected, the strongest sea breeze occurs in April when the cooling advection from the north is negligible and the solar heating is at its first maximum, with which the cloudiness of the rainy season does not yet interfere. It is remarkable that northeasterly winds dominate in the months of July and August, in the middle of the rainy season, although they are slightly weaker than the sea breeze (north and northeast 49%, 2.5 m/sec; south and southwest 26.5%, 2.7 m/sec. The nocturnal westerly is a local orographic effect which is less conspicuous at the airport station. The mean scalar wind is stronger in the afternoon (12h00–18h00) than in the morning (06h00–12h00) with the exception of November and December, when the sea breeze effect is usually reduced to a lessening of the north wind.

Barometric pressure

It is customary to present the mean pressure conditions for January and July. However, in Central America and the West Indies these two months represent maxima, so that some additional information must be given to indicate the minima.

The climatic tables at the end of this chapter offer monthly means of the atmospheric pressure for many stations. When one supplements them by maritime data from the Northern Hemisphere charts of normal pressure (U.S. WEATHER BUREAU, 1952), one can devise maps of departures from the annual mean. Such maps (not published) show the pressure variation of this region to be composed of two components. One of them can be presented in form of a sine curve with its maximum in December in the western part, January in the eastern part of the region, apparently produced by the monsoonal effect of the North American land mass. The other effect emanates from approximately 25°N 60°W and produces a sharp maximum in July and an equally sharp minimum in October. The two pressure oscillations can be separated statistically and their amplitudes can be computed. In the northwestern part of the Gulf of Mexico the amplitude of the continental pressure wave is more or less twice as high as the oceanic one. Along a line running from Florida over Grand Cayman to Colombia, both waves are of the same in-

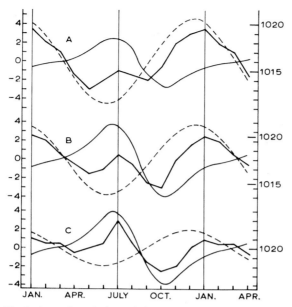

Fig.4. Annual variation of barometric pressure. $A = 25°N\ 95°W$ (centre of the Gulf of Mexico); $B = 25°N\ 80°W$ (southern tip of Florida); $C = 30°N\ 65°W$ (south of Bermuda Islands). Bold solid curves = original data; thin solid curves = maritime component; dashed curves = sinusoidal annual variation (continental).

tensity. Eastward of this line the amplitude of the oceanic pressure wave rises to the double of the amplitude of the monsoonal wave. Fig.4 gives some examples.

After WRAIGHT (1951) all coastal stations of Cuba (including Pinar del Rio which is largely influenced by marine conditions) show a slight maximum in March. It is his opinion that the light depression between the January and March maxima may be due to occasional cyclonic storms and to a short period of disturbances over the Gulf of Mexico and the Caribbean. No corresponding increase of rainfall has been found in Cuba.

The atmospheric pressure displays a well expressed double wave every day with minima near 04h00 and 16h00 local time (L.T.), and maxima near 10h00 and 22h00 L.T. The data of San Salvador (PORTIG, 1956) and of Cuba (WRAIGHT, 1951) show in all seasons the 16h00 minimum to be much deeper than that in the early morning. The two maxima,

TABLE IV

ANNUAL VARIATION OF THE DOUBLE DIURNAL PRESSURE WAVE IN SAN SALVADOR, C.A., BASED ON DATA OF FOUR YEARS[1]

Local time	Jan.	Feb.	Mar.	Apr.	May	June	July	Aug.	Sept.	Oct.	Nov.	Dec.	Year
03–04h first minimum, N	4.3	4.0	3.3	3.1	2.7	3.2	3.9	3.6	2.7	2.8	3.9	4.4	933.5
09–10h first maximum, X	6.8	6.0	5.5	4.9	4.1	4.5	5.1	4.8	4.2	4.5	5.7	6.2	935.2
15–16h second minimum, n	3.0	1.9	1.3	1.2	1.1	1.8	2.5	2.0	1.1	1.6	2.4	2.7	931.9
22–23h second maximum, x	6.1	5.1	4.5	4.4	4.2	4.6	5.3	5.2	4.2	4.4	5.4	5.7	935.0

[1] Units mbar minus 930, or mbar (year).

411

however, are practically equal when averaged over all months, and they have an annual variation with the 10h00 maximum being smaller than the 22h00 maximum during May through August (Table IV).

Cloudiness

It is customary that climatological treatises offer values of mean cloudiness. The following objections may be raised against this procedure.

The cloudiness of tropical stations displays a prominent diurnal variation. This implies that: (*1*) stations can be compared only when their hours of observation are equal which is—for historical reasons—normally not the case in our region; and that (*2*) observations of certain hours are not representative for the mean cloudiness of the day. The region is also more often than not under a canopy of cirrus. Even an excellent observer cannot decide where the cirrus ends and where the blue sky begins, and it is anyway not established how thin a cirrus layer has to be to become negligible.

The difficulties for the observer increase (especially when altocumulus lenticularis is involved) when the clouds form or dissipate rapidly. Neither the present cloud code nor the *International Cloud Atlas* (W.M.O., 1956) help to describe the state of the sky adequately in such cases.

Whether clouds can be seen depends largely on the time of the day. Cirrus clouds are best visible during only a few minutes at sunrise and sunset. Other clouds may be overlooked or misinterpreted at night. Because of the long tropical nights and the strong diurnal variation of cloudiness these conditions add to the difficulties of establishing proper and meaningful cloud statistics.

Tropical clouds show a strong tendency to respond to geographical features. This becomes obvious in satellite photographs where coastlines and mountain ranges are readily seen in the cloud patterns. In consideration of the many unassessed factors which affect cloud statistics, Table V, presenting some monthly means of cloudiness, will not be discussed. Further data of mean cloudiness are listed in the climatic tables at the end of this chapter.

The objections to cloud statistics in the tropics apply to a lesser degree to observations made aboard ships. Since many observers contribute to the statistics of one ocean field the effects of the individualities of the observers cancel out while local effects are either absent or greatly reduced. This does not eliminate the difficulties of interpreting the data physically but it does make the data mutually comparable.

Table VI informs about the annual variation of cloudiness in some ocean fields. The minimum of the annual range is found along a line running from Trinidad through point *L* (Fig.6) to the Yucatan Straits, and from there to the Mississippi delta. The largest annual range is observed in the Pacific Ocean off the coast of Central America.

Minima of mean annual cloudiness are found along the north coast of Colombia, in the bay between the two Haitian peninsulas, and north and west off Yucatan. Maxima are found in the western Gulf of Mexico and along the Central American east coast from Belize, British Honduras, to Turbo, Colombia. This corresponds fairly well to the distribution of rainfall, see pp.418–445.

TABLE V

MEAN CLOUDINESS IN TENTHS

Station	Hour	Jan.	Feb.	Mar.	Apr.	May	June	July	Aug.	Sept.	Oct.	Nov.	Dec.	Year
Bridgetown	09h	6.9	6.9	7.0	7.6	7.6	7.7	7.7	7.6	7.7	7.8	7.3	7.1	7.4
(Barbados)[1]	15h	7.1	7.0	7.0	7.2	7.2	7.5	7.4	7.4	7.3	7.2	7.2	7.1	7.2
Port au Prince	07h	1.4	1.9	2.0	2.4	3.3	2.8	2.1	2.2	2.6	2.7	2.4	1.5	2.3
(Haiti)[2]	13h	2.0	2.6	2.4	3.2	4.7	4.6	3.9	3.9	4.0	3.9	3.2	2.2	3.4
	19h	4.2	5.4	5.9	7.1	7.0	6.5	6.1	6.9	7.2	7.0	5.5	3.7	6.0
Piarco	08h	4.2	3.9	4.1	4.5	5.1	5.4	4.3	4.2	4.3	4.4	4.2	4.1	4.4
(Trinidad)[2]	14h	5.4	5.5	5.7	6.1	6.3	6.3	5.8	6.0	5.9	6.1	5.8	5.3	5.9
	20h	3.1	2.4	2.8	3.3	3.6	4.4	3.2	3.1	3.2	3.5	3.5	5.0	5.9
Kingston	07h	3.1	2.9	2.3	3.4	4.6	5.3	4.5	4.7	5.6	5.2	4.8	3.6	4.2
(Jamaica)[2]	13h	4.0	4.2	3.8	4.6	5.6	5.6	5.2	5.4	6.1	5.9	5.3	4.4	5.0
	19h	2.5	2.8	2.8	4.2	5.5	5.5	4.9	5.3	5.5	4.8	3.9	2.9	4.2
Nassau	04h	3.4	3.3	3.4	3.8	3.6	4.4	4.0	3.8	4.4	4.3	3.8	2.9	3.8
(Bahamas)[2]	10h	5.0	5.3	5.9	6.4	6.6	7.8	7.1	7.1	7.4	6.6	5.6	5.0	6.4
	16h	5.0	4.9	5.3	5.9	6.3	7.4	7.4	7.4	7.4	6.4	5.6	5.1	6.1
Havana	(*)b	4.8	4.4	3.9	4.0	4.6	5.3	4.4	4.7	5.1	5.8	5.7	5.3	4.8
(Cuba)[2]	c	4.7	4.5	4.0	4.3	5.5	6.7	6.3	6.6	6.7	6.4	5.4	5.0	5.5
	d	3.7	3.4	2.8	2.7	3.6	4.4	3.8	4.0	4.1	4.6	4.0	3.9	3.7
Fort de France	06h	4.9	5.2	5.2	5.3	5.4	6.1	6.3	6.0	5.9	6.0	5.8	5.2	5.6
(Martinique)[2]	10h	5.9	6.0	6.1	6.3	6.5	6.9	6.8	6.7	6.7	6.7	6.3	6.1	6.4
	16h	5.6	5.6	5.8	5.9	6.3	6.8	6.7	6.6	6.7	6.7	6.0	5.7	6.2
Belize[2]	07h	7.3	6.8	6.5	7.3	7.6	7.8	7.5	7.5	7.8	7.7	7.5	7.5	7.4
	17h	6.1	5.4	5.0	5.6	7.2	7.6	7.5	7.1	7.9	7.3	7.3	7.1	6.8
Guatemala City[3]	07h	6.3	6.2	6.0	7.2	7.8	8.7	9.2	9.2	8.4	8.4	7.5	6.4	7.6
	14h	4.8	5.0	5.7	7.1	8.0	9.1	8.3	8.9	8.6	8.6	7.2	5.1	7.2

[1] From *Sailing Directions for the West Indies*, vol. II, H.O. Publ. No.129, 6th ed., 1949.
[2] From *Aviation Meteorology of the West Indies*, Meteorological Office, London, MO621f, 1959, 85 pp.
[3] From *Weather Summary Central America*, H.O. Publ. No.531, 1948.
(*)b = average of 08h, 10h and 12h; c = average of 14h, 16h and 18h; d = average of 20h and 22h.

TABLE VI

MEAN MONTHLY CLOUDINESS IN TENTHS OVER SELECTED OCEAN AREAS[1]

Position	Jan.	Feb.	Mar.	Apr.	May	June	July	Aug.	Sept.	Oct.	Nov.	Dec.	Year
20–22°N 90–92°W (B)	3.9	3.5	3.1	2.9	3.9	4.5	3.9	4.0	5.0	4.5	4.5	4.6	4.0
	2.1	2.6	2.3	1.9	2.6	2.6	2.9	2.8	4.0	3.9	3.5	3.3	2.9
24–26°N 84–86°W (C)	4.7	4.7	4.3	4.3	4.0	4.5	4.7	4.6	5.3	5.1	5.1	5.2	4.7
	3.8	3.9	3.4	3.2	2.8	3.1	3.3	3.2	4.0	4.0	4.2	4.2	3.6
14–16°N 64–66°W (K)	4.1	4.3	3.9	4.3	4.8	5.2	4.6	4.4	4.8	4.3	4.3	4.0	4.4
	3.6	3.5	3.2	3.4	3.4	3.7	3.4	3.1	3.6	3.4	3.4	3.2	3.4
14–16°N 72–74°W (L)	4.1	4.0	4.0	4.1	5.0	4.7	4.7	4.0	4.7	4.9	4.7	4.3	4.4
	3.4	3.5	3.2	3.0	3.7	3.4	3.5	3.0	3.3	3.5	3.6	3.5	3.4
18–20°N 80–82°W (M)	3.9	3.9	3.1	3.8	4.4	5.2	4.7	4.8	5.4	5.2	5.3	4.8	4.5
	3.6	3.2	2.6	2.8	2.6	3.5	3.4	3.3	3.7	3.9	4.1	3.8	3.4
10–12°N 78–80°W (N)	5.1	5.1	5.5	6.0	6.6	6.8	6.7	6.4	6.2	6.3	6.0	5.2	6.0
	4.1	4.3	4.6	4.6	4.7	4.8	5.0	4.6	4.3	4.6	4.2	4.3	4.5
10–12°N 86–88°W (north of P)	1.9	1.8	1.8	3.4	6.1	6.6	6.5	6.7	7.0	6.3	4.6	3.1	4.6
	1.2	1.0	1.0	1.7	3.9	4.5	4.0	4.7	5.3	4.6	2.9	2.4	3.2

[1] Upper line, total cloud coverage; lower line, coverage with low clouds. Capital letters in parentheses refer to Fig.6.
Data compilation through courtesy National Weather Records Center, Asheville, N.C. (unpublished).

Sunshine

Sunshine records give a fair estimate of the daytime shielding effect of the clouds. As does the human observer, the sunshine recorder notices thick cirrus but disregards medium or thin cirrus.

The climatic tables at the end of this chapter give mean sunshine duration in hours for eight stations. Table VII presents the same data expressed in percent of the maximal possible sunshine. (One station has been added from an older source.) Most stations show a tendency to have their sunshine extremes a month or so earlier than their rainfall extremes.

Particularities at certain stations deserve special attention. As can be seen in the chapter on precipitation, May is the month in which, in most parts of the region, significant changes of the rainfall pattern occur. Curaçao does not participate in such a change, yet the sunshine has a well developed minimum in May.

San Salvador has practically the same rainfall total in each of the summer months and 20 days with measurable rain in each of them. The sunshine, however, has a well expressed maximum in the middle of the rainy season, i.e., in July and August. For more details see Table XV and the text on p.426.

The sunshine regimes at the two ends of the Panama Canal are very similar while the rainfall regimes have substantial differences. The Atlantic end has much more rain and slightly more sunshine than the Pacific end. (See Tables XXXIII, XXXIV and LV).

Temperature

Since our region is tropical and predominantly maritime, temperature changes are small. This facet would make it desirable to obtain more accurate observations than those of regions with larger variations. This is, however, certainly not the case. This author has made measurements that showed that the temperature measurement in a well ventilated meteorological shelter depends on the cloud cover. And PROHASKA (1963) pointed to the fact that daily means computed from hourly observations are up to 1.5°C lower than the average of the maximum and minimum.

TABLE VII

MEAN MONTHLY SUNSHINE IN PERCENT OF THE POSSIBLE

Station	Jan.	Feb.	Mar.	Apr.	May	June	July	Aug.	Sept.	Oct.	Nov.	Dec.	Year
Pointe-à-Pitre	65	68	69	64	58	59	57	59	57	60	62	63	62
Fort de France	68	71	69	65	62	55	56	65	60	61	64	64	63
Curaçao	75	84	76	71	57	70	75	76	76	74	71	67	73
San Juan, P.R.	63	69	68	64	60	61	64	67	62	64	62	53	65
Port au Prince	79	79	72	68	63	63	67	68	64	62	73	74	69
Nassau, Bah.	70	72	71	68	64	58	65	58	59	58	69	66	65
San Salvador, C.A.	85	85	79	65	56	45	60	65	49	56	77	84	67
Cristóbal, C.Z.	64	72	72	74	51	45	38	41	46	48	38	51	53
Balboa, C.Z.	61	76	75	72	50	46	35	46	46	44	37	59	54
Average*	70	**75**	71	68	58	*55*	57	**61**	*58*	59	62	65	63.4

* Numbers in bold face are relative maxima, numbers in italics are relative minima.

Annual variation of temperature

Two types of annual variation can be distinguished. In most of the region we find the summer–winter variation of higher latitudes, though with high tropical temperatures and a small range. In Central America (with the exception of the Atlantic coasts of Honduras and northern Nicaragua) the monsoon type dominates with the highest temperatures before the onset of the summer rains. Fig.5 presents the annual temperature variation for three stations.

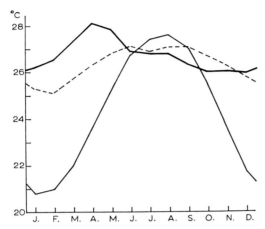

Fig.5. Annual temperature variation. Bold curve: Santa Cruz Porrillo, 13°26′N 88°46′W (14 years); thin curve: Nassau, 25°03′N 77°28′W (1946–1960); dashed curve: Seawell Airport, 13°07′N 59°29′W (1956–1960, reduced to fit the annual variation of the mean extreme temperatures of the 35 year record of Bridgetown).

Throughout the entire region (except Bermuda) the average temperature (at sea level) of the coldest month is higher than 19°C, that of the warmest less than 29°C, except for a few places on the Pacific side of Central America where the mean April temperature rises up to 31°C (e.g., in Amapala, an Hondurian island).

The annual temperature range decreases from 8°C in the northern Bahamas to 1.6°C in the Canal Zone. Yucatan and most of Honduras have a range of more than 4°C, dropping to less than 3°C at the coasts. The temperature range of stations at higher elevations is generally of the same magnitude as that of nearby lowland stations (Table VIII).

TABLE VIII

MONTHLY MEAN TEMPERATURES (°C) OF TWO COSTA RICAN STATIONS

Station	Jan.	Feb.	Mar.	Apr.	May	June	July	Aug.	Sept.	Oct.	Nov.	Dec.	Year	Temp. range
Villa Mills[1]	6.0	6.9	7.8	8.1	8.1	8.1	7.8	7.9	7.6	7.0	6.5	6.2	7.3	2.1
Orotina[2]	27.2	27.4	28.4	28.3	27.1	25.6	25.4	26.0	25.6	25.5	25.7	26.1	26.5	3.0

[1] 9°34′N 83°41′W, elev.3,000 m; 16 years of record.

[2] 9°53′N 84°31′W, elev.223 m; 9 years of record.

The vertical temperature gradient between these two stations varies between 6.5°C/km in June and July, and 7.6°C/km in January.

415

Temperature variation with altitude

The historical grouping in "tierra caliente" (warm land, i.e., land at low elevations), "tierra templada" (mild land, i.e., land at moderate elevations), and "tierra fría" (cold land, i.e., land at high elevations) has neither a scientific nor practical justification.

While most of the West Indies lack temperature measuring stations at higher elevations, WERNSTEDT'S (1961) data collection makes it possible to study the average decline of temperature with height in Central America. In the dry months January and April, the dispersion of the measured monthly means allows some doubts as to the representativeness of an ideal height–temperature curve (Table IX). Less uncertainty exists with respect to the curves of the moist months July and October. Table IX contains also the mean lapse rates computed from the temperature data. Their minimum near the 2,000-m level indicates the mean position of the trade-wind inversion. For Martinique, a mean lapse rate of 6°C/km is reported (SERVICE MÉTÉOROLOGIQUE DU GROUPE ANTILLES–GUYANE, 1962). REED (1926) mentions 7°C/km for maximum temperatures and 5.5°C/km for minimum temperatures measured in Jamaica up to 1,600 km. ROUSE (1966) reports for Barbados an annual variation from 7.7°C/km in September to 11.1°C/km in March, between 60 and 326 m elevation.

TABLE IX

TEMPERATURE (°C) AS A FUNCTION OF ELEVATION IN CENTRAL AMERICA[1]

Elev. (m)	Jan.	Apr.	July	Oct.	Lapse rate per kilometer			
					Jan.	April	July	Oct.
3,000	6.0	10.4	8.4	6.9				
					10.0	7.0	9.4	8.2
2,500	11.0	13.9	13.1	11.0				
					7.8	5.4	6.8	7.6
2,000	14.9	16.6	16.5	14.8				
					4.8	5.6	5.0	7.6
1,500	17.3	19.4	19.1	18.6				
					6.0	9.0	6.0	5.8
1,000	20.3	23.9	22.1	21.5				
					6.4	5.4	6.0	5.2
500	23.5	26.6	25.1	24.1				
					8.6	4.6	5.8	5.2
M.S.L.	27.8	28.9	28.0	26.7				

[1] Computation based on data published in WERNSTEDT, 1961.

According to HOFFMANN'S charts (1960), the absolute temperature extremes depend almost completely on the elevation. He lists absolute temperature maxima of slightly above 40°C along most of the Central American coast of the Pacific Ocean and in the interior of Yucatan, and 35°–40°C for almost all other low lying areas. In consideration of the difficulties in obtaining reliable temperature measurements, no more details will be discussed.

The distribution of the absolute temperature minima shows the effects of the winterly cold air invasions from the North American continent. Southeast of a line running from the straits between Puerto Rico and Hispaniola ("Mona Passage") to the Nicoya peninsula at the Pacific coast of Costa Rica, the temperature of the lowlands never dropped below 15°C. Northwest of this line the absolute temperature minima at sea level occurred with approximately 7°C in western Cuba and most of Yucatan as well as in northern Honduras. Such low temperatures usually occur after a cold air intrusion from the north has begun to slow down, and night-time cooling adds to the effect of advection.

HOFFMANN (1960) reports the highest mountains of Guatemala and of Costa Rica to have experienced air temperatures slightly below freezing.

For soil temperatures the reader is referred to the *Boletín Meteorológico de El Salvador*.

Diurnal temperature variation

Although reduced by the nearness of the ocean, the diurnal temperature range exceeds the annual temperature range. The climatic tables at the end of this chapter give numerical values. In general terms one can say that the dry season has a larger daily temperature variation than the rainy season, and coastal stations, especially when they lie on the windward side have smaller variations than inland stations. Table X illustrates these differences through juxtaposition of two Salvadorean stations, one of which lies directly at the coast, the other 55 km away in a wide valley. Incidentally, the latter (San Andrés, the city) has the largest diurnal range of the entire region, as far as the sparse data allow such a statement. The station Acajutla, though only a few metres from the beach, has almost double the diurnal variation typical for most stations of the Antilles. (See the climatic tables at the end of this chapter.)

In an unpublished investigation this author listed for San Salvador–Ilopango (airport) the highest and lowest temperatures observed at every full hour during the period 1955–1959. It turned out that the highest temperatures are less dependent on the season than the lowest (see Table XI).

Moisture

Available statistics of atmospheric humidity are as unrevealing as cloud statistics. This is mainly due to the heterogeneity of available data. The publications reviewed for this work

TABLE X

DIURNAL TEMPERATURE RANGE (°C)

Station	Jan.	Feb.	Mar.	Apr.	May	June	July	Aug.	Sept.	Oct.	Nov.	Dec.	Year
Acajutla[1]	10.5	10.2	9.0	8.3	8.1	8.0	8.9	9.1	8.6	8.4	9.3	10.6	9.0
San Andrés[2]	18.1	18.7	18.3	16.6	13.9	12.0	12.7	12.8	11.9	12.3	14.8	16.9	14.9
							rainy						

[1] Coastal station, 13°36'N 89°50'W, 8 years of record.
[2] Inland station at 55 km from Acajutla, 13°49'N 89°24'W, 14 years of record.

TABLE XI

LOWEST AND HIGHEST TEMPERATURES OBSERVED AT CERTAIN HOURS IN SAN SALVADOR–ILOPANGO

Time 90°W:	00h	05h or 06h*[1]	12h	14h or 15h*[2]	18h
October–January	12.8–24.5	9.2–21.4	21.2–30.2	21.6–32.6	18.4–27.2
February–May	15.1–25.6	12.7–22.6	23.8–32.9	22.6–34.2	21.1–31.3
June–September	18.2–23.6	16.8–21.9	21.5–30.6	20.9–32.6	18.6–30.4

*[1] Whichever is lower.
*[2] Whichever is higher.
First line: time of occasional strong northers.
Second line: time of weaker northers and many calms.
Third line: rainy season.

present averages of the relative humidity in seventeen different combinations of hours but only one of them contains a night hour. The magnitude of the biases can be studied with the following example. On January 4, 1957, a sunny, windy winter day, the relative humidity in San Salvador oscillated between 40 and 73% with a true average of 61.5%. The seventeen procedures to select representative hours would yield values ranging from 40 to 72%. Most of the publications contain observations of two hours. Averaging the pairs of hours actually used in our region would give us values ranging from 54% (08h00 and 12h00) to 66.5% (06h00 and 10h00). This example which deals with a typical day of the season shows that the published data do not allow comparison of the moisture conditions of different parts of the region.

The comparability would be better if dew points, vapor pressure, or mixing ratio would be published; all these measures have much smaller diurnal variabilities than the relative humidity. Substituting the dew point in our example, we find that on that day the dew point oscillated only between 12° and 15°C with a true average of 13.3°C. The dew points of the published hours would range from 12° to 15°C, and the published combinations of hours would yield averages between 12.5° and 14°C, i.e., close to the true mean.

Due to the strong influence the oceans have on the climate of the West Indies, the moisture always remains rather high. In Central America, however, it can drop in a few minutes to values so low that a hygrometer has difficulty in recording them. In San Salvador, minima below 20%, or below 7°C dew point temperature, have been observed in January through April between 14h00 and 15h00 local time. They were obviously the consequence of rapidly subsiding upper air masses.

The dew point can be crucial in the identification of tropical air masses, especially in the dry season (PORTIG, 1959). But since the climatological summaries (with one exception: Puerto Rico and the Virgin Islands) do not mention them (although they are measured and transmitted every hour from every international airport) no presentation can be given here.

Rainfall

Introduction

The most important meteorological element in the tropics is the rainfall. Although Central America and the West Indies are practically always under the influence of maritime air masses, the local and, partly, seasonal variations of precipitation are surprisingly high. As far as the rainfall totals are concerned, most of them can easily be attributed to orographic effects. However, we also find significant differences in the timing of the rainfall seasons over short distances.

Sea

There are no measurements of the rainfall amounts from the open sea. There are many reasons to believe that the rainfall over the oceans is less than over adjacent land and islands. SKEETE (1931), for example, found that the rainfall on the windward part of the rather flat island of Barbados is 15–20% less than the average for the entire island. He attributes this to the fact that the rain in that part falls from clouds generated at sea and blown over the island while the rain in the other parts stems from clouds produced by convection over the land. (Reference is further made to the paragraphs on annual rainfall totals, especially Cuba and Hispaniola pp.433–434 and p.435.)

Counts of rainfall frequency can be presented here through courtesy of the U.S. National Weather Records Center, Asheville, N.C. These counts have some obvious shortcomings that cannot be discussed here nor can they be easily eliminated. They show, however, some consistency in time and space which allows drawing conclusions that will help to understand the rainfall patterns of the land stations as a part of larger circulation systems.

Mean annual rainfall frequency on sea

The map in Fig.6 shows lines of annual rainfall frequency in percent of all ship observations used. Remarkable features are the minima along the Caribbean coasts of Venezuela and Colombia as well as north and west of Yucatan, and the maxima in the western parts of the gulfs of the Mosquitoes and of Campeche, along the Lesser Antilles, and—in the Pacific Ocean—along 5°N. The decrease of rainfall across the peninsula of Yucatan is well documented (TREWARTHA, 1961, p.70). Fig.6 shows that the Yucatan dryness is, though less famous, more intense than the Curaçao dryness.

It is remarkable that the increase of rainfall activity along the Lesser Antilles is not confined to the islands but is also evident at sea, even upstream from the islands.

Annual variation of rainfall frequency on sea

The insets of Fig.6 inform about the seasonal variation of the rainfall frequencies at the locations indicated by letters in the map. Since these diagrams have been derived from rigorously smoothed data they are valid also for the surroundings of the $2° \times 2°$ fields for which they stand. The maps (Fig.8,19) that indicate rainfall regimes for land and island areas show some suggestions of how to divide the ocean areas into climatic regions.

All maritime areas have their rainfall minimum in the first half of the year, and their highest maximum (except *J*, off the Guiana coast) in the second. This is especially conspicuous for area *O* which is situated to the south from the intertropical rainfall belt. Concluding from the rainfall conditions over Africa one would rather expect either two equally spaced and equally high rainfall maxima at O, or the main maximum in the first half of the year, and the minimum in the second. Data published by ALPERT (1946a) reveal that south of the rainfall belt the annual variation of the rainfall frequencies is substantially different from that of the rainfall totals. He gives the figures reproduced in Table XII for Seymour Island, 0°28′S 90°18′W, i.e., southwest of field O. The last line of the table indicates the weather that accounts for the difference of the annual variations of frequencies and totals.

TABLE XII

RAINFALL CONDITIONS AT SEYMOUR ISLAND

	Jan.	Feb.	Mar.	Apr.	May	June	July	Aug.	Sept.	Oct.	Nov.	Dec.	Year
Rainfall total (mm)	21	35	27	17	0.1	0.2	0.2	0.2	0.1	0.2	0.1	0.1	101
No. of rain days	8	9	6	6	4	4	9	8	7	2	4	6	73
Weather	E wind with showers						SSE wind with drizzle						

This difference may also account for the slight discrepancy between the mean position of the Intertropical Convergence Zone (I.T.C.Z.) (U.S. WEATHER BUREAU, 1938; ALPERT, 1945, 1946b) and the position of maximum rainfall frequency (see Table XIII). FLOHN (1957) published a thorough discussion on the asymmetry of the Atlantic I.T.C.Z. He had the advantage of much more and better observations at his disposal. ALPERT (1945, p.432) mentions the possibility of the coexistence of a shallow I.T.C.Z. convergence and an active convergence aloft at another latitude.

TABLE XIII

ANNUAL VARIATION OF THE LATITUDE OF THE I.T.C.Z. AND OF THE ZONE OF MAXIMUM RAIN FREQUENCY OVER THE PACIFIC OCEAN ADJACENT TO CENTRAL AMERICA

	Jan.	Feb.	Mar.	Apr.	May	June	July	Aug.	Sept.	Oct.	Nov.	Dec.
Mean I.T.C.Z.	5	3	3	5	8	9	8–9	8	8	8–9	7	6
Zone of maximum rainfall frequency	2	3	4	5	6	6	5	5	6	6	5	3

While the Pacific rainfall zone moves only a short distance, its intensity varies considerably in the course of the year. At all longitudes east of 90°W we find that the rainfall frequency in March (locally February) does not exceed 9% at any northern latitude. It exceeds 30% somewhere along these longitudes in one or several summer months. If one accepts the assumption that the rainfall frequencies are indicative for the rainfall totals, one can show that the rainfall maxima in Central America do not indicate the presence of the I.T.C.Z. during those months over Central America. Table XIV presents rainfall frequencies along 91°W for different latitudes over sea. By different symbols it

TABLE XIV

RAINFALL FREQUENCIES (%) ALONG 91°W AT DIFFERENT LATITUDES OVER SEA

Lat. (°N)	Jan.	Feb.	Mar.	Apr.	May	June	July	Aug.	Sept.	Oct.	Nov.	Dec.	Year
12–14	1	1	0−	1	8	11+	9−	10	14+	12	4	0−	5.6
10–12	×	1	0−	2	7	10+	9−	10	13+	11	4	0−	5.6
8–10	1	×	0−	2	6	9	10	11	13+	12	7	2	6.2
6–8	11	7	5−	6	9	*14*	*18*	20+	19	*18*	16	14	13.1
4–6	**25+**	**18**	**10−**	**11**	**15**	**16**	**19**	**25+**	**24**	**21**	**19−**	**24**	**19.2**
2–4	*21+*	*15*	8	7−	9	10	14	19	*20+*	15	14−	*19*	*14.3*
Eq.–2	13	9	3	2−	3	4	7	12	13+	8	7−	14+	7.5

Legend: × = more than 0 but less than 0.5; − = relative minimum for that latitude; + = relative maximum for that latitude; bold face = largest value for that month; italics = second largest value for that month.

is marked *where* the highest amount of each month occurs, and *when* each latitude has its maxima in the course of the year. One sees that the highest local rainfall frequencies stay at any time considerably south of Central America at that longitude.

In spring, when the northernmost part of the region presented in Fig.6 has a rainfall minimum (fields *A*, *B*, *C*, *D*, *H*), other portions already have their first maximum. This first peak occurs in May in the central Caribbean Sea and in the region from the central Bahamas eastward (*E*, *G*, and *L*). In some places it is surprisingly strong and followed by a minimum in June or July (*G* and *K*). Another area with a rainfall maximum is centered around field *S* in the Pacific Ocean. From these two areas the rains spread and produce June maxima in a small area around the northern Lesser Antilles, an area off the Guiana coast, and a large area extending from the 8th parallel over most of Central America, and over most of Cuba to the northern Bahamas (*D*, (*E*), *J*, *M*, *N*, *P* and *Q*).

In the eastern Caribbean, on parts of the Atlantic seaboard of Central America, and off the Pacific coast of Colombia, the first summer rain maximum is delayed until July (*K*, *I* and *R*).

There is a zone extending from Hispaniola eastward and northward where both occur, a spring maximum in May or June, and a summer maximum in July or August (*K*, *I* and *G* lie on the border line of this zone).

The entire region has a rainfall maximum in fall. It begins in September in the west (*A*, *B*, *C*, *Q* and *S*) and moves from there to the east where the fields around *D*, *E*, *F*, *I*, *J*, *L*, *M* and *P* show it in October, and *G*, *H*, *K* and *N* in November. In some places the fall maximum is delayed until December, such as in the field 24°–26°N 58°–62°W and at the coast of Venezuela.

There are also some areas which have a winter rain maximum in addition to a fall maximum, such as the Lesser Antilles including the seas off the Guiana coast, the Pacific Ocean right north of the Equator (see also Table XIV), and most of the area north of 24°N (*A*, *B*, *C*, *D*, *E*, *I*, *J*, *K*, *O*, *R* and *S*). Considering this multitude of rainfall patterns over sea, it is not surprising to find over land areas very complicated conditions with little or no possibility to explain them with the presently available information.

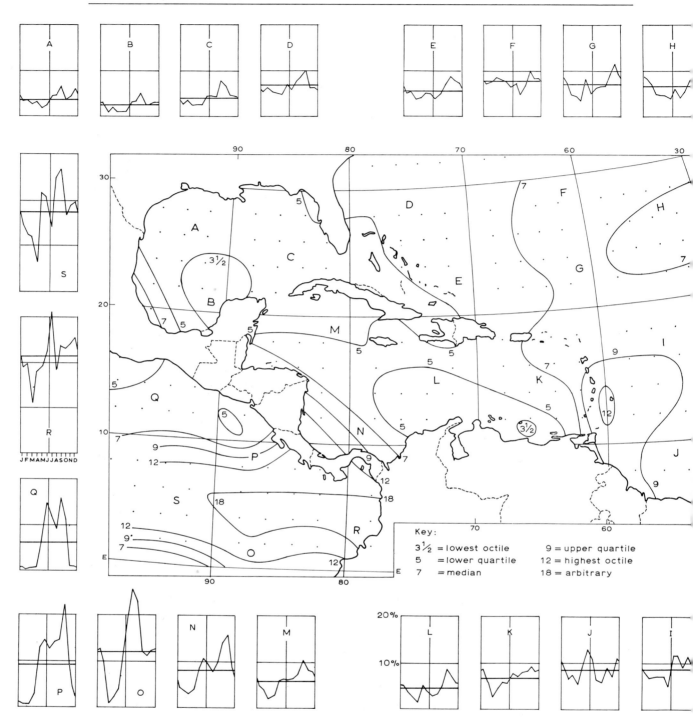

Fig.6. Rainfall frequency in percent of all observations made in 2° × 2° fields aboard ships. The body of the figure depicts the mean annual rainfall frequency. The letters refer to the insets which give the annual variation of the respective fields. In the insets the vertical lines at the left and at the right correspond to January 1, the middle line to July 1. The bold horizontal line denotes the annual mean. All insets are drawn to the same scale (see inset *L*).

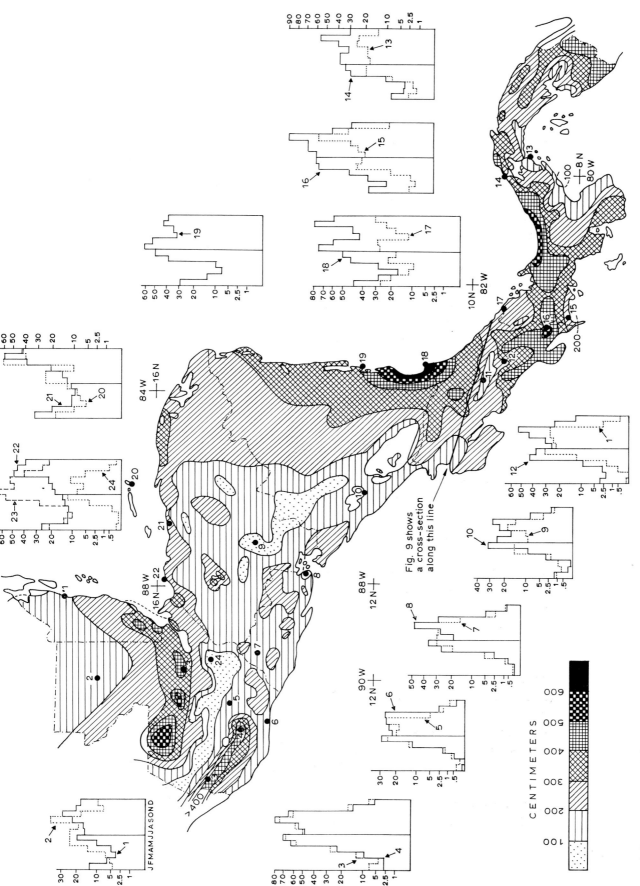

Fig.7. Mean annual rainfall (mm) in Central America. The straight line through Costa Rica is the base of the cross-section presented as Fig.9. For explanation of numbers see Fig.1.

Central America

Annual rainfall total of Central America

The chart of annual rainfalls presented here as Fig.7, though far from being definite, is an improvement over its predecessor (PORTIG, 1965).

The mean annual rainfall totals of Central America vary in a surprisingly wide range. There is a wide belt of steppe conditions through Guatemala and Honduras into north-western Nicaragua, in which the annual totals stay below 1,000 mm. There are indications also that parts of the Pacific coast west of the Bay of Fonseca (HASTENRATH and LESSMANN, 1963), and southwest of the city of Panama are very close to only 1,000 mm. Two types of areas with excessive rains contrast to this dryness. The first of them is found in the mountains of Guatemala, its southern (coastal) as well as its northern range, and in the mountains crossing the political border between Costa Rica and Panama. The carefully analyzed map of the annual rainfall of Costa Rica (Servicio Meteorológico Nacional de Costa Rica, 1965) shows that the large rainfall totals are concentrated on the tops of lesser mountains and on the slopes of higher mountains while the crests above approximately 2,500 m receive less than half of what falls on the slopes.

The other type of area with excessive rains exists along some portions of the coasts such as between Belize and Puerto Barrios, near San Juan del Norte (Greytown), at the Atlantic coast of Panama near 8.7°N 81°W, and—to a lesser degree—at the extreme southeast corner of Panama as well as at the south coast of Costa Rica.

The reasons for the strong contrasts of precipitation totals are only partly understood. For a discussion of possible causes the reader is referred to this author's previous paper (PORTIG, 1965). It suffices here to state that, according to TROLL (1952), the dryness of the interior of Central America is due to mountain-valley winds rather than to shielding effects of the mountains. The dryness of the narrow strip along the Pacific coast west of the Bay of Fonseca may partly be due to the same katabatic effect enhanced by LAHEY's (1958) friction effect at coasts running parallel to the mean wind flow.

Most of the rain-bringing disturbances come from the Caribbean so that the rainfall at the Atlantic seaboard is much higher than at the Pacific (compare insets *13* and *14* of Fig.7). The mountains of the Azuero peninsula may, in addition, have a shielding effect against disturbances of Pacific origin.

Annual variation of rainfall in Central America

The insets of Fig.7 show that the annual variation of rainfall *amounts* in Central America corresponds fairly well to that of the rainfall *frequencies* over the surrounding waters. (In order to make the variations during the dry season more evident a non-linear scale has been used in the insets.)

While the rainfall totals are, to a large amount, decided by the local characteristics of the stations (especially elevation and exposure to rain-bringing winds), the regime is rather due to macro-scale developments and, hence, can be found throughout relatively large areas. (Compare, for example, insets *4* and *5* in Fig.7). Therefore, instead of the customary monthly rainfall maps, a chart of rainfall regimes is presented in Fig.8. It must be read in connection with Table XV where typical monthly rainfall totals are given ex-

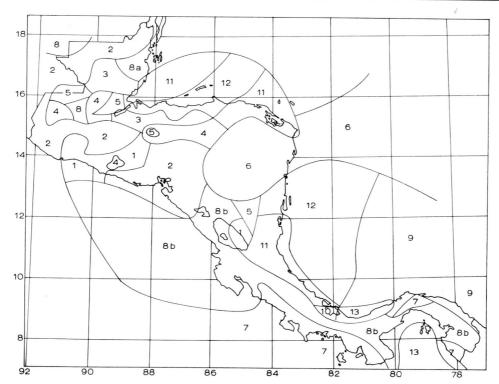

Fig.8. Rainfall regimes of Central America. See Table XV for explanation of the numbers.

TABLE XV*

RAINFALL REGIMES OF CENTRAL AMERICA—MONTHLY RAINFALL IN TENTHS OF PERCENT OF THE ANNUAL TOTAL

Regimes	Jan.	Feb.	Mar.	Apr.	May	June	July	Aug.	Sept.	Oct.	Nov.	Dec.
1	1	1	3	26	101	182	174	166	196	126	20	4
2	13	9	11	22	131	182	133	112	200	134	33	20
3	52	32	26	29	75	132	113	98	147	130	84	80
4	12	8	2	22	82	164	183	162	190	113	44	18
5	35	24	14	26	74	155	177	155	141	94	63	42
6	52	18	2	32	82	143	170	117	113	125	78	68
7	16	10	9	31	115	117	103	112	118	205	113	51
8a	in British Honduras:											
	54	29	16	37	64	116	104	104	141	149	110	75
8b	elsewhere:											
	14	7	6	24	117	141	107	108	138	182	109	47
9	36	14	10	24	113	118	110	108	131	125	147	64
10	68	41	52	75	78	111	95	96	94	86	102	102
11	56	33	11	41	65	155	125	111	78	95	114	116
12	78	44	28	28	52	64	96	83	78	134	195	120
13	39	23	36	68	94	94	96	112	83	101	142	112
Pacific, (1) + (2) + (7) + (8b):												
	11	7	7	26	115	156	129	125	163	162	69	30
Atlantic, (6) + (8a) + (9) + (11) + (12):												
	47	23	15	40	84	125	121	110	109	119	118	87

* To be used with Fig.8.

Remarks: 1 has a poorly, 2 a well expressed summer minimum; 3 is similar to 1 and 2 but with more rain in winter; 4 is typical for mountainous areas; it is similar to 5; 7 has a strong, isolated maximum in October; compare with 12.

pressed in percent of the annual. The percentages are based on all available stations considered typical for the respective regime. In some data-sparse areas Henry's (HENRY et al., 1963–1966, 7th report) and URRUTIA's (1964) monthly maps have been used for Fig.8 and Table XV. Where the regime boundaries arrive at the coast they have been extended over the waters; they then represent frequency percentages rather than the variation of the totals.

With exception of regimes 5 and 9, all of Central America has two rainfall maxima. The main dry season occurs in winter or early spring and is much more intense in the Pacific part of Central America than in the Atlantic part. Table XV does not show the full extent of this difference since the contrast of percentage is enhanced by the generally higher totals of the Atlantic region. This is borne out by the example given in Table XVI. The six driest consecutive months produce 5.5 % of the annual rain in regime 1 which is found only at the Pacific side of Central America, and 29.4 % in region 12 on the Atlantic side. The change from the Pacific to the Atlantic rainfall patterns occurs in many places abruptly. An example is shown in form of a cross-section through Costa Rica (Fig.9; for location see Fig.7). The figure based on the data of 43 stations illustrates the annual variation of rainfall by means of isopleths of rainfall totals. Minima are designated by *N*, maxima by *X*. It can readily be seen that the Atlantic side of Costa Rica has four seasons of almost equal length (regime 11). The pattern changes at the eastern slope of the mountains, whereas their western slope is not reflected in the rainfall regime (regime 8). The secondary Pacific minimum, the "veranillo" (little summer) occurs at the same time as does the Atlantic maximum. The main Pacific maximum in September–October coincides with the secondary Atlantic minimum, and the Pacific primary minimum lasts as long as one and a half of the seasons on the Atlantic side.

In order to prevent unwarranted generalization, it has to be mentioned that right north of the center of the cross-section depicted in Fig.9, and within the visual range of some of its stations, there are three stations on the Pacific side of the mountains with typical signs of the Atlantic regime. The insufficiency of the station network does not allow an explanation of this interesting phenomenon.

According to Fig.10 (taken from PORTIG, 1965) the rainfall of Belize can be divided into three additive components. If the two summer components are added omitting the com-

TABLE XVI

MONTHLY RAINFALL DURING THE DRY SEASON AT THE ATLANTIC AND PACIFIC COASTS OF HONDURAS

	Jan.	Feb.	Mar.	Apr.	May
	Total in mm:				
Puerto Cortés[1]	263	123	122	114	134
Amapala[2]	2.3	0.2	11.4	23.1	292
	% of annual total:				
Puerto Cortés	9.2	4.3	4.3	4.0	4.7
Amapala	0.1	0.0	0.5	1.1	13.7

[1] 15.8°N 87.9°W, at the Caribbean north coast.
[2] 13.3°N 87.6°W, island in the Bay of Fonseca.

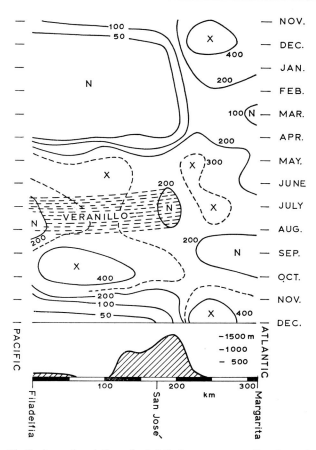

Fig.9. Annual variation of rainfall along a cross-section through Costa Rica (cf. Fig.7). Isohyets labelled in millimetres; X = maximum, N = minimum.

ponent S_3, a rainfall curve results that is characteristic of the Pacific side of Central America (e.g., regimes 1 and 2), with a deep minimum in February. The addition of S_3 shifts the minimum to March and makes it less intense (curve I and regime 8). S_3 is attributed to the cold air invasions called "northers" (Spanish: nortes) which are discussed below in the section on weather.

Closer study of the regimes reveals some interesting details. The summer and fall extremes of rainfall do not move with the sun to the north or south but rather in a zonal direction. The curve of regime 5, which has only one maximum, can easily be split into two bell-shaped curves of equal height with the maxima near July 1 and September 20. In this respect it is very similar to the neighboring regimes, with the difference that the bell-shaped curves are of different steepness. In regime 5 the steepness is so small that the two curves merge into one; in the neighboring regimes the steepness is such that the identity of each peak is easily recognized.

While in this case it can be shown that the difference of rainfall curves is nothing but the variation of the same basic pattern (which probably means "of the same physical causes"), the very similar curves 8a and 8b reveal, when analyzed, quite different basic curves. True, both consist of three components, one with its peak in late spring, the second in fall, and the third in winter. But the times of the occurrence of the respective peaks as

427

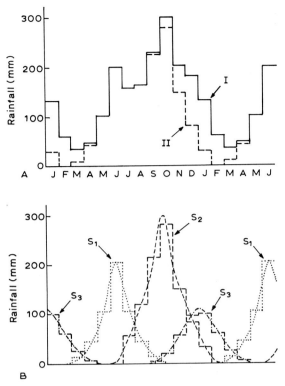

Fig.10. Annual variation of rainfall of Belize, British Honduras (from PORTIG, 1965). A. I = original data; $II = I$ minus S_3; B. Split of the original data into three symmetrical components, S_1, S_2, and S_3.

well as their amounts differ considerably, more than, for example, between regimes 2 and 4. Returning to the non-analyzed data of Table XV, one notices at its bottom the average values for the regimes that border only the Pacific Ocean, and of those bordering only the Caribbean Sea. In spite of the smoothing effect of each averaging one still finds the seasons well expressed on the Pacific side (range 15.7% of the annual total), while the Atlantic regimes show less annual variation (11.0%) as far as percentages are concerned. The secondary rainfall minimum in mid-summer is well recognized by the man in the street, also in the area of regime 1 where the average amounts drop only slightly. (Because of the dispersion around the mean there must have been many individual years in which the mid-summer totals were the highest of the year.) The reason is that the rains tend to fall at night during this period. This is brought out by the strong sunshine maximum in August which is much more impressive than the simultaneous rainfall minimum (Table XVII).

TABLE XVII

RAINFALL TOTALS, SUNSHINE, AND PRESSURE IN SAN SALVADOR, C.A.

	May	June	July	Aug.	Sept.	Oct.
Rainfall (mm)	187	321	314	299	318	232
Sunshine (%)	56	45	60	65	49	56
Pressure (mbar)	932.8	933.1	933.7	933.5	932.8	932.7
No. of days with rain	13.3	19.6	20.3	19.9	20.0	15.8

The lessening of the rains in mid-summer is commonly attributed to the strengthening of the Atlantic anticyclone and a subsequent tendency toward increased subsidence with drying of the middle and higher troposphere. Since the pressure is—on the average—not fully in phase with the rain, and since the pressure extremes occur throughout the entire region at the same time while this is not the case with the rains, this explanation is not satisfactory. Hourly rainfall data of San Salvador show that one ought to consider the peculiarities of the two peaks rather than the lessening in between. *All* months have a rainfall maximum between sunset and midnight, but June and September have an additional maximum in the early afternoon. This was even true in the years 1954–1957 when the monthly totals of July and August exceeded those of June and September. In other words, although the secondary rainfall minimum was replaced by the primary maximum, the hourly distribution produced the subjective impression of a rainfall reduction in mid-summer (Table XVIII). In those years of an abnormal mid-summer maximum of rainfall the atmospheric pressure had its maximum exactly as in normal years, i.e., in July. This is another argument against the hypothesis that the pressure maximum reduces the rainfall totals.

The displacement of the June rainfall peak to July, mentioned in the previous paragraph, was only one of several changes in the annual rainfall pattern. Fig.11, taken from PORTIG (1965) and extended to 1970, presents overlapping 11-year monthly means of the rainfall of San Salvador. They show many decade-long displacements of rainfall extremes. Not only did the rainfall peaks change their times of occurrence but also the rainfall minima. The driest month in the years ca. 1910–1929 was January. In the 1930's it was replaced by March, after which it switched to February (with fourteen Februaries without a single drop of rain), and more recently the driest month is again January.

Discussions of climatic changes in other parts of our region, or material for such studies can be found in CHAPEL (1927), JONES (1939), BRENNAN (1942), MILLÁS (1943), WRAIGHT (1951), ALAKA (p.495 of this volume) and in the mostly unpublished data by the Panama Canal Company.

TABLE XVIII

FOUR-HOURLY RAINFALL TOTALS (mm) AT SAN SALVADOR, C.A. (1954–1957)

Time 90°W	June	July	Aug.	Sept.
00–04h	32	37	58	63
04–08h	13	5	3	14
08–12h	12	6	3	34
12–16h	50	24	31	42
16–20h	45	146	82	43
20–24h	54	166	142	96
Total	206	384	319	292
Normal total	321	314	299	318

Number of days with at least a trace of rain in Central America

It is useful to note that the number of days with rain is different from the rainfall frequency discussed above in context with ship observations. The former take into account the

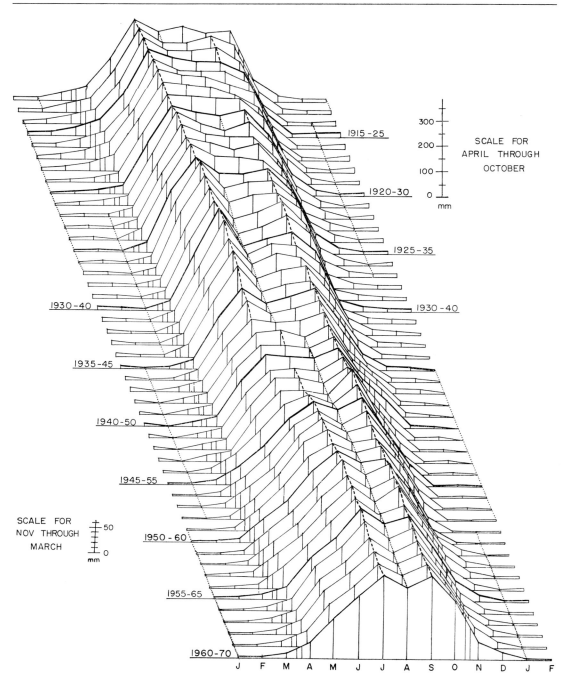

Fig.11. Eleven-year moving averages of rainfall at San Salvador, El Salvador. Minima are connected by dotted lines, maxima by dashed lines. The height of the curves over their respective base line indicates the mean rainfall total during the eleven years marked at the left or right end of the base line. The height scale is different for the dry and rainy seasons. (From PORTIG, 1965.)

full day, the latter only the hours of observation. In the former it has no effect whether rain falls several times on the same day, or whether the rain is hours-long or a brief shower. In the latter the same storm is counted as often as an observation happens to be made as long as it lasts, whereas rain between the observations is not counted however heavy it is.

The number of days with rain in an individual month can—despite well expressed dry and rainy seasons—deviate considerably from the mean number. Table XIX presents, as an example, the highest and smallest numbers observed in San Salvador in 40 years. For comparison, Table XX is added to present the extreme monthly totals for the same station and the same period. The tables show that the (relative) variability of rainfall totals is greater yet than that of the number of days with rain. The amount of rain fallen during a certain number of days can vary substantially. During the six days listed in Table XIX as the September minimum, 305 mm of rain were recorded, whereas the minimum total of 184 mm was spread over 15 days.

TABLE XIX

NUMBER OF DAYS WITH AT LEAST A TRACE OF RAIN AT SAN SALVADOR, C.A.*

	Jan.	Feb.	Mar.	Apr.	May	June	July	Aug.	Sept.	Oct.	Nov.	Dec.	Year
Maximum	6	7	5	16	23	28	30	27	28	23	13	7	167
Average	1.6	1.3	1.7	5.5	14.5	21.0	21.2	20.8	21.0	16.7	4.7	2.4	132.2
Minimum	0	0	0	0	8	11	8	5	6	3	0	0	46

* Data from PORTIG, 1961.

TABLE XX

RAINFALL TOTALS (mm) AT SAN SALVADOR—COMPARABLE WITH TABLE XIX*

	Jan.	Feb.	Mar.	Apr.	May	June	July	Aug.	Sept.	Oct.	Nov.	Dec.	Year
Maximum	40	35	80	511	380	617	492	471	611	505	124	37	2,284
Minimum	0	0	0	0	69	147	117	82	184	25	0	0	1,029

* Data from PORTIG, 1961.

It can be noted that the annual minimum of Table XIX is only slightly greater than the sum of the monthly values. This is so because all monthly minima of the rainy season occurred in one year, 1944.

CROW and COBB (1962) report that there was only one day in either of the entire rainy seasons of 1960 and 1961 when it did not rain at one place at least, within a banana plantation of 180 km² size, in the sea breeze zone at the Pacific coast near Puerto Armuelles in west Panama.

A study of rainfall frequencies in the sense of maritime observations would shed much light on the rain-producing mechanisms and on details of the circulation. Also, for days with rain there are hardly more data available in published form than those tabulated at the end of this chapter. Only for San Salvador do two publications (LESSMANN, 1956; SERVICIO METEOROLÓGICO NACIONAL DE EL SALVADOR, 1959) go beyond monthly sums and present frequencies of rainy and dry spells and intensities, or daily measurements, respectively.

The latter, daily rainfall totals for a long period, can be used to compute frequency distributions. This has been previously done by this author (PORTIG, 1961) for three seasons.

The rainfall frequencies around the two annual peaks show some strong similarities, i.e., maxima and minima at almost the same amounts. In spite of this the spread of the distributions is different. This is borne out by the percentiles at either end of the distributions. (Table XXI). The typical weather distributions that cause the wider spread in June and September will be discussed in the section on weather (pp.447–451). The paragraphs on pp.410–412 also contribute to an understanding of this facet.

Our knowledge of climates would certainly be improved if rainfall frequency distributions were computed for other parts of the region also.

TABLE XXI

SOME PERCENTILES OF SALVADOR RAINFALL DISTRIBUTIONS (mm)

	N_5	Percentiles					N_{100}
		25	50	75	95	98	
June + September	558	2.3	9.0	17.7	47.2	67.0	13
July + August	487	3.0	9.5	18.0	39.7	56.0	2

N_5 = number of days with less than 5 mm.
N_{100} = number of days with more than 100 mm.

Islands near Central America

There are few observing stations on islands around Central America, and their records are too short to be very useful with exception of Swan Island which has longer, though not yet fully satisfactory, records. Table XXII presents the available mean monthly rainfall totals.

TABLE XXII

MEAN MONTHLY RAINFALL TOTALS (mm) OF ISLAND STATIONS NEAR CENTRAL AMERICA

	Jan.	Feb.	Mar.	Apr.	May	June	July	Aug.	Sept.	Oct.	Nov.	Dec.	Year
Swan Island[1]	85	36	22	19	81	160	118	94	136	251	190	142	1,334
Guanaja[2]	244	125	59	97	77	127	259	180	124	497	427	457	2,641
San Andrés[3]			driest						rainiest				1,824
Isla del Rey[4]	20	0	0.3	92	245	320	263	297	275	651	257	139	2,559
Isla San José[5]	0	0	0	56	168	363	236	365	363	665	222	107	2,545

[1] 17°24′N 83°56′W, 1941–1964, data (partly published) from National Weather Records Center, Asheville, N.C.
[2] 16°28′N 85°54′W, 1951–1961, data from *Almanaque Hondureño* (Servicio Meteorológico Nacional de Honduras, Tegucigalpa).
[3] 12°32′N 81°33′W, 1932–1941, annual mean allegedly varying between 500 and 4,100 mm.
[4] 8°28′N 78° 56′W, 7 years, two locations (M. A. Fradel, personal communication, 1964).
[5] 8°16′N 79°6′W, one year only (M. A. Fradel, personal communication, 1964).

The rainfall of Swan Island is remarkably low although it lies directly in the path of hurricanes, one of which caused most severe damage in September 1955 and destroyed the weather station. According to ship observations (Fig.6), Swan Island lies near an axis of minimum rainfall frequency from where it increases to the northeast as well as to the southwest. Inset *M* of Fig.6 corresponds well to the island's numbers-of-days-with-rainfall which have another annual variation than the totals, the latter matching regime 8 described on p.425, Table XV (Table XXIII).

TABLE XXIII

SOME MONTHLY RAINFALL PARAMETERS AT SWAN ISLAND

	Jan.	Feb.	Mar.	Apr.	May	June	July	Aug.	Sept.	Oct.	Nov.	Dec.
N	15	8	6	5	9	14	13	13	14	18	16	16
R	85	36	22	19	81	160	118	94	136	251	190	142
R/N	5.6	4.5	3.7	3.8	9.0	11.4	9.1	7.2	9.7	14.0	11.9	8.9

N = mean number of days with rain; R = mean rainfall total in mm.

Guanaja's rainfall has much similarity with that at the continental coast south of it.
The *highest* reported annual rainfall total of San Andrés is in the order of magnitude of the *normal* total at the coast of Nicaragua. Its alleged minimum competes with the lowest normals in the dry belt of Central America: severe drought conditions are reported to occur from time to time on San Andrés (PARSONS, 1956)—although the island lies near the path of the November hurricanes.
Both stations that contributed to the listed Isla del Rey data (M. Fradel, personal communication, 1963) show a very strong maximum in October and complete dryness in February and most of March. They resemble the regimes 7 and 8b (Table XV), more than the ship observations near and south of the island would suggest. San José Island is close to Isla del Rey and its very short record matches well.

Annual rainfall totals of islands other than those near Central America

Cuba

The annual rainfall totals of Cuba (Fig.12) are surprisingly small. The minima, all of them in the eastern part of the island, drop below 1,000 mm. Most of the dry areas are lined up along the coast including the windward coasts. One station, Cayo Guano del Este, situated at 21°41′N 81°2′W on a flat island approximately 40 km off the coast, reports 957 mm (WERNSTEDT, 1961) as an average total of eight years.
The maxima, found in mountainous terrain, appear not to exceed 2,000 mm.

Jamaica

Also, most of Jamaica has but little rain, in various areas less than 1,000 mm/year. Much

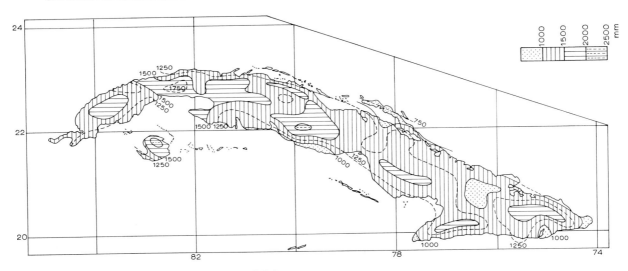

Fig.12. Mean annual rainfall of Cuba.

rainfall, however, is recorded in the mountains of the eastern part of the island (Fig.13), with the highest average, 7,820 mm, recorded at Bowden Pen. Fig.14 is a cross-section of annual rainfall totals along the meridian 76°25′W, running from coast to coast across the mountains.

As in Cuba and in parts of Central America, the rainfall totals drop off toward the coasts. The Jamaican cloudbursts are described in the section on weather, pp.449–450.

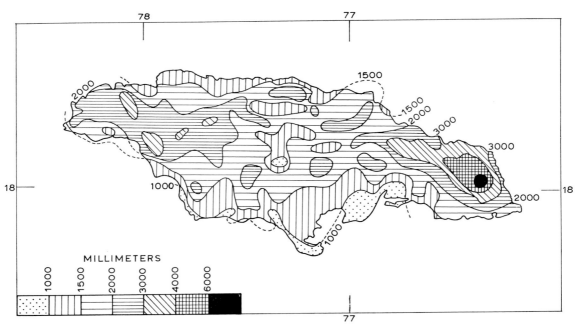

Fig.13. Mean annual rainfall of Jamaica.

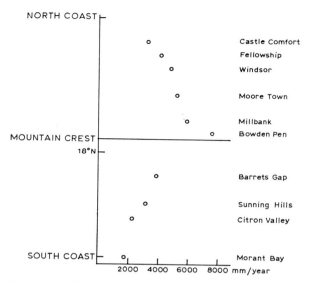

Fig.14. Meridional section of the annual rainfall through eastern Jamaica along 76°25′W.

Hispaniola

Again we find that dryness is more characteristic than wetness (Fig.15), but the decrease of rainfall totals toward the coasts is not quite so typical as for the islands discussed before. Some low-lying places in the interior have less than 500 mm annually, and almost the entire east (windward) and west coasts have mean annual rainfall totals below 1,000 mm. These rains are more evenly spread over the year than comparable totals in Central America, leaving the rainy season less wet, and the dry season not quite as dry as there. Two stations taken from WERNSTEDT's (1961) collection may illustrate this.

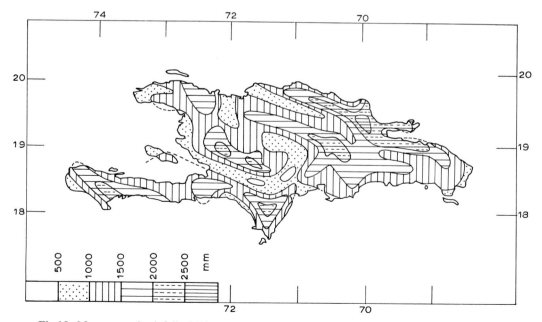

Fig.15. Mean annual rainfall of Hispaniola.

435

Chiquimula in Guatemala (14°48′N 89°42′W) has an annual rainfall total of 636 mm, 19 mm of which (3 %) fall in the six consecutive driest months. Padre las Casas in the Dominican Republic (18°44′N 70°56′W) has an annual total of 716 mm, and 197 mm (24 %) fall in the six driest consecutive months.

Annual variation of rain in the area 16–27°N 68–85°W

Fig.16 and Table XXIV inform about rainfall regimes in the northwestern portion of the West Indies. As done above for Central America and the surrounding waters, the monthly rainfall totals are expressed in tenths of percent of the annual. This eliminates the effect of elevation on the totals, and it allows to present totals over land and frequencies over sea in the same way (though basic objections can be raised against doing so). We find one rainfall maximum in May or June, and another in September, October, or November. They are separated by a minimum which in almost all places occurs in July. The listed regimes show variations of this general theme. The last line of Table XXIV allows a brief glimpse on the climatic conditions of the Gulf of Mexico; it suggests that the Gulf is not a climatic extension or appendix of the Caribbean Sea but a climatic region in its own right.

There is no area over sea and only one regime (11) in Central America where the first maximum of the rainy season is higher than the second. There are, however, many such areas in the Greater Antilles (regimes 14, 15, 17, 23, 24, 26, 27, 28, 30, and 31). In most of the regimes listed in Table XXIV the first rainfall maximum is very pronounced and fol-

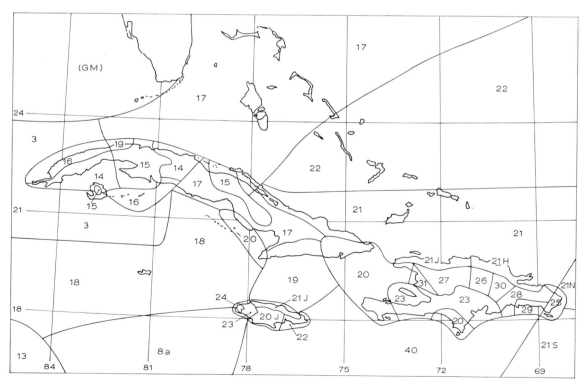

Fig.16. Rainfall regimes of the northwestern portion of the West Indies. See Table XXIV for explanation of the numbers.

TABLE XXIV*

MONTHLY RAINFALL IN TENTHS OF PERCENT OF THE ANNUAL TOTAL IN RAINFALL REGIMES OF THE NORTH-WESTERN PORTION OF THE WEST INDIES

Regime	Jan.	Feb.	Mar.	Apr.	May	June	July	Aug.	Sept.	Oct.	Nov.	Dec.
14	35	27	36	54	136	163	117	118	141	117	36	20
15	25	28	38	59	136	166	116	120	138	109	43	22
16	45	34	40	41	110	136	96	125	166	131	48	28
17	27	25	23	56	134	157	118	116	125	146	47	26
18	33	26	29	55	130	139	79	89	147	177	58	38
19	42	34	36	59	144	124	97	102	132	152	50	28
20	28	37	39	68	129	105	70	99	124	204	64	33
20J	33	37	40	77	128	87	70	104	128	168	86	44
21	70	44	35	67	122	81	43	56	93	139	161	91
21H	85	80	46	67	108	66	75	69	78	77	128	121
21J	80	68	46	65	101	73	49	64	81	120	145	108
22	38	36	24	45	103	102	70	101	138	176	113	54
23	24	33	43	87	157	96	86	112	130	137	66	29
24	30	28	41	82	136	136	111	119	112	117	59	29
25	62	47	35	62	95	98	97	80	102	115	112	95
26	68	57	42	82	168	80	69	71	90	100	93	80
27	21	27	30	80	175	130	85	105	129	121	64	33
28	34	40	31	57	147	127	120	107	121	107	67	42
29	30	28	24	53	143	96	103	111	141	135	96	43
30	63	61	44	72	147	90	108	94	83	83	81	74
31	9	18	18	50	134	166	137	144	147	123	40	14
Aver.	42	39	35	64	132	115	92	100	121	131	79	50
GM (as a contrast)	61	88	55	53	39	75	78	90	169	112	89	90

* To be used with Fig.16.
GM = average of the entire Gulf of Mexico.

lowed by a sharp drop (e.g., in 14, 15, 26), even in areas where the second maximum is higher (e.g., in 16, 18, 21). In most of the areas mentioned in this paragraph the first rainfall maximum occurs in May, i.e., at the same time when the neighbouring Gulf of Mexico has its primary minimum.

The second rainfall maximum appears in September in west and central Cuba (14, 15, 16) and in small portions of Hispaniola (27, 28, 29, 31). From there it moves to the northeast and to the southeast where it occurs in October, in regime 21 (with the local modifications 21H, 21J, 21N, and 21S; see Tables XXIV and XXVI) further delayed until November. This statement concerning the migration of the rainfall maximum also holds true when one considers rainfall totals instead of percentages. Surrounded by areas with an October or November maximum we find region 30 with the last maximum of the year in July. The situation is similar to that of regime 5 in Central America (Fig.8 and Fig.16).

The regimes 18, 19, 20, 23, 24, and 31 show an especially short transition from the rainy to the dry season. In this respect they resemble the regimes 7 and 8 in Central America. These are areas with or near southwest coasts which tend to have a smaller rainfall range (percentage of the wettest month minus that of the driest) in the northeast than in the southwest. This can be observed in the gross features of the entire region as well as in each of its subdivisions.

Puerto Rico and the Lesser Antilles

Annual rainfall totals

Fig.17–20 show the mean annual rainfall of Puerto Rico, the Guadeloupe Islands, Martinique and Margarita, respectively. The isohyets reflect primarily the elevation with

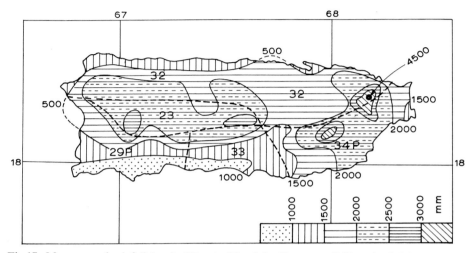

Fig.17. Mean annual rainfall (mm) of Puerto Rico (after SMEDLEY, 1961) and rainfall regimes. (Consult Table XXVI for explanation of the regime numbers.)

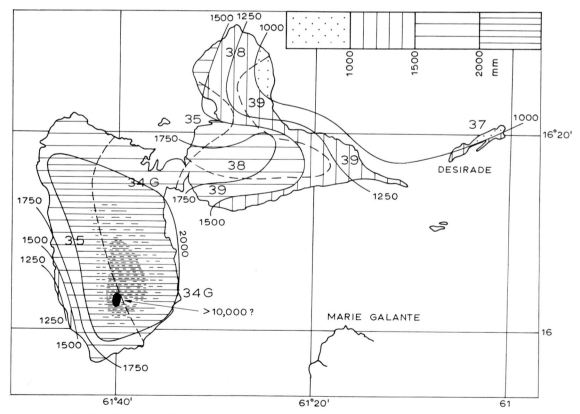

Fig.18. Mean annual rainfall of the Guadeloupe Islands (after CAMPAN, 1959) and rainfall regimes. (Consult Table XXVI for explanation of the regime numbers.)

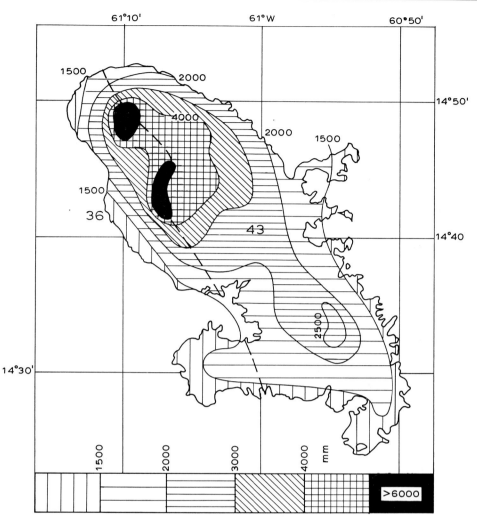

Fig.19. Mean annual rainfall of Martinique (after SERVICE MÉTÉOROLOGIQUE DU GROUPE ANTILLES–GUYANE, 1962) and rainfall regimes. (Consult Table XXVI for explanation of the regime numbers.)

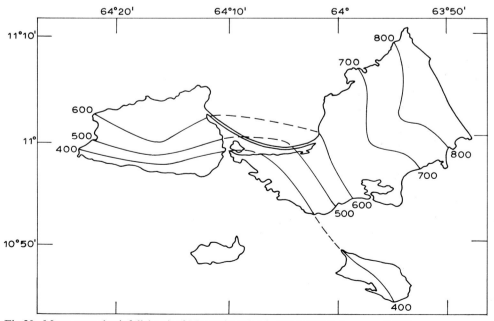

Fig.20. Mean annual rainfall (mm) of Margarita Island. (After GOLDBRUNNER, 1962.)

TABLE XXV

ANNUAL RAINFALL TOTALS

Station	Source	Rainfall (mm)	Years of observ.
Virgin Gorda, Spanish Town	Reed, 1926	920	12
Tortola, Roadtown	Reed, 1926	1,300	24
Vieques Island, Santa Maria	Ward and Brooks, 1934	1,231	26
St. John, American Hill	U.S. Weath. Bur., P.R.	1,300	13
St. John, Cruz Bay	U.S. Weath. Bur., P.R.	610	30 w.b.[4]
St. Thomas, Bonne Esperance[1]	U.S. Weath. Bur., P.R.	1,330	26 w.b.
St. Thomas, Dorothea[1]	U.S. Weath. Bur., P.R.	980	12 w.b.
St. Croix, Annaly[2]	U.S. Weath. Bur., P.R.	1,360	20 w.b.
St. Croix, Bethlehem New Works[2]	U.S. Weath. Bur., P.R.	717	9
St. Maarten, Philipsburg	Braak, 1935	1,083	52
St. Barthélemy, Gustavia		879	6
Saba, Bottom	Braak, 1935	1,124	41
Saba, Windwardside	Braak, 1935	1,055	18
St. Eustatius, Pleasures	Braak, 1935	1,197	6
St. Eustatius, Bengalen	Braak, 1935	924	15
Barbuda, Codrington	Reed, 1926	1,030	14
St. Kitts, Lower Canada	Reed, 1926	1,130	18
St. Kitts, Basseterre	Reed, 1926	1,250	56
St. Kitts, Brotherson Estate	Reed, 1926	2,130	5
Antigua, St. John's		1,240	61
Montserrat, Plymouth	Meteorol. Off., London, 1958	1,620	15
Marie Galante, Grand Bourg	Reed, 1926	1,450	23
Dominica, Shawford	Reed, 1926	4,660	27
Dominica, Batalie[3]	Reed, 1926	1,750	5
St. Lucia, Uplyme	Ward and Brooks, 1934	3,404	15
St. Lucia, Moule à Chique	Ward and Brooks, 1934	1,168	20
St. Vincent, Kingstown	Alexander, 1958	2,300	24
St. Vincent, Bayabou	Reed, 1926	1,740	6
Barbados, Bridgetown	Reed, 1926	1,230	78
Barbados, Dunscombe	Reed, 1926	2,182	10
Carriacou, Top Hill	Reed, 1926	1,250	10
Grenada, Grand Etang	Ward and Brooks, 1934	3,731	10
Grenada, Point Saline	Ward and Brooks, 1934	990	10
Tobago, Roxborough Estate	Reed, 1926	2,280	12
Tobago, Government Farm	Reed, 1926	1,160	13
Trinidad, Blanchisseuse	Reed, 1926	2,150	29
Trinidad, Port of Spain	Meteorol. Off., London, 1958	1,370	32
Margarita, La Asunción	Alexander, 1958	535	20
Margarita, Salamanca	Lahey, 1958	1,300	5
Bonaire, Rinçon	Braak, 1935	527	29
Bonaire, Kralendijk	Braak, 1935	511	29
Curaçao, St. Thomas College	Braak, 1935	707	7
Curaçao, Knip	Braak, 1935	380	6
Aruba, Oranjestad	Braak, 1935	447	33

[1] Five more stations on St. Thomas range between the two mentioned.
[2] Eighteen more stations on St. Croix range between the two mentioned.
[3] Five more stations on Dominica range between the two mentioned.
[4] w.b. = with breaks.

the smallest totals at the coasts. Guadeloupe is the only area with sufficiently detailed data where the windward (northeast and east) coasts receive less rain than the leeward coasts. The areas of maximum rainfall are expected to be very close to the highest elevation; the totals at these points are not known. SMEDLEY (1961) suggests at least 4,500 mm for Puerto Rico; CAMPAN (1959) has fragmentary indications that 10,000 mm is a good estimate for the highest point of Guadeloupe; and the official map for Martinique shows isohyets of 8,000 mm. From these maxima the annual rainfall totals drop sharply to the nearby coasts where averages near or even below 1,000 mm are well documented.

Table XXV lists annual rainfall totals of stations in the Lesser Antilles. The values depend greatly on the elevation of the stations and on their exposure to the rain-bringing trade winds.

The light decrease of rainfall north of 17°N which can be inferred from Table XXV, is attributed (BLUME, 1962) to the geological structure of those islands which consist of horizontally stratified lime rather than of volcanic formations as do the islands from St. Kitts southward.

Another sharp decrease characterizes the islands off the coast of Venezuela (Fig.21). The stations in this dry belt show a large variability which is typical for most dry climates. ALEXANDER (1958) reports that the annual rainfall of La Asunción, Margarita Island, oscillated in 20 years between 96 mm (1924) and 1,532 mm (1938). More to the west the rainfall totals remain low but the variability decreases.

The dry area can be considered as delineated by the 5% line of Fig.6. If there should happen to be a definite boundary with the areas of more abundant rainfall, such boundary could only be found through statistical evaluation of satellite photographs. Such a study has not as yet been made.

LAHEY (1958) made an intense investigation as to the causes of the dryness. He found that

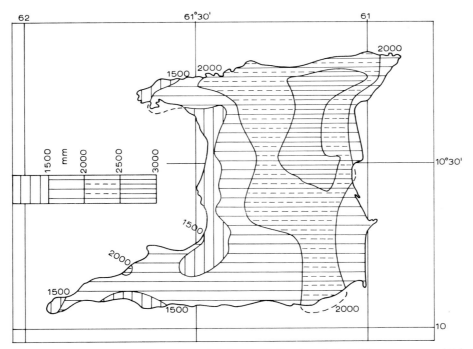

Fig.21. Mean annual rainfall of Trinidad. (After A. K. Powis and C. A. Thompson, unpublished.)

441

the combination of a hot land surface bordering a cooler ocean with a mountainous coast that parallels the main wind flow is the prerequisite for the suppression of rain. Differential frictional and turbulent stresses are exerted in such a way that there is divergence in the lower layers of the (rather moist) air in places where it flows parallel and close to the coast. This divergence near the ground is complemented by a decrease of the form stresses which act on the slightly higher layer that crosses the coast line from the south under a small angle. Both effects result in subsidence. These dynamic processes are coupled with thermal modifications. The air over the water surface is cooled, or at least not warmed, while the air of the next higher stratum has moved over and around the heated mountains of Venezuela. Heated also by the mentioned subsidence upon leaving the continent, the thermodynamical stability of the air over the coastal waters is greater than in maritime areas farther away. Lahey considers the negative temperature anomaly of the surface water as a consequence of the mentioned processes rather than as a primary cause of the dryness. Rains can fall when a synoptic weather pattern interrupts the described interplay of rain-preventing controls.

As one of many examples for the strong influence of topography on rainfall totals, we quote the *Sailing Directions for the West Indies*, (1950, vol.II, p.14): ". . . the influence of even relatively low mountains in the northeast trade wind belt is found on the island of Trinidad. On the north part of that island the mean annual rainfall drops from 3,000 to 1,500 mm in a space of less than 16 km between a mountain crest and a lowland on its leeward side. The mountains on the north side are in no case higher than 1,000 m, but the rainfall varies from 3,500 mm in the middle of the mountain range near the northern edge of the island to less than 1,500 mm along most of the western shore!" (See Fig.21.)

At higher elevations not only the amount of rainfall is higher than in the lowlands but also the number of days with rain is increased, especially in the dry season. Camp Jacob, Guadeloupe, elevation 540 m, reports 3,560 mm of rain fallen on 274 days annually (METEOROLOGICAL OFFICE, LONDON, 1958). The driest month, February, has 18 days with rain and a total of 154 mm. Neufchâteau, at 250 m upstream of Camp Jacob, reports even 4,414 mm annually, and 173 mm in the driest month (March). Marquisat at the nearby windward coast has only 2,061 and 68 mm, respectively (CAMPAN, 1959). The reader is also referred to Fig.14, which presents another example.

Annual variation of rainfall

Puerto Rico can be divided into five rainfall regimes (Fig.17 and Tables XXV, XXVIII). All of them have a rainfall maximum in May and another in September or October. In the northern and central portions of the island the former is the larger one. The minimum of rainfall can occur in January (regimes 23 and 29P, Fig.16), in March (32 and 34P), or in both (33), but not in February although February is the shortest month. (The data are not adjusted for equal length of month.)

The oceanic areas north and south of Puerto Rico (21N and 21S) have both their main maximum in November. Both maritime areas have a minimum, or are close to a minimum, in September when the entire coast of Puerto Rico and most of its interior have a maximum. Also the beginning of the rainfall season is over water different from the land in between as indicated through the May and July values of the regimes 21N, 21S, 23, 29P, 32, 33, and 34P (Table XXVI).

TABLE XXVI*

RAINFALL REGIMES OF THE EASTERN PORTION OF THE WEST INDIES—MONTHLY RAINFALL IN TENTHS OF PERCENT OF THE ANNUAL TOTAL OVER LAND, OR OF THE ANNUAL FREQUENCY OVER WATER

Regime	Jan.	Feb.	Mar.	Apr.	May	June	July	Aug.	Sept.	Oct.	Nov.	Dec.
21N	93	76	54	62	89	82	87	82	75	88	115	97
21S	95	72	47	54	71	84	90	84	99	100	120	89
29P	27	34	41	60	122	79	93	126	150	137	89	42
32	68	49	43	66	124	89	89	102	101	96	95	78
33	23	32	19	50	123	91	80	124	167	154	102	35
34G	57	36	36	50	101	77	86	109	140	126	103	78
34P	52	40	31	55	115	103	97	105	130	117	91	64
35	61	39	37	44	58	96	108	120	130	121	107	79
36	69	54	41	54	63	98	116	108	103	100	113	85
37	45	42	34	65	60	87	80	117	144	131	118	77
38	55	46	44	36	78	74	81	112	143	154	90	77
39	50	32	36	46	81	74	78	115	129	163	118	78
40	85	70	54	40	90	74	68	70	104	139	109	95
41	65	45	42	55	82	74	82	101	117	126	126	85
42	108	64	58	54	79	84	76	82	82	103	126	84
43	79	64	46	51	74	90	94	97	101	101	125	78
44	106	56	42	40	29	45	54	58	65	142	202	163
45	87	58	44	38	74	54	59	54	91	148	156	137
46	110	65	45	44	44	79	109	92	69	90	121	132

* To be used with Fig.17–20.

Although there are differences from island to island, it appears appropriate to unite the islands between Puerto Rico and Guadeloupe into one regime (41). For data of individual islands the specific literature should be consulted (REED, 1926; U.S. NAVY HYDROGRAPHIC OFFICE, 1949; METEOROLOGICAL OFFICE, LONDON, 1959; CAMPAN, 1959.) The user of such data should, however, be aware that the published numbers are subject to change with more observations becoming available. Regime characteristics derived from, and devised for, a number of stations will change less in this process.

The islands of regime 41 show an annual variation of the rainfall which resembles that of the rainfall frequencies north of Puerto Rico (21), but the amplitude is greater at the islands. However, the maritime areas directly adjacent to regime 41, 21S and especially 42, show the same phenomenon described above for Puerto Rico: the open seas have their extremes later than the islands.

CAMPAN (1959) and THÉVENEAU (1966) have published enough data to divide the Guadeloupe islands into climatic zones according to their annual variation of rainfall (Fig.18 and Table XXVI). There can be doubts about the representativeness of regime 37 (island of Désirade) which is based only on two stations on the south coast. It also can be disputed whether the distinction between regimes 38 and 39 is justified. The climates of Guadeloupe are rendered different from those of their surroundings by their larger autumn maximum of precipitation.

The east and west sides of Dominica and Martinique (south of Guadeloupe) are separated by elevations of more than 1,000 m in spite of their small horizontal distance. The east sides of the islands have only one long rainy season (regime 43, compare with 35—Barbados, see below) with the maximum at its end, in November. The leeward (west) side of the islands has two rainfall peaks, the larger of them in July–August, the other in November (Table XXVII).

TABLE XXVII

MONTHLY RAINFALL (mm) OF TWO STATIONS ON MARTINIQUE

Station	Jan.	Feb.	Mar.	Apr.	May	June	July	Aug.	Sept.	Oct.	Nov.	Dec.	Year
Basse Pointe[1]	79	88	74	109	146	141	186	217	231	246	343	143	2,003
St. Pierre[2]	135	149	85	90	86	247	252	288	225	243	241	112	2,153

[1] 14°52′N 61°7′W, coast, regime 43   } distance: 18 km.
[2] 14°45′N 62°11′W, coast, regime 36 }

The westerly stations have their rainfall minimum much later than the easterly ones. This late minimum together with the early maximum results in a short transition from the dry to the rainy season. This is contrary to nearby Guadeloupe where the long transition occurs on the west side (regime 35) and a short one on the rest of the island (34G, 38, 39) (THÉVENEAU, 1966).

Barbados has simpler rainfall conditions, probably because of the absence of high elevations. SKEETE (1931) condensed the 40-year records of 160 rain gauges into 16 areal means. All of them show with little variation a broad minimum in February–March, and a likewise broad maximum in September, partly with a secondary maximum in November. The Barbados data have been combined with those of western Guadeloupe and listed as regime 35. The center of Barbados has the highest (2,100 mm), the south and the southeast coasts the lowest (1,000 mm) rainfall totals. (LaSEUR, 1966, presents a map of the annual rainfall of Barbados.)

The islands south of Martinique, from St. Lucia to Trinidad, have basically the same rainfall regime (36) as the western half of Martinique where the long rainy season has a slight lessening in September or October. The broad minimum is centred in March. The form of the rainfall curve indicates that the time of rainfall extremes varies considerably from year to year. This is in sharp contrast to the Greater Antilles where the extremes occur almost always in the same months. (See, for example, regime 20 of Table XXIII in which the extremes stand out very well in spite of the smoothing effect of the averaging.) In the dry region west of Trinidad, the long dry season with the rainfall minimum in April or May is well distinguished from the rainy season which occurs very late in the year. In the eastern part of the dry zone (regime 46) there is still a July maximum as in Trinidad; the secondary minimum in September is, however, much stronger in 46 than in 36.

More to the west, the July or August maximum disappears gradually; it is barely detectable on Curaçao whereas the November maximum is outstanding. Climatic Table XXXVII for Curaçao (at the end of this chapter) shows that the number of days with thunderstorms is greatest one month earlier, in October. It also shows that the weak mid-summer maximum of rainfall totals is well expressed in the number of days with rain while it is completely absent in the number of thunderstorm days.

The rainfall minimum in regime 44 is delayed until April. The reader is recommended to compare regime 44 (Table XXVI) with regime 30 (Table XXIV), the northeastern part of the interior of Hispaniola, which lies almost exactly north of Curaçao and whose annual rainfall variation is almost opposite to that of regime 44.

As far as the second quarter of the year is concerned, this contrast between the northern and southern part of the Caribbean Sea becomes obvious through juxtaposition of regimes 20, 40, and 45. At the same longitude we find a strong rainfall maximum of 12.9 % in regime 20 (peninsula Barahona, Dominican Republic), a moderate maximum of 9.0 % in regime 40 at approximately 17°N, and a weak, almost doubtful, maximum of 7.4 % in regime 45 near 15°N.

Thunderstorms

For the reader from extratropical regions it seems to be natural that rains in the tropics fall in form of thundershowers. However, it is true only within certain limitations. This author is glad to have lived in an area where practically every rainfall (with exception of "temporals", see p.448) was accompanied by a frequently grandiose display of lightning and thunder. The beauty of a thunderstorm is most evident in zones where they approach the observer from the east bathed in the evening in many colors of the setting sun. It seems appropriate to mention that occasionally a very rare variety of "thunder" storms can be seen. The author once observed for several hours intense fireworks of horizontal lightning covering the elevation angle from the zenith to the horizon, facing the south. No thunder was heard nor was rain observed or reported.

The thunderstorm statistics show some striking discrepancies. The same publication (METEOROLOGICAL OFFICE, LONDON, 1959) lists for Kingston, Jamaica, in September (October in parentheses) 25 (24) days with thunderstorm together with 10 (12) days with more than 0.25 mm rain, and for Palisadoes, Kingston's airport, 13 (11) days with thunder and/or lightning, and 8 (9) days with more than 1 mm rain. REED (1926) lists for Kingston 12 (5) days with thunderstorms and 9 (11) days with rain.

PORTIG (1965) mentions another impressive example of bias in thunderstorm data. The statistics published for San Salvador by U.S. HYDROGRAPHIC OFFICE (1948) are based on observations made between morning and early afternoon and cover only approximately one quarter of the thunderstorms since the other three quarters occurred when no weather observer was on duty.

With all reservations one may conclude that there are two regions of maximum thunderstorm occurrence in Central America, one in the Gulf of Fonseca (116 storm days a year on Amapala), and the other in the center of the Canal Zone (196 at Madden Dam). From there the frequency decreases to a questionable three at Belize and a reliable 37 at Chimax, central Guatemala (McCULLOUGH, 1956; PORTIG, 1965).

With the same reservations one may state that smaller islands have fewer thunderstorms than larger islands, and that the southeastern part of the Caribbean Sea has fewer thunderstorms than the northwestern part and the Bahamas. Since thunderstorms, cost-free and delivered by nature, are an essential means of gaining insights into certain atmospheric processes, efforts should be made to improve thunderstorm statistics and their publication.

While the absolute numbers of thunderstorms in Central America and the West Indies are very much in doubt, their annual distribution is better established. It can be derived from the data that the maximum thunderstorm frequency is approximately one month earlier than the autumnal rainfall maximum. In the few cases in which the thunderstorm frequency has a secondary maximum in mid-summer, it is somewhat later than the corresponding rainfall maximum. Note, however, that not every May or June rainfall maximum is followed by a thunderstorm maximum.

Table XXVIII shows the monthly rainfall totals and monthly mean number of days with thunderstorms averaged for two Caribbean regions. For obvious reasons the actual amounts are not listed but the percentages with respect to the annual sum.

TABLE XXVIII

RAINFALL TOTALS (% OF ANNUAL TOTAL) AND NUMBER OF DAYS WITH THUNDERSTORMS[1]

Month	15°–25°N 70°–85°W (11 stations)		10°–20°N 55°–70°W (16 stations)	
	rainfall total (%)	number of days with thunderst. (%)	rainfall total (%)	number of days with thunderst. (%)
Jan.	5	1	6	1
Feb.	4	1	3	0
Mar.	4	2	4	0
Apr.	6	4	5	1
May	11	10	7	7
June	11	13	9	11
July	7	16	10	14
Aug.	9	18	11	18
Sept.	13	17	12	21
Oct.	14	12	13	16
Nov.	10	4	12	8
Dec.	6	2	8	3

[1] The same stations have been used for the rainfall and thunderstorm observations.

Visibility

Fog occurs practically never at sea and at coastal stations. Inland stations occasionally have shallow fog in the morning that is rapidly burned by the rising sun.

The dry season has many days with haze. As for the Caribbean Sea it is possible that the haze originates from the remainders of West African sand storms. In winter the trajectories lead from the Sahara directly into the West Indies and suggest such a cause–effect relationship. It is worth mentioning that the author saw considerable deposits of sand aboard ship when travelling at 50–80 km from the African coast near Dakar.

In Central America dust is stirred up in the dry season by the northers (see p.447) and reduces the visibility considerably. The dust has a sharp upper surface, probably the trade wind inversion, through which the mountains penetrate into clear air with unlimited visibility. In satellite pictures the mountains stand out black against a whitish diffuse background. The dust can be a real hazard for aircraft.

The range of visibility increases from the east to the west through our region. This means that Central America, especially on its Pacific side, has the worst but also the best visibilities found. Frequently in the dry season, and at any time of the year for the duration of heavy showers the author has had difficulty seeing nearby elevations, but has observed from the same point the cloud system of hurricane Hilda, whose center was passing by at a distance of 600 km. In cases as the latter, vision seems to be limited only by the curvature of the earth.

Weather

Introduction

Climate is the essence of weather, and therefore consideration must be given to the weather-types encountered in Central America and the Caribbean. One may distinguish between synoptic situations (German: Grosswetterlagen) and local developments.

"Grosswetterlagen"

The following paragraphs follow to some extent the text in *Aviation Meteorology of the West Indies* (METEOROLOGICAL OFFICE, LONDON, 1959) where the reader will find more details, some weather maps, and a good bibliography.

In winter and in spring, i.e., in the dry season, there is generally fine weather with steadily blowing trade winds. This weather is from time to time interrupted by outbreaks of cold air from the North American continent. Of course, their intensity and frequency decreases from the northwest to the southeast.

Where a front trails against mountains, long lasting rain and drizzle are typical; most of the winter rain of the north shores of Honduras and the Greater Antilles is due to this phase of weather development. On the Pacific side of Central America the cold fronts, with few exceptions, pass dry because of the foehn effect. The cold north wind, well familiar to the man in the street, is called norte, norther, or tehuantepeco. Locally it can become destructively strong, but in many places the norther is a welcome disruption of the monotony of the dry season (WRAIGHT, 1951). The lowest temperatures occur through nocturnal radiation after the wind has subsided.

The first norther(s) of the season is (are) generally a marked change of weather, while late in the season the northers become less and less distinct. For this reason there can be sometimes doubts as to what weather situation should be called a norther, and it is not possible to give exact numbers of northers per season. General estimates speak of 30–40 (U.S. HYDROGRAPHIC OFFICE, 1949). For Cuba, WRAIGHT (1951) mentions 25 per season and MILLÁS (1946) only 10 (Table XXIX).

CHAPEL (1927) defines a norther by the wind rather than through the weather map, and gives a table with daily means of wind, pressure, and rainfall during six northers. Their averages are presented in Table XXX.

While moderate or even no rain falls where the cold front passes, heavy downpours can occur where the front stalls. This occurs in some places more often than in others. Places with frequent heavy rainfall due to cold fronts can be found in the Caribbean coastal

447

TABLE XXIX

NUMBER OF NORTHERS IN HAVANA (after MILLÁS, 1946)

Oct.	Nov.	Dec.	Jan.	Feb.	Mar.	Apr.	May
0.2	1.2	1.4	2.2	2.3	2.1	0.6	0.1

TABLE XXX

AVERAGE WIND SPEED, BAROMETRIC PRESSURE, AND RAINFALL DURING SIX NORTHERS AT COLON, C.Z. (After CHAPEL, 1927)

Days:	−3	−2	−1	N	+1	+2	+3
Wind velocity (m/sec)	4.1	4.2	5.0	8.6	8.6	7.2	6.2
Pressure (mbar)	1010.2	1010.5	1011.1	1011.5	1011.3	1010.8	1010.8
Rainfall (mm)	21	17	6	32	50	25	10

N = day with onset of norther.

area of southeast Nicaragua (PORTIG, 1965, p.74), and in parts of Jamaica (see next section pp.449–450).

THEVENEAU (1961) reports that typical northers do not occur at Martinique but that the trough preceding the cold front can develop a convergence line running approximately east–west which may affect the island.

The rainy season has more "Grosswetterlagen" than the dry season. In the weather chart they become manifest as hurricanes, tropical depressions, easterly waves, pulsations of the trades, and displacements of the Intertropical Convergence Zone (I.T.C.Z.).

Hurricanes occur in most of the Caribbean Sea as well as in the Pacific Ocean along the coast of Central America. They are discussed in another part of this volume. Here it should only be emphasized (RIEHL 1954, p.83) that hurricanes are not responsible for the autumnal rainfall maximum. These rains, even when they derive from hurricanes, are considered beneficial in many locations.

Tropical lows or depressions occur at varying intensities all over our region, and also occasionally in the dry season.

Accrual of data and experience have led to confusion and subsequent new research concerning waves in the easterlies. There are at least two genetically different types of these waves (RIEHL, 1954; FREEMAN et al., 1964; MERRITT, 1964).

Studies by SIMPSON and GARSTANG (1965) and by HUBERT (1965) show that the present knowledge of tropical weather is insufficient, and that new models are needed.

Local developments

The most important local development, as far as human life is concerned, is the "temporal". It can be disputed whether the attribute "local" is appropriate, but these long-lasting rains with no or only little electrical activity are typical for two areas, the north coast of Honduras, and the Pacific coast of all of Central America. The temporals of Honduras occur during invasions of cold air from the north.

Although structure and origin of the temporals of the Pacific coast have not been sufficiently explored, it is known that they are quite different from the Honduran type. At least frequently, perhaps always, the Pacific temporals are tropical depressions slowly trailing along the coast. They seem to be akin to hurricanes, and transformations from "temporal"-lows into hurricanes, and vice versa, can occur (LESSMANN, 1962, p.8).

TIROS pictures seem to indicate that a temporal has the spiral bands and the size of a hurricane, but the wind is (with few exceptions) light to moderate, and any lightning and thunder display that may occur in a hurricane, is almost completely absent in a Pacific temporal. The economical impact of a temporal comes from the large amount of rain on a relatively large area, causing floods, land slides, and other water damage.

Fig.22. TIROS picture of a typical Central American norther. Rain is reported by all stations in north Honduras.

It is not possible to clearly distinguish between temporals and temporal-like situations ("tiempo atemporalado" as the man in the street says), but a fair estimate calls for one or two temporals a year for each location on the Pacific coast of Central America with the greatest probability of occurrence in September–October, and a secondary maximum in June.

Heavy downpours, more local in character than temporals, are typical, among many other places, for Jamaica. The Plumb Point observer wrote on May 12, 1916: "There was a tremendous downpour or cloudburst at the lighthouse. This heavy deluge was confined to quite a small area, and it was plainly seen from Kingston. The cloudburst lasted only 15 minutes, but at the end of that time the raingauge that holds 200 mm had over-flowed. The cloudburst was followed by a small whirlwind." (quoted after THEAMAN, 1943). Jamaican cloudbursts show a distinct annual variation. The number of reported days with more than 300 mm is largest in November, and second largest in May. These downpours are not always isolated in time and/or space. Until recently (PAULHUS, 1965) the Silver Hill Plantation held the world record of five-day rains (Table XXXI). On November 4, 1909, a cold front protruding from a depression over Labrador reached Jamaica and became stationary. On November 7 an anticyclone over Maine pushed a broad swift current of cold air towards the Caribbean. On November 8 the cold front was clearly south of Jamaica, and a tropical storm formed at the north coast of Panama. This storm moved slowly northeastward, and the heavy rains in Jamaica ceased when the storm was east of the island on November 12 (UNITED STATES WEATHER BUREAU, 1909). While SMEDLEY (1961) contends that most heavy rainfalls on Puerto Rico and the Virgin Islands stem from hurricanes, he reports of stationary thunderstorms that pro-duced 370 mm in three days and caused the May 1960 flood in the Virgin Islands. He also mentions cold fronts as flood-producing factors.

TABLE XXXI

24-HOUR RAINFALL TOTALS (mm) IN EASTERN JAMAICA IN NOVEMBER 1909

	Radnor	Farm Hill	Silver Hill Plant	
Nov. 4	—	—	165	
Nov. 5	—	—	525	
Nov. 6	—	600	765	
Nov. 7	—	720	675	2,865 mm in 5 days
Nov. 8	630	530	450	
Nov. 9	610	565	450	
Nov. 10	700	—	200	
Nov. 11	—	—	150	
Totals	1,940	2,415	3,380	

In spite of the temporary occurrence of heavy rain there is a serious danger of droughts in many locations of the region where the *average* totals of rain are sufficient. SMEDLEY (1961) mentions the almost permanent water shortage on St. Thomas caused by the interaction of steep slopes with a high evaporation rate. Small islands such as this induce the trade winds to release their moisture. However, much of it is wasted, falling downwind of the islands into the ocean. But droughts occur also on larger islands such as in spring 1958 on Puerto Rico near San Juan (SMEDLEY, 1961).

A local phenomenon encountered in many places is the sea breeze. In some places and at some times it is so strong that it overcomes the large-scale wind regime, i.e., it may blow from the west in a typical trade wind area. CROW and COBB (1962) report a sea breeze front at the Pacific coast of West Panama penetrating inland up to 30 km. They describe that the retiring maritime air has also front-like effects which after dark continue over sea. This is in concurrence with the author's observations in El Salvador.

In Central America the sea breeze is so well developed in the rainy season that it has been confounded in older climatographies with a monsoon (McBryde, 1942).

Squalls are frequent, especially in waters to the west of islands whose coasts are backed by mountains (U.S. Hydrographic Office, 1949). Preceded by sultry weather with light, variable winds, the squalls rush from under thunderclouds toward lower elevations. Usually one speaks of "white squalls" or "black squalls", but at the south coast of Cuba they are so typical that they have a special name, "bayamo". According to Wraight (1951) the convective storms in Cuba, especially in May, are called "turbanos". Thunderstorms over water quite frequently generate waterspouts, sometimes several of them from the same cloud (*Aviation Meteorology of the West Indies*).

The more dangerous brother of the waterspout, the tornado, is fortunately less frequent than the waterspout. Quate (1962) reports of some tornadoes and briefly describes the synoptic situation in west Panama at the time of occurrence. Also Alpert (1955) describes tornadoes in Panama. Several funnel clouds and damage typical for smaller tornadoes were reported to this author during his stay in El Salvador.

Hail is rare even in places with many thunderstorms. Alexander (1903) and Smedley (1961) mention hail on Puerto Rico, the former also on the Bahamas (1906). This author observed several hail falls in El Salvador, one of them moderate, the others light. McBryde (1942) mentions several hail falls in Pacific Guatemala, also two snowfalls on elevations of 3,400 m and higher. Lemons (1942) lists the Panama Canal Zone among the places from where reports on hail have become known.

Some minor phenomena should not be omitted. Haloes are frequent at all times of the year, also in the dry season, documenting the high frequency of cirrostratus. Of course, rainbows are frequent in the rainy season. This writer photographed at sunrise a rainbow without rainclouds above it. (This was the second case in four years when rain fell without a cloud.) Thunderstorms are frequently preceded by iridescent altocumuli lenticularis that change their form and amount very rapidly. This precursory sign appears up to six hours before the thunderstorm begins to develop and aids in short-range forecasting.

Appendix—Climatic tables

The climatic tables XXXII–LV are shown on pp.455–478 and are arranged from south to north.

References

Admiralty Hydrographic Department, 1957. *West Indies Pilot*, 3. London, 5th ed., 312 pp.
Alexander, C. S., 1958. *The Geography of Margarita and Adjacent Islands, Venezuela*. University of California Press, Santa Monica, Calif.
Alexander, W. H., 1903. Hailstorms in Puerto Rico. *Monthly Weather Rev.*, 31: 233–234.
Alexander, W. H., 1906. Climatology of Puerto Rico from 1857 to 1905. *Monthly Weather Rev.*, 34: 315–324.

ALPERT, L., 1941. The areal distribution of mean annual rainfall over the island of Hispaniola. *Monthly Weather Rev.*, 69:201–205.

ALPERT, L., 1945. The Intertropical Convergence Zone of the Eastern Pacific region. *Bull. Am. Meteorol. Soc.*, 26: 426–432.

ALPERT, L., 1946a. Notes on the weather and climate of Seymour Island, Galapagos Archipelago. *Bull. Am. Meteorol. Soc.*, 27: 200.

ALPERT, L., 1946b. The Intertropical Convergence Zone of the Eastern Pacific region. *Bull. Am. Meteorol. Soc.*, 27: 15–29.

ALPERT, L., 1946c. *Contributions to the Climate of the Tropical Eastern Pacific Ocean.* Dissertation, Clark University, Worcester, Mass.

ALPERT, L., 1955. Panama tornadoes. *Bull. Am. Meteorol. Soc.*, 36:171–172.

BLISS, E. W., 1930. A study of rainfall in the West Indies. *Quart. J. Roy. Meteorol. Soc.*, 56: 67–72.

BLUME, H., 1962. Beiträge zur Klimatologie Westindiens. *Erdkunde*, 16: 271–288.

BRAAK, C., 1935. Het climaat van Nederlandsch West Indië. *Mededel. Verhandl.*, 36/39: 120 pp.

BRENNAN, J. F., 1942. *The Rainfall of Jamaica from about 1870 to the End of 1939.* Government Printer, Kingston, Jamaica, 86 pp.

BRENNAN, J. F., 1943. *Meteorology of Jamaica.* Government Printer, Kingston, Jamaica, 32 pp.

BROOKS, C. E. P., 1918. The meteorology of Belize, British Honduras, 1888–1917. *Quart. J. Roy. Meteorol. Soc.*, 44: 305–306.

CAMPAN, G., 1959. Note sur la climatologie des Antilles et de la Guyane Française. *Monographies Météorol. Natl.*, 15:26 pp.

CHAPEL, L. T., 1927. Winds and storms on the isthmus of Panama. *Monthly Weather Rev.*, 55:519–530.

CHAPEL, L. T., 1934. The significance of air movements across the equator in relation to development and early movement of tropical cyclones. *Monthly Weather Rev.*, 62: 433–438.

CHERBONNIER, J., 1963. *Essai de Classification des Différents Climats de la Martinique d'après les Méthodes de Thornthwaite.* Service Météorologique du Groupe Antilles–Guyane, Fort-de-France, Martinique, 28 pp.

CROW, L. W. and COBB, G., 1942. *Life Cycles of Tropical Cumulus in Southwestern Panama.* Final Rept., Natl. Science Foundation, Contract NSF C-184, 31 pp.

DEUTSCHES HYDROGRAPHISCHES INSTITUT, 1958. *Westindien-Handbuch.* Hammerich and Lesser, Hamburg, 3rd. ed., 386 pp.

FLOHN, H., 1957. Studien zur Dynamik der äquatorialen Atmosphäre. *Beitr. Phys. Atmos.*, 30:18–46.

FORS, A. L., 1952. Observaciones sobre el efecto de los vientos fuertes en la producción y el rendimiento de las cañas. *Bol. Ofic. Asoc. Téc. Azucareros Cuba*, 11:71–75.

FREEMAN JR., C. F., GRAVES, L. F., HANNA, P., PORTIG, W. H., ARMIJO, L. and BURGMEIER, J., 1964. The effects of internal waves in the easterlies on mesoscale weather developments. *U.S. Army Electron. Command, Interim Rept.*, 51 pp.

GOLDBRUNNER, A. W., 1962. El clima de la isla de Margarita. *Mem. Soc. Cienc. Nat. La Salle*, 22:146–155.

HALL, M., 1923. *The Rainfall of Jamaica from about 1870 to the End of 1919.* Government Printer, Kingston, Jamaica, 21 pp. (Abstract by A. J. HENRY in *Monthly Weather Rev.*, 47:207–208.)

HASTENRATH, S. and LESSMANN, H., 1963. Mapa preliminar de la producción anual de lluvia en El Salvador. *Serv. Meteorol. Nacl. El Salvador, Publ. Téc.*, 3:18 pp.

HENRY, W. K., GRIFFITHS, J. F., COBB, G., et al., 1963–1966. *Research on Tropical Rainfall Patterns and Associated Mesoscale Systems.* Texas A & M University, College Station, Texas, 11 reports.

HOFFMANN, G., 1960. Die mittleren jährlichen und absoluten Extremtemperaturen der Erde. *Meteorol. Abhandl. Inst. Meteorol. Phys. Freien Univ. Berlin*, 8, pt.1, 97 pp.; pt.2, 18 tables, 47 maps; pt.3, 161 pp.

HUBERT, L. F., 1965. Three-dimensional field of motion in a Pacific tropical disturbance. *Tech. Conf. Hurricanes Tropical Meteorol.*, 4th, Miami, Fla., presented paper. (Abstract in *Bull. Am. Meteorol. Soc.*, 46: 585.)

JAMAICA METEOROLOGICAL OFFICE, 1961. *Jamaica Weather Report 1960.* (Contains 30-year rainfall averages for 103 stations.)

KLOSTER, W., 1922. Bewölkungs-, Niederschlags- und Gewitterverhältnisse der westindischen Gewässer und der angrenzenden Landmassen. *Arch. Deut. Seewarte*, 40(1): 67 pp.

LAHEY, J. F., 1958. On the origin of the dry climate in northern South America and the southern Caribbean. *Univ. Wisc., Dept. Meteorol., Sci. Rept.*, 10:290 pp.

LASEUR, N., 1966. On the distribution of rainfall probability over Barbados. *Proc. Army Conf. Tropical Meteorol.*, Ft. Monmouth, 1966, pp.36–39.

LEMONS, H., 1942. Hail in high and low latitudes. *Bull. Am. Meteorol. Soc.*, 23:61.

LESSMANN, H., 1956. Características de la lluvia en El Salvador. *Bol. Lluvia El Salvador*, 1956, pp. 56–63.

LESSMANN, H., 1961. El clima en la cuenca del Río Grande de San Miguel. *Serv. Meteorol. Nacl. El Salvador, Publ. Tec.*, 1:57 pp.

MACKY, W. A., 1957. The rainfall of Bermuda. *Meteorol. Office Bermuda, Tech. Note*, 8:58 pp.

McBRYDE, F. W., 1942. Studies in Guatemalan meteorology. *Bull. Am. Meteorol. Soc.*, 23:254–263, 400–406.

McCULLOUGH, C. R., 1956. *Terrain Study of the Panama Canal Zone*. North Carolina State College, Dept. of Engineering Research, Raleigh, N.C., 267 pp.

MERRITT, E. S., 1964. Easterly waves and perturbations, a reappraisal. *J. Appl. Meteorol.*, 3:367–382.

METEOROLOGICAL OFFICE, LONDON, 1959. Aviation meteorology of the West Indies. *Meteorol. Office, London, Meteorol. Rept.*, 22 (MO 621f):85 pp.

MILLAS, J. C., 1937. Algunas diferencias entre años tormentosos y años no-tormentosos en la porción occidental de Cuba. *Bol. Obs. Nacl.*, 3(2): 194–205.

MILLAS, J. C., 1943. *La Lluvia en La Habana*. Servicio Meteorológico de Cuba, La Habana, 34 pp.

MILLAS, J. C., 1945. Los brisotes en La Habana. *Bol. Obs. Nacl.*, 4(1): 73–83.

MILLAS, J. C., 1946. Los Nortes en La Habana. *Arch. Inst. Nacl. Hidrol. Climatol. Médicas*, 1:29–44.

PAULHUS, J. L. H., 1965. Indian Ocean and Taiwan rainfalls set new records. *Monthly Weather Rev.*, 93: 331–335.

PAWLING, J. W., 1956. *The Climate of Central America*. Thesis, Clark University, Worcester, Mass. (unpublished).

PORTIG, W. H., 1956. La variación de la doble onda diurna de la presión atmosférica en San Salvador. *Comun. Inst. Tropical Univ. El Salvador*, 5: 171–178.

PORTIG, W. H., 1957. Wettervorhersage-Regeln für El Salvador. *Meteorol. Rundschau*, 10:107–108.

PORTIG, W. H., 1958. Über die Trockenzeit von San Salvador, Mittelamerika. *Meteorol. Rundschau*, 11: 80–85.

PORTIG, W. H., 1959. Air masses in Central America. *Bull. Am. Meteorol. Soc.*, 40:301–304.

PORTIG, W. H., 1960. Beiträge zur Meteorologie Mittelamerikas (insbesondere El Salvadors). *Deut. Wetterdienst, Seewetteramt, Einzelveröffentl.*, 28:24 + 16 pp.

PORTIG, W. H., 1961. Some climatological data of Salvador, Central America. *Weather*, 16:103–112.

PORTIG, W. H., 1965. Central American rainfall. *Geograph. Rev.*, 55:68–90.

PORTIG, W. H., 1966. Atmospheric water budget over tropical areas. *U.S. Army Electron. Command, Ft. Monmouth, Tech. Rept.*, ECOM-01549-F: 92 pp.

PROHASKA, F., 1963. Die Verfälschung der Temperaturmittel in den Tropen bei ihrer Berechnung aus den Extremwerten. *Meteorol. Rundschau*, 16: 145–149.

QUATE, B. E., 1962. *A Pictorial Report on Project "Wind Control"*. Weather Engineers, Inc., Soquel, Calif., 66 pp.

REED, W. W., 1926. Climatological data for the West Indian Islands. *Monthly Weather Rev.*, 54: 133–160.

REYES, R. L., 1964. Resumen climatológico aeronáutico para el Aeropuerto Internacional de Ilopango, parte No. 1, Modelo "B", Promedio mensual de días en que se dan casos de simultaneidad de velocidad y dirección del viento dentro de límites determinados. *Serv. Meteorol. Nacl. El Salvador, Publ. Téc.*, 4, 29 pp.

RIEHL, H., 1954. *Tropical Meteorology*. McGraw-Hill, New York, N.Y., 392 pp.

ROUSE, W. R. and WATTS, D., 1966. *Two studies in Barbadian Climatology*. McGill University, Montreal, 65 pp.

SALIVIA, L., 1950. *Historia de los Temporales de Puerto Rico (1508–1949)*. San Juan, Puerto Rico.

SANTAMARIA, L., 1945. Promedios de las lluvias de algunos lugares de Cuba (1922–1942). *Bol. Obs. Nacl.*, 4(1):91–94.

SAPPER, K., 1932. Klimakunde von Mittelamerika. In: *Handbuch der Klimatologie*, 2. Band, 2. Teil. Borntraeger, Berlin, pp.1–74.

SERVICE MÉTÉOROLOGIQUE DU GROUPE ANTILLES–GUYANE, 1962. *Quelques Élements du Climat de la Martinique*. 6 pp. (mimeographed).

SERVICIO METEOROLÓGICO NACIONAL DE EL SALVADOR, 1954–1957. *Boletín Meteorológico de El Salvador*.

SERVICIO METEOROLÓGICO NACIONAL DE EL SALVADOR, 1959. *Anexo al Boletín Meteorológico de El Salvador.* ("Réseau Mondial" data and daily rainfall totals 1918–1957.)

SERVICIO METEOROLÓGICO NACIONAL DE HONDURAS, 1960. *El viento de superficie en Honduras*. Tegucigalpa, 3 pp. (mimeographed).

SERVICIO METEOROLÓGICO NACIONAL DE HONDURAS, 1962. *Precipitación máxima Caída en 24 Horas y Número de Días con Precipitación de 0.01 Pulgadas y Más*. Tegucigalpa, 53 pp. (mimeographed).

SIMPSON, J. and GARSTANG, M., 1965. *A Study of a Non-deepening Tropical Disturbance*. Florida State University, Department of Meteorology, Tallahassee, Fla., 48 pp.

SKEETE, C. C., 1931. Barbados rainfall. *Barbados Dept. Sci. Agr. Pamphlet*, 9:3–18.

SMEDLEY, D., 1961. Puerto Rico and the Virgin Islands. In: *Climatography of the United States*. U.S. Weather Bureau, Washington, D.C., pp.1–20.

STATISTIEK- EN PLANBUREAU, CURAÇAO, 1960–61. *Meteorologische Waarnemingen in de Nederlandse Antillen*. The Hague.

THEAMAN, J. R., 1943. *Tropical Downpours of Jamaica. Maximum Precipitation for Various Periods*. Theaman, Indianapolis.

THÉVENEAU, A., 1961. *Les Saisons et les Types de Temps aux Antilles Françaises*. Fort-de-France, Martinique, 1:10 pp., 2:11 pp.

THÉVENEAU, A., 1966. *Le Climat de la Guadeloupe*. Direction de la Météorologie Nationale, Service Météorologique du Groupe Antilles-Guyane, Fort-de-France, Martinique, 1:57 pp., 2:58 pp.

THORP, J., 1941. Climate and settlement in Puerto Rico and the Hawaiian Islands. In: GOVE HAMBIDGE (Editor), *Climate and Man, Yearbook of Agriculture*. Washington D.C., pp.217–226.

TREWARTHA, G. T., 1961. *The Earth's Problem Climates*. The University of Wisconsin Press, Madison, Wisc., 334 pp.

TROLL, C., 1952. Die Lokalwinde der Tropengebirge und ihr Einfluss auf Niederschlag und Vegetation. *Bonner Geograph. Abhandl.*, 9:124–182.

UNITED STATES HYDROGRAPHIC OFFICE, 1948. *Weather Summary: Central America*. U.S. Govt. Printing Office, Washington, D.C., H.O. Publ. 531, 64 pp.

UNITED STATES HYDROGRAPHIC OFFICE, 1949. *Sailing Directions for the West Indies*. U.S. Govt. Printing Office, Washington, D.C., H.O. Publ. 129, 250 pp.

UNITED STATES HYDROGRAPHIC OFFICE, 1955. *Atlas of Pilot Charts of the South Atlantic and Central American Waters*. U.S. Govt. Printing Office, Washington, D.C., H.O. Publ., 576, 65 pp. (Does not contain the Central American waters in the Pacific Ocean.)

UNITED STATES WEATHER BUREAU, 1909. Rainfall in Jamaica. *Monthly Weather Rev.*, 37: 988.

UNITED STATES WEATHER BUREAU, 1955. *Rainfall Intensity–Duration–Frequency Curves for Selected Stations in the United States, Alaska, Hawaiian Islands, and Puerto Rico*. U.S. Govt. Printing Office, Washington, D.C., U.S.W.B. Techn. Pap., 25, 53 pp.

UNITED STATES WEATHER BUREAU, 1961. *Generalized Estimates of Probable Maximum Precipitation and Rainfall Frequency Data for Puerto Rico and Virgin Islands for Areas to 400 Sq. miles, Durations to 24 h, and Return Periods from 1 to 100 Years*. U.S. Govt. Printing Office, Washington, D.C., U.S.W.B. Tech. Pap., 42, 94 pp., 130 figs.

URRUTIA, E. C., 1964. *Atlas Climatológico de Guatemala*. Ministerio de Agricultura Observatorio Nacional, Guatemala. (Monthly and annual maps of temperature and precipitation.)

WAGNER, L. P., 1958. *Nicoya, a Cultural Geography*. University of California Press, Santa Monica, Calif.

WARD, R. DE C. and BROOKS, C. F., 1934. Climatology of the West Indies. In: W. KÖPPEN and R. GEIGER (Editors), *Handbuch der Klimatologie*. Borntraeger, Berlin, 2(1): 1–47.

WEHEKIND, L. and SMITH, G. W., 1955. *Trinidad Rainfall 1933–1952, and Rainfall Reliability in Trinidad*. Port of Spain, Trinidad.

WERNSTEDT, F. L., 1961. *World Climatic Data, 2. Latin America and the Caribbean*. Pennsylvania State University, College Park, Penn., 88 pp.

WEST INDIES METEOROLOGICAL SERVICE, 1951–57. *Climatological summaries 1951–1957*. Port of Spain, Trinidad.

WISE, H., 1958. *Análisis de caída de lluvia en cincuenta estaciones de Honduras, 1913–1957*. Servicio Meteorológico de Honduras, Tegucigalpa.

WRAIGHT, J., 1951. *A Climatic Survey of Cuba*. Dissertation, Clark University, Worcester, Mass., 161 pp. (unpublished).

TABLE XXXII

CLIMATIC TABLE FOR JAQUE (1951–60)
Latitude 7°32′N, longitude 78°10′W, elevation 5 m

Month	Temperature (°C)				Mean precip. (mm)	Number of days with		Preval. wind direction
	daily mean	mean daily range	absol. extremes			precip. ⩾1 mm	thunder-storm	
			max.	min.				
Jan.	26.1	8.9	33.9	18.9	8	4	0	
Feb.	26.1	10.0	35.0	18.9	3	<0.5	0	NNE
Mar.	26.7	7.8	33.9	20.0	28	3	3	
Apr.	26.7	6.4	35.0	21.1	91	10	9	
May	26.1	6.4	31.7	20.0	622	22	12	NE
June	26.1	6.4	31.7	21.1	411	23	9	
July	25.6	6.1	30.6	20.0	432	25	22	
Aug.	25.6	6.4	31.7	20.0	409	24	22	NNE
Sept.	25.6	6.4	30.6	20.0	353	21	21	
Oct.	25.0	5.6	30.6	20.0	955	26	9	
Nov.	25.6	6.1	30.6	20.0	696	23	9	NNE
Dec.	25.6	6.4	32.8	20.0	305	18	9	
Annual	25.9	6.9			4,313	199	125	

TABLE XXXIII

CLIMATIC TABLE FOR BALBOA (1951–60)*
Latitude 8°58′N, longitude 79°33′W, elevation 31 m

Month	Mean sea level press. (mbar)	Temperature (°C)				Precipitation (mm)	
		daily mean	mean daily range	extremes		mean	max. in 24 h
				max.	min.		
Jan.	1010.0	26.6	8.9	33.3	18.9	38	37
Feb.	1010.3	26.9	9.4	34.4	17.8	16	48
Mar.	1010.3	27.6	9.4	35.6	18.9	17	30
Apr.	1009.9	27.9	8.9	35.0	20.0	76	74
May	1009.6	27.2	6.4	35.0	21.1	198	184
June	1009.7	26.8	6.4	33.9	21.7	203	107
July	1009.8	26.9	7.2	33.9	21.1	183	138
Aug.	1010.1	26.8	6.4	33.9	20.0	190	112
Sept.	1009.7	26.6	6.1	33.3	21.1	193	130
Oct.	1010.3	26.2	6.1	33.3	20.6	254	125
Nov.	1009.8	26.2	7.2	33.3	20.6	249	116
Dec.	1010.2	26.6	7.8	33.3	20.0	130	66
Annual	1010.1	26.9	7.5			1,747	184

Month	Mean evap.[1] (mm)	Number of days with			Mean cloud-iness (tenths)	Mean sun-shine (h)	Wind	
		precip. ⩾0.1 mm	thunder-storm	fog			preval. direct.	mean speed (m/sec)
Jan.	128	6	1	0	6.2	218	NW	3.7
Feb.	138	3	<0.5	0	5.2	249	NW	4.3
Mar.	164	<0.5	<0.5	0	6.1	278	NNW	4.2
Apr.	148	5	4	0	6.9	271	NNW	3.8
May	106	17	10	0	8.2	195	NW	2.7
June	80	17	12	0	8.7	172	NW	2.1
July	81	20	12	0	9.2	139	NW	2.3
Aug.	80	20	14	0	9.1	133	NW	2.3
Sept.	77	15	13	1	8.7	170	NW	2.1
Oct.	80	22	11	1	8.8	158	NW	2.2
Nov.	78	24	8	1	8.0	158	NW	2.3
Dec.	112	12	3	0	7.1	214	NW	2.9
Annual	1,272	162	88	3	7.6	2,355	NW	2.9

* See also Table LV.
[1] 8 km northwest of Balbao.

TABLE XXXIV

CLIMATIC TABLE FOR CRISTOBAL (COLON) (1951–60)*
Latitude 9°21′N, longitude 79°55′W, elevation 12 m

Month	Mean sea level press. (mbar)	Temperature (°C)				Mean vap. press. (mbar)	Precipitation (mm)		
		daily mean	mean daily range	extremes			mean[1]	mean[2]	max. in 24 h
				max.	min.				
Jan.	1011.5	26.8	6.4	31.7	17.8		70	85	77
Feb.	1010.8	26.8	6.1	32.8	17.8		38	40	113
Mar.	1010.8	27.0	6.1	33.9	17.8		40	42	106
Apr.	1010.5	27.2	6.1	33.9	18.9		94	105	157
May	1009.6	26.8	6.7	33.9	21.1		315	324	154
June	1010.1	26.4	7.2	35.0	21.1		314	307	137
July	1010.2	26.6	6.4	33.9	21.1		389	406	166
Aug.	1010.3	26.6	6.4	37.2	20.0		384	387	135
Sept.	1009.7	25.6	7.8	35.0	20.0		321	321	87
Oct.	1010.0	26.5	7.8	33.9	20.0		432	397	173
Nov.	1009.9	26.1	7.2	33.9	20.0		646	569	343
Dec.	1010.6	26.4	6.4	33.9	20.0		381	302	130
Annual	1010.3	26.6	6.7			27.7	3,424	3,285	343

Month	Mean evap. (mm)	Number of days with			Mean cloud-iness (tenths)	Mean sun-shine (h)	Wind	
		precip. ≥0.1 mm	thunder-storm	fog			preval. direct.	mean speed (m/sec)
Jan.	117	12	0	0	6.4	248	NE	5.5
Feb.	129	11	0	0	5.9	235	N	5.9
Mar.	147	12	<0.5	0	6.0	268	N	5.7
Apr.	145	12	2	0	6.1	243	N	5.4
May	108	21	12	0	8.0	188	N	3.4
June	93	22	15	0	8.3	156	NW	2.5
July	82	24	16	0	8.7	154	NNE	3.2
Aug.	85	26	16	0	8.5	156	N	3.0
Sept.	85	22	15	0	8.2	171	NW	2.4
Oct.	89	22	14	0	8.4	164	S	2.3
Nov.	80	25	8	0	8.5	148	N	2.9
Dec.	104	22	2	0	7.7	206	NE	4.5
Annual	1,264	231	100	0	7.6	2,337	N	3.9

* See also Table LV.
[1] For standard period 1931–1960.
[2] For a period of 103 years ending in 1973.
[3] For the period 1908–1965.

TABLE XXXV

CLIMATIC TABLE FOR SAN JOSÉ, COSTA RICA (1951–60)
Latitude 9°56′N, longitude 84°08′W, elevation 1,120 m

Month	Mean sea level press. (mbar)	Temperature (°C)			Precipitation[1] (mm)		Relat. humid. (%)	Days with precip.[1] ⩾0.1 mm
		daily mean	extremes		mean	max. in 24 h		
			max.	min.				
Jan.	1010.7	19.0	30.5	9.4	8	12	80	3
Feb.	1010.1	19.3	31.1	10.5	5	8	80	1
Mar.	1009.0	20.3	32.7	10.0	10	12	80	2
Apr.	1008.6	21.0	31.6	11.7	37	25	79	7
May	1008.1	21.4	31.0	12.2	244	83	84	19
June	1008.3	21.2	33.2	13.9	284	88	86	22
July	1008.9	20.6	28.9	12.2	230	123	86	23
Aug.	1008.8	20.8	29.4	13.3	233	60	85	24
Sept.	1008.4	20.9	30.0	13.3	342	60	86	24
Oct.	1008.7	20.6	29.4	12.7	333	75	88	25
Nov.	1009.5	19.9	28.7	11.1	172	75	84	14
Dec.	1010.6	19.3	30.5	9.4	46	22	82	6
Annual	1009.1	20.4			1,944	123	83	170

[1] Period of record 1931–60.

TABLE XXXVI

CLIMATIC TABLE FOR PIARCO (1951–60)
Latitude 10°37′N, longitude 61°21′W, elevation 14 m

Month	Mean sea level press. (mbar)	Temperature (°C)				Mean vap. press. (mbar)	Precipitation (mm)	
		daily mean	mean daily range	extremes			mean	max. in 24 h
				max.	min.			
Jan.	1013.2	24.5	9.4	31.1	16.1	24.4	77	58
Feb.	1013.3	24.7	10.0	32.1	16.7	24.4	61	46
Mar.	1012.7	25.4	10.5	32.7	16.7	24.0	27	25
Apr.	1012.7	26.3	10.0	33.9	17.2	24.9	71	71
May	1012.8	26.6	8.3	34.4	19.4	27.6	129	46
June	1013.6	26.1	7.2	35.5	18.9	28.4	269	73
July	1013.7	25.9	7.8	32.7	18.3	28.3	243	89
Aug.	1012.8	26.1	8.9	33.2	18.9	28.2	213	119
Sept.	1012.0	26.2	9.4	33.9	19.4	28.3	144	61
Oct.	1011.4	25.9	8.9	32.7	20.5	28.2	151	58
Nov.	1011.1	25.4	9.4	32.7	18.3	27.7	212	79
Dec.	1012.0	24.8	9.4	31.6	16.7	26.3	153	109
Annual	1012.6	25.7				26.7	1,750	119

Month	Number of days with			Mean cloudiness (tenths)			Preval. wind direct.
	precip. ⩾1 mm	thunder- storm	fog	08h	14h	20h	
Jan.	13	0	1	4.2	5.4	3.1	E
Feb.	8	0	0.5	3.9	5.5	2.4	E
Mar.	8	0	0.1	4.1	5.7	2.8	E
Apr.	6	0.2	0.5	4.5	6.1	3.3	E
May	15	1	1	5.1	6.3	3.6	E
June	22	4	0.7	5.4	6.3	4.4	E
July	21	4	2	4.3	5.8	3.2	E
Aug.	20	8	2	4.2	6.0	3.1	E
Sept.	14	7	0.7	4.3	5.9	3.2	E
Oct.	15	8	1	4.4	6.1	3.5	E
Nov.	15	2	2	4.2	5.8	3.5	E
Dec.	15	1	1	4.1	5.3	3.0	E
Annual	172	35	12	4.4	5.9	3.3	E

TABLE XXXVII

CLIMATIC TABLE FOR MANAGUA (1951–60)
Latitude 12°08′N, longitude 86°11′W, elevation 56 m

Month	Mean sta. press. (mbar)	Temperature (°C)				Precipitation (mm)	
		daily mean	mean daily range	extremes		mean	max. in 24 h
				max.	min.		
Jan.	1005.5	26.3	10.6	31.0	20.4	4	9
Feb.	1005.1	27.2	11.5	32.1	20.6	1	2
Mar.	1004.3	28.6	11.9	33.6	21.7	5	29
Apr.	1004.0	29.3	11.7	34.3	22.6	6	23
May	1003.9	29.4	10.6	34.0	23.4	76	92
June	1004.0	27.2	8.4	31.4	23.0	296	119
July	1004.8	26.9	8.3	30.9	22.6	134	89
Aug.	1004.1	27.2	9.0	31.4	22.4	130	60
Sept.	1003.5	26.9	9.1	31.3	22.2	182	119
Oct.	1003.3	26.5	8.7	30.8	22.1	243	108
Nov.	1004.1	26.3	9.7	30.6	20.9	59	45
Dec.	1005.1	26.1	10.8	30.8	20.0	6	9
Annual	1004.3	27.3	10.0	31.8	21.8	1,142	119

Month	Mean evap. (mm)	Days with precip. ⩾1 mm	Mean cloud-iness (tenths)	Mean sun-shine (h)	Wind	
					preval. direct.	mean speed (m/sec)
Jan.	286	2.5	4.8	217	E	4.3
Feb.	297	0.8	4.4	214	E	4.8
Mar.	392	1.0	4.4	250	E	4.8
Apr.	380	1.0	5.5	223	E	4.7
May	325	5.7	5.8	193	E	4.1
June	141	21.5	8.4	128	E	3.1
July	166	19.7	8.2	142	E	3.5
Aug.	170	17.0	7.9	174	E	3.4
Sept.	125	20.0	8.1	178	E	2.7
Oct.	114	19.3	8.1	181	E	2.3
Nov.	144	9.9	6.2	195	E	2.8
Dec.	231	2.4	5.0	215	E	3.5
Annual	2,771	120.8	6.4	2,310	E	3.7

TABLE XXXVIII

CLIMATIC TABLE FOR CURAÇAO (WILLEMSTAD AND DR. PLESMAN AIRPORT COMBINED) (1951–60)
Latitude 12°12′N, longitude 68°58′W, elevation 8 m

Month	Mean sea level press. (mbar)	Temperature (°C)				Mean vap. press. (mbar)	Precipitation (mm)	
		daily mean	mean daily range	extremes			mean	max. in 24 h
				max.	min.			
Jan.	1013.0	26.2	5	30.9	19.0	25.3	68	75
Feb.	1013.1	26.1	5	31.1	19.9	24.8	31	24
Mar.	1012.4	26.5	5	32.2	20.0	25.1	14	33
Apr.	1011.8	27.1	5	33.4	21.4	26.4	12	33
May	1011.6	27.7	5	35.6	22.2	27.6	18	43
June	1012.3	28.0	5	32.9	21.0	28.0	26	98
July	1012.7	28.0	6	33.8	22.6	28.0	34	42
Aug.	1011.7	28.4	6	34.9	20.4	28.4	48	64
Sept.	1010.8	28.8	6	35.8	20.4	29.1	31	91
Oct.	1010.3	28.3	6	35.2	20.0	29.1	67	119
Nov.	1010.8	27.8	5	33.5	20.0	28.8	98	157
Dec.	1011.9	26.9	5	32.7	20.1	27.2	85	104
Annual	1011.9	27.5	5	35.8	19.0	27.3	532	157

Month	Number of days with			Mean cloudiness (tenths)	Mean sunshine (h)	Wind	
	precip. ⩾0.1 mm	thunderstorm	fog			preval. direct.	mean speed (m/sec)
Jan.	9.1	0.3	0	3.6	266	E	7.3
Feb.	4.6	0	0	3.2	272	E	8.0
Mar.	2.3	0	0	3.4	282	E	8.0
Apr.	2.3	0.2	0	4.3	263	E	8.2
May	2.8	0.6	0	5.0	217	E	8.2
June	3.6	0.5	0	4.6	266	E	8.5
July	5.4	1.2	0	4.2	297	E	8.1
Aug.	5.6	2.4	0	4.0	296	E	7.6
Sept.	3.9	2.8	0	3.8	280	E	7.3
Oct.	6.6	5.5	0	4.2	273	E	6.3
Nov.	10.2	4.4	0	4.2	245	E	6.1
Dec.	9.5	1.0	0	3.9	242	E	6.7
Annual	65.9	18.9	0	4.0	3,199	E	7.5

TABLE XXXIX

CLIMATIC TABLE FOR BRIDGETOWN (AND SEAWELL AIRFIELD, IN PARENTHESES) (1951–60)
Latitude 13°08′N (13°04′N), longitude 59°36′W(59°30′W), elevation 55 m (60 m)

Month	Mean sea level press. (mbar)	Temperature (°C)				Mean vap. press. (mbar)	Precipitation (mm)	
		daily mean	mean daily range	extremes			mean	max. in 24 h
				max.	min.			
Jan.	1014.2	25.2	7.2	30.5	16.1	(24.6)	66	63
Feb.	1014.7	25.1	7.8	30.5	16.1	(24.9)	28	23 (48)
Mar.	1014.7	25.9	8.3	31.6	16.6	(24.1)	33	23
Apr.	1014.0	26.3	7.8	31.6	17.7	(25.1)	36	71
May	1014.1	26.8	7.8	32.8	18.9	(27.5)	58	63
June	1014.8	27.1	7.2	32.2	19.4	(28.4)	112	94
July	1014.7	26.8	6.7	32.2	20.0	(28.8)	147	66
Aug.	1013.7	27.1	7.2	35.0	20.6	(29.3)	147	89
Sept.	1013.0	27.1	7.2	32.8	19.4	(29.7)	170	94
Oct.	1012.5	26.7	7.2	33.2	19.4	(29.3)	178	107
Nov.	1012.1	26.3	6.7	31.6	18.9	(28.2)	205	152
Dec.	1013.1	25.9	6.7	31.1	17.7	(26.5)	96	74
Annual	1013.8	26.3				(27.2)	1,276	152

Month	Number of days with			Wind		Mean cloudiness (tenths)	
	precip. ⩾0.1 mm	thunder-storm	fog	preval. direct.	mean speed (m/sec)	08h00	14h00
Jan.	13	0 (0.1)	(0)	(ENE)	4.8	(5.6)	(5.6)
Feb.	8	0 (0.1)	(0)	(ENE)	5.4	(5.4)	(5.2)
Mar.	8	0 (0)	(0)	(E)	5.5	(5.9)	(5.5)
Apr.	7	1 (0)	(0)	(E)	5.3	(6.3)	(6.1)
May	9	5 (0.2)	(0)	(E)	5.4	(6.5)	(6.5)
June	14	9 (1)	(0)	(E)	5.7	(6.5)	(6.6)
July	18	11 (2)	(0)	(E)	5.2	(6.1)	(6.1)
Aug.	16	19 (3)	(0.1)	(E)	4.4	(5.9)	(5.9)
Sept.	15	17 (4)	(0.1)	(E)	3.6	(6.3)	(6.5)
Oct.	15	15 (2)	(0)	(E)	3.3	(6.2)	(6.8)
Nov.	16	1 (1)	(0)	(E)	3.8	(5.4)	(5.8)
Dec.	14	3 (0.1)	(0)	(ENE)	4.4	(5.5)	(5.2)
Annual	153	81 13	(0)	(E)	4.7	(6.0)	(6.0)

TABLE XL

CLIMATIC TABLE FOR SAN SALVADOR, C.A. (1931–60)
Latitude 13°43′N, longitude 89°12′W, elevation 700 m

Month	Mean sta. press. (mbar)	Temperature (°C)					Mean vap. press. (mbar)	Precipitation (mm)	
		daily mean	mean daily range	extremes				mean	max. in 24 h
				max.	min.				
Jan.	934.0	22.1	13.3	29.2	15.9		13.0	5	18
Feb.	934.0	22.4	15.0	30.8	15.8		12.9	3	24
Mar.	933.3	23.5	15.0	32.0	17.0		13.8	8	32
Apr.	933.1	24.2	13.6	32.1	18.5		15.5	60	88
May	932.8	23.7	12.0	30.7	18.7		17.1	190	120
June	933.1	23.1	10.6	29.3	18.7		17.5	322	205
July	933.7	22.9	11.5	29.6	18.1		17.1	304	95
Aug.	933.5	23.0	11.6	29.9	18.3		17.1	297	165
Sept.	932.8	22.5	10.7	29.1	18.4		17.4	325	175
Oct.	932.7	22.4	10.0	28.3	18.3		16.7	220	170
Nov.	933.3	22.0	11.2	28.3	17.1		14.6	35	54
Dec.	933.7	22.0	12.1	28.4	16.3		13.6	7	46
Annual	933.3	22.8	12.2	29.8	17.6		15.5	1,775	205

Month	Mean evap.[1] (mm)	Number of days with			Mean cloud-iness (tenths)	Mean sun-shine (h)	Wind	
		precip. ⩾1 mm	thunder-storm	fog			preval. direct.	mean speed (m/sec)
Jan.	214	1.2	0.5	0	2.4	301	N	3.4
Feb.	223	0.9	0.6	0.2	2.8	277	N	2.8
Mar.	236	1.3	1.1	0	3.7	294	SW	2.8
Apr.	189	4.5	5.8	0.7	5.4	243	SW	2.4
May	146	13.3	11.5	0	6.6	220	W	2.0
June	87	19.6	17.1	1.0	7.6	174	SW	1.7
July	87	20.3	21.8	1.3	7.4	239	N	1.8
Aug.	87	19.9	21.6	0	6.8	257	W	1.8
Sept.	63	20.0	16.6	1.0	7.7	180	SW	1.7
Oct.	90	15.8	14.2	0.7	6.8	211	N	1.9
Nov.	151	3.9	3.0	0	4.2	267	N	2.7
Dec.	202	1.8	2.0	0	2.8	294	N	3.2
Annual	1,775	122.5	115.8	4.9	5.4	2,957	N	2.4

[1] Relative values obtained with Piche instrument in the shadow.

TABLE XLI

CLIMATIC TABLE FOR TEGUCIGALPA (TONCONTÍN) (1951–60)
Latitude 14°04′N, longitude 87°13′W, elevation 1,007 m

Month	Mean sea level press. (mbar)	Temperature (°C)				Precipitation (mm)	
		daily mean	mean daily range	extremes		mean	max. in 24 h
				max.	min.		
Jan.	1012.8	19.3	11.2	24.9	13.7	12	21
Feb.	1011.8	20.5	12.9	26.9	14.0	2	8
Mar.	1010.2	21.9	14.3	29.0	14.7	1	7
Apr.	1008.9	23.3	13.4	29.9	16.5	26	93
May	1008.9	23.6	11.8	29.6	17.8	180	109
June	1009.7	23.1	9.7	27.9	18.2	177	50
July	1010.8	22.5	9.7	27.3	17.6	70	59
Aug.	1010.4	22.8	11.0	28.3	17.3	74	62
Sept.	1009.2	22.9	11.0	28.4	17.4	151	172
Oct.	1009.7	22.1	8.5	26.8	17.3	87	79
Nov.	1011.3	20.7	9.8	25.6	15.8	38	54
Dec.	1012.4	19.6	10.1	24.8	14.7	14	15
Annual	1010.5	21.9	11.1			832	172

Month	Number of days with		Wind	
	precip. ⩾0.1 mm	thunder-storm	preval. direct.	mean speed (m/sec)
Jan.	5	0	N	5.7
Feb.	3	0	N	5.1
Mar.	2	1	N	4.6
Apr.	2	4	N	4.6
May	9	9	N	4.1
June	13	11	N	4.1
July	11	6	NE	4.1
Aug.	11	8	N	4.1
Sept.	16	11	E	4.1
Oct.	13	5	N	4.1
Nov.	8	1	N	5.1
Dec.	7	0	N	5.1
Annual	100	56	N	4.6

TABLE XLII

CLIMATIC TABLE FOR FORT-DE-FRANCE (1935–60)
Latitude 14°35′N, longitude 61°12′W, elevation 144 m

Month	Mean sea level press.[1] (mbar)	Temperature (°C)				Mean vap. press. (mbar)	Precipitation[2] (mm)	
		daily mean	mean daily range	extremes			mean	max. in 24 h
				max.	min.			
Jan.	1014.9	23.5	5.5	30.2	17.8	22.8	96	45
Feb.	1014.6	23.5	5.6	30.0	17.3	22.4	68	63
Mar.	1014.3	24.0	6.1	32.6	18.6	22.5	58	42
Apr.	1014.0	24.7	6.1	32.8	18.9	24.1	82	78
May	1013.7	25.4	5.8	32.5	19.9	26.0	126	123
June	1014.7	25.7	5.2	31.5	20.0	26.8	160	76
July	1014.4	25.6	5.0	31.3	19.5	27.3	214	71
Aug.	1014.1	26.0	5.5	33.0	20.3	27.6	227	103
Sept.	1013.0	25.9	5.7	32.9	20.6	27.6	232	238
Oct.	1012.7	25.6	5.6	32.0	20.2	27.4	221	152
Nov.	1012.7	25.2	5.4	31.8	19.8	26.5	230	133
Dec.	1013.2	24.2	5.1	31.0	17.4	24.6	126	86
Annual	1013.8	24.9	5.6	33.0		25.5	1,840	238

Month	Mean evap. (mm)	Number of days with			Mean cloud-iness (tenths)	Mean sun-shine (h)	Wind	
		precip.[2] ⩾1 mm	thunder-storm	fog			preval. direct.	mean speed (m/sec)
Jan.	132	21	0	0	4	237	ENE	6.3
Feb.	126	17	0	0	4	227	ENE	6.7
Mar.	153	18	0	0	4	257	ENE	6.8
Apr.	139	19	0	0	5	245	ENE	6.5
May	131	21	1	0	5	242	ENE	6.4
June	121	24	3	0	5	211	ENE	7.0
July	108	27	6	0	5	224	ENE	6.8
Aug.	106	25	6	0	5	256	ENE	5.6
Sept.	93	23	9	0	5	219	ENE	4.9
Oct.	95	23	7	0	5	223	ENE	5.0
Nov.	91	22	4	0	5	220	NE	5.1
Dec.	113	23	1	0	4	226	ENE	6.1
Annual	1,408	263	37	0	5	2,787	ENE	6.1

[1] Period of record 1947–60.
[2] Period of record 1931–60.

TABLE XLIII

CLIMATIC TABLE FOR GUATEMALA CITY (OBSERVATORIO) (1931–60)
Latitude 15°29′N, longitude 90°16′W, elevation 1,300 m

| Month | Mean sta. press. (mbar) | Temperature (°C) | | | | Mean precip. (mm) | Relat. humid. (%) | Number of days with precip. ≥0.25 mm |
| | | daily mean | mean daily range | extremes | | | | |
				max.	min.			
Jan.	854.4	16.3	11.1	30.0	5.0	3	70	4
Feb.	853.9	17.0	12.8	29.5	6.1	2	68	2
Mar.	853.2	18.4	13.3	30.0	5.0	7	69	3
Apr.	853.2	19.5	13.3	32.2	8.3	19	70	5
May	852.8	19.6	13.3	31.6	11.1	141	76	15
June	853.2	18.7	11.1	30.0	11.1	265	83	23
July	854.0	18.5	10.0	28.8	10.6	211	81	21
Aug.	853.7	18.7	10.6	28.3	11.1	187	79	21
Sept.	852.7	18.3	10.6	27.8	12.2	257	83	22
Oct.	853.0	17.7	8.9	27.8	10.0	159	82	18
Nov.	853.9	16.7	9.5	28.3	6.7	23	76	7
Dec.	854.2	16.3	9.5	28.3	5.0	7	73	4
Annual	853.5	18.0	11.2			1,281	76	145

TABLE XLIV

CLIMATIC TABLE FOR TELA (1951–60)
Latitude 15°43′N, longitude 87°29′W, elevation 3 m

Month	Mean sea level press. (mbar)	Temperature (°C)				Precipitation (mm)	
		daily mean	mean daily range	extremes		mean	max. in 24 h
				max.	min.		
Jan.	1015.5	23.2	8.3	27.4	19.1	255	129
Feb.	1014.1	23.7	7.9	28.2	19.3	146	128
Mar.	1012.1	25.1	9.3	29.7	20.4	72	75
Apr.	1011.4	26.2	9.0	30.7	21.7	90	254
May	1010.6	26.8	8.6	31.2	22.6	103	83
June	1010.4	27.1	8.7	31.6	22.9	122	101
July	1012.3	26.7	8.4	30.9	22.5	181	90
Aug.	1011.7	26.8	8.6	31.1	22.5	240	116
Sept.	1010.0	27.2	8.9	31.6	22.7	207	153
Oct.	1011.2	25.9	8.4	30.1	21.7	405	187
Nov.	1013.6	24.5	8.2	28.6	20.4	410	317
Dec.	1015.1	23.5	7.8	27.4	19.4	356	279
Annual	1012.3	25.6	8.5			2,587	317

Month	Number of days with		Wind	
	precip. ≥0.1 mm	thunder-storm	preval. direct.	mean speed (m/sec)
Jan.	15	0	NE	3.1
Feb.	11	0	NE	3.1
Mar.	6	1	NE	3.1
Apr.	5	2	NE	3.1
May	6	5	NNE	3.1
June	12	15	NNE	3.1
July	17	20	NE	3.1
Aug.	17	18	NNE	3.1
Sept.	15	18	NW	2.6
Oct.	19	9	SE	2.6
Nov.	15	3	NW	2.6
Dec.	18	0	NW	2.6
Annual	156	91	NE	

TABLE XLV

CLIMATIC TABLE FOR POINTE-À-PITRE (1951–60)
Latitude 16°15′N, longitude 61°31′W, elevation 7 m

Month	Mean sea level press. (mbar)	Temperature (°C)				Mean vap. press. (mbar)	Precipitation[1] (mm)	
		daily mean	mean daily range	extremes max.	min.		mean	max. in 24 h
Jan.	1015.2	23.4	9.3	29.9	13.5	23.1	91	51
Feb.	1015.4	23.5	9.3	30.0	13.0	22.6	66	58
Mar.	1015.1	24.0	9.7	31.3	13.9	23.1	66	60
Apr.	1014.5	26.0	8.9	31.8	15.7	24.5	86	46
May	1014.3	26.0	7.8	32.0	17.8	26.3	135	121
June	1015.5	26.8	7.5	32.6	18.9	27.4	148	115
July	1015.5	26.7	7.7	32.2	19.6	27.7	179	86
Aug.	1014.1	26.8	8.0	32.4	19.8	28.4	244	109
Sept.	1013.0	26.7	8.0	32.5	19.5	28.4	196	113
Oct.	1012.7	26.1	8.1	32.0	19.0	27.8	229	81
Nov.	1012.3	26.3	8.5	31.3	17.2	26.6	231	123
Dec.	1013.9	24.3	8.6	30.4	15.6	24.4	143	103
Annual	1014.3	25.5	8.5			25.9	1,814	123

Month	Mean evap. (mm)	Number of days with			Mean cloud-iness (tenths)	Mean sun-shine (h)	Wind	
		precip.[1] ≥1 mm	thunder-storm	fog			preval. direct.	mean speed (m/sec)
Jan.	104	19	0	0	4	223	E	1.9
Feb.	117	16	0	0	5	219	E	2.4
Mar.	141	16	0	0	5	257	E	2.5
Apr.	138	16	0	0	5	239	E	2.5
May	126	21	2	0	5	231	E	2.4
June	129	21	4	0	5	227	E	2.7
July	124	25	6	0	5	231	E	2.5
Aug.	107	23	7	0	5	232	E	2.0
Sept.	89	22	10	0	5	210	E	1.9
Oct.	88	22	8	0	5	220	E	1.7
Nov.	82	23	5	0	5	210	E	1.7
Dec.	90	25	1	0	5	220	E	2.0
Annual	1,335	249	43	0	5	2,719	E	2.2

[1] Period 1931–60.

TABLE XLVI

CLIMATIC TABLE FOR SWAN ISLAND (1931–60)
Latitude 17°24′N, longitude 83°56′W, elevation 9 m

Month	Mean sea level press.[1] (mbar)	Temperature (°C)				Mean relat. humid.[2] (%)	Precipitation (mm)	
		daily mean	mean daily range	extremes			mean	max. in 24 h
				max.	min.			
Jan.	1015.2	25.7	4.8	28.1	23.3	75	87	75
Feb.	1014.4	25.9	5.2	28.5	23.3	75	29	79
Mar.	1013.7	26.6	5.7	29.2	23.5	74	16	51
Apr.	1013.1	27.4	5.3	30.1	24.8	75	21	30
May	1012.3	27.9	5.5	30.7	25.2	77	84	114
June	1012.3	28.0	4.9	30.5	25.6	82	164	128
July	1013.6	27.9	5.4	30.6	25.2	80	107	210
Aug.	1012.5	28.3	5.6	31.1	25.5	79	93	192
Sept.	1011.0	28.3	5.7	31.2	25.5	80	132	110
Oct.	1011.1	27.7	5.5	30.4	24.9	79	246	272
Nov.	1012.7	26.6	5.2	29.2	24.0	77	188	229
Dec.	1014.1	25.8	4.8	28.2	23.4	75	143	104
Annual	1013.0	27.2	5.3			77	1,311	272

Month	Number of days with			Mean cloud- iness (tenths)	Wind	
	precip. ⩾0.1 mm	thunder- storm	fog		preval. direct.	mean speed (m/sec)
Jan.	15	0.1	0	5.3	E	4.7
Feb.	8	0.1	0	4.5	E	4.7
Mar.	6	0	0	4.1	SE	4.8
Apr.	5	0.1	0	5.1	SE	4.7
May	9	1.9	0	6.2	E	4.5
June	14	6.0	0	7.9	E	4.7
July	13	9.0	0	6.5	E	4.2
Aug.	13	6.3	0	5.9	E	3.8
Sept.	14	7.7	0	6.9	SE	3.6
Oct.	18	5.6	0	6.8	NE	3.9
Nov.	16	2.5	0	6.9	NE	4.4
Dec.	16	1.1	0	5.5	E	4.4
Annual	147	40.4	0	6.0	E	4.4

[1] Period of record 1941–60.
[2] Period of record 1951–60.

TABLE XLVII

CLIMATIC TABLE FOR BELIZE (1951–60)
Latitude 17°31′N, longitude 88°11′W, elevation 5 m

Month	Mean sea level press. (mbar)	Temperature (°C)				Mean vap. press. (mbar)	Precipitation (mm)	
		daily mean	mean daily range	extremes			mean	max. in 24 h
				max.	min.			
Jan.	1016	23.3	7.8	32.1	9.4	26.1	137	107
Feb.	1015	23.8	7.2	33.8	9.4	26.7	61	74
Mar.	1014	25.0	7.2	34.9	12.2	28.8	38	66
Apr.	1013	26.6	6.7	36.0	15.0	30.4	56	79
May	1012	27.2	6.7	35.5	15.5	31.9	109	150
June	1011	27.2	6.7	36.0	17.7	33.1	198	158
July	1013	27.2	6.7	34.9	16.7	32.3	163	132
Aug.	1012	27.2	7.2	35.5	15.5	32.7	170	157
Sept.	1011	27.2	7.2	36.0	15.5	31.9	244	209
Oct.	1011	25.6	7.8	35.5	14.4	29.6	305	130
Nov.	1014	23.8	8.3	34.9	11.1	28.3	226	249
Dec.	1016	23.3	7.2	33.2	9.4	26.7	186	122
Annual	1013	25.6	7.2			29.9	1,893	249

Month	Number of days with		Preval. wind direct.	Mean cloudiness (tenths)	
	precip. ≥0.1 mm	fog		07h00	17h00
Jan.	12	0.1	NW	7.3	6.1
Feb.	6	0	ESE	6.8	5.4
Mar.	4	0.1	ESE	6.5	5.0
Apr.	5	0	ESE	7.3	5.6
May	7	0	E	7.6	7.2
June	13	0	E	7.8	7.6
July	15	0	E	7.5	7.5
Aug.	14	0.1	E	7.5	7.1
Sept.	15	0	ESE	7.8	7.9
Oct.	16	0	NW	7.7	7.3
Nov.	12	0.1	NW	7.5	7.3
Dec.	14	0.4	NW	7.5	7.1
Annual	133	0.8		7.4	6.8

TABLE XLVIII

CLIMATIC TABLE FOR KINGSTON (1951–60)
Latitude 17°58′N, longitude 76°48′W, elevation 8 m

Month	Mean sea level press. (mbar)	Temperature (°C)				Mean vap. press.* (mbar)	Precipitation (mm)	
		daily mean	mean daily range	extremes			mean	max. in 24 h
				max.	min.			
Jan.	1015.2	25.4	10.5	33.9	13.9	23.8	20	46
Feb.	1014.9	25.4	10.5	33.2	15.0	24.0	18	46
Mar.	1014.4	25.9	10.0	33.9	14.4	24.6	10	89
Apr.	1013.6	26.7	9.4	33.9	17.2	26.5	37	46
May	1012.8	27.6	8.3	34.4	18.9	28.7	138	235
June	1013.5	28.0	8.3	35.0	20.0	29.1	114	238
July	1014.3	28.3	9.4	35.5	18.9	29.2	51	111
Aug.	1013.3	28.5	9.4	36.1	20.0	29.9	92	124
Sept.	1012.0	28.2	8.9	35.5	20.0	30.0	86	162
Oct.	1011.8	27.6	8.3	35.5	18.3	29.1	168	280
Nov.	1012.7	27.1	8.9	35.5	16.3	27.3	52	226
Dec.	1014.1	26.1	10.0	35.5	13.9	25.6	25	122
Annual	1013.5	27.1	9.3			27.3	811	280

Month	Number of days with			Mean cloudiness (tenths)		Wind	
	precip. ≥1 mm	thunder-storm	fog	07h	15h	preval. direct.	mean speed (m/sec)
Jan.	3	1	0	3.3	5.2	N, SE	3.0
Feb.	3	1	0	2.8	5.6	N, SE	3.3
Mar.	3	2	0	2.8	5.7	SE	3.5
Apr.	4	1	0	3.0	6.5	SE	3.6
May	6	7	0	4.3	7.0	SE	3.8
June	4	5	0	5.1	7.2	SE	4.4
July	3	14	0	4.2	6.6	SE	4.2
Aug.	7	22	0	4.1	7.4	SE	3.7
Sept.	7	25	0	4.8	7.8	SE	3.0
Oct.	9	24	0	4.8	7.9	SE	2.9
Nov.	6	11	0	4.6	6.8	N	2.7
Dec.	2	2	0	3.4	5.3	N	2.7
Annual	57	115	0	3.9	6.6	SE	3.4

* Mean vapor pressure measured at Palisadoes, 17°56′N 76°47′W.

TABLE XLIX

CLIMATIC TABLE FOR SAN JUAN, P.R. (1931–60)
Latitude 18°28'N, longitude 66°06'W, elevation 15 m

Month	Mean sta. press. (mbar)	Temperature (°C)				Mean vap. press.[1] (mbar)	Precipitation (mm)	
		daily mean	mean daily range	extremes			mean	max. in 24 h
				max.	min.			
Jan.	1014.2	23.9	5.3	32.2	17.2	24.2	105	115
Feb.	1014.2	23.9	5.5	32.8	16.7	23.7	69	123
Mar.	1013.6	24.4	5.6	35.6	17.2	24.0	53	134
Apr.	1012.9	25.0	5.4	33.9	18.3	25.2	99	171
May	1012.5	26.0	5.2	34.4	18.9	27.4	182	136
June	1013.6	26.6	5.0	33.9	18.9	28.6	148	164
July	1013.9	26.7	4.6	33.3	21.1	29.1	153	103
Aug.	1012.5	27.1	5.1	34.4	20.0	29.5	161	251
Sept.	1011.2	27.1	5.7	34.4	20.1	29.5	153	216
Oct.	1010.2	26.8	5.9	35.6	20.0	28.8	133	98
Nov.	1010.3	25.9	5.4	33.9	18.3	27.5	154	205
Dec.	1012.5	24.8	5.2	32.2	16.7	25.5	124	268
Annual	1012.6	25.7	5.3			26.9	1,534	268

Month	Mean evap.[1] (mm)	Number of days with			Mean cloud-iness (tenths)	Mean sun-shine (h)	Wind	
		precip. ⩾1 mm	thunder-storm	fog			preval. direct.	mean speed (m/sec)
Jan.	150	16	<0.5	0	5.2	217	E	3.9
Feb.	160	10	<0.5	0	5.1	222	E	4.0
Mar.	202	10	<0.5	0	5.0	255	E	4.1
Apr.	185	10	2	0	6.0	241	E	3.8
May	186	12	5	0	6.9	239	E	3.6
June	184	13	6	0	7.1	240	E	3.8
July	202	15	7	0	6.2	259	E	4.2
Aug.	198	16	7	0	6.2	266	E	3.9
Sept.	160	14	10	0	6.6	227	E	3.3
Oct.	150	15	8	0	6.7	233	SE	3.0
Nov.	137	16	3	0	5.9	210	SE	3.2
Dec.	146	16	1	0	5.7	218	ESE	3.5
Annual	2,060	163	49	0	6.1	2,827	E	3.7

[1] Period of record 1951–60.

TABLE L

CLIMATIC TABLE FOR SANTO DOMINGO (1951–60)
Latitude 18°29′N, longitude 69°54′W, elevation 19 m

Month	Mean sea level press. (mbar)	Temperature (°C)				Precipitation (mm)		Relat. humid. (%)	
		daily mean	mean daily range	extremes		mean	max. in 24 h	06h00	11h30
				max.	min.				
Jan.	1016.9	23.9	10.0	32.7	15.0	47	60	91	64
Feb.	1016.9	24.0	10.5	31.1	15.5	45	37	88	58
Mar.	1015.9	24.5	9.4	32.1	15.5	45	74	90	60
Apr.	1015.5	25.3	8.9	32.1	16.3	65	225	90	62
May	1014.7	25.9	8.3	31.6	18.3	190	124	89	65
June	1016.1	26.5	8.9	32.1	19.4	175	137	90	66
July	1016.6	26.7	8.9	32.1	20.0	158	145	90	66
Aug.	1015.0	27.0	8.3	32.1	17.8	147	101	90	66
Sept.	1013.8	26.8	8.9	32.7	20.0	168	508	91	66
Oct.	1013.0	26.4	8.3	33.2	18.9	165	117	92	66
Nov.	1013.2	26.9	8.9	33.2	16.1	113	155	92	66
Dec.	1015.4	24.6	10.0	32.7	16.3	67	37	91	66
Annual	1015.0	25.6	9.1			1,386	508	90.3	64.7

Month	Number of days with		Mean cloud- iness[1] (tenths)	Wind	
	precip. ⩾1 mm	fog		preval. direct.	mean speed (m/sec)
Jan.	7	0	4.4	N	3.9
Feb.	6	0	4.5	N	3.7
Mar.	5	0	4.3	N	4.0
Apr.	7	0	5.1	N	3.9
May	11	0	5.2	N	3.7
June	12	0	5.0	N	3.8
July	11	0	4.6	N	3.9
Aug.	11	0	4.8	N	3.6
Sept.	11	0	4.5	N	3.5
Oct.	11	0.1	4.4	N	3.7
Nov.	10	0	4.3	N	4.2
Dec.	8	0	4.7	N	4.3
Annual	110	0.1	4.6	N	3.9

[1] At 06h00, 12h00 and 19h00.

TABLE LI

CLIMATIC TABLE FOR PUNTA MAISÍ

Latitude 20°16′N, longitude 74°09′W, elevation: coast

Month	Mean sea level press. (mbar)	Temperature (°C)				Mean precip. (mm)	Wind	
		daily mean	mean daily range	extremes			preval. direct.	mean speed (m/sec)
				max.	min.			
Jan.	1017.1	24.1	6.1	29.3	16.6	52	NE	9.3
Feb.	1016.6	24.2	6.2	29.7	17.2	34	ENE	10.3
Mar.	1015.8	25.0	6.6	31.4	17.5	42	ENE	10.3
Apr.	1015.4	25.9	6.6	32.2	19.2	33	ENE	10.3
May	1014.5	26.7	6.2	32.7	20.1	75	ENE	7.7
June	1015.0	27.7	6.9	33.1	21.1	51	E	7.7
July	1016.0	28.2	6.6	32.9	22.3	15	E	9.8
Aug.	1014.2	28.6	6.7	33.9	23.8	44	E	8.2
Sept.	1013.2	27.8	7.2	33.5	21.2	59	E	8.2
Oct.	1012.5	27.1	7.1	32.8	19.7	198	NE	6.2
Nov.	1014.0	25.8	6.5	31.3	18.9	112	NE	8.7
Dec.	1015.6	24.8	6.3	29.8	17.5	69	NE	8.7
Annual	1015.2	26.3	6.6			783	ENE	8.7

TABLE LII

CLIMATIC TABLE FOR CAMAGÜEY (1951–60)
Latitude 21°24′N, longitude 77°55′W, elevation 127 m

Month	Mean sea level press. (mbar)	Temperature (°C)					Precipitation (mm)	
		daily mean	mean daily range	extremes			mean	max. in 24 h
				max.	min.			
Jan.	1017.0	22.0	12.9	32.4	10.4		30	50
Feb.	1016.4	23.0	12.7	33.2	11.2		37	66
Mar.	1015.3	24.5	14.1	34.6	12.1		55	300
Apr.	1015.0	25.3	13.4	35.2	14.5		86	112
May	1014.5	26.1	12.2	34.6	16.1		180	149
June	1014.8	26.7	11.1	34.2	19.9		230	119
July	1015.8	27.3	12.0	34.8	19.2		134	102
Aug.	1014.4	27.6	12.2	35.4	19.1		170	107
Sept.	1012.8	26.9	11.2	34.4	19.2		187	100
Oct.	1012.2	25.9	10.6	33.7	17.5		170	105
Nov.	1014.4	24.0	11.0	33.7	14.2		69	188
Dec.	1017.0	22.6	11.4	31.4	11.2		37	86
Annual	1015.0	25.1	12.1				1,385	300

Month	Number of days with		Mean cloud-iness (tenths)	Wind	
	precip. ⩾0.1 mm	fog		preval. direct.	mean speed (m(sec)
Jan.	5	1.0	3.7	NE	3.6
Feb.	9	0.3	3.4	ENE	3.6
Mar.	5	0.6	3.1	ENE	3.9
Apr.	6	0.2	4.0	ENE	4.2
May	12	1.5	4.6	ENE	3.3
June	15	0.7	4.7	E	3.1
July	11	0	4.2	ENE	3.6
Aug.	11	0.3	3.2	ENE	3.1
Sept.	13	1.2	3.8	E	3.1
Oct.	12	1.0	3.8	ENE	3.1
Nov.	9	0	4.5	NE	3.6
Dec.	5	0.3	3.6	NE	3.9
Annual	113	7.1	3.9	ENE	3.5

TABLE LIII

CLIMATIC TABLE FOR PASO REAL (1951–60)
Latitude 22°35′N, longitude 83°20′W, elevation 41 m

Month	Mean sea level press. (mbar)	Temperature (°C)				Mean precip. (mm)	Wind	
		daily mean	mean daily range	extremes			preval. direct.	mean speed (m/sec)
				max.	min.			
Jan.	1018.3	21.3	11.1	30.0	9.1	50	NE	2.5
Feb.	1016 8	22 4	11.3	30.7	11.1	43	ENE	2.5
Mar.	1015.6	23.4	11.7	32.2	11.1	52	ENE	3.1
Apr.	1015.7	24.9	11.3	33.3	14.5	78	ENE	3.1
May	1014.6	26.1	10.7	33.9	16.8	205	ENE	2.2
June	1014.4	27.0	9.6	34.3	19.8	237	ENE	1.9
July	1015.6	27.3	10.0	34.1	20.3	187	NE	2.5
Aug.	1014.8	27.7	10.0	34.8	20.5	219	NE	1.9
Sept.	1013.7	27.0	9.4	33.9	20.5	212	NE	2.2
Oct.	1014.1	25.7	10.5	33.0	16.7	183	NE	2.5
Nov.	1015.4	23.7	10.6	31.5	12.6	40	NE	2.5
Dec.	1017.6	21.7	10.9	30.2	9.5	17	NE	2.5
Annual	1015.3	24.9	10.7			1,523	NE	2.5

TABLE LIV

CLIMATIC TABLE FOR NASSAU (INTERNAT. AIRPORT) (1951–60)
Latitude 25°03′N, longitude 77°28′W, elevation 3 m

Month	Mean sea level press. (mbar)	Temperature (°C)				Mean vap. press. (mbar)	Precipitation (mm)		Mean relat. humid. (%)
		daily mean[1]	mean daily range	extremes[2]			mean[4]	max. in 24 h	
				max.	min.				
Jan.	1019.7	20.3	7.4	27.9	10.7	19.2	36	19	71
Feb.	1018.7	20.9	7.9	28.7	11.6	19.4	43	48	80
Mar.	1017.7	22.2	7.9	29.2	12.1	20.8	45	30	74
Apr.	1017.1	23.6	7.4	30.2	15.2	22.8	78	131	77
May	1016.1	25.2	7.5	31.2	17.2	25.1	117	78	78
June	1016.2	26.7	7.3	32.4	20.2	27.7	159	124	79
July	1018.1	27.4	7.2	32.6	21.4	27.9	150	61	82
Aug.	1016.2	27.7	7.5	33.3	21.7	29.2	135	58	79
Sept.	1014.6	27.1	7.2	32.5	21.4	28.8	165	58	82
Oct.	1014.3	25.5	6.8	31.7	17.9	26.2	164	64	80
Nov.	1017.0	23.5	7.3	30.0	14.9	22.9	85	40	80
Dec.	1019.0	21.6	7.3	28.4	12.7	20.4	39	106	79
Annual	1017.1	24.3	7.4	34.7[3]	8.1[3]	24.2	1,216	131	78

Month	Number of days with			Mean cloud-iness (tenths)	Mean sun-shine (h)	Preval. wind direct.
	precip. ⩾1 mm	thunder-storm	fog			
Jan.	6	0.5	0.4	3.7	234	NE
Feb.	6	0.6	0.4	3.7	229	NE
Mar.	5	1.3	0.5	3.7	263	NE
Apr.	6	2.4	0.2	4.6	259	NE
May	11	5.7	0.1	4.5	264	SE
June	14	9.7	0.1	5.2	234	SE
July	16	12.4	0	4.9	271	SE
Aug.	16	14.5	0	5.0	256	SE
Sept.	17	10.2	0.1	5.1	215	NE
Oct.	15	4.9	0	4.9	210	NE
Nov.	9	1.6	0.3	3.9	226	NE
Dec.	9	0.6	0.8	3.7	218	NE
Annual	130	64.4	2.9	4.4	2,879	NE

[1] From 8 observations daily.
[2] Mean of monthly extremes of 15 years.
[3] Absolute extreme.
[4] Period 1931–60.

TABLE LV*

RADIATION DATA (ly/day) FOR FT. SHERMAN, CANAL ZONE, 9°22′N 79°57′W (4.5 km WNW of Colon)

	Jan.	Feb.	Mar.	Apr.	May	June	July	Aug.	Sep.	Oct.	Nov.	Dec.	Year
Mean global radiation													
A. 1958–1974	461	504	510	504	414	364	351	365	381	382	340	391	414
B. 1958–1964	487	512	529	507	418	361	346	377	381	399	350	410	423
C. 1968–1974	438	484	500	495	405	356	341	355	369	364	337	373	401
C−B	−49	−28	−29	−12	−13	−4	−5	−22	−12	−35	−13	−37	−22
(C−B)/A × 100	−11	−6	−6	−2	−3	−1	−1	−6	−3	−10	−4	−10	−5
Mean net radiation													
A. 1964–1974	209	259	272	260	192	151	154	155	171	150	112	168	188
B. 1964–1968	232	272	290	272	199	163	154	151	176	162	103	173	196
C. 1970–1974	191	253	260	249	190	139	162	161	171	133	120	158	182
C−B	−41	−19	−30	−23	−9	−24	+8	+10	−5	−29	+17	−15	−13
(C−B)/A × 100	−20	−7	−11	−9	−5	−16	+5	+6	−3	−19	+15	−9	−7

* Added in proof, supplement to Table XXXIV.

Instruments. Eppley pyranometer, 0.4–20 μm, entire sky; Belfort net radiometer, all wavelengths, one horizontal plate looking up, the other looking down into grass.

Note. The change shown in the second to fifth lines of either part of the table is not reflected in the duration of sunshine. It is, however, confirmed by other instruments (global radiation), some located at other places of the Canal Zone. Hence, it is suggested that there was an increase of turbidity of the atmosphere. This change was not evenly spread over the total period; it was rather irregular, but with the same irregularities at both coasts, Atlantic and Pacific.

Chapter 8

Climatology of Atlantic Tropical Storms and Hurricanes

M. A. ALAKA[1]

Classification of disturbances in the tropics

Corresponding to the typhoons of the western Pacific Ocean, the most severe tropical storms of the Atlantic Ocean[2] are the hurricanes. These violent storms originate from weaker circulations and pass through several increasingly intense phases before they ultimately attain hurricane status. To make the present study meaningful we shall adopt the following terminology most commonly used in the Atlantic area, but unfortunately not universally adopted.

The weakest stage is the *tropical disturbance* in which rotary circulation is slight or absent at the surface, but possibly better developed aloft. There are no closed surface isobars and no strong winds. Disturbances of this type are common in the tropics. Next in intensity is the *tropical depression* which is associated with one or more closed surface isobars and with sustained winds equal to or less than Beaufort force 7 (32–38 miles/h). When the winds are of Beaufort force 8–11 (39–72 miles/h) we have a *tropical storm*[3]. Finally, the strongest stage, with winds of Beaufort force 12 ($\geqq$ 73 miles/h), is the *hurricane*.

Sources of information

Although the first recorded account of Atlantic hurricanes dates back to the time of Columbus, very little has been written about storm occurrences, in the centuries immediately following, to provide a basis for meaningful climatological studies of these storms. According to TANNEHILL (1938), who summarized known records of early West Indian hurricanes, only sixteen hurricanes were listed by Poëy as having occurred in the 16th century. Of these, two were probably not authentic tropical cyclones (GENTRY, 1963). Judging from present-day frequency, the number of storms was more nearly 800.

---

[1] *Editor's note*: Dr. Alaka's original manuscript was submitted in July, 1965, but its publication had to be delayed by circumstances beyond his and the editor's control. We are grateful to Mr. R. M. DeAngelis (Marine Climatology Branch, National Oceanographic Data Center) who in 1973 provided the information necessary to update tables and figures.

[2] For the purpose of this study the Atlantic Ocean is considered to include the Caribbean Sea and the Gulf of Mexico.

[3] Frequently the terms "tropical storms" and "tropical cyclones" have been used interchangeably in the literature. For the purpose of this article the term "tropical cyclones" is used to denote tropical disturbances of all intensities while "tropical storms" comprise only those with winds between 39 and 72 miles/h.

Toward the last quarter of the 19th century, records of tropical cyclone occurrences became sufficiently complete so that a systematic climatology of hurricanes began to be established. The records have been summarized, among others, by MITCHELL (1924), COLÓN (1953), and DUNN (1956). One of the most complete summaries for the period 1886–1958, was prepared in the Office of Climatology of the U.S. Weather Bureau by CRY et al. (1959) who scrutinized and combined available records including earlier works by GARRIOTT (1900), FASSIG (1913), MITCHELL (1924, 1932), CLINE (1926) and TANNE-HILL (1956). This work was recently brought up to date by CRY (1965).

The above works, in addition to many others to which reference is made in the text, represent the sources on which the material presented in this study is based.

Hurricane formation

The birthplace of Atlantic hurricanes

Gradually, with increased traffic over the tropical Atlantic and, more recently, as a result of systematic air reconnaissance and the advent of meteorological satellites, the number of undetected tropical storms in the Atlantic has become vanishingly small and the positioning of these storms has become more and more accurate. It is a fact, nevertheless, that long-period records on which any climatological studies can be based, perforce suffer from a lack of homogeneity and deteriorating reliability as we go backward in time.

But aside from the uneven quality of the data, the exact locality where a hurricane originated is not always easy to determine. As mentioned earlier, hurricanes develop in pre-existing disturbances, waves or shearlines. Sometimes the intensification is rapid; at other times it is slow. It is not uncommon for an easterly wave to travel several thousand miles before it eventually develops into a warm-core vortex of hurricane intensity. DUNN and MILLER (1964) ask: "Where would it be said that the hurricane formed? When the initial disturbance began to intensify, when the tropical storm reached hurricane intensity, or perhaps at some other point in its life history?"

The literature contains many conflicting statements about the birthplace of even such important and recent storms as hurricane Connie of 1955. NAMIAS and DUNN (1955) contend that this storm, though first reported on August 4 at about 16.6°N 48.0°W, actually developed earlier off North Africa. Similarly the birthplace of the September 1938 storm has been variously located by different authors at longitudes 68°W, 50°W, 37°W and 5°W (HAGGARD, 1958).

COLÓN (1953) used the beginning of published tracks as an indication of storm formation. He pointed out that the beginning of the track was not necessarily the point of formation, since the initial disturbance usually existed for some time prior to intensification. Moreover, in the earlier years of the record disturbances of hurricane intensity east of the Lesser Antilles could well remain undetected for several days. The results of a similar, more recent and comprehensive analysis by HOPE and NEUMANN (1971) are shown in Fig.1. To quote the authors: "The count of all tropical storms or hurricanes passing through each $2\frac{1}{2}°$ latitude-longitude box has been determined, analyzed by computer contouring routine, and drawn by an electric plotter."

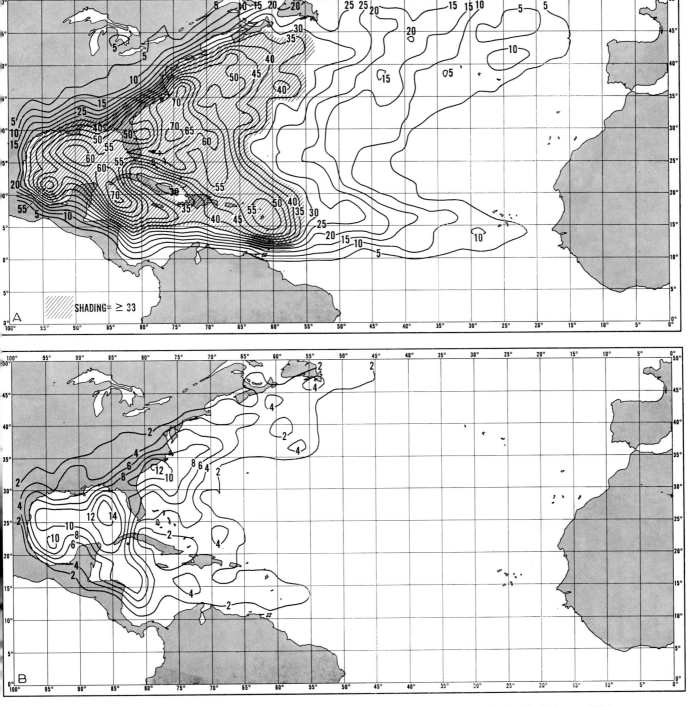

Fig.1. Number of tropical storms and hurricanes passing through $2\frac{1}{2}°$ latitude–longitude boxes, 1886–1968. A. Entire season, 1 May–30 November; B. 1 May–15 July; C. 16 July–20 September; D. 21 September–30 November.

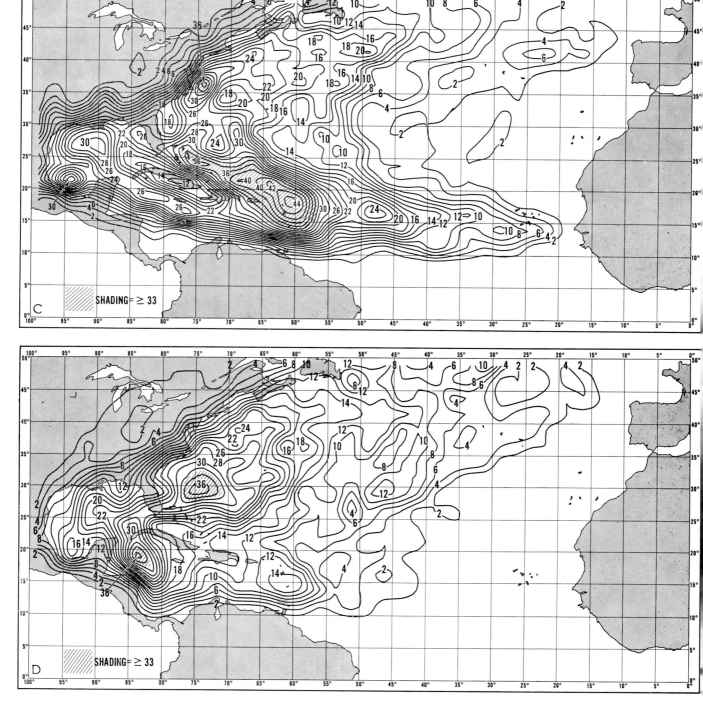

Fig.1. C, D (legend on p. 479)

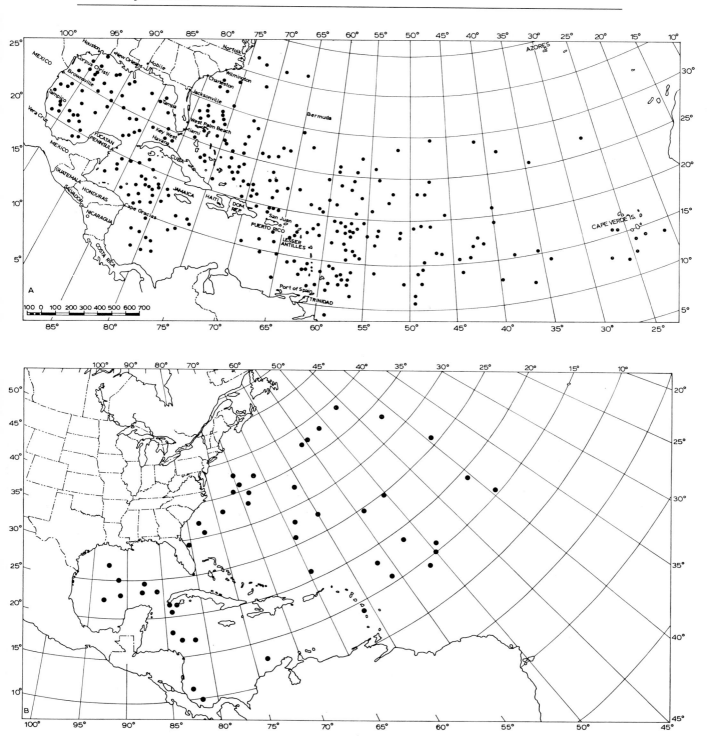

Fig.2. Points at which Atlantic tropical storms reached hurricane intensity. A. 1901–1963; B. 1964–1972.

DUNN (1956) devised a different procedure to determine the place of formation of Atlantic hurricanes. He plotted the points at which each storm reached hurricane intensity, using only those storms during the period 1901–1955 for which he was reasonably confident that this point could be established accurately. The result of this procedure, extended to include the years 1956–1972, is shown in Fig.2 which we shall discuss in terms of the following climatological factors which are relevant to hurricane formation.

Availability of latent energy

The formation of hurricanes cannot be explained directly in terms of a simple buoyant convective process. Indeed it has been established that the horizontal scale of circulations produced by thermal buoyancy is different from the observed scale of tropical cyclones (YANAI, 1964). Nevertheless the availability of latent energy plays a fundamental role in hurricane formation since it indubitably constitutes the principal energy source of these storms. Thus, land areas are not only immediately excluded as likely sites for hurricane development but are indeed known to be effective graveyards for these storms.

Nor do the tropical oceans always endow the atmosphere above them with large quantities of latent energy available for release. When air is lifted beyond its condensation level latent heat is released so that the potential temperature is increased while the equivalent potential temperature remains constant. If the resulting temperature of the rising air particle is higher than that of the surrounding air, it is accelerated upward by its buoyancy. If we neglect friction we may write the expression for this acceleration as:

$$\frac{\mathrm{d}w}{\mathrm{d}t} = g\frac{T'-T}{T} \tag{1}$$

where w denotes the upward motion, T' the temperature of the rising air and T the temperature of the ambient atmosphere. Eq.1, integrated over the path of the rising air particle from sea level to the top of the atmosphere, gives the total latent energy (E_1) which may be converted into the kinetic energy of the rising particle. Thus:

$$E_1 = \tfrac{1}{2}w^2 = g\int_0^z \frac{T'-T}{T}\,\mathrm{d}z = -R\int_{p_0}^p (T'-T)\,\frac{\mathrm{d}p}{p} \tag{2}$$

PALMÉN (1948) has found a remarkable variation, both seasonal and areal, in the availability of latent energy over the Atlantic. Fig.3 shows the large seasonal range at Swan Island (17°N 84°W), in the difference of temperature between a parcel of ascending air and the ambient air at various isobaric levels.

Assuming that the surface air has the same temperature as the ocean and a relative humidity of 85%, Palmén considered as a latent instability index the difference between the mean 300 mbar temperature and the temperature of an air parcel lifted adiabatically to its condensation level and then pseudo-adiabatically to 300 mbar. Palmén's results are shown in Fig.4.

Fig.4A, for February, shows only a very weak instability over the West Indies and surrounding seas while Fig.4B, for September, reveals a large area where the rising air

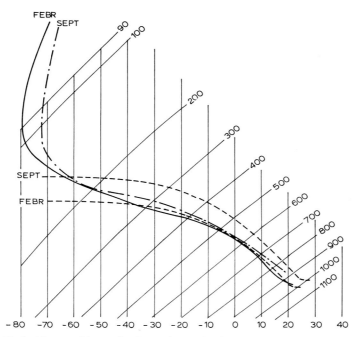

Fig.3. Mean tephigram for September and February at Swan Island (17°N 84°W) showing the difference between an ascending air mass and the ambient air at different isobaric levels.

particle at 300 mbar has a temperature in excess of the surrounding atmosphere. The figures may explain why Atlantic hurricanes are exceedingly rare during the cold season. Fig.4B, which shows the latent instability decreasing eastward, may also explain the lower tendency for hurricanes to form in the eastern Atlantic.

The above considerations led Palmén to suggest that hurricanes can form only in regions where the mean surface water temperature is above 26 or 27°C. Fig.5 delineates the area in February and September where the water temperature in the North Atlantic exceeds 27°C. Comparison of the size of this area during these two months again explains why winter hurricanes are unlikely in the Atlantic. In February, 27°C water is confined to an area within 10°N of the Equator, while in September its northern edge slopes from about 15°N in the eastern Atlantic to about 34°N in the western Atlantic. The boundary runs almost north–south in mid-Atlantic, thus marking the western half as being more propitious for hurricane development than the eastern half, as is clearly depicted by Fig.1 and 2. Fig.6, adapted from an earlier figure published by PALMÉN (1948), shows that water temperatures in the warmest season are appreciably lower south than north of the Equator. This may explain why no storms of hurricane intensity have been known to occur in the South Atlantic.

Vorticity due to earth's rotation

The change of relative vorticity (ζ) along an air trajectory is given with sufficient approximation by the following simplified form of the vorticity equation:

$$\frac{\mathrm{d}\zeta}{\mathrm{d}t} = -(\zeta + 2\omega \sin \varphi)\, \mathrm{div_H}\, V \tag{3}$$

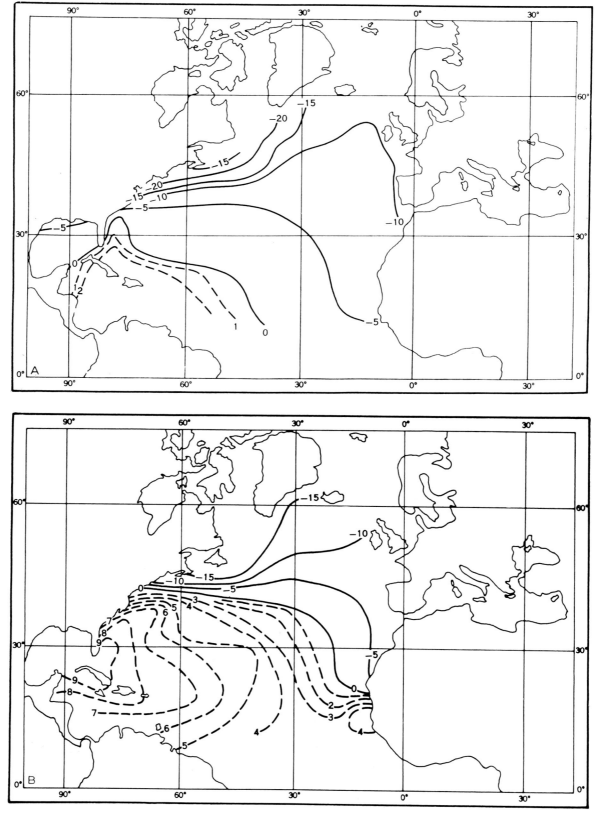

Fig.4. Difference in temperature (°C) between air lifted adiabatically from the sea surface to the 300-mbar level and the air at the same level over the North Atlantic Ocean. A. February. B. September. (After PALMÉN, 1948.)

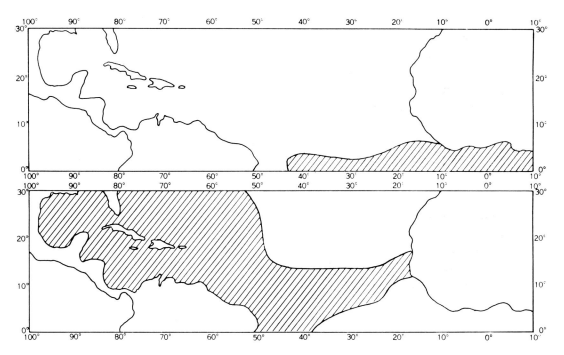

Fig.5. Delineation of the areas in the North Atlantic where the mean surface water temperature exceeds 27°C: in February (top) and September (bottom).

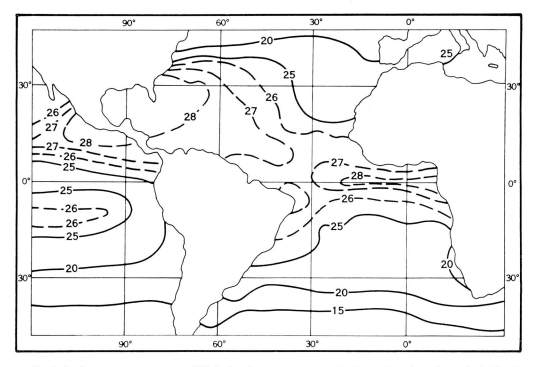

Fig.6. Surface water temperature (°C) during the warmest season in the north and south tropical Atlantic Ocean. (Adapted from PALMÉN, 1948.)

487

where ω is the earth's rotation, φ the latitude and $\mathrm{div_H}\ V$ the horizontal wind divergence. PALMÉN (1948) has argued that since a hurricane represents an area of high cyclonic vorticity produced by the convergence of air and, according to eq.3, the effectiveness of the convergence in increasing the vorticity varies directly with the initial total vorticity $(\zeta + 2\omega\sin\varphi)$, and since moreover the initial relative vorticity in the tropics is not usually very great, then effective cyclogenesis must depend on a comparatively large value of the Coriolis term.

According to this reasoning, hurricanes cannot occur very close to the Equator—a conclusion which is amply confirmed by Fig.2 where no hurricanes are reported to have occurred Equator-ward of 9°N. This observation should not be confused with the fact that hurricanes are known to form in disturbances which originate in the equatorial zone. Indeed out of the 242 hurricanes considered by DUNN (1956) approximately 40% developed in perturbations moving away from the Intertropical Convergence Zone (I.T.C.Z.); but almost all cases of development occurred when the perturbation was a considerable distance from this zone, well in the trade-wind current. A majority of the hurricanes developed in easterly waves.

The upper-air flow

RIEHL (1948) was among the first to emphasize the importance for hurricane development of mass divergence in the upper troposphere associated with migratory upper anticyclonic vortices. This feature places the poleward limit of the tropical belt which is propitious for hurricane development well Equator-ward of the subtropical jet stream. Fig.2 indeed shows that very few tropical disturbances developed north of 30° of latitude and these are confined to the western Atlantic.

The low-level flow

Wind constancy (q) is defined as ($V_\mathrm{r}/V_\mathrm{s}$) $\times$ 100, where V_r is the magnitude of the resultant mean wind vector and V_s is the scalar mean wind speed. If the wind always blew from one direction the value of (q) would be 100%. The body of the trades is characterized by high values of (q) while the doldrums have minimum values. CRY (1965) has plotted cases of tropical storm development on maps with isolines of wind constancy. Almost all cases of development occurred where the climatological wind constancy was higher than 80%.

Fig.2 shows a concentration of cases of development just to the east of Swan Island and another just east of the Leeward and Windward Islands. A relative void is indicated between Hispaniola and Venezuela, in agreement with the observation by MITCHELL (1924) that "in every instance, the first evidence of storm development, although rather obscure in some cases, was found either over the western third of the Caribbean Sea (west of longitude 78°W) or to the east of the eastern limits of the Caribbean Sea." DUNN and MILLER (1964) ascribe the comparatively infrequent development of tropical cyclones in this area to the significant divergence in the low-level flow as the easterly trade winds are diverted into the semi-permanent low over the Amazon valley. These authors, however, note that two recent severe hurricanes, Hazel of 1954 and Janet of 1955, attained great intensity in this area.

488

The total number of tropical storms originating in different areas of the Atlantic during the period 1901–1972 are given in the upper portion of Table I. Comparable figures for locations where tropical cyclones first attained hurricane intensity are given in the lower section of the table.

TABLE I

NORTH ATLANTIC TROPICAL CYCLONE AND HURRICANE ORIGINS BY AREAS, 1901–1972

	East of 50°W	50°–60°W	60°–70°W	70°–80°W	80°–90°W	90°–100°W	Total
Tropical cyclones with origins south of 20°N:							
Total	98	93	28	50	71	18	358
Percent	27.4	26.0	7.8	14.0	19.8	5.0	
Tropical cyclones with origins north of 20°N:							
Total	16	30	36	63	45	32	222
Percent	7.2	13.5	16.2	28.4	20,3	14.4	
Hurricanes developing south of 20°N:							
Total	31	40	27	17	29	1	145
Percent	21.4	27.6	18.6	11.7	20.0	0.7	
Hurricanes developing north of 20°N:							
Total	13	26	37	51	31	34	192
Percent	6.8	13.5	19.3	26.6	16.1	17.7	

Frequency of Atlantic hurricanes

Seasonal variation

Atlantic hurricanes occur mostly during the six months from June through to November. These months are usually considered to constitute the hurricane season. According to Table II, which shows the percentage frequency of all hurricanes and tropical storms occurring in the North Atlantic during each month of the year in the period 1886–1972, only 1.4% occurred out of season.

TABLE II

PERCENTAGE MONTHLY FREQUENCY OF ATLANTIC TROPICAL STORMS AND HURRICANES FROM 1881–1972

	Feb.	Mar.	Apr.	May	June	July	Aug.	Sept.	Oct.	Nov.	Dec.
Hurricanes	0	1	0	3	21	30	117	153	76	14	2
Percent	0	0.2	0	0.7	5.0	7.2	28.1	36.7	18.2	3.4	0.5
Storms	1	0	0	8	26	24	45	85	80	18	2
Percent	0.3	0	0	2.8	9.0	8.3	15.6	29.4	27.7	6.2	0.7
Total	1	1	0	11	47	54	162	238	156	32	4
Percent	0.1	0.1	0	1.6	6.7	7.6	22.9	33.7	22.1	4.5	0.6

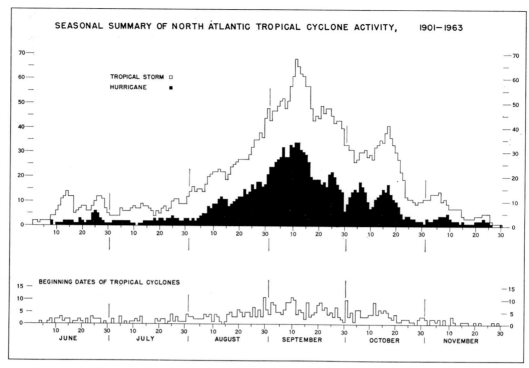

Fig.7. Total daily frequency of occurrence of North Atlantic tropical cyclones and dates of beginning of tropical cyclone, 1901–1963. Shading refers to frequency of hurricanes.

Within the season, storm activity is far from uniform as can readily be seen from Fig.7 (CRY, 1965) which shows the total number of tropical storms and hurricanes which were in existence on each calendar day in the months June–November during the period 1901–

TABLE III

OBSERVED FREQUENCIES OF NORTH ATLANTIC TROPICAL CYCLONE OCCURRENCES, 1901–1972

Number of storms	June	July	Aug.	Sept.	Oct.	Nov.	Dec.–May
All tropical cyclones:							
0	41	40	14	1	13	47	59
1	24	22	17	18	21	24	13
2	5	7	21	13	22	1	0
3	2	2	12	16	12	0	0
4	0	1	5	16	2	0	0
5	0	0	0	4	1	0	0
6	0	0	2	3	1	0	0
7	0	0	1	1	0	0	0
Hurricanes:							
0	56	53	22	9	25	60	68
1	15	15	21	30	35	12	4
2	1	3	18	11	11	0	0
3	0	3	9	15	0	0	0
4	0	0	1	5	1	0	0
5	0	0	1	2	0	0	0

1963. Starting with the low-level activity of June and July, there is a sharp increase beginning early in August. This culminates in a peak in mid-September, followed by a decline to a lower plateau in October, a small increase to a secondary peak in mid-October and finally a sharp decrease to the low-level activity of late October and November. All told, about 79% of all hurricanes from 1886 to 1961 occurred during the three months, August–October.

The proportion of tropical storms which reach hurricane intensity also follows a seasonal trend. According to Table II nearly 56% of all tropical storms intensified into hurricanes in July, 72% in August, 64% in September, and 49% in October.

The observed frequency with which a given number of hurricanes occurred in different months is given in Table III. The probability of occurrences of tropical storms and hurricanes, based on observations during the period 1901–1972, is given in Table IV.

While the above clear-cut seasonal trend applies to the North Atlantic taken as a whole, it does not obtain to an equal degree in all areas. This is due to the fact that there is a marked seasonal shift in the main cyclogenetic areas from the western Caribbean and Yucatan in June, eastward to the Bahamas and even to Cape Verde in mid-season, and then gradually back to the western Caribbean in October and November. This is clearly demonstrated by Fig.8 which shows the points of origin of North Atlantic tropical storms and hurricanes during different months of the hurricane season. CRY and HAGGARD (1962) divided the North Atlantic into six cyclogenetic zones delineated in Fig.9. The daily distribution of tropical storms and hurricanes originating in each zone is shown in Fig.10. The daily contribution of each hurricane or tropical storm is credited to the zone where the tropical cyclone was first tracked. The early season activity is seen to consist almost entirely of tropical storms forming in the western Caribbean and the Gulf of Mexico. Only three tropical cyclones have been first noted in June outside this area. In July, the most genetic areas are the Lesser Antilles and the southwestern Atlantic. During August and the first half of September the Lesser Antilles and the southwestern Atlantic spawn the greatest number of hurricanes while in October the western Caribbean again becomes the primary genetic zone.

TABLE IV

OBSERVED PROBABILITIES OF NORTH ATLANTIC TROPICAL CYCLONE OCCURRENCES, 1901–1972

	June	July	Aug.	Sept.	Oct.	Nov.	Other months
All tropical cyclones:							
At least 1	0.43	0.44	0.81	0.99	0.82	0.35	0.18
2 or more	0.10	0.14	0.57	0.74	0.53	0.01	0
3 or more	0.03	0.04	0.28	0.56	0.22	0	0
4 or more	0	0.01	0.11	0.33	0.06	0	0
Hurricanes:							
At least 1	0.22	0.29	0.69	0.88	0.65	0.17	0.06
2 or more	0.01	0.08	0.40	0.46	0.17	0	0
3 or more	0	0.04	0.15	0.30	0.01	0	0
4 or more	0	0	0.03	0.10	0.01	0	0

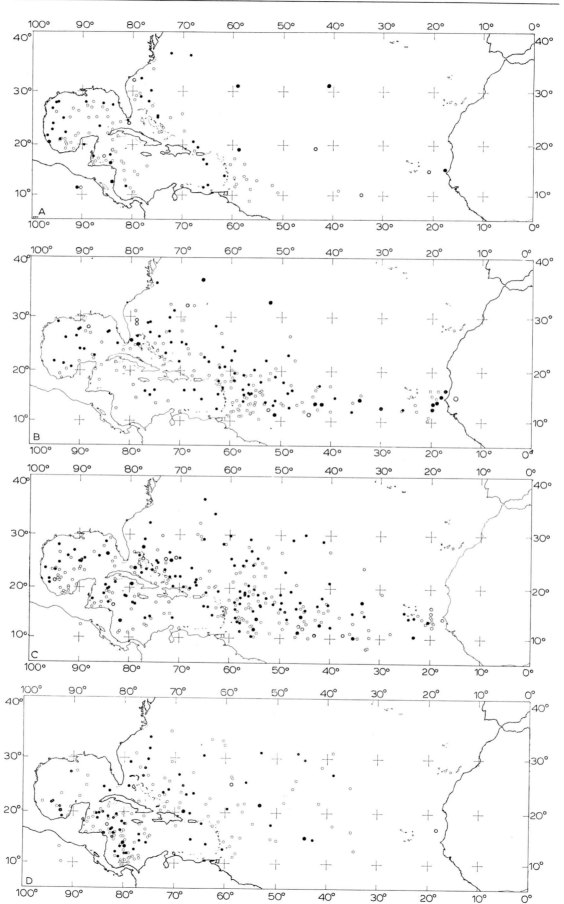

Fig.8. Points of tropical storm origin (open circles) and hurricanes (black dots) during the period 1901–1972 in: A. June–July; B. August; C. September; D. October. (Courtesy of G. W. Cry.)

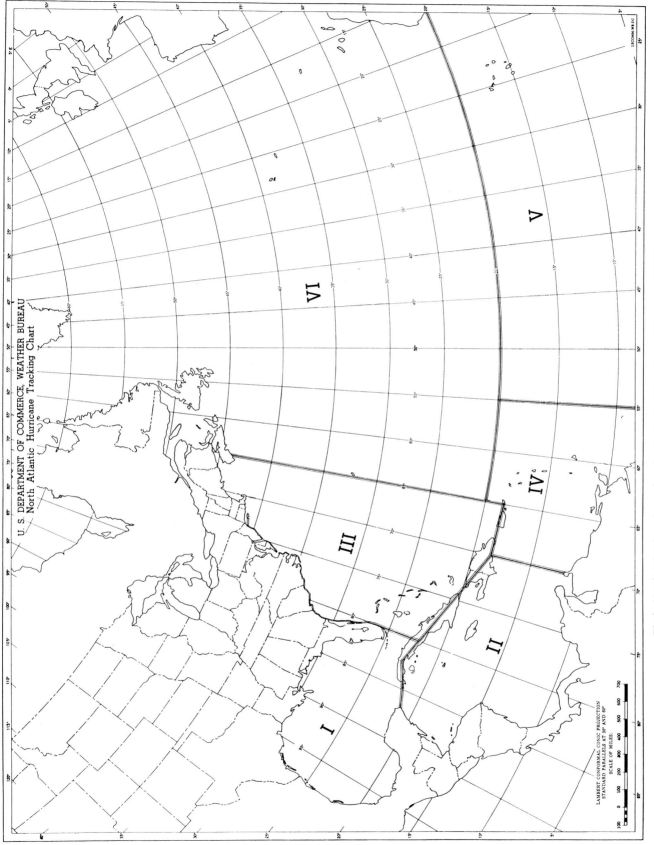

Fig.9. North Atlantic regions referred to in Fig.10. (After CRY and HAGGARD, 1962.)

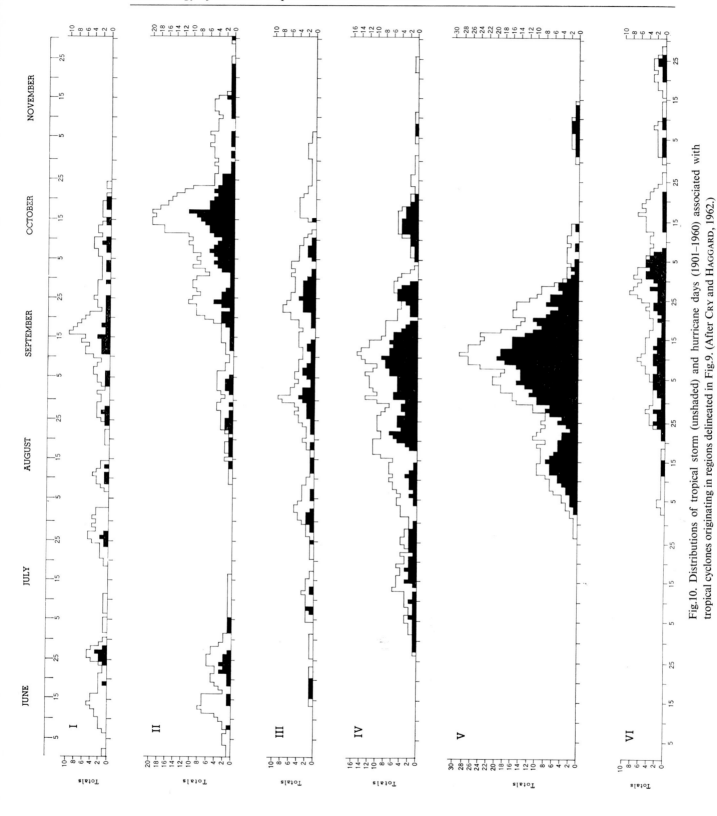

Fig.10. Distributions of tropical storm (unshaded) and hurricane days (1901–1960) associated with tropical cyclones originating in regions delineated in Fig.9. (After CRY and HAGGARD, 1962.)

Secular variations and trends

The frequency of occurrence of tropical cyclones varies considerably from year to year. During the period 1871–1964, there have been as many as 21 tropical storms and as few as one storm in any one year. The corresponding seasonal maximum and minimum for hurricanes are respectively eleven and zero. The high annual variability demonstrated in Fig.11 completely masks any secular trends. It is to be noted, however, that while the median number of storms for the whole period is eight, it is only four in the period 1910–1930 and ten from 1931–1957. Clearly, this latter period was characterized by a distinctly

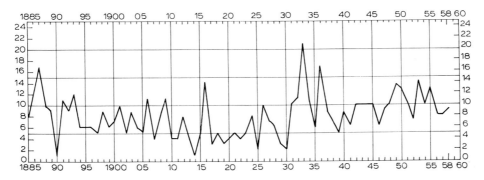

Fig.11. Annual variation in the number of tropical cyclones, 1885–1958. (After DUNN and MILLER, 1964.)

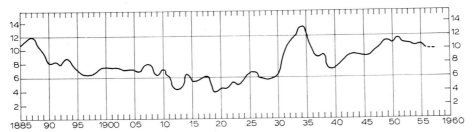

Fig.12. Five-year running average of tropical cyclone frequencies. (After DUNN and MILLER, 1964.)

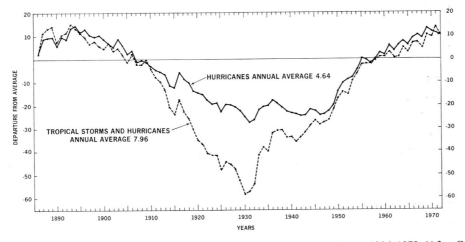

Fig.13. Cumulative departure from annual North Atlantic tropical cyclones, 1886–1972. (After CRY et al., 1959; R. M. DeAngelis, personal communication to the editor, 1973.)

495

higher level of tropical storm activity than the former. This is revealed by Fig.12 in which running averages have been computed from Fig.11. Fig.13 shows the cumulative departure from the *annual* average of storm occurrences. Beginning in 1894, a steady downward progression of activity occurred, becoming more pronounced in 1910 and continuing until 1930. During the period 1931–1936 a sharp increase in the level of activity occurred and a generally high level has been maintained since. The upward trend is, however, small when compared to the year to year variability. During the years 1901–1963, it amounts to 16 % of the total variability for tropical cyclones and 22 % of the variability for hurricanes (CRY, 1965).

The causes underlying the variability of hurricane frequencies from season to season are not sufficiently known. The abnormal poleward displacement of the planetary westerlies which NAMIAS and DUNN (1955) associated with the formation of hurricanes Connie and Diane of 1955, has been observed to be generally favorable for hurricane development. WILLETT (1955) and DUNN and MILLER (1964) have suggested a possible relation between the upward trend in the level of hurricane activity and long-term changes in temperature which might influence the patterns of the general circulation and hence the potential for tropical cyclogenesis.

Hurricane motion

Types of storm movements

When paths of individual tropical storms occurring over many years are plotted on a chart, as was done by TANNEHILL (1938) and others, there results a hopelessly tangled skein with no noticeable pattern. If, however, only the paths for a given season of a given year or for a particular month are plotted, a more closely knit family of tracks appears with the individual storms following similar though never identical courses (NAMIAS, 1955a).

For many years it has been known that the movement of tropical cyclones is strongly influenced by the great centres of action. The recurvature of the storms follows a path which evokes the shape of the isobars around the western periphery of the subtropical high. There is clear evidence that hurricanes are steered by the current in which the vortex is embedded. These currents are in turn influenced both by the small-scale circulation of the vortex and the broad-scale circulation over a large area. The planetary extent of the broad-scale effect was demonstrated by KLEIN and WINSTON (1947) who traced the failure of the September 1947 hurricane to recurve, as predicted, to dynamic anti-cyclogenesis over the eastern United States which was brought about by a chain of events originating in the Pacific Ocean a few days earlier.

Three general types of storm movements may be distinguished: westward, recurving and eastward. Most tropical storms develop in a deep easterly current and move initially in a westerly direction. Many of these continue their westward motion at low latitudes until they pass inland and dissipate over the United States or Mexico. The majority of storms drift northward as they are carried by the easterly currents of lower latitudes and eventually become dominated by the westerly current of higher latitudes. These are the recurving storms. Finally, a number of tropical storms which develop in the Gulf of Mexico, the

ANNUAL FREQUENCY OF NORTH ATLANTIC TROPICAL CYCLONES, 1871-1972

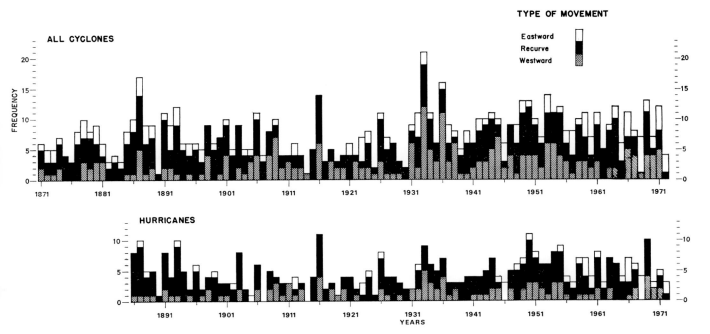

Fig.14. Annual frequency of North Atlantic tropical cyclones, 1871–1972, and hurricanes, 1886–1972. General movement is indicated by shading. (After CRY, 1965; R. M. DeAngelis, personal communication to the editor, 1973.)

Caribbean or, occasionally, over the Atlantic are steered throughout their life span by the westerly current in which they form. The annual frequency of tropical storms and hurricanes of the above three categories is given in Fig.14. Other statistical characteristics of storm movement during different periods are given in Table V.[1] Tracks for each individual storm since 1871 have been published by CRY et al. (1959).

Persistence and recurvature

COLÓN (1953) has computed for each 5° longitude–latitude square the percentage frequency with which storms have moved along each of the sixteen principal directions. He also calculated the probability of a displacement along the modal direction with the mean speed of motion and the probability that the track in the next 24 h would not deviate by more than 10° from that of the previous 24 h. He found that linear extrapolation verified at least 80% of the time over most of the region south of 20°, especially in mid-season. Extrapolation was least accurate early and late in the season and near the mean position of the subtropical ridge, where the storm motion vacillates between the influence of the easterlies and westerlies. This would be the region of maximum frequency of recurvature. Fig.15, which shows the locus of points of highest recurvature frequency, reflects the seasonal shift of the subtropical ridge (RIEHL, 1954).

---

[1] Occasionally, storms move in an erratic fashion—they remain stagnant, describe loops or change course abruptly. These short-period, erratic movements have not been considered in compiling Table V.

TABLE V

NORTH ATLANTIC TROPICAL CYCLONES, GENERALIZED MOVEMENT CHARACTERISTICS
(After CRY, 1965; R. M. De Angelis, personal communication to the editor, 1973)

	Westward[1]			Recurved			Eastward[2]			Total	
	number	percent	annual mean	number	percent	annual mean	number	percent	annual mean	number	annual mean
All tropical cyclones:											
1871–1885	18	20.2	1.2	49	55.1	3.3	22	24.7	1.5	89	6.0
1886–1900	21	17.1	1.4	77	62.6	5.1	25	20.3	1.7	123	8.2
1901–1963	186	37.2	3.0	241	48.2	3.8	73	14.6	1.2	500	8.0
1964–1972	25	31.3	2.8	28	35.0	3.1	27	33.7	3.0	80	8.9
Cyclones with hurricane intensity:											
1886–1900	13	16.7	0.9	60	76.9	4.0	5	6.4	0.3	78	5.2
1901–1963	86	29.7	1.4	176	60.7	2.8	28	9.6	0.4	290	4.6
1964–1972	11	22.9	1.2	23	47.9	2.5	14	29.2	1.6	48	5.3

[1] Always in easterlies.
[2] Always in westerlies.

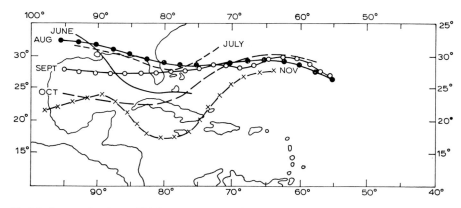

Fig.15. Locus of points of highest recurvature frequency for different months of the hurricane season. (After COLÓN, 1953.)

Tropical cyclones affecting the United States

The incidence of tropical cyclones or their effects on various sectors of the eastern and southern United States is shown in Fig.16 (CRY, 1965). The lower graph represents the number of tropical depressions (unshaded), tropical storms (hatched), and hurricanes (shaded) which passed inland over the continental United States during the period 1901–1972. In the upper six graphs, storms which did not move inland but nevertheless affected the mainland by bringing high wind, rain or high tides are also included.

The regions most vulnerable to tropical cyclone activity are Florida, Texas, the middle Gulf Coast, and the Carolinas. Florida, which experienced a high level of strong activity

■ HURRICANES *(Winds 73 m.p.h. or over)* ▨ TROPICAL STORMS *(Winds 39-72 m.p.h.)* ☐ DEPRESSIONS

☐ CENTER MOVED INLAND IN INDICATED AREA ○ CENTER REMAINED OFFSHORE OR MOVED INLAND IN ANOTHER AREA

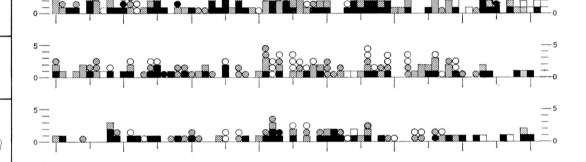

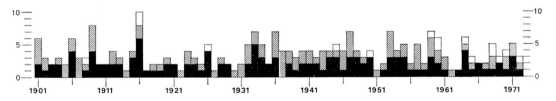

ANNUAL
NUMBER
OF TROPICAL
CYCLONE
CENTERS
PASSING
INLAND

Fig.16. Annual frequencies of tropical cyclones and hurricanes passing inland or significantly affecting various sections of the United States, 1901–1972. (For 1901–1963 after CRY, 1965; for 1964–1972 after R. M. DeAngelis, personal communication to the editor, 1973.)

in the 1940's, had a respite during the 1950's and early 1960's. A resurgence occurred in 1964 when no less than four storms crossed the peninsula. The Carolinas, on the other hand, experienced widespread tropical cyclone effects primarily during the 1950's. Finally, the Gulf Coast region, after being highly vulnerable in the early years of this century, has experienced fewer severe tropical cyclones during the last 20 years.

The middle Atlantic states mostly suffer from storms moving parallel to the coast. Only two hurricanes, one in 1903 and one in 1923, made a landfall on this section of the United States. In contrast, five severe hurricanes have crossed inland over the northeastern United States since 1938.

The observation that years of maximum threat to New England from tropical storms tend to occur in clusters has been ascribed by NAMIAS (1955b) to the year to year persistence of general circulation anomalies. The planetary interdependence of these anomalies may have been responsible for the fact that in the fall of 1954 there was an unusual westward displacement of tropical cyclone tracks, north of 35°N, both along the east coast of North America and over east Asia.

Structural aspects and related phenomena

The wind field

The highest winds in Atlantic hurricanes have not yet been directly measured. At sea, such direct measurements are virtually unobtainable because ships are warned to avoid the paths of severe hurricanes. Since 1944, when routine hurricane reconnaissance flights began and thus good estimates of hurricane winds became available, speeds exceeding 70 m/sec (150 miles/h) or more have frequently been reported.

Over land, direct measurements of the strongest winds associated with hurricanes are no less difficult to come by since even the sturdiest wind instruments have been known to collapse by their onslaught. Even so, anemometers have registered some notably high winds associated with hurricanes. Before it collapsed, the instrument at the airport terminal building at Chetumal, Mexico, recorded 78 m/sec (175 miles/h) during the passage of hurricane Janet of 1955. The wind, however, continued to increase and was estimated to have exceeded 90 m/sec (200 miles/h). Other examples of extremely strong winds measured in hurricanes are described by DUNN and MILLER (1964). Table VI gives a short summary of these records which refer to winds sustained over a period of one or more minutes. Extreme gusts are known to exceed these sustained winds by 30–50%. Additional details and references on tropical cyclone winds are given in KUTSCHENREUTER (1957) and WINSTON (1959).

The pressure field

Clearly, the devastatingly strong hurricane winds must be associated with a rapid decrease of pressure toward the centre of the storm. CLINE (1926) found an average pressure gradient of 0.42 mbar/km (0.02 inch/mile) in the storms he studied. But, as DUNN (1951) noted, these storms must have been mostly past their prime, as is usual with hurricanes on the Gulf Coast (except occasionally in June) and must have sustained the characteristic spread of isobars which occurs in older storms. The most extreme gradients are usually found when the hurricane is in the lower latitudes during the immature and early mature stages. Some fantastic pressure changes have been reported by ships near hurricanes. The S.S. Virginia, for instance, experienced a pressure drop of 45.4 mbar (1.34 inch) in 20 min in the central Caribbean on September 20, 1943. This is equivalent to a pressure gradient of 2.3 mbar/km (0.11 inch/mile). Table VII (DUNN and MILLER,

1964) gives some record low barometer readings in Atlantic hurricanes as well as in typhoons in the Western Pacific and in cyclones in the Indian Ocean and near Australia. In Fig.17 the accumulated frequency of hurricanes occurring in various sectors of the Atlantic is plotted against the central pressure.

TABLE VI

SOME EXTREME WINDS RECORDED IN ATLANTIC HURRICANES

Date	Place	Wind speed (miles/h)	Duration (min)
Sept. 18, 1926	Miami Beach, Fla.	138 (132 true)	2
		128 (123 true)	5
Sept. 13, 1928	San Juan, P.R.	150 (135 true)	5
		160 (144 true)	1
Oct. 18, 1944	Havana, Cuba	163	—
Sept. 17, 1947	Hillsboro Lighthouse, Fla.	155 (extreme)	1
		121 (maximum)	
Aug. 26, 1949	Jupiter Lighthouse, Fla.	132	
		153 (extreme)	
Sept. 27–28, 1955	Chetumal, Mexico	175*	—
Sept. 10, 1960	Central Keys, Fla.	180	—
Sept. 11, 1961	Port Lavaca, Texas	175*	—
Sept. 6, 1965	Great Abaco, Bahama Is.	178	—
Aug. 17, 1969	Boothville, La.	172 (gust)	—
		200 (estimated)	
Aug. 3, 1970	Aransas Pass, Texas	130	

* Wind-measuring equipment disabled at speed indicated. Highest wind probably higher.

TABLE VII

SELECTED LOW-PRESSURE READINGS IN TROPICAL CYCLONES

Location	Date	Pressure (mbar)
1. Eye of typhoon Ida (18.9°N 135.3°E)	Sept. 24, 1958	877
2. *SS Saporoea* (460 miles E of Luzon)	Aug. 18, 1927	887
3. Matecumbe Key (Florida Keys)	Sept. 2, 1935	892
4. Eye of typhoon Marge (700 miles NW of Guam)	Aug. 15, 1951	895
5. Eye of hurricane Camille (Gulf of Mexico)	Aug. 16, 1969	905
6. Basilan (Frank Helm Bay, Philippines)	Sept. 25, 1905	909
7. Cossak (Australia)	June 7, 1881	914
8. Chetumal (Janet) (Mexico)	Sept. 27–28, 1955	914
9. *SS Phemius* (western Caribbean)	Nov. 5, 1932	915
10. *SS Laisang* (26.7°N 123°E)	Aug. 2, 1901	915

Rainfall and floods

Hurricanes derive their energy from the latent heat of water vapor. By some mechanism, which is not yet fully understood, the heat released by the condensation of the water

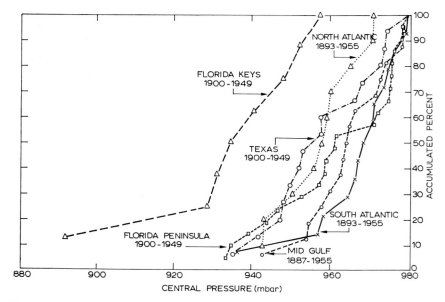

Fig.17. Accumulated frequency of occurrence of hurricanes vs. central pressure (mbar) in different regions.

vapor is transformed into the mechanical energy of the storm. Since condensation is also a prime requisite for rainfall it is not surprising that some of the world's heaviest rainfall has occurred in connection with hurricanes. The greatest amounts are frequently measured when storms stall or when orographic effects are pronounced. Since both these factors occur after a storm has moved inland and started to dissipate, decaying storms can produce more rainfall in a given area than more intense storms with a fast rate of motion. Table VIII lists some record rainfalls measured in Atlantic hurricanes. These measurements, however, underestimate the true amounts because of the loss from rain gauges which may amount to 50% when the wind reaches 22.5 m/sec (50 miles/h) or more (DUNN, 1951).

Yet hurricanes do not always produce heavy rain. An extreme example of storms of this

TABLE VIII

SOME HEAVY RAINFALL TOTALS ASSOCIATED WITH ATLANTIC HURRICANES

Date	Place	Rainfall (mm)
June 15–16, 1886	Alexandria, La.	543.6
Nov. 1909 (4 days)	Silver Hill, Jamaica	2,451.1
July 14–15, 1916	Altapass, N.C.	564.4
Sept. 9–10, 1921	Taylor, Texas	590.0
Sept. 13–14, 1928	Adjuntas, P.R.	751.8
Aug. 6–10, 1940 (5 days)	Crowley, La.	501.9
Aug. 6–10, 1940 (5 days)	Abbeville, La.	804.2
Aug. 6–10, 1940 (5 days)	Lafayette, La.	753.1
Oct. 3–8, 1963 (6 days)	Tacajo, Cuba	2,024.9
Aug. 19, 1969 (8 hours)	Massies Mill, Va.	685.8

category is one whose centre passed within 20 km of Miami on October 6, 1941. Only a meagre nine mm of rain were collected in a locality where the winds reached 55 m/sec (123 miles/h).

SCHONER and MOLANSKY (1956) have compiled some representative rainfall totals as well as 12, 18 or 24 h amounts from hurricanes entering the United States at various points along the Atlantic and Gulf coasts. A climatological summary of the mean areal average and percentage frequency of precipitation amounts following passage of storms across the Gulf coast is given in Fig.18 and 19. These figures show that the distribution of rainfall is not symmetrical with respect to the centre of the storm and are in qualitative agreement with the findings of CLINE (1926), who noted that in storms along the Gulf coast the greatest rainfall intensity occurred about 100–130 km (60–80 miles) in front of the centre and mostly to the right of the hurricane track.

The above distribution is typical of hurricanes in the mature stage or after they have recurved. In immature storms, those observed in low latitudes and stationary storms, the rainfall is more or less symmetrical with some tendency for a maximum to occur to the rear of the centre.

Radar observations have established that hurricane rainfall is concentrated along a series of spiral bands. Fig.20, prepared by ROCKNEY (1956), shows a schematic picture of the appearance on radar of four major east-coast hurricanes. The concentration of the more solid bands in the forward semi-circle and the lack of precipitation to the rear, which is typical of these storms, is in evidence.

The torrential rains from hurricanes often result in floods: some minor, others devastating. In the period since 1886 more than 60 floods due to rains from hurricanes and other

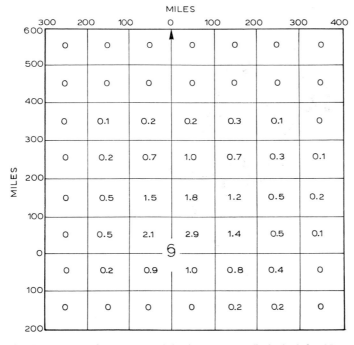

Fig.18. Mean areal average precipitation amounts (in inches) for 35 tropical storms entering the Texas coast. All values are located relative to the position of storm as it crossed the coast and its motion (in direction of arrow) during the 24 h following coastal crossing. (After SCHONER, 1957.)

MILES

300	200	100	0	100	200	300	400
O	03	03	O	O	O	O
O	03	03	03	03	O	O
O	11	23	23	29	11	O
03	20	66	66	63	34	11
03	46	94	97	86	46	20
03	43	100	100	97	49	09
O	20	77	80	60	43	03
O	O	O	03	18	18	O

(Vertical axis labelled MILES: 600, 500, 400, 300, 200, 100, 0, 100, 200)

Fig.19. Percentage frequencies of areal 24-h precipitation occurrence for 35 tropical storms entering the Texas coast. All percentages are located relative to the position of storm as it crossed the coast and its motion (in direction of arrow) during the 24 h following coastal crossing. (After SCHONER, 1957.)

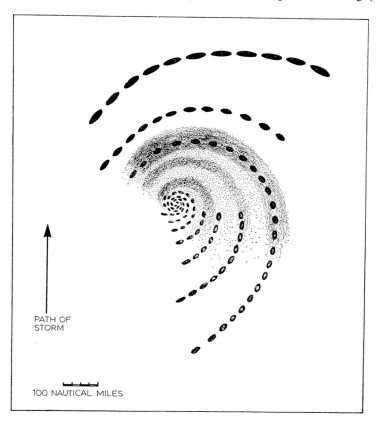

PATH OF
STORM

100 NAUTICAL MILES

Fig.20. Schematic illustration of the typical appearance of four major east coast hurricanes on radar. (After ROCKNEY, 1956.)

tropical cyclones are known to have affected the United States. Nearly every coastal state from Texas to Maine and even some interior states have been affected. Some of the most severe of these floods occurred in connection with hurricanes in the decaying stage, as in the case of hurricane Diane which caused nearly 200 deaths and about one billion dollars in damage.

Elsewhere, in the West Indies and Central America, floods from hurricanes have been equally frequent and have killed thousands of people.

The storm surge

The storm surge is a rapid rise in the water produced by onshore hurricane winds and falling barometric pressures. The surge may be as little as three or four feet or less or it may be as much as fifteen to twenty feet or more. The most destructive situations occur when the surge coincides with high astronomical tides.

The study of storm surges is not readily amenable to a climatological approach because of their dependence on a multitude of factors and their great variability within short distances. At least five distinct processes associated with passage of a storm have been

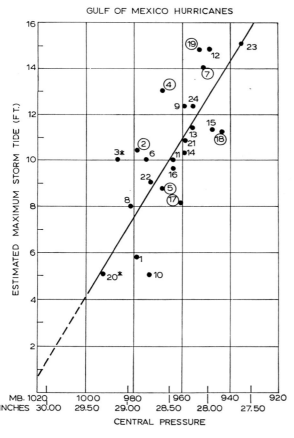

Fig.21. Regression of maximum storm tide height on central pressure for hurricanes entering the United States Gulf of Mexico coastline west of Tallahassee, Fla. (After HOOVER, 1957.) The numbers in the graph refer to 24 individual hurricanes of the years 1893–1950, listed in Hoover's table II.

recognized to alter the level in tidewater regions (HARRIS, 1963). These are (*1*) the pressure effect; (*2*) the wind effect; (*3*) the effect of the earth's rotation; (*4*) the effect of waves; and (*5*) the effect of rainfall. In addition such factors as the shape of the coastline and the topography of the continental shelf are also highly relevant.

In view of this multiplicity of causal factors it would at first glance appear hopeless to correlate any one of them with the observed water level. A correlation between peak water levels and minimum pressures associated with hurricanes has, however, been found (CONNER et al., 1957; HOOVER, 1957); Fig.21 illustrates this relationship. Some notably high tides associated with Gulf of Mexico hurricanes are listed in Table IX. Detailed data and discussions of water levels in 27 hurricanes affecting the Atlantic and Gulf coasts during the years 1926–1961 were given by HARRIS (1963).

Another characteristic which has been found to obtain in the generality of hurricanes is that the peak surge occurs to the right of the storm track. HOOVER (1957) found that the peak occurred on the average about 15–30 km to the right of the track. However, peak surges have been observed as far as 65 km to the right and, occasionally, to the left of the hurricane track.

TABLE IX

HIGHEST TIDES AND LOWEST CENTRAL PRESSURES OF GULF OF MEXICO HURRICANES

Date	Location of highest tide on open coast	Lowest pressure (mbar)	Maximum tide height (ft.)*
Sept. 8, 1900	Galveston, Texas	936	14.5
Sept. 27, 1906	Fort Barrancas, Fla.	965	10.8
July 21, 1909	Galveston, Texas	959	10.0
Aug. 16, 1915	High Island, Texas	953	13.9
Sept. 14, 1919	Port Aransas, Texas	948	11.1
Aug. 25, 1926	Timbalier Bay, La.	959	10.0
Sept. 5, 1933	Brownsville, Texas	949	13.0
Aug. 30, 1942	Matagorda, Texas	951	14.8
Sept. 19, 1947	Biloxi, Miss.	968	11.1
Sept. 11, 1961	Port Lavaca, Texas	935	16.6
Aug. 17, 1969	Pass Christian, Miss.	905	24.2

* Above mean sea level

Hurricane tornadoes

Although TANNEHILL (1956) has noted that few authentic records existed of tornado activity in hurricanes, and despite the fact that tornadoes are not normally considered tropical phenomena, there is increasing evidence of tornado occurrences associated with hurricanes. In a study of reported occurrences over the years, MALKIN and GALWAY (1953) found that tornadoes in hurricanes were observed only in the forward semi-circle or along the periphery of the storm. SADOWSKI (1962), on the other hand, showed that the 16 tornadoes which were reported in hurricane Carla of 1962, favoured a location to the right of the hurricane track. In a study of a larger sample of tornado occurrences over the last ten years PEARSON and SADOWSKI (1965) showed that both the above results were partially correct. Their finding illustrated in Fig.22 indicates the right front quadrant as the preferred location for tornado occurrences.

There is no indisputable evidence that tornadoes form during a preferred time of day. Nor, apparently, are they associated with hurricane winds. Out of a sample of 16 studied by SADOWSKI (1962), none formed within hurricane force wind, but 13 formed in areas of gale wind.

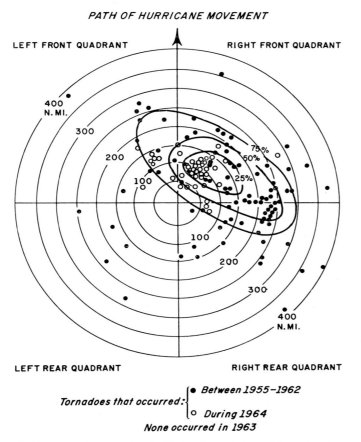

Fig.22. Tornadoes associated with hurricanes, 1955–1964. (After PEARSON and SADOWSKI, 1965.)

Acknowledgement

The author wishes to express his warm thanks to Mr. G. Cry of the Laboratory of Climatology, U.S. Weather Bureau, who made available the revised edition of *Technical Paper* no.55 prior to its publication.

References

ALAKA, M. A., 1968. Climatology of Atlantic tropical storms and hurricanes. *Weather Bur.*, *ESSA*, *Tech. Rept.*, WB-6, 18 pp.

CLINE, I. M., 1926. *Tropical Cyclones*. Macmillan, New York, N.Y., 301 pp.

COLÓN, J. A., 1953. A study of hurricane tracks for forecasting purposes. *Monthly Weather Rev.*, 81:53–66.

CONNER, W. C., KRAFT, R. H. and HARRIS, D. L., 1957. Empirical methods for forecasting the maximum storm tide due to hurricanes and other tropical storms. *Monthly Weather Rev.*, 85:113–116.

CRY, G. W., 1965. Tropical cyclones of the North Atlantic Ocean. *U.S. Weather Bur., Tech. Paper*, 55: 148 pp.

CRY, G. W. and HAGGARD, W. H., 1962. North Atlantic tropical cyclone activity, 1901–1960. *Monthly Weather Rev.*, 90:341–349.

CRY, G. W., HAGGARD, W. H. and WHITE, H. S., 1959. North Atlantic tropical cyclones—tracks and frequencies of hurricanes and tropical storms 1886–1958. *U.S. Weather Bur., Tech. Paper*, 36:214 pp.

DUNN, G. E., 1951. Tropical cyclones. In: T. F. MALONE (Editor), *Compendium of Meteorology*. American Meteorological Society, Boston, Mass., pp.887–901.

DUNN, G. E., 1956. Areas of hurricane development. *Monthly Weather Rev.*, 84:47–51.

DUNN, G. E. and MILLER, B. I., 1964. *Atlantic Hurricanes*. Louisiana State Univ. Press, Baton Rouge, La., 377 pp.

FASSIG, O. L., 1913. Hurricanes of the West Indies. *U.S. Weather Bur., Bull.*, X: 28 pp.

GARRIOTT, E. B., 1900. West Indian hurricanes. *U.S. Weather Bur., Bull.*, H:69 pp.

GENTRY, R. C., 1963. Historical survey and climatology of hurricanes and tropical storms. In: *Proceedings of the Inter-Regional Seminar on Tropical Cyclones in Tokyo, 18–31 January, 1962—Japan Meteorol. Agency, Tech. Rept.*, 21:17–29.

HAGGARD, W. H., 1958. The birthplace of North Atlantic tropical storms. *Monthly Weather Rev.*, 86: 397–404.

HARRIS, D. L., 1963. Characteristics of the hurricane storm surge. *U.S. Weather Bur., Tech. Paper*, 48: 139 pp.

HOOVER, R. A., 1957. Empirical relationships of the central pressures in hurricanes to the maximum surge and storm tide. *Monthly Weather Rev.*, 85:167–174.

HOPE, J. R. and NEUMANN, C. J., 1971. Computer methods applied to Atlantic area tropical storm and hurricane climatology. *Mariners Weather Log*, 15(5):272–278.

KLEIN, W. H. and WINSTON, J. S., 1947. The path of the Atlantic hurricane of September 1947 in relation to the hemispheric circulation. *Bull. Am. Meteorol. Soc.*, 28:447–452.

KUTSCHENREUTER, P. H., 1957. Survey of meteorological factors pertinent to reduction of loss of life and property in hurricane situations. *U.S. Weather Bur., Natl. Hurricane Res. Project, Rept.*, 5:87 pp.

MALKIN, W. and GALWAY, J. G., 1953. Tornadoes associated with hurricanes—as illustrated by Franconia, Va., tornado, September 1, 1952. *Monthly Weather Rev.*, 81:299–303.

MITCHELL, C. L., 1924. West Indian hurricanes and other tropical cyclones of the North Atlantic Ocean. *Monthly Weather Rev., Suppl.*, 24:47 pp.

MITCHELL, C. L., 1932. West Indian hurricanes and other tropical cyclones of the North Atlantic Ocean. *Monthly Weather Rev.*, 60:253.

NAMIAS, J., 1955a. Long range factors affecting the genesis and paths of tropical cyclones. In: *Proceedings of The UNESCO Symposium on Typhoons, 9–12 November, 1954, Tokyo*. The Japanese National Commission for Unesco (with the co-operation of the Organizing Committee for the Unesco Symposium on Typhoons 1955), pp.213–219.

NAMIAS, J., 1955b. Secular fluctuations in vulnerability to tropical cyclones in and off New England. *Monthly Weather Rev.*, 83:155–162.

NAMIAS, J. and DUNN, C. E., 1955. The weather and circulation of August 1955—including the climatological background for hurricanes Connie and Diane. *Monthly Weather Rev.*, 83:163–170.

PALMÉN, E., 1948. On the formation and structure of tropical hurricanes. *Geophysica*, 3:26–38.

PEARSON, A. D. and SADOWSKI, A. F., 1965. Hurricane-induced tornadoes and their distribution. *Monthly Weather Rev.*, 93:461–464.

RIEHL, H., 1948. On the formation of typhoons. *J. Meteorol.*, 5:247–264.

RIEHL, H., 1954. *Tropical Meteorology*. McGraw-Hill, New York, N.Y., 392 pp.

ROCKNEY, V. D., 1956. Hurricane detection by radar and other means. In: *Proceedings of the Tropical Cyclone Symposium, Brisbane, December 1956*. Bureau of Meteorology, Melbourne, pp.179–197.

SADOWSKI, A., 1962. Tornadoes associated with hurricane Carla, 1961. *Monthly Weather Rev.*, 90:514–516.

SCHONER, R. W., 1957. Frequency and distribution of areal rainfall averages associated with tropical storms entering the coast of the United States (Texas to Maine). *U.S. Weather Bur., Res. Rept.*, 60 pp. (unpublished).

SCHONER, R. W. and MOLANSKY, S., 1956. Rainfall associated with hurricanes (and other tropical disturbances). *U.S. Weather Bur., Natl. Hurricane Res. Project, Rept.* 3:305 pp.

TANNEHILL, I. R., 1938. *Hurricanes—Their Nature and History*. Princeton Univ. Press, Princeton, N.J., 257 pp.

TANNEHILL, I. R., 1956. *Hurricanes—Their Nature and History, Particularly Those of the West Indies and the Southern Coasts of the United States*. Princeton Univ. Press, Princeton, N.J., 308 pp.

WILLETT, H. C., 1955. A study of the tropical hurricane along the Atlantic and Gulf Coasts of the United States. *Inter-Regional Insurance Conference, New York City, 1955*, 63 pp.

WINSTON, J. S. (Editor), 1959. Hurricane forecasting. *U.S. Weather Bur., Forecasting Guide*, 3:108 pp.

YANAI, M., 1964. Formation of tropical cyclones. *Rev. Geophys.*, 2:367–414.

Reference Index

ADMIRALTY HYDROGRAPHIC DEPARTMENT, 451
ALAKA, M. A., 507
ALBRECHT, F., 11, 71
ALDAZ, L., 269
ALEWIJNSE, C. R. J., d'HAMECOURT, J. L. A. and
 FUNG LOY, L. W., 313, 376
ALEXANDER, L. S., 440, 441, 451
ALEXANDER, W. H., 451
ALMEIDA SIMÕES, R. M., 267
ALMEYDA, E., 118, 120, 121, 129
ALMEYDA, E. and SAEZ, F., 113, 127, 129
ALPERT, L., 200, 372, 376, 420, 451, 452
AMERICAN METEOROLOGICAL SOCIETY, 71
ANGOT, A., 240, 267
ANONYMOUS, 117, 118, 129
ANTEZANA PARDO, O., 72
ARDISSONE, R., 72
ARMIJO, L., see FREEMAN JR., C. F. et al.
ARNETT, J. S., 115, 126, 129
ATKINSON, G. D. and SADLER, J. C., 329, 361, 376

BAILEY, S. I., 178, 200
BALAGOT, V. F., see FLORES, J. W. and BALAGOT, V. F.
BARROS, L., 269
BARROS, L. F., 267
BATTIONE CHIARINO, J. A., 72
BECKER, B. K., 269
BERGEIRO, J. M., 71, 72
BERNARDES, L. M. C., 255, 258, 268
BERNHARDT, F. and PHILIPPS, H., 327, 376
BERTONE, M., 66, 72
BEZERRA DOS SANTOS, L., 268
BJERKNES, J., 170, 200
BLANCHARD, W., 334, 376
BLISS, E. W., 452
BLUME, H., 441, 452
BOFFI, J. A., 72
BÖGEL, R., 41, 72
BOER, G. J., see NEWELL, R. E. et al.
BORZACOV, V., see KNOCHE, W. et al.
BOWMAN, I., 192, 193, 200
BRAAK, C., 309, 311, 312, 313, 316, 317, 318, 355, 376,
 440, 452
BRANDT, B., 200
BRENNAN, J. F., 429, 452
BRINKMANN, W. L. F., WEINMAN, J. A. and RIBEIRO,
 M. N. G., 346, 377
BROOKS, C. E. P., 452
BROOKS, C. F., see WARD, R. DE C. and BROOKS, C. F.
BUDYKO, M. I., 377

BULTOT, F. and GRIFFITHS, J. F., 199, 200
BURGMEIER, J., see FREEMAN JR., C. F. et al.
BURGOS, J. J., 49, 72
BURGOS, J. J. and VIDAL, A. L., 72
BURMEISTER, H., 72

CAMPAN, G., 303, 305, 377, 430, 441, 442, 443, 452
CATALANO, L. R., 72
CAVIEDES, C. L., 130, 171, 200
CEDANO, L. V., 163, 197, 200
CHAPEL, L. T., 429, 447, 448, 452
CHERBONNIER, J., 452
CHIOZZA, E. M. and GONZÁLEZ VAN DONSELAR, Z., 71
CLAYTON, F. L., see CLAYTON, H. H. and CLAYTON,
 F. L.
CLAYTON, H. H. and CLAYTON, F. L., 379
CLEARE, L. D., 325, 377
CLINE, I. M., 480, 500, 503, 507
COBB, G., see CROW, L. W. and COBB, G.,
 see also HENRY, W. K. et al.
COILE, R. C. and CULMSEE, W., 163, 193, 200
COLÓN, J. A., 480, 497, 498, 508
COLQUI, B. S., 62, 72
COLQUI, B. S., see RAFFO, J. M. et al.
CONNER, W. C., KRAFT, R. H. and HARRIS, D. L., 506,
 508
CONRAD, V. and POLLAK, L. W., 237, 240, 268
CONWAY, W. M., 192, 200
CORNEJO, T. A., 168, 200
COUSSIRAT ARAÚJO, C., 267
CROW, L. W. and COBB, G., 431, 450, 452
CRUTCHER, H. L., see TALJAARD, J. J. et al.
 see also VAN LOON, H. et al.
CRY, G. W., 480, 488, 490, 496, 498, 499, 508
CRY, G. W. and HAGGARD, W. H., 491, 493, 494, 508
CRY, G. W., HAGGARD, W. H. and WHITE, H. S., 472,
 495, 497, 508
CULMSEE, W., see COILE, R. C. and CULMSEE, W.
CZAJKA, W., 71
CZAJKA, W. and VERVOORST, F., 72

DAVIS, L. G., see KELLEY, J. I. et al.
DAVIS, W. G., 71
DAY, P. C., 373, 377
DEAN, G. A., 269, 325, 377
DEAN, G. A., NUNES, G. S. and NUNES, H. T., 269
DE FINA, L. et al., 72
DE GARCÍA, M. T. U., see PROHASKA, F. and DE
 GARCÍA, M. T. U.

DE LEVIN, M. M., 71
DELGADO DE CARVALHO, C. M., 255, 267
DEPARTAMENTO DE IRRIGACIÓN, SECCIÓN DE CLIMA E INGENIERÍA AGRÍCOLA, 379
DEPARTAMENTO DE IRRIGACIÓN, SECCIÓN DE METEO-ROLOGÍA Y AFOROS, 379
DEPARTAMENTO NACIONAL DE METEOROLOGÍA, 269
DEUTSCHES HYDROGRAPHISCHES INSTITUT, 452
D'HAMECOURT, J. L. A., *see* ALEWIJNSE, C. P. J. et al.
DIRECCIÓN GENERAL DE METEOROLOGÍA, 71
DIRECCIÓN DE PLANIFICACIÓN AGROPECUARÍA, 377
DREWES, A. T., *see* DREWES, W. V. and DREWES, A. T.
DREWES, W. V. and DREWES, A. T., 158, 200
DUFFIE, J. A., see LÖF, G. O. G. et al.
DUNN, G. E., 480, 484, 488, 500, 502, 508
DUNN, G. E. and MILLER, B. I., 480, 488, 495, 496, 500, 508
DUNN, C. E., *see* NAMIAS, J. and DUNN, C. E.

ELLENBERG, H., 200
ESCOBAR VALLEJO, I., 200

FARIÑA SANCHEZ, T., 72
FASSIG, O. L., 480, 508
FEDERACIÓN NACIONAL DE CAFETEROS DE COLOMBIA, 379
FLOHN, H., 190, 200, 310, 333, 350, 363, 368, 377, 420, 452
FLORES, J. W. and BALAGOT, V. F., 348. 377
FORS, A. L., 452
FOSBERG, F. R., GARNIER, B. J. and KÜCHLER, A. W., 259, 267
FOUGEROUZE, J., 303, 377
FREEMAN JR., C. F., GRAVES, L. F., HANNA, P., PORTIG, W. H., ARMIJO, L. and BURGMEIER, J., 448, 452
FUENZALIDA, H., 127, 129
FUGLISTER, F. C., 352, 377
FUNG LOY, L. W., *see* ALEWIJNSE C. P. J. et al.

GAIGNARD, R., 72
GALVÃO, M. C., 268
GALVÃO, M. V., 268, 269
GALWAY, J. G., *see* MALKIN, W. and GALWAY, J. G.
GARNIER, B. J., *see* FOSBERG, F. R. et al.
GARRIOTT, E. B., 480, 508
GARSTANG, M., *see* SIMPSON, J. and GARSTANG, M.
GEIGER, R., 367, 377
GENTRY, R. C., 479, 508
GOLDBRUNNER, A. W., 345, 346, 347, 377, 379, 439, 452
GONZALEZ, E., 379
GONZÁLEZ VAN DONSELAR, Z., *see* CHIOZZA, E. M. and GONZÁLEZ VAN DONSELAR, Z.
GORDON, D., *see* MINTZ, Y. and GORDON, D.
GRANDOSO, H. N. and NUÑEZ, J., 14, 58, 72
GRANDOSO, H. N., *see* IRIBARNE, J. V. and GRANDOSO, H. N.

GRAVES, L. F., *see* FREEMAN JR., C. F. et al.
GRAVES, M. E., 167, 200
GRIFFITHS, J. F., *see* BULTOT, F. and GRIFFITHS, J. F.
    see also HENRY, W. K. et al.
GROSH, R. C., 309, 377
GUERRA, I. A. L., 268
GUIMARÃES, F. M. S., 268
GUNTHER, E. R., 200
GUTIERREZ, M. M., 370, 377
GUTMANN, G. J. and SCHWERDTFEGER, W., 8, 11, 193, 200

HAGGARD, W. H., 480, 508
HAGGARD, W. H., *see* CRY, G. W. and HAGGARD, W. H.
    see also CRY, G. W. et al.
HAHN, J., 178, 188, 200
HALL, M., 452
HANNA, P., *see* FREEMAN JR., C. F. et al.
HARRIS, D. L., 506, 508
HARRIS, D. L., *see* CONNER, W. C. et al.
HASTENRATH, ST., 9, 11, 200
HASTENRATH, ST. and LESSMANN, H., 424, 452
HAUTHAL, R., 200
HAVLIK, D., *see* WEISCHET, W. and HAVLIK, D.
HEINSHEIMER, J. J., 66, 72
HENRY, W. K., GRIFFITHS, J. F., COBB, G. et al., 426, 452
HERINGER, E. P., *see* RIZZINI, C. T. and HERINGER, E. P.
HEROLD, L., 129
HOFFMANN, G., 416, 417, 452
HOOVER, R. A., 505, 506, 508
HOPE, J. R. and NEUMANN, C. J., 480, 508
HOSLER, C. L., *see* KELLEY, J. I. et al.
HOWELL, W. E., 161, 200, 309, 377
HOWELL, W. E., *see also* LÓPEZ, M. F. and HOWELL, W. E.
HOYT, D. V., *see* VAN LOON, H. et al.
HUBERT, L. F., 448, 452
HUSCHKE, R. E., 296, 302, 335, 339, 377

IM THURN, E. F., 324, 377
INSTITUTO GEOGRÁFICO "AGUSTIN CODAZZI", 379
IRIBARNE, J. V. and GRANDOSO, H. N., 69, 72

JAHR, A., 338, 377
JAMAICA METEOROLOGICAL OFFICE, 452
JEFFERSON, M., 113, 129
JENNE, R. L., *see* TALJAARD, J. J. et al.
    see also VAN LOON, H. et al.
JOHANNSON, O. V., 240, 268
JOHNSON, A. M., 201

KELLEY, J. I., HOSLER, C. L., DAVIS, L. G. and KELLEY, N. D., 356, 377
KELLEY, N. D., *see* KELLEY, J. I. et al.

KENDREW, W. G., 113, 114, 129
KESSLER, A., 198, 201
KESSLER, A. and MONHEIM, F., 195, 196, 197, 198, 201
KIDSON, J. W., *see* NEWELL, R. E. et al.
KING, C. A. M., 201
KINZL, H. and WAGNER, A., 201
KLEIN, W. H. and WINSTON, J. S., 496, 508
KLOSTER, W., 452
KNOCH, K., 1, 11, 71, 113, 114, 129, 163, 170, 193, 201, 203, 215
KNOCHE, W., 114, 129
KNOCHE, W., BORZACOV, V. et al., 16, 69, 71
KRAFT, R. H., *see* CONNER, W. C. et al.
KRISHNAMURTI, T. N., *see* YOUNG, J. A. et al.
KÜCHLER, A. W., *see* FOSBERG, F. R. et al.
KUHN, F., 71
KUTSCHENREUTER, P. H., 500, 508

LABITZKE, K., *see* VAN LOON, H. et al.
LABORATORIO DE FISICA COSMICA, UNIVERSIDAD MAYOR DE SAN ANDRES, 193, 195, 201
LAHEY, J. F., 349, 351, 352, 377, 424, 440, 441, 452
LAMB, H. H., 115, 129
LANDSBERG, H. E., 32, 72, 339, 377
LaSEUR, N., 444, 452
LEMONS, H., 451, 452
LENSELINK, H. J. and VAN DER WEERT, R., 319, 377
LESSMANN, H., 431, 449, 452, 453
LESSMANN, H., *see* HASTENRATH, S. and LESSMANN, H.
LETTAU, H., 167, 190, 191, 201
LETTAU, H. and SCHWERDTFEGER, W., 190, 201
LETTAU, H. H., 310, 351, 357, 368, 377
LETTAU, H. H. and LETTAU, K., 338, 354, 377
LETTAU, K., *see* LETTAU, H. H. and LETTAU, K.
LIETH, H., *see* WALTER, H. and LIETH, H.
LINDZEN, R. S., *see* YOUNG, J. A. et al.
LIST, R. J., 199, 201, 327, 377
LLIBOUTRY, L., 9, 11, 127, 129
LÖF, G. O. G., DUFFIE, J. A. and SMITH, C. O., 327, 377
LONDON, J., *see* VAN LOON, H. et al.
LÓPEZ, M. E., 373, 377
LÓPEZ, M. E. and HOWELL, W. E., 333, 360, 361, 363, 377
LÓPEZ, M. F. and HOWELL, W. E., 379
LÜCKEFEDT, W., *see* RIEHL, L. and LÜCKEFEDT, W.

MACHADO, F. P., 268
MACKY, W. A., 453
MADEJSKI, M. E., *see* RAFFO, J. M. et al.
MAHRT, L. J. and SCHWERDTFEGER, W., 190, 201
MALKIN, W. and GALWAY, J. G., 407, 508
MALKUS, J. S., 311, 351, 355, 378
MARCHETTI, A. A., 66, 72, 73
MARKSON, R. and NELSON, R., 201
MARTIN, D. W., *see* SCHWERDTFEGER, W. and MARTIN, D. W.
MARTIN DE MOUSSY, J. A., 71

MATTOS, J. N. B., 268
McBRYDE, F. W., 451, 453
McCULLOUGH, C. R., 445, 453
MERRITT, E. S., 448, 453
METEOROLOGICAL OFFICE, AIR MINISTRY, 379
METEOROLOGICAL OFFICE, LONDON, 440, 442, 443, 445, 447, 453
METEOROLOGICAL OFFICE OF GREAT BRITAIN, 113, 129
MILLAS, J. C., 429, 447, 448, 453
MILLER, A., 120, 121, 129
MILLER, B. I., *see* DUNN, G. E. and MILLER, B. I.
MILLER, D. B., 297, 338, 378
MINISTERIE VAN OPENBARE WERKEN EN VERKEER, 308, 313, 378, 379
MINISTRY OF WORKS AND COMMUNICATIONS, METEOROLOGICAL DIVISION, 379
MINTZ, Y. and GORDON, D., 266, 268
MITCHELL, C. L., 480, 508
MOLANSKY, S., *see* SCHONER, R. W. and MOLANSKY, S.
MONHEIM, F., *see* KESSLER, A. and MONHEIM, F.
MONTEIRO, C. A. F., 268, 269
MORANDI, L., 73
MORIZE, H., 255, 268
MOSTAJO, B., 201
MURPHY, R. C., 201, 375, 378

NAMIAS, J., 496, 500, 508
NAMIAS, J. and DUNN, C. E., 480, 496, 508
NELEMANS, B., 379
NELSON, R., *see* MARKSON, R. and NELSON, R.
NEUMANN, C. J., *see* HOPE, J. R. and NEUMANN, C. J.
NEWELL, R. E., KIDSON, J. W., VINCENT, D. G. and BOER, G. J., 269
NEWTON, C. W., *see* VAN LOON, H. et al.
NIMER, E., 269
NUNES, G. S., *see* DEAN, G. A. et al.
NUNES, H. T., *see* DEAN, G. A. et al.
NUÑEZ, J., *see* GRANDOSO, H. N. and NUÑEZ, J.

OBSERVATORIO METEOROLÓGICO NACIONAL, 378
OFICINA METEOROLÓGICA, 71
OFICINA METEOROLÓGICA DE CHILE, 113, 118, 119, 124, 127, 129, 130
OLASCOAGA, M. J., 73
ORTOLANÍ, M., 201

PAES DE CAMARGO, A., 255, 288
PALMÉN, E., 484, 485, 486, 487, 488, 508
PAPADAKIS, J., 73
PAULHUS, J. L. H., 450, 453
PAWLING, J. W., 453
PEARSON, A. D. and SADOWSKI, A. F., 507, 508
PHILIPPS, H., *see* BERNHARDT, F. and PHILIPPS, H.
PINTO, M. M., *see* RIZZINI, C. T. and PINTO, M. M.
POLLAK, L. W., *see* CONRAD, V. and POLLAK, L. W.
PORTIG, W. H., 320, 348, 349, 353, 372, 378, 405, 411, 418, 424, 426, 429, 430, 431, 432, 445, 448, 453

PORTIG, W. H., *see* FREEMAN JR., C. F. et al.
PRADA ESTRADA, R., 71
PREGO, A. J., 64, 73
PROHASKA, F., 9, 11, 22, 47, 60, 73, 179, 167, 180, 181, 201, 414, 453
PROHASKA, F. and DE GARCÍA, M. T. U., 41, 73
PROHASKA, F., *see* SCHWERDTFEGER, W. and PROHASKA, F.
PROHASKA, F. J., 120, 130, 374, 378

QUATE, B. E., 451, 453

RAFFO, J. M., COLQUI, B. S. and MADEJSKI, M. E., 66, 73
RAMAGE, C. S., 334, 335, 378
RAMIA, M., 333, 334, 378
RATISBONA, L., *see* SERRA, A. and RATISBONA, L.
RECK, H., 192, 201
REED, W. W., 163, 201, 379, 440, 443, 445, 453
REYES, R. L., 453
RIBEIRO, M. N. G., *see* BRINKMAN, W. L. F. et al.
RICHARDS, P. W., 299, 321, 378
RICO NEGRETTI, J. J., 201
RIEHL, H., 266, 268, 344, 348, 350, 378, 448, 453, 488, 497, 508
RIEHL, H. and LÜCKEFEDT, W., 346, 378
RIZZINI, C. T. and HERINGER, E. P., 268
RIZZINI, C. T. and PINTO, M. M., 260, 268
ROCKNEY, V. D., 503, 504, 508
ROESSEL, H., 201
RÖHL, E., 330, 338, 339, 345, 378
ROUSE, W. R. and WATTS, D., 448, 453
ROHMEDER, W., 73
ROHMEDER, W., *see* WILHELMY, H. and ROHMEDER, W.
RÖNICKE, G., 69, 73

SAA, G. E. and VALDEZ, M., 120, 130
SADLER, J. C., *see* ATKINSON, G. D. and SADLER, J. C.
SADOWSKI, A., 506, 507, 508
SADOWSKI, A. F., *see* PEARSON, A. D. and SADOWSKI, A. F.
SAEZ, F., *see* ALMEYDA, E. and SAEZ, F.
SALIVIA, L., 453
SAMPAIO FERRAZ, J., 263, 264, 268
SANTAMARIA, L., 453
SAPPER, K., 453
SASAMORI, T., *see* VAN LOON, H. et al.
SCHMIDT, J. C. J., 255, 266
SCHMIDT, R. D., 370, 373, 378
SCHMITHÜSEN, J., 130
SCHONER, R. W., 503, 504, 508
SCHONER, R. W. and MOLANSKY, S., 508
SCHOTT, G., 117, 130
SCHRÖDER, R., 379
SCHÜTTE, K., 169, 201
SCHWEIGGER, E., 117, 119, 130, 169, 170, 187, 201
SCHWERDTFEGER, W., 3, 7, 11, 21, 22, 58, 65, 66, 68,
73, 121, 125, 127, 128, 130, 193, 201
SCHWERDTFEGER, W. and MARTIN D. W., 5, 11
SCHWERDTFEGER, W. and PROHASKA, F., 7, 11, 15, 73
SCHWERDTFEGER, W. and VASINO, C. J., 73
SCHWERDTFEGER, W. et al., 37, 73
SCHWERDTFEGER, W., *see* GUTMAN, G. J. and SCHWERDTFEGER, W.
see also LETTAU, H. and SCHWERDTFEGER, W.
MAHRT, L. J. and SCHWERDTFEGER, W.
SEARS, A. E., 170, 201
SECOND WEATHER GROUP, 324, 378
SEREBRENICK, S., 255, 258, 268
SERRA, A., 240, 241, 255, 264, 268
SERRA, A. and RATISBONA, L., 264, 268, 269
SERVICE MÉTÉOROLOGIQUE DU GROUPE ANTILLES–GUYANE FRANÇAISE, 379
SERVICE MÉTÉOROLOGIQUE DU GROUPE ANTILLES–GUYANE, 406, 416, 439, 453
SERVICIO METEOROLÓGICO DEL URUGUAY, 71
SERVICIO METEOROLÓGICO NACIONAL, BUENOS AIRES, 70, 71, 72
SERVICIO METEOROLÓGICO NACIONAL DE EL SALVADOR, 408, 431, 453
SERVICIO METEOROLÓGICO NACIONAL DE HONDURAS, 408, 453
SERVICO DE METEOROLOGÍA Y COMMUNICACIONES, SECCIÓN DE CLIMATOLOGÍA, 378
SETZER, J., 255, 269
SIMPSON, J. and GARSTANG, M., 448, 453
SIMPSON, R. H., 346, 378
SKEETE, C. C., 418, 444, 453
SMEDLEY, D., 438, 441, 450, 451, 454
SMITH, C. O., *see* LÖF, G. O. G. et al.
STATISTIEK- EN PLANBUREAU, CURAÇAO, 409, 454
STENZ, E., 201
STONE, R. G., 320, 378
STRANG, D. M. D., 269

TALJAARD, J. J., VAN LOON, H., CRUTCHER, H. L. and JENNE, R. L., 194, 199, 201
TALJAARD, J. J., *see* VAN LOON, H. et al.
TANNEHILL, I. R., 479, 480, 496, 508, 509
THEAMAN, J. R., 450, 454
THÉVENEAU, A., 443, 444, 448, 454
THORNTHWAITE ASSOCIATES, 72
THORNTHWAITE, G. W., 258, 269
THORP, J., 454
TRAUTWINE, JOHN C., 371, 373, 378
TREWARTHA, G. T., 114, 115, 119, 130, 379, 419, 454
TREWARTHA, J. T., 201
TROJER, H., 365, 373, 378
TROLL, C., 11, 201, 424, 454

URRUTIA, E. C., 454
U.S. DEPARTMENT OF COMMERCE, ESSA, 379
U.S. HYDROGRAPHIC OFFICE, 410, 443, 445, 447, 451, 454
U.S. NAVAL WEATHER SERVICE, 379

U.S. NAVY, 5, 11
U.S. NAVY, CHIEF OF NAVAL OPERATIONS, 374, 378
U.S. WEATHER BUREAU, 65, 73, 302, 323, 355, 370, 374, 378, 379, 410, 420, 440, 450, 454

VALDEZ, M., *see* SAA, G. E. and VALDEZ, M.
VAN DER WEERT, R., *see* LENSELINK, H. J. and VAN DER WEERT, R.
VAN LOON, H., 1, 7, 12, 130
VAN LOON, H., TALJAARD, J. J., JENNE, R. L. and CRUTCHER, H. L., 7, 12, 201
VAN LOON, H., TALJAARD, J. J., SASAMORI, T., LONDON, J., HOYT, D. V., LABITZKE, K. and NEWTON, C. W., 12
VAN LOON, H., *see* TALJAARD, J. J. et al.
VASINO, C. J., *see* SCHWERDTFEGER, W. and VASINO, C. J.
VERVOORST, F., *see* CZAJKA, W. and VERVOORST, F.
VIDAL, A. L., *see* BURGOS, J. J. and VIDAL, A. L.
VINCENT, D. G., *see* NEWELL, R. E. et al.
VON TSCHUDI, J. J., 192, 202
VULQUIN, A., 269
VUORELA, L. A., 269

WAGNER, A., *see* KINZL, H. and WAGNER, A.
WAGNER, L. P., 454
WALKER, H. O., 199, 202

WALTER, H., and LIETH, H., 114, 130
WARD, R. DE C. and BROOKS, C. F., 440, 454
WATTS, D., *see* ROUSE, W. R. and WATTS, D.
WATTS, I. E. M., 335, 348, 350, 378
WEHEKIND, L. and SMITH, G. W., 454
WEINMAN, J. A., *see* BRINKMANN, W. L. F. et al.
WEISCHET, W. and HAVLIK, D., 202
WELCH, M. M., 12
WERNSTEDT, F. L., 379, 416, 433, 435, 454
WEST, ROBERT C., 373, 375, 378
WEST INDIES METEOROLOGICAL SERVICE, 454
WEXLER, R., 353, 378
WHITE, H. S., *see* CRY, G. W. et al.
WILHELMY, H., 371, 378
WILHELMY, H. and ROHMEDER, W., 72
WILLETT, H. C., 496, 509
WINSTON, J. S., 500, 509
WINSTON, J. S., *see* KLEIN, W. H. and WINSTON, J. S.
WISE, H., 454
WÖLCKEN, K., 63, 68, 73
WRAIGHT, J., 411, 429, 447, 451, 454

YANAI, M., 484, 509
YOUNG, J. A., KRISHNAMURTI, T. N. and LINDZEN, R. S., 348, 378

ZIMMERSCHIED, W., 160, 202, 203

Geographical Index

Abbeville, 502
Acajutla, 407, 408, 417
Aconcagua, Mount, 17, 125
— river, *see* Rio Aconcagua
Aconquija, 62
Adjuntas, 502
Afobaka, 309
Africa, 180, 185, 192, 198, 251, 334, 419, 446, 480
Alegrete, 231, 232, 238, 239, 241, 246, 247, 250, 256, 261, 292
Alexandria, 502
Alliance, 308
Altapass, 502
Alto Tapajós, 261, 278
Amapa, 245, 249
Amapala, 406, 407, 408, 415, 426, 445
Amazon Basin, 120, 154, 155, 156, 157, 158, 184, 219, 221, 222, 223, 225, 226, 228, 229, 230, 231, 233, 234, 236, 237, 238, 240, 241, 242, 243, 244, 245, 248, 249, 250, 251, 253, 254, 255, 259, 260, 261, 262, 265, 266, 298, 324, 325, 336, 337, 338, 346, 347, 359, 360, 368, 369–370, 376, 488
Amazon plains, *see* Planicie Amazonica
Amazon river, *see* Amazonas (river)
Amazonas (river), 147, 148, 220, 229, 314, 326, 346, 358, 359
— (state), 52
Ambato, 150, 164, 183
American Hill, 440
Ananea, 196
Ancón, 372
Andagoya, 300, 370, 371, 373, 374, 394
Andalgalá, 26. 34, 49, 50
Andes (Cordillera), 2, 5, 6, 8, 9, 13, 14, 16, 17, 21, 22, 25, 27, 28, 29, 36, 37, 38, 41, 44, 48, 50, 52, 58, 60, 62, 65, 67, 113, 117, 120, 121, 122, 123, 124, 125, 126, 127, 128, 147, 148, 150, 151, 153, 154, 155, 156–165, 167, 168, 169, 172, 176–179, 183, 184, 185, 188, 189, 190, 191, 192, 193, 224, 261. 296, 338, 358, 359, 360, 361–368, 376
Andes, Colombian, 361–368
Andes, Patagonian, 44, 58
Andes, Venezuelan, 339
Angel Gallardo, 50
Angola, 169
Antabamba, 193
Antarctica, 190
Antauta, 195
Antigua, 407, 440
Antofagasta, 7, 8, 39, 40, 41, 59, 116, 118, 119, 131, 134, 194

Apolo, 150, 152, 196, 214
Approuague river, 303
Aracataca, 360
Aransas Pass, 501
Arapa, 196
Araquaia, 231
Arauca, 300, 368, 389
Arenales, 8
Arequipa, 150, 152, 153, 158, 163, 165, 169, 175, 178, 179, 183, 185, 187, 188, 196, 199, 215
Argentina, 5, 13–112, 118, 120, 121, 125, 127, 128, 135, 137, 149, 153
Argentine Islands, 6
Arica, 39, 40, 59, 118, 119, 131, 132, 165
Artigas, 39, 40, 59, 61, 108
Aruba, 407, 440
Asia, 334, 500
Asunción, 38, 39, 40, 48, 59, 106
Atacama, 58
Atlantic Ocean, 7, 20, 21, 50, 51, 52, 63, 64, 65, 126, 221, 222, 228, 230, 233, 250, 251, 253, 259, 261, 407, 410, 420, 425, 426, 479–507
Atlantic Ocean, North, 4, 222, 223, 225, 243, 248, 266, 295, 298, 308, 321, 323, 329, 334, 351, 357, 485, 486, 487, 489, 490, 491, 493, 495, 497, 498, 500, 502
Atlantic Ocean, South, 3, 5, 58, 222, 223, 225, 227, 228, 243, 248, 250, 251, 265, 266, 302, 323, 485, 502
Atlantic Ocean, West, 2, 302, 488
Australia, 180, 501
Ayaviri, 196
Ayoayo, 196
Azangaro, 196
Azores, 225, 483
Azuero Peninsula, 424
Azul, 26

Bahamas, 407, 413, 414, 415, 421, 445, 451, 491, 501
Bahía, 180, 229
Bahía Blanca, 39, 40, 59
Bahía de Puno, 197
Bahía Felix, 126
Bahía Negra, 42, 52, 53
Baia de Todos os Santos, 226, 229, 244, 267
Bajo Calima, 373
Balboa, 407, 414, 456
Balmaceda, 127
Barahona Peninsula, 407, 445
Barbados, 406, 407, 413, 416, 419, 440, 444

Barbuda, 407, 440
Barcelona (Venezuela), 300, 327, 328, 344, 350, 358, 385
Bariloche, *see* San Carlos de Bariloche
Barquisimeto, 332, 352
Barra do Corda, 231, 232, 246, 247, 249, 261, 275
Barrancabermeja, 300, 367, 389
Barranquilla, 300, 361, 381, 405, 407
Barrets Gap, 435
Basilan, 501
Basseterre, 440
Batalie, 407, 440
Bayabou, 407, 440
Bay of Fonseca, 424, 426
Belém, 231, 232, 246, 247, 249, 256, 261, 265, 273
Belize (city), 406, 407, 412, 413, 424, 426, 428, 470
Belize (country), *see* British Honduras
Bellavista, 150, 154, 165, 171, 172, 180
Belo Horizonte, 231, 232, 241, 246, 247, 249, 250, 261, 288
Benguela Current, 233
Berbice river, 300, 321
Bermuda Islands, 406, 411, 415, 483
Biloxi, 506
Bio Bio river, *see* Rio Bio Bio
Blanchisseuse, 407, 440
Blonay, 368
Bluefields, 407
Boa Vista do Rio Branco, 223, 246, 247, 249, 266
Bogotá, 8, 300, 359, 365, 366, 396
Bolivia, 26, 54, 59, 62, 117, 118, 119, 120, 121, 147–218
Bolivian Chaco, 148, 156, 174
Bonaire, 407, 440
Bonne Espérance, 440
Boothville, 501
Borborema, *see* Planalto da Borborema
Bottom, 440
Bowden Pen, 434, 435
Brazil, 57, 59, 63, 64, 149, 171, 219–293, 299, 321, 326, 336, 337
Brazil Current, 19, 221, 225, 233
Bridgetown, 407, 413, 415, 440, 462
British Honduras, 407, 412, 425, 428
Brotherson Estate, 440
Brownsville, 483, 506
Bucay, 150, 165
Buenaventura, 372, 373, 374, 375
Buenos Aires (province), 18, 19, 25, 38, 42, 45, 52, 53, 54, 63, 66, 67, 68, 69, 92, 93, 96
Buenos Aires (town), 19, 36, 39, 40, 47, 49, 53, 59, 61, 68, 91

Caa Yarí, 67
Cabanillas, 196
Cabo Blanco, 42, 52, 54
Cabo Virgenes, 15, 42, 52, 54
Caceres, 238, 239
Cachi, 8
Caetite, 231, 261, 282

Caiapos Mountains, *see* Serra do Caiapos
Cajamarca, 150, 152, 161, 170, 186, 199, 211
Cajota, 196
Calacoto, 196
Calama, 118
Calamarca, 196
Calchaquí Valley, 22
Calen Fjord, 66
Callao, 150, 165, 170, 180, 181
Caluma, 150
Camagüey, 407, 475
Camarones, 15, 45
Camiri, 150, 152, 174
Campinas, 261, 289
Camp Jacob, 442
Cañadón León, 61
Canal Beagle, 127
Canal Zone, *see* Panama Canal Zone
Cañar, 150, 183
Cañete, 150, 177
Capachica, 196
Capazo, 196
Cape Gracias, 483
Cape Horn, 7
Cape of São Roque, 221
Cape Verde Islands, 483, 491
Caracas, 300, 343, 383
Caracol, 407
Cariamanga, 152
Carolinas, 498, 499
Carpish range, 156
Carriacou, 407, 440
Caribbean Sea, 265, 295, 329, 330, 335, 340, 345, 347, 348, 349, 352, 353, 357, 358, 359, 360, 361, 365, 371, 405, 407, 411, 421, 424, 436, 445, 446, 447, 448, 450, 479, 488, 491, 497, 500, 501
Cartagena, 349
Casaracra, 150
Casilda, 26
Castelar, 52
Castle Comfort, 435
Catacamas, 406, 407
Catamarca (province), 21, 26, 27, 28, 34–35, 52, 58, 60
— (town), 26, 39, 40, 43, 59, 83
Cayenne, 300, 302, 307, 395
Cayo Guano del Este, 407, 433
Central America, 348, 405–478, 505
Central Keys, 501
Central Valley (Chile), 122, 123, 124, 125
Ceres, 26, 61, 84
Cerro de Pasco, 150, 152, 157, 158
Cerro Ojos del Salado, 29
Chacaltayao, 196
Chachani, Mount, 165, 187
Chaco (province), 52, 54, 56, 64, 79
—, *see* Gran Chaco *and* Bolivian Chaco
Challapalca, 196
Challapata, 152
Chañi, 8, 62
Chapada Diamantina, 221, 228, 235, 238, 253, 254

Chapada do Araripe, 245
Chapada dos Parecis, 230
Charaña, 118, 150, 152, 164, 175, 177, 196
Charleston, 483
Cherrapunji, 370
Chetumal, 500, 501
Chiclayo, 150, 152, 171, 182, 210
Chico river, *see* Rio Chico
Chile, 22, 25, 39, 40, 57, 58, 59, 65, 113–145, 149, 166
Chillan, 39, 40, 59
Chilligua, 196
Chiloé, Isla, 39, 40, 59, 117, 123, 126
Chimax, 445
Chimborazo, 8, 147
Chimbote, 150, 152, 169, 171, 187
Chinchiná, 365, 368
Chiquimula, 407
Chiriboga, 150, 164, 179
Chone, 150
Chosica, 150, 168, 185
Chos Malal, 39, 40, 59, 95
Chubut province, 48, 99, 100, 101
— river, *see* Rio Chubut
—, town, 25, 30
Chuquibambillao, 196
Cipoletti, 43, 55, 97
Citron Valley, 435
Ciudad Bolivar, 333, 335, 344
— Mutis, 372
Coalaque, 196
Cobija, 152
Cocuy, 8
Codrington, 440
Colombia, 149, 190, 295, 296, 298, 300, 326, 329, 337,
    348, 349, 353–376, 380, 381, 387, 389, 391, 394,
    396, 398, 402, 405, 407, 410, 412, 419, 421
Colombian Altiplano, 366, 367
Colón, 406, 407, 448, 457, 478
Colonia Alvear, 50
Colonia Tovar, 300, 384
Colorado river, *see* Rio Colorado
Columna, 8
Comodoro Rivadavia, 15, 30, 38, 39, 40, 41, 52, 57,
    59, 101
Conceicão do Araguaia, 246, 247, 250
Concepción (Bolivia), 152
— (Chile), 39, 40, 59, 114, 122, 123, 129, 131, 140
— (Paraguay) 39, 40, 59
Concordia, 68, 85
Constitución, 123, 124
Copacabana, 196
Copiapó, 39, 40, 59, 115, 117, 118, 120, 121
— river, *see* Rio Copiapó
Cordillera Azul, 155
— Blanca, 8
— Central, 362, 363
— de la Costa, 326, 331, 332, 333, 345, 350, 352, 354,
    358
— de Mérida, 326, 332, 333, 334, 352
— Huayabamba, 8

— Macarena, 369
— Occidental, 361, 365, 371, 373
— Oriental, 156, 360, 362, 369
— Real, 8, 147, 157
Córdoba, province, 19, 36, 38, 52, 53, 55, 58, 64, 69
—, town, 26, 34, 39, 40, 59, 86
Coro, 327, 328, 344, 354, 355, 356, 357, 358, 361
Corpus Christi, 483
Corrida de Cori, 22, 27, 29, 36, 41, 47–48, 50, 53, 59,
    118, 121, 131, 135
Corrientes, 20, 39, 40, 48, 49, 51, 57, 59, 61, 67, 81
Corumbá, 48, 241, 261, 286
Cossak, 501
Costa Rica, 407, 415, 417, 423, 424, 426, 427, 458, 483
Cotopaxi, 147, 154, 158, 161
Cristóbal, 407, 414, 457
Cristo Redentor, 29, 43, 44, 46, 47, 48, 50, 61, 125, 131,
    137
Crowley, 502
Crucero, 196
Cruta, 406, 407, 408
Cruz Bay, 440
Cruzeiro do Sul, 246, 247, 249, 250
Cuba, 406, 407, 411, 413, 417, 419, 421, 433, 434, 437,
    447, 451, 483, 501, 502
Cuchilla Grande, 63
Cuenca, 152
Cuiabá, 231, 246, 247, 249, 250, 261, 285
Cumbal, 8
Curaçao, 335, 336, 349, 350, 351, 355, 356, 406, 407,
    409, 414, 419, 440, 444, 445, 461
Curicó, 39, 40, 59, 122
Curitiba, 231, 232, 241, 246, 247, 250, 261, 290
Cutuco, 407
Cuyo, 17–21, 25, 28–29, 30, 34, 35, 38, 45, 46, 48, 49,
    52, 55, 56, 58, 63, 66, 68
Cuzco, 150, 152, 154, 158, 162, 186, 199, 213

Dakar, 446
Darwin Cordillera, 66
Depressão Paulista, 249
— Central, 263
Desaguadero, 17, 196
Desaix, 406, 407
Désirade, 407, 443
Desolación, Isla, 39, 40, 59
Diamantina plateau, *see* Chapada Diamantina
Dominica, 407, 440, 443
Dominican Republic, 445, 483
Dorothea, 440
Drake Passage, 7, 37
Dunscombe, 440
Durazno, 39, 40, 59

Ecuador, 147–218, 371, 372
El Alto, 194, 195, 199, 215
El Belen, 196
El Corazón, 150, 164

El Misti, 150, 158, 165, 175, 178, 187
El Niño, 169–171, 182, 191–192
El Porvenir, 150
El Progreso, 150, 154, 196
El Salvador, 407, 408, 409, 410, 417, 430, 451, 483
Embalse de Guárico, 334
Entre Rios, 18, 38, 51, 52, 55, 67, 68, 85
Equatorial Current, 221
Esmeraldas, 148, 150, 166, 372
Espigao Mestre, 248
Espinal, 300, 398
Esquel, 15
Esquilache, 196
Essequibo river, 321
Europe, 47, 161
Evangelistas, 5, 39, 40, 59, 114, 125, 126, 127, 128, 131, 144
Ezeiza, 52

Falkland Current, 19, 38
Falkland Islands, 39, 40, 59
Famatina, 8
Farm Hill, 450
Faro Cabo Blanco, *see* Cabo Blanco
Faro Cabo Virgenes, *see* Cabo Virgenes
Faro Punta Delgada, *see* Punta Delgada
Fellowship, 435
Fernando Noronha, 230, 231, 232, 241, 246, 247, 249
Fiambala, 17
Flores, 407
Florianópolis, 238, 241
Florida, 407, 410, 411, 498, 501, 502, 505, 506
— Keys, 501, 502
Fonseca Bay, 407
Formosa, Minas Gerais, 231, 232, 246, 247, 249, 261, 280
—, province, 21, 44, 48, 51, 64, 75
Fortaleza, 231, 232, 246, 247, 249
Fort Barrancas, 506
Fort-de-France, 407, 413, 414, 465
Fortin Mercedes, 50
Fort Sherman, 478
Frank Helm Bay, 501
French Guiana, 298, 299–305, 324, 395, 400
Frutillar, 124

Galapagos, Islas, 147, 149, 153, 154, 165, 171, 180, 182, 184, 188, 205
Gallegos, 39, 40, 59
— river, *see* Rio Gallegos
Galveston, 506
Georgetown, 300, 307, 323, 390
Gobernador Gregores, 15, 16, 26, 30
Golfo, *see also* Gulf
Golfo de Peñas, 126
— — Venezuela, 326, 340, 347, 353
— San Matias, 16
Government Farm (Tobago), 407, 440

Grajau, 220
Gran Chaco, 13, 17, 21, 22, 27, 28, 35, 37, 38, 41, 43, 47, 48, 49, 51, 52, 55, 57, 58, 60, 62, 63, 66, 67, 69
Grand Bourg, 440
— Cayman, 410
— Etang, 440
— Turks, 406, 407
Great Abaco, 501
Greater Antilles, 407, 436, 444, 447
Grenada, 407, 440
Greytown, 407, 424
Grytviken, 61
Guabito, 407
Guadeloupe, 406, 407, 441, 442, 443, 444
— Islands, 438, 443
Guafo, Isla, 39, 40, 59
Gualeguaychú, 68
Guam, 501
Guanaja, 407, 408, 432, 433
Guaqui, 196
Guarello, Isla, 65, 127
Guatemala, 407, 417, 424, 445, 451, 483
— City, 407, 413, 466
Guayaquil, 148, 149, 150, 152, 166, 180, 183, 184, 187, 209
Guayas, 166
Guemes, 49, 50
Guiana, French, *see* French Guiana
Guianas, 261, 296, 302, 306, 307, 314, 419, 421
Gulf Coast, 498, 499, 500, 503, 506
— of Anoud, 122
— — Campeche, 419
— — Fonseca, 445
— — Guayaquil, 166
— — Mexico, 264, 407, 410, 411, 412, 436, 437, 479, 491, 501, 502, 505, 506
— — Panamá, 372, 374, 375
— — Paria, 340
— — San Jorge, 54
— — the Mosquitoes, 419
Gustavia, 406, 407, 440
Guyana, 295, 298, 300, 308, 309, 321–326, 390, 391, 401
— Highlands, 296, 321, 326, 328, 329, 337–338, 341, 346, 347

Haiti, 407, 412, 413, 483
Havana, 407, 413, 448, 483, 501
Hawaii, 370
High Island, 506
Hispaniola, 407, 417, 419, 421, 435–436, 437, 445, 488
Honduras, 406, 407, 408, 410, 415, 424, 426, 447, 449, 483
Houston, 483
Huancané, 196
Huancayo, 9, 150, 163, 167, 182, 183, 193, 199, 212
Huánuco, 150, 152, 156, 157
Huaraya, 196
Huascarán, 147

Hudson, 8
Humboldt Current, 148, 150, 169, 172, 173, 174, 180

Iauareté, 239, 240
Ibarra, 150, 159, 182
Ica, 150, 187
Ichuña, 196
Iguape, 232
Ilave, 196
Île de Cayenne, 304
Ilo, 196
Imata, 150, 175, 176, 196
Imperatriz, 239
India, 334, 370
Indian Ocean, 501
Indonesia, 348
Iquique, 39, 40, 59, 118, 119, 131, 133
Iquitos, 150, 152, 154, 155, 158, 175, 185, 210
Isla de Diego, 39, 40, 59
— del Rey, 407, 432, 433
— de Margarita, 340, 349, 350, 353
— Desolación, 39, 40, 59
— Guafo, 39, 40, 59
— San Cristobal, 154, 184, 205
— San Felix, 39, 40, 59
— San José, 407, 432, 433
— Santa Cruz, 154, 165, 180
Islas Juan Fernández, 5, 39, 40, 59
Itajai, 230, 232
Izobamba, 150, 161, 182, 183, 186, 199, 204

Jachal, 29
Jacksonville, 483
Jama, 150, 171
Jamaica, 407, 413, 416, 433–435, 445, 448, 449, 450, 483, 502
Jaque, 407, 455
Jauja, 150
Jequitinhonha, 243
João Pessoa, 238, 239
Joazeiro, 231, 232
Jose C. Paz, 50
Juan Fernández, Islas, 5, 39, 40, 59
Jujuy, province, 22, 25, 34–35, 44, 51, 52, 53, 55, 57, 60, 67, 74
Juli, 196
Juliaca, 150, 152, 163, 177
Junín, 39, 40, 59, 92

Kabelstation, 313
Kauai, 370
Key West, 483
Kingston, 407, 413, 445, 450, 471
Kingstown, 407, 440
Knip, 440
Kralendijk, 440

La Asunción, 407, 440, 441
Labrador, 450
La Casualidad, 27, 29, 36, 48, 77
La Ceiba, 407, 408
Lafayette, 502
La Florida, 300, 368, 402
La Galera, 196
Lago Argentino, 15, 46, 66, 127
Lago Buenos Aires, 127, 128
Lago del Toro, 127
Lago de Maracaibo, 353
Lago Llanquihue, 122
Lago Poopo, 177
Lago Quillen, 67
Lago San Martin, 127, 128
Lago Titicaca, 147, 150, 163, 186, 193, 195–199
Lago Viedma, 127
La Gran Sabana, 337, 341
La Guajira Península, 355, 360, 376
Lake Bangweulu, 199
Lake Kivu, 199
Lake Maracaibo, 407
Lake Victoria, 199
Lake, *see* also Lago
Lambrama, 193
Lamentin, 406, 407
Lampa, 196
La Orchila, 340, 349, 355, 356
La Palma, 372
La Pampa, 18, 35, 64, 94
La Paz, 148, 149, 150, 152, 154, 158, 159, 164, 177, 187, 193, 195, 196, 199, 215, 216
La Poma, 22
La Quiaca, 22, 23, 25, 27, 32, 33–35, 36, 43, 44, 45, 46, 47, 48, 54, 55, 56, 57, 61, 67, 68, 74, 164, 194, 195
La Rioja, province, 21, 22, 25, 26, 28, 29, 34–35, 38, 58
La Rioja, town, 26, 27, 34, 40, 59
La Serena, 39, 40, 59, 114, 118, 121, 131, 136
Las Flores, 39, 40, 59, 93
Las Lomitas, 21, 40, 59, 75
La Soledad, 150, 154
Las Piedras, 300, 380
Latacunga, 150, 161
Lebu peninsula, 123
Leeward Islands, 488
Lesser Antilles, 407, 419, 421, 438–445, 480, 483, 491
Lethem, 401
Lima, 8, 9, 147, 148, 149, 150, 152, 153, 165, 167, 168, 169, 171, 172, 179, 180, 181, 182, 184, 185, 186, 187, 194, 212
Limbani, 196
Lita, 150
Litoral Sudeste, 221
Llailio, 196
Llaima, 8
Llanos, 298, 309, 324, 326, 329, 333, 334–337, 340, 341, 342, 344, 346, 347, 350, 352, 357, 359, 366, 368–369
Llanquihue, Lake, *see* Lago Llanquihue

Lloró, 370
Llullaillaco, 62
Loa river, *see* Rio Loa
Lobitos, 150, 170, 372
Locumba, 196
Loreto, 80
Los Angeles (Chile), 39, 40, 59, 123
Lower Canada, 440
Luanda, 169
Luzon, 501

Macachín, 39, 40, 94
Macusani, 196
Madden Dam, 407, 445
Maine, 450, 505
Maipo, 8
Maiquetia, 342, 343, 345, 349, 350, 351, 354, 355, 356
Malacatos, 150, 173
Managua, 407, 460
Manaus, 231, 232, 241, 261, 272, 336
Mañazo, 196
Manta, 150, 152, 166, 171, 372
Mantiqueira, *see* Serra da Mantiqueira
Mapocho river, *see* Rio Mapocho
Maquinchao, 26, 30
Maracaibo, 300, 326, 332, 333, 336, 338, 340, 344, 352, 382
Maracaju, 230, 243
Maracay, 343
Mar Chiquita, 39, 40, 59
Mar del Plata, 19, 20, 39, 40, 42, 43, 44, 45, 57, 59, 61, 67, 96
Margarita, 407, 438, 439, 440, 441
Marie Galante, 407, 440
Maripasoula, 300, 304, 400
Mariscal Estigarribia, 21, 40, 43, 48, 61, 104
Marquisat, 442
Martinique, 406, 407, 413, 416, 438, 439, 441, 443, 444, 448
Massies Mill, 502
Matagorda, 506
Matecumbe Key, 501
Mato Grosso, 26, 230, 233, 236, 242
Matucana, 150, 179
Maule river, *see* Rio Maule
Mazocruz, 196
Medellín, 300, 366, 391
Meio Norte, 220, 229, 230, 236, 242, 243, 244, 248, 259, 261
Melo, 39, 40, 59
Mendoza, province, 25, 28, 29, 30, 32, 33, 48, 68, 69
Mendoza, town, 18, 29, 39, 40, 56, 59, 88
Mercedes, 8
Mérida, 300, 327, 328, 333, 336, 338, 341, 342, 386
Meridional plateau, *see* Planalto Meridional
Merume Mountains, 321
Mesa de Guanipa, 344
Mexico, 483, 496, 500, 501
Miami, 483, 503

Miami Beach, 501
Milagro, 150
Millbank, 435
Misiones, 20, 25, 42, 44, 48, 49, 51, 53, 54, 57, 58, 63, 64, 67, 68, 80
Mississippi delta, 412
Mitchell Fjord, 66
Mobile, 483
Moho, 196
Mollendo, 150, 152
Mona Passage, 407, 417
Montserrat, 407, 440
Montevideo, 19, 32, 39, 40, 45, 46, 59, 68, 111
Monts Tumuc-Humac, 303
Moore Town, 435
Moquegua, 196
Morant Bay, 435
Moreno glacier, 66
Morococha, 150
Moule à Chique, 407, 440
Mt. Waialeale, 370
Moyobamba, 150, 152, 186
Muisne, 150, 152, 171
Muñanio, 196

Nahuel Huapi, 46
Nassau, 406, 407, 413, 414, 415, 477
Natal, 230
Netherlands West Indies, 336
Neuquén, province, 16, 25, 30, 33, 34, 35, 45, 53, 95
Neuquén, town, 39, 40, 59, 61
New Amsterdam, 300, 391
New England, 161, 500
New Orleans, 483
Nicaragua, 180, 407, 415, 424, 433, 448, 483
Nicoya Peninsula, 407, 417
Nieuw Nickerie, 310
Norfolk, 483
North America, 38, 410, 417, 447
Nuñoa, 196

Oiapoque, 223, 265
Ojos del Salado, 8
Olinda, 231, 232, 246, 247, 250, 277
Ollague, 118, 121
Oploca, 150, 152, 174
Oranjestad, 440
Orotina, 407, 415
Oroya, 150, 152, 158, 162, 184, 185
Oruro, 150, 152, 158, 164, 175, 177, 187, 217
Osorno, 8, 39, 40, 59, 123
Otora, 196
Oyapock river, 303

Pachamama, 150, 161
Pacific Ocean, 6, 7, 9, 13, 50, 65, 113, 126, 171, 172, 188, 222, 224, 355, 358, 371, 374, 375, 407, 410,

412, 419, 420, 421, 424, 425, 426, 448, 449, 479, 496
— —, East, 2, 13
— —, North, 295
— —, South, 3, 5, 189
— —, Southeast, 5
—, Western, 501
Padre las Casas, 407
Pakaraima Mountains, 321
Palisadoes, 471
Pampa de La Joya, 189, 190
— del Castillo, 38
— de Majes, 150
Panamá, 371, 372, 407, 424, 431, 450, 451
— Canal, 7, 374, 414
— — Zone, 415, 445, 451, 478
— City, 407, 424
Pantanal Matogrossense, 221, 230, 233, 236, 237, 238, 240, 241, 242, 244, 249, 258, 261
Paraguay, 13–112, 149, 186
Paraguay river, *see* Rio Paraguay
Paraiba river, 221, 238, 243
Paramaribo, 300, 307, 308, 309, 310, 311, 312, 313, 316, 317, 318, 319, 320, 392
Paraná, 39, 40, 59
— river, *see* Rio Paraná
— valley, 221
Parnaiba, 220
Pass Christian, 506
Paso de la Patria, 81
— — los Libres, 26
— — los Toros, 109
— Real, 407, 476
Pastaza, 150
Patacamaya, 196
Patagonia, 13, 14–17, 18, 19, 25, 27, 30–31, 33, 34, 35, 36, 37, 38, 42, 45, 46, 47, 48, 49, 50, 51, 52, 53, 54, 55, 56, 57, 58, 60, 63, 64, 65, 66, 67, 68, 127, 128
Patagonian Andes, 44, 58
Paulista plateau, *see* Planalto Paulista
Paysandú, 39, 40, 59
Peñas, 196
Peninsula de Paraguaná, 349, 356
— — Paria, 349
— — Taitao, 39, 40, 59
— Valdés, 47, 53, 54
Pergamino, 50
Perito Moreno, 15
Peru, 9, 58, 115, 117, 119, 147–218, 354, 372
— Current, 117, 189
Philippines, 348, 501
Philipsburg, 440
Piarco, 407, 413, 459
Piaui Mountains, 220
Pichacani, 196
Pichanal, 67
Pichilingue, 150, 166, 183, 187, 207
Pilar, 36, 39, 40, 59
Pinar del Rio, 407, 411
Pindapoy, 67

Piracicaba, 231
Pirapora, 246, 247, 249
Pisco, 150, 152, 169, 170, 185, 187
— Bay, 170, 187
Piura, 147, 150, 152, 169, 171
Pizacoma, 196
Planalto Central, 220, 221, 226, 228, 229, 238, 240, 243, 244, 248, 249, 253, 254, 261
— Centro-Oriental, 221, 235, 237, 242, 243, 244, 245, 249, 254, 258
— da Borborema, 220, 228, 230, 232, 233, 237, 238, 240, 242, 243, 251, 253, 258, 263, 267
— da Mantiqueira, 230, 254
— Meridional, 221, 230, 232, 233, 235, 236, 237, 238, 242, 243, 244, 245, 246, 254, 261, 263
— Paulista, 221, 238, 251, 261, 263
Planicie Amazonica, 220, 233, 238, 261
— do Nordeste, 233
— Riograndense, 221, 229, 230, 233, 237, 238, 240, 241, 242, 244, 245, 250, 254, 255, 261, 263
Playas, 150, 171
Plumb Point, 449
Plymouth (Montserrat), 440
Pointe-à-Pitre, 407, 414, 468
Point Saline, 440
Poopo, 150
Porlamar, 340, 350, 358
Port Aransas, 506
— au Prince, 406, 407, 413, 414
— Lavaca, 501, 506
— of Spain, 407, 440, 483
— Stanley, 7, 61
Porto Alegre, 261, 293
— Nacional, 231, 232, 261, 279
Portoviejo, 150, 166, 180, 183, 184, 187, 206
Posadas, 39, 40, 59, 61, 67, 80
Potrerillos, 118
Presidencia Roque Saenz Peña, 21, 34, 39, 40, 59, 79
Pucallpa, 150, 152, 154, 155, 158, 174, 186
Pucaráo, 195
Puente del Inca, 29
Puerto Aisen, 127
— Armuelles, 407, 431
— Ayacucho, 300, 333, 335, 336, 337, 341, 343, 344, 368, 393
— Baquerizo, 150, 154, 171, 180, 184, 205
— Barrios, 407, 424
— Bolívar, 150, 171
— Casado, 40, 59, 105
— Chicama, 150, 170, 187
— Cortés, 407, 426
— Deseado, 15, 16
— Guarani, 39, 40, 59
— Ila, 150, 152
— Maldonado, 150, 152, 213
— Montt, 8, 39, 40, 59, 114, 115, 116, 122, 131, 142
— Natales, 128
— Presidente Franco, 21, 107
— Rico, 407, 417, 418, 438–445, 450, 451, 483, 502
— Santa Cruz, 15, 102

Puna plateau, 157
Puno, 150, 175, 177, 196, 197
Punta Areñas, 39, 40, 59, 61, 114, 131, 145
— de Atacama, 67
— del Agua, 22, 25, 27, 61
— — Este, 19, 21, 26, 39, 40, 42, 43, 44, 45, 48, 54, 57, 59, 67, 68, 112
— Delgada, 54
— Maisí, 407, 474
Punto Deseado, 42
Putina, 196
Putumayo, 150, 152, 174
Puyo, 150, 155, 156, 183, 184, 185, 208

Quibdó, 365, 370, 371, 373, 375
Quintero, 8, 116
Quito, 148, 150, 152, 154, 158, 159, 160, 161, 163, 164, 174, 179, 183, 186, 199, 203
Quixeramobim, 240, 261, 276

Radnor, 450
Raizet, 406, 407
Ramirez, 39, 40, 59
Recife, 241, 250, 261, 277
Região Leste, 230
Remanso, 246, 247, 249, 250, 261, 281
Riberalta, 150, 152, 173, 175
Rinçon, 440
Rio Aconcagua, 39, 40, 59, 122, 123
— Alto Cuyuni, 337
— Apure, 326, 347
— Apurimac, 161
— Arauca, 333, 341
— Aroa, 332
— Atrato, 359, 370, 371, 373, 375, 376
— Beni, 157
— Bio Bio, 39, 40, 59, 122, 123, 124
— Branco, 229, 230, 248
— Cauca, 360, 361, 362, 363, 364, 365, 367
— Chicama, 161
— Chico, 39, 40, 59
— Chubut, 39, 40, 59
— Coata, 196
— Colorado, 13, 14, 17, 39, 40, 59, 64
— Copiapó, 39, 40, 59, 117, 119
— de Janeiro, 231, 232, 246, 247, 251, 261, 291
— de la Plata, 19, 21, 34, 35, 38, 48, 51, 52, 53, 54, 57, 58, 67, 68
— Desaguadero, 197, 198
— Deseado, 37
— Doce, 231, 243, 249
— Esmeraldas, 154
— Gallegos, 15, 16, 38, 39, 40, 52, 59
— Grande (Tierra del Fuego), 38
— — do Sul, 237
— — plains, *see* Planicie Riograndense
— Guayas, 148
— Huancané, 196

— Ilave, 196
— Limay, 37
— Loa, 121
— Magdalena, 359, 360, 361, 362, 363, 364, 365, 366, 367, 376
— Mantaro, 161
— Mapocho, 124
— Marañon, 161
— Masparro, 347
— Maule, 39, 40, 59, 124
— Meta, 326, 333
— Mira, 164
— Nechí, 365
— Negro, Argentina, 13, 14, 16, 17, 33, 39, 40, 48, 59, 64
— —, Brazil, 231, 232, 244, 245, 248, 249, 251, 259, 326, 337, 358
— —, province, 25, 34, 35, 45, 97
— Orinoco, 261, 295, 326, 331, 333, 334, 337, 341, 347, 358, 368, 369
— Paraguay, 13, 38, 39, 40, 48, 50, 51, 52, 58, 59, 219, 220, 221, 228, 229, 230, 243, 244
— Paraná, 13, 39, 40, 42, 50, 51, 52, 53, 58, 59, 219, 220, 228, 229, 230, 236, 237, 241, 242, 243, 244, 249, 256, 258, 261
— Paucartambo, 157
— Pita, 150, 175
— Ramis, 196
— Rimac, 168, 185
— San Juan, Argentina, 22
— — —, Colombia, 371, 376
— Santa Cruz, 39, 40, 59
— Solimões, 259
— Suchez, 196
— Tocuyo, 332, 347
— Tuy, 333, 347, 350
— Unare, 333, 350
— Urubamba, 161
— Uruguay, 39, 40, 51, 59, 230, 263
— Yaracuy, 332, 347
Riobamba, 150, 152, 183
River, *see* Rio
Road Town, 407, 440
Roboré, 152
Rochambeau, 300, 304, 395
Rosario, 39, 40, 59, 89
Roxborough Estate, 407, 440

Saba, 407, 440
Sahara, 185, 446
St. Barthélémy, 407, 440
— Croix, 407, 440
— Eustatius, 407, 440
— Georges, 304
— Ignatius, 300, 401
— John, 407, 440
— John's, 407, 440
— Kitts, 407, 440, 441
— Laurent du Maroni, 304

St. Lucia, 440, 444
— Maarten, 407, 440
— Martin, *see* St. Maarten
— Thomas, 407, 440, 450
— — College, 440
— Vincent, 407, 440
Sajama, 8
Salado, 17
Salamanca, 407, 440
Salinas, 150, 152, 164, 166, 171, 179, 180
Salinas-Imbabura, 150
Salta, province, 27, 34–35, 44, 45, 51, 52, 53, 54, 55, 57, 60, 67, 77
—, town, 26, 29, 34, 39, 40, 54, 59, 76
Salto, 39, 40, 59
— Grande, 85
Salvador (Bahía), 180, 231, 232, 241, 246, 247, 250, 261, 283
San Andrés, El Salvador, 407, 408, 417
— —, island, 407, 432, 433
— — Osuna, 407
— Antonio, Venezuela, 335, 337, 352
— — de los Cobres, 29
— Carlos de Bariloche, 39, 40, 43, 57, 59, 61, 67, 98
— — — Rio Negro, 300, 331, 337, 341, 401
— Felipe, 59
— Felix, Isla, 39, 40, 59
— Fernando, Chile, 124
— —, Venezuela, 300, 327, 328, 333, 335, 336, 337, 341, 344, 388
— — de Atabapo, 341
— Ignacio, 152
— José, 407, 458
San Juan, Argentina, 26, 39, 40, 59, 61, 87
— — del Norte, 407, 424
— —, Peru, 150, 152, 214
— —, province, 22, 28, 29, 33, 57, 58, 60, 63, 66, 67, 68
— —, Puerto Rico, 405, 406, 407, 414, 450, 472, 483, 501
— — river, *see* Rio San Juan
— Julian, 15
— Lorenzo, 150, 152, 165, 183
— Luis, province, 19
— —, town, 26, 28, 39, 40, 59, 61, 90
— Miguel, 407, 408
— Pedro, 39, 40, 59, 114, 126, 128, 131, 143
— — Sula, 407, 408
— Rafael, glacier, 9
— Ramón, 150, 176
— Salvador, 407, 408, 409, 410, 411, 414, 418, 428, 429, 430, 431, 432, 445, 463
Santa Ana, 150
— Clara, 372
— Cruz, Argentine province, 30, 67, 102
— —, Argentine town, 39, 40, 59
— —, Bolivian department, 148
— —, Bolivian town, 150, 152, 156, 174, 186, 216
— — Porrillo, 407, 408, 415
— — river, *see* Rio Santa Cruz
— Domingo, 407

— Elena, 300, 327, 328, 332, 333, 335, 337, 341, 344, 347, 397
— — peninsula, 166, 180
— Fé, province, 18, 55, 63, 67, 84, 89
— —, town, 42
— Lucia, Lesser Antilles, 407
— —, Peru, 193, 196
— Maria, Argentina, 22
— —, Vieques Island, 440
— Marta, 8, 360, 361
— Rosa, Argentina, 50
— — de Copán, 407, 408
— —, Peru, 196
— Tecla, 407, 408
— Teresa, 407
Santarem, 261, 274
Santiago de Chile, 39, 40, 59, 123, 124, 129, 131, 139
— del Estero, province, 21, 26, 34, 48, 61, 64
— — —, town, 39, 40, 59, 82
Santo Domingo, 473
Santos, 231, 232, 238, 239
São Francisco river, 220, 221, 228, 229, 230, 236, 237, 240, 241, 243, 244, 247, 249, 250, 253, 258, 259, 261
— — valley, 220, 221
— Jeronimo, 230
— Paulo, 7, 8, 57, 244, 255
— Rogue, 250
Sarmiento, 15, 26, 30, 38, 39, 40, 41, 48, 59, 61, 100
Segovia Highlands, 332, 352, 354
Sena Madureira, 261, 271
Sepacuité, 407
Seridó, 258, 259
Serra da Canastra, 230, 245, 253
— — Ibiapada, 228
— — Mantiqueira, 221, 238, 245, 253, 259
— de Ibiapaba, 220, 253
— de Maracaju, 228
— do Caiapos, 230, 253
— — Espinhaço, 221, 228, 238, 254
— — Mar, 237, 238, 243, 253
— — Parecis, 253
— Geral, 233, 235, 237, 243, 253, 263
— Lombard, 303
Serranía de Baudó, 371
Seymour Island, 150, 165, 171, 420
Shawford, 407, 440
Sierra de Chañi, 58
— — Perijá, 332
— Nevada de Santa Marta 360
Sierro do Mar, 57
Silver Hill, 450, 502
Sina, 196
Sipaliwini, 300, 309, 310, 319, 403
Sorata, 196
Soto, Isla, 195, 196
Spanish Town, 407, 440
Strait of Magellan, 47, 117, 126
Sucre, 150, 152, 157, 163, 193
Sunning Hills, 435

Surinam, 295, 298, 299, 300, 303, 305–321, 324, 325, 329, 340, 351, 392, 399, 403
Swan Island, 407, 432, 433, 469, 484, 485, 488

Tabiazo, 150, 154
Tacagua, 150, 177
Tacajo, 502
Tacna, 150, 152, 169, 217
Tacuarembó, 39, 40, 59
Tafelberg, 300, 310, 319, 399
Taisha, 150, 152, 156
Taitao peninsula, *see* Peninsula de Taitao
Talara, 150, 152, 169, 171, 181
Talca, 39, 40, 59, 123
Talenga, 150, 157
Tallahassee, 505
Tamarugal Pampa, 119
Tambopata, 196
Tampa, 483
Tampico, 483
Tapajos, 220
Tapanahoni river, 308
Taperinha, 274
Taquili, Isla, 195, 196
Taraco, 196
Tarija, 150, 152, 157, 187
Tarma, 150, 184
Taylor, 502
Tegucigalpa, 407, 408, 464
Tela, 407, 408, 467
Temuco, 39, 40, 59, 122
Tena, 150, 152, 154, 155, 161, 186
Texas, 498, 501, 502, 503, 504, 505, 506
Texis Junction, 407
Tierra del Fuego, 7, 8, 13, 14–17, 30, 31, 32, 33, 34, 35, 37, 38, 44, 46, 47, 48, 49, 51, 53, 54, 55, 60, 65, 66, 67, 68, 103, 126, 127, 128
Timbalier Bay, 506
Tingo María, 150, 152, 155, 156, 175, 176, 186, 211
Tiputini, 150, 175
Tiquina, 196
Titicaca Basin, 163, 175, 177, 178, 195–199
Tobago, 407, 440
Tocantins, 220, 231, 236
Tolima, 8
Toncontín, 464
Top Hill, 440
Torata, 196
Toroya, 196
Tortola, 407, 440
Treinta y Tres, 39, 40, 57, 59, 61, 110
Trelew, 15, 39, 40, 59, 61, 99
Trenque Lauquen, 61
Tres Esquinas, 369, 370
— Lagoas, 231, 232, 241, 246, 247, 249, 250, 261, 287
Trinidad, Bolivia, 150, 152
—, Lesser Antilles, 407, 412, 413, 440, 441, 442, 444, 483
Tronador, 8

Trujillo, 150, 170, 182
Tucumán, province, 25, 34–35, 51, 53, 54, 57, 58, 60, 67, 68
—, town, 26, 27, 34, 39, 40, 55, 59, 78
Tulcán, 150, 159, 183
Tumaco, 300, 372, 374, 394
Tumbes, 170
Tumeremo, 337
Tupungato, 8
Turbo, 300, 373, 374, 375, 387, 407, 412

Uaupés, 231, 232, 241, 246, 249, 261, 270, 335, 337
Umayo, 196
Uplyme, 407, 440
Upsala glacier, 66
Uribia, 300, 380
Uruguay, 13–112
— river, *see* Rio Uruguay
U.S.A., 32, 496, 498, 499, 500, 505
Ushuaia, 7, 8, 15, 25, 30, 33, 35, 38, 39, 40, 42, 45, 46, 56, 59, 61, 103, 116

Valdés peninsula, *see* Peninsula Valdés
Valdivia, 39, 40, 59, 114, 123, 131, 141
Valentin, 8
Vallenar, 39, 40, 59, 118
Valparaíso, 39, 40, 59, 114, 117, 123, 124, 131, 138
Velluda, 8
Venezuela, 4, 295, 296, 298, 300, 309, 321, 324, 326–358, 361, 368, 369, 380, 382, 383, 384, 385, 386, 388, 393, 397, 401, 407, 419, 421, 441, 442, 488
Viacha, 196
Victorica, 26
Vera Cruz, 483
Viedma glacier, 66
— Lake, 66
Vieques Island, 407, 440
Villa Mills, 407, 415
Villarrica, 39, 40, 59
Villavicencio, Argentine, 29
—, Colombia, 300, 368, 398
Virgin Gorda, 407, 440
— Islands, 407, 418, 450
Visayas, 348
Vitor, 150
Vitoria, 223, 261, 284
Vizcachani, 196
Volcan, 8

West Indies, 410, 416, 418, 436, 437, 442, 443, 446, 447, 479, 484, 505
— Palm Beach, 483
Willemstad, 355, 356, 461
Wilmington, 483
Windsor, 435
Windward Islands, 488
Windwardside, 440

Yacuíba, 150, 152, 156, 174, 175, 176, 186, 218
Yaupi, 150, 152, 157, 158
Yauricocha, 150, 158
Yeguas, 8
Yucatan, 407, 412, 415, 416, 417, 419, 483, 491
— Straits, 407, 412
Yuncan, 150, 157
Yungas, 156, 157, 176

Yunguyo, 196
Yúrac, 150, 155

Zacapa, 407
Zamora, 150, 156
Zanderij, 309, 310, 315, 316, 317, 319
Zaruma, 150

Subject Index

Airmasses, AR* 27, 38, 43, 60; CHI 116; BRA 225; CAM 418
Anticyclones, blocking, AR 14, 48, 65
—, position, 3
—, subtropical, 5; AR 13, 17, 18; CHI 115, 117; BRA 222, 225, 265
Anticyclonic ridge over the Andes, 7, 194
Aridity, CHI 117–122
Atmospheric circulation, SAM 1–7

Birthplace of Atlantic hurricanes, 480
Brazil current, APU 19; BRA 225

Caatinga, BRA 221
Calms, frequency, AR 18; EC 186; BRA 229; CAM 408
Cerrado, BRA 221, 259
Clear days, number, AR 29, 31, 32
Climate diagrams, AR 43; BRA 256
— types, BRA 225; VE 328, 338; CO 359
Climatic factors, BRA 219
— strata, VE 338
— tables, APU 74–112; CHI 131–145; PBE 203–218; BRA 270–293; NSA 380–403; CAM 455–477
— zones, BRA 257
Cloud observations in the tropics, CAM 412
— regime of Patagonia, AR 31
Cloudiness, AR 23, 24, 31; PE 181; BRA 229; CAM 412
— and elevation, AR 31
— and hurricanes, 504
— and land areas, CAM 412
—, mean, SAM 297
— and ocean areas, CAM 413
— tables, see Climatic tables
Cloudy days, number, AR 31, 32
Cold fronts, VE 345
— —, crossing equator, NSA 346
— pool disturbances, NSA 345
Constancy of the tradewinds, 488
Continental troughline, AR 17
Continentality, AR 17, 44; BRA 240
Crónicas Potosinas, PBE 163

Desert, coastal, — theory, 188–192
— conditions, CHI 121
Dew-point temperature, SU 316
— —, diurnal variation, SU 316
— —, monthly means, SE 319
Discharge of river Rimac, PE 168
— of Lake Titicaca, 197
Disturbances, classification of tropical, 479
Drought, BRA 263, 264
Dry zone, S. Caribbean, 348–357
Dunes, AR 64; PBE 188, 189; NSA 333
Dust, see Haze
— storms, AR 23, 64; PBE 186

Easterly waves, NSA 347, 488
Ekman drift, 189, 190
El Niño, duration, BPE 169
— —, precipitation, PBE 170, 171
— —, pressure, PBE 182
— —, temperature, PBE 169, 182
— —, theory, 191
Equatorial current, BRA 221
— front, BRA 221
— trough, BRA 223; NSA 296, 334, 346
Evaporation, SAM 11; APU 74–112; BRA 270–293; SU 305–319; NSA 382–403; CAM 456–472
—, high Andes, 9; AR 62
—, Lake Titicaca, 196
Evaporimeter, types used, APU 70–71; BRA 267

Falkland (Malvinas) current, APU 19, 38
Firn line, AR 62
Floods, BRA 263, 264; VE 334, 345
— from hurricanes, 501
— from sudestadas, APU 66
Foehn winds in the Andes, AR 22
— in Venezuela, 352
Fog, CHI 115, 119, 121, 124; PE 167; 168; BRA 253; CAM 446
—, number of days with, see Climatic tables
— and precipitation, CHI 119, 120; PE 168
Friagem, BRA 260–262
Fronts, AR 63; CHI 115, 117; BRA 226, 227, 262

---

* *Abbreviations:* APU = Argentina (AR), Paraguay (PA), Uruguay (UR); CHI = Chile; PBE = Peru (PE), Bolivia (BO), Ecuador (EC); BRA = Brazil; NSA = Northern South America; French Guiana (FG), Surinam (SU), Guyana (GU), Venezuela (VE), Colombia (CO); CAM = Central America (incl. Caribbean islands); SAM = South America.

Frost, occurrence, BRA 263
— free days, number, APU 48, 49

Garua, PE 167, 168
Glaciation in the Andes, 8, 9; AR 60, 66; CHI 127;
   PBE 158
Grosswetterlagen, CAM 447
Groundfrost, APU 48

Hail, AR 69; CAM 451
Haze, BRA 254; VE 334
— and sandstorms, CAM 446
—, number of days with, BRA 270–293
Heat low, 2; AR 63
High pressure systems, *see* Anticyclones
Humboldt current, CHI 177; PBE 148, 172, 180
Humidity, *see also* Vapor pressure
—, relative, AR 50; PE 166; BRA 244; NSA 305; 319,
   325; CAM 417
— —, tables, PBE 203–218; NSA 380–403
Hurricanes, 479–509
— and surface water temperature, 485
—, areas of origin of, 489, 492
—, floods caused by, 501
—, formation of, 480
—, frequency of affecting U.S., 499
—, — of occurrence, 489, 494, 495, 496
—, movement of, 496–498
—, number of, 481, 482
—, persistence of, 497
—, precipitation, 501
—, pressure field, 500
—, radar appearance of, 504
—, recurvature of, 498
—, storm surges of, 505
—, structure of, 500
—, wind field of, 500

Igapo, BRA 221
Inflow Lake Titicaca, 197
Insolation, BRA 231
—, annual variation of, BRA 232
— in the high subtropical Andes, 9; AR 62
Inversions, AR 38, 41; CHI 116; PBE 172, 178, 180
—, tradewinds, BRA 225; CAM 446
Irrigation, APU 47, 63; PE 153; VE 334
I.T.C. (Intertropical Convergence Zone), 2, 3; PBE
   161; BRA 224, 265; NSA 298, 323, 337, 346;
   CAM 420, 448

Jetstream, average position, 7
—, coastal low level, 189–191; NSA 354, 355, 361
Jungle, NSA 299

Land and sea breeze, AR 19, 20; CHI 119; NSA 311;

   CAM 410, 431, 451
Latent energy of tropical storms, 484
Low pressure systems, APU 13, 21, 63

Meteorological equator, NSA 301
Moisture, *see* Humidity
— index of Thornthwaite, BRA 257
Mountain and valley winds, AR 22; PBE 186
Monsoon-like phenomena, NSA 336, 337, 341

Nevoa seca, BRA 254
Northerns, CAM 427, 447–449

Oasis effect, APU 47

Pampero, AR 13, 47
Páramos, PBE 172; NSA 367
Patagonian ice fields, AR 66; CHI 127
Penitents, 9
Peru current, *see* Humboldt current
Precipitation along Pacific coast, NSA 372
—, annual, frequency over sea, CAM 419
—, — totals, APU 58, 59; CHI 59; BRA 246; FG
   301; SU 307; GU 322; VE 331; CO 364; CAM
   423
—, — — of Lake Titicaca, 196
—, — — of Lesser Antilles, CAM 440, 441
—, — —, section through Costa Rica, CAM 427
—, — —, section through eastern Jamaica, CAM 435
—, — variation, APU 61; CHI 114, 121; PBE 152,
   159, 164; BRA 246, 247; NSA 301, 306, 322, 330,
   337, 363; CAM 424, 436, 443
— characteristics, CHI 118, 125
—, cloudbursts, CAM 449, 450
—, coefficient of variation, APU 66; NSA 304
—, diurnal variation, PBE 160, 162; NSA 309, 311,
   351, 365; CAM 429
— divide, CHI 113; PBE 163
—, effect of Andes, CHI 126
—, hurricane, 501, 502
—, —, areal averages, 503
— intensity, APU 68; BRA 245, 248; NSA 344;
   CAM 433
—, large values, APU 65, 66; CHI 127; NSA 370;
   CAM 450
—, latitudinal differences, CHI 121; PBE 158; BRA
   256
—, max. in 24 h, *see* Climatic tables
—, monsoonal character, GU 324, 325
—, number of days with, *see* Climatic tables
— percent in summer half-year, PBE 150, 151
—, percentiles, CAM 432
— producing processes, AR 63; VE 344
— regimes, APU 60–68; CHI 114, 118, 123; BRA 248
—, seasonal distribution, APU 61; CHI 121, 125;
   PBE 152, 157, 164; BRA 246; NSA 302, 308,
   322, 363

— tables, *see* Climatic tables
—, time of maximum, SU 312
Pressure, annual mean hourly values, Quito, EC 160
—, —, BRA 223; NSA 336; CAM 411
— centers, BRA 222
— changes from summer to winter 4, 5
—, diurnal variation, SU 317
— field, hurricanes, 500
— gradient, meridional, APU 15, 20; CHI 126
— in hurricanes, lowest values, 501, 502, 506
— in mountainous regions, 199
—, semi-annual variation of, 7
—, semi-diurnal variation of, NSA 351; CAM 411
— tables for sea or station level, *see* Climatic tables
Puna, AR 69; PBE 172
Puna brava, PBE 172

Radiation, global, APU 36; PBE 181, 182; NSA 327, 342, 359; CAM 478
—, —, semi-annual variation, NSA 342
Rainfall, *see* Precipitation
Rain-front, AR 60; CHI 174; BRA 221; NSA 299, 300, 307
Relative humidity, *see* Humidity
Relief of Brazil, 221
Resultant winds, AR 19; FG 302; SU 313; GU 323; VE 335
— —, Altiplano, 195
— —, upper air at the Tropic, 7

Salares, AR 60
Savanna, NSA 295, 299, 321
Sea breeze, *see* Land and sea breeze
Semi-arid zones, APU 64
Snow, AR 60, 62; CHI 120; BRA 263; VE 338
— line, 8, 9; AR 60, 62; PBE 158, 172, 176
Soil temperatures, AR 49, 50; SU 318
Squalls, CAM 451
Steppe, AR 14, 57, 65; NSA 299
Storm surge, 505
Storminess, NSA 347
Streamlines, mean, CAM 408
Sudestada, APU 13, 58
Sunshine, APU 32, 34, 70, 74–111; PBE 181, 182, 212; BRA 230–232, 272–293; NSA 310, 382–403; CAM 414, 428, 456–477

Temperature, annual range, BRA 240, NSA 317–319
—, — variation, APU 36, 37, 42; CHI 123; PBE 171, 172, 177, 178; BRA 232, 238, 239; FC 305; SU 316; VE 338; 343; CAM 414, 415
— anomalies, Brazil, 237, 238
— —, Galapagos Is., 180
— —, Lima, 180
—, change with height, APU 38, 41; CHI 116; PBE 172, 173; CO 367; CAM 416, 485, 486
—, diurnal variation, AR 44–46; CHI 120, 121; BRA

240; SU 316; VE 343; CAM 417
— gradient, meridional, AR 37
—, hourly values, NSA 342, 343
—, interdiurnal variation, AR 46, BRA 241
— inversion, AR 38, 41; CHI 116; PBE 172, 178, 180
—, January isotherms, APU + CHI 39, BRA 234
—, July isotherms, APU + CHI 40, 235
—, max. and min. extremes, APU 47; PBE 174–176; BRA 236, 237; *see also* Climatic tables
—, time of maximum, AR 45; BRA 241
—, water, surface, SU 314; 485, 487
Temporal, CAM 448, 449
Thunderstorms, APU 60, 69; PBE 163, 192; BRA 251–253; NSA 304, 313; CAM 445, 446, 451
—, number of days with, *see* Climatic tables
—, over subtropical Andes, 192–195
Tides, highest in hurricanes, 506
Tornadoes, CAM 451
— assoc. with hurricanes, 507
Tradewinds, PBE 170; BRA 225, 226; NSA 298, 352, 361; CAM 405, 406, 410
—, constancy of, 488
Travelogs, 1, 192
Tropical cyclones, 479–509
— —, annual variation, 495
— —, areas of origin, 489
— —, dates of beginning, 490
— —, frequency, 490
— —, probability of occurrence, 491
—, discontinuity, BRA 227
— storms, 479–509
— —, areal 24-h precipitation, 504
— —, latent energy available, 484
— —, number of, 481, 482
— —, points of origin, 492
Troughline, continental, AR 17
—, equatorial, BRA 225, 312; NSA 368

Upwelling, 190, 191; NSA 352

Vapor pressure, APU 50, 51, 74–112; CHI 131–145; BRA 242, 270–293; NSA 320, 326; CAM 455–477
Visibility, CAM 446–447

Warm days and nights, BRA 242
Water budget of Lake Titicaca, 194–198
Waterspouts, CAM 451
Wind, surface, direction and speed, APU 14, 16–19, 23; CHI 119, 125, 127; PBE 184, 187; BRA 226–229; NSA 374; CAM 406; *see also* Climatic tables
—, —, Altiplano, 195
— erosion, AR 64
— regime, AR 17, 22; BRA 229
— speed, diurnal variation, APU 16; NSA 355; CAM 409
— — extremes, BRA 229

Wind *(continued)*
——— in the high Andes, AR 22; CHI 121, 125, 126
——— and hurricanes, 501
—, thermal, 189

Zero degree isotherm, elevation, APU 41
Zonal atmospheric flow, 6
———, summer maximum, 7
Zonda, AR 13, 22